INSECT SYMBIOSIS

CONTEMPORARY TOPICS in ENTOMOLOGY SERIES

THOMAS A. MILLER EDITOR

INSECT SYMBIOSIS

Edited by

Kostas Bourtzis

Thomas A. Miller

CRC PRESS

Boca Raton London New York Washington, D.C.

Cover: The genome sequences of the *Buchnera aphidicola* symbionts of *Schizaphis graminum* and *Acyrthosiphon pisum* have been determined. A comparison of the two genomes shows that no rearrangements or gene acquisitions have occurred in the past 50 to 70 million years, despite high levels of nucleotide-sequence divergence. This is the first time that whole-genome evolution for microbes has been calibrated with respect to time. The analysis has shown that *B. aphidicola* have the most stable genomes characterized to date. (Photograph courtesy of Ola Lundström, Department of Molecular Evolution, Uppsala University, Uppsala, Sweden.)

Senior Editor: John Sulzycki
Production Editor: Christine Andreasen
Project Coordinator: Erika Dery
Marketing Manager: Nadja English

Library of Congress Cataloging-in-Publication Data

Insect symbiosis / edited by Kostas Bourtzis and Thomas A. Miller.
p. cm. -- (Contemporary topics in entomology)
Includes bibliographical references (p.).
ISBN 0-8493-1286-8
1. Insects--Ecology. 2. Symbiosis. I. Bourtzis, Kostas. II. Miller, Thomas A. III. Series.

QL496.4 .I57 2003
595.71785--dc21 2002038796

International Standard Book Number 0-8493-1286-8
Library of Congress Card Number 2002038796
Printed in the United States of America 1 2 3 4 5 6 7 8 9 0
Printed on acid-free paper

Dedication

This book is dedicated to Frank F. Richards, M.D., Professor (Emeritus) of Internal Medicine, Yale University School of Medicine. Frank's creative drive and imagination in molecular parasitology led to the development of the paratransgenesis approach described in Chapter 6 by Frank's collaborator and friend Ravi Durvasula. This approach is currently being used to develop a cure for Pierce's disease in California grape vineyards.

Photo by Soo-ok.

Frank has influenced the work of Peter Hotez, Serap Aksoy, Scott O'Neill, Ben Beard, and many others. His tireless support of parasitology and immunology has led to many exciting projects that have influenced colleagues all over the world. His advice has benefited the Rockefeller Foundation and the National Institutes of Health.

Those of us who are fortunate enough to interact with Frank appreciate his unflagging enthusiasm and positive attitude. His laboratory door is always open and his advice is golden. Although Frank has left day-to-day operations at Yale, he remains keenly interested in following the projects he began and the individuals he touched.

Tom Miller

Foreword

When Charles Darwin looked at the ground, he saw billions of earthworms joined in a feast that could bury the ruins of past civilizations. When he looked at a bee crawling across an orchid, he saw an economy of sex and food reaching back millions of years. Darwin had a kind of biological x-ray vision that let him see the hidden episodes of life's history that others missed. And yet the creatures so lovingly profiled in *Insect Symbiosis* fell completely within Darwin's blind spot. Given how common, powerful, and fascinating they are, this oversight comes as a particularly big jolt.

It is not as if Darwin was ignorant about creatures that make other creatures their hosts. He was born over a century after Leeuwenhoek discovered microscopic organisms swimming in the scum of his teeth. Darwin kept up to date with the reports from Louis Pasteur, showing, among other things, that silkworms were infected by microbes (which would later turn out to be microsporidia). But Darwin did not consider these symbionts very important to evolution. He sometimes mentioned "a particle of small-pox matter" or "the contagious matter of scarlet fever," but only as an analogy to illustrate the particles he thought carried heredity. He thought parasitic wasps that devoured their caterpillar hosts alive summed up nature's cruelty. But he did not muse over how those wasps had evolved, or what effect they had on their hosts' evolution.

It is hard to prove the cause of silence. I believe in this case Darwin's social background played a role. He was a Victorian gentleman with a fortune built on railroads and Wedgwood china. Victorian gentlemen typically saw history as progress, from simple to complex, from crude to sophisticated, from barbarians to — well, to Victorian gentlemen. Darwin's triumph was to challenge this ubiquitous belief with his theory of evolution by natural selection. It required no built-in progress to guide life's history, only the natural competition and variation found in every generation. But the idea that parasites and other symbionts might have an important evolutionary history — that seems to have been too much even for Darwin.

Darwin's acolytes were not so silent on the subject. In 1879 Ray Lankester wrote an essay called "Degeneration: A Chapter in Darwinism." He observed with a mix of fear and loathing how parasites could lose the complex anatomy their forebears had taken millions of years to evolve. The crustacean *Sacculina*, for example, turned from an animal to a hairy sac barely distinguishable from a plant. "Let the parasitic life once be secured," Lankester warned, "and away go legs, jaws, eyes, and ears; the active, highly-gifted crab may become a mere sac, absorbing nourishment and laying eggs." Progress had a dark counterpart in degeneration, both for highly gifted crabs and highly gifted humans. Parasites degenerated, Lankester warned, "just as an active healthy man sometimes degenerates when he becomes suddenly possessed of a fortune; or as Rome degenerated when possessed of the riches of the ancient world."

The 20th century did not do much for the reputation of parasites and other symbionts in evolutionary circles. For the most part, it was an age of hosts. The genes of hosts were the only ones that mattered in the history of life, as they were handed down from one generation to the next, mutated, and altered fitness. But by the end of the 20th century, as genetics gave way to genomics, it was becoming inescapably clear that the hidden passengers were just as important. Parasites have adapted to their hosts with staggering sophistication, able in many cases to control their physiology, reproduction, and even their brains. Animals and plants simply would not exist without the help of hordes of bacteria, fungi, and other symbionts that have the biochemical skills their hosts lack.

These revelations are having an impact as subtle as a slap in the face. Take river blindness, a horrible disease that has robbed the sight of millions of people. Scientists have long known that when infected black flies bite a human, they transmit nematodes that settle in their host's skin, producing thousands of babies that crawl around the body, scarring the eyes in the process. Only in 2002 did scientists establish that a symbiotic bacterium in the nematode's skin, *Wolbachia*, actually triggered the immune reaction that causes the blindness. Antibiotics that kill *Wolbachia* may be able to cure the disease. *Wolbachia*, of course, is no stranger to the pages of this book. It has made many insect species its host, and as a result it may be the most common form of infectious bacteria on Earth. The insights from the research described in *Insect Symbiosis* may help turn *Wolbachia* into a symbiotic insecticide, allowing entomologists to control pests that carry diseases and destroy crops.

At the same time, the insights found in *Insect Symbiosis* challenge some of the basic vocabulary we use to describe life. In their chapter on microsporidia, for example, Agnew et al. describe them as "degenerate eukaryotes," using the term Lankester used with such anxiety over 120 years ago. It is true that microsporidia lack many of the genes their ancestors had and cannot make the things those genes code for. But microsporidia are hardly evolutionary backsliders. As Agnew and his colleagues demonstrate, they adapt nimbly to the challenges posed by their hosts. And having a reduced genome does not make them degenerate so much as sleek and efficient. They are not bogged down with the "junk DNA" that clutters our own cells. Even the genes they have retained have shrunk — succinct commands in place of hazy instructions. And while they may not have the organelles they would need to live on their own, they have the ability to seize the organelles of their host. They even go so far as to puncture those organelles with gap junctions so they can extract the molecules they create more efficiently. You could even say that microsporidia have not lost the genes for these organelles. Their host's genes have become their own.

Symbionts are turning evolutionary biology into a symphony. The evolutionary fate of a host depends not only on its own genome, but also on the hidden agendas of mutualists, parasites, commensalists, and other passengers who fall somewhere in between on the spectrum of coexistence. Darwin would be astonished to see what was hiding in his blind spot: the tangled bank he described in *The Origin of Species* has been carried within.

Carl Zimmer

Insect Symbiosis: A Personal Journey

I studied biology and completed my doctorate in the Department of Biology, University of Patras, Greece under the supervision of Professors Antigoni Zacharopoulou and Vassilios Marmaras, to whom I am grateful for introducing me to the biological sciences. In January 1991, I moved to the Institute of Molecular Biology and Biotechnology (IMBB) in Crete, Greece, to do a postdoc in the lab of Professor Charalambos (Babis) Savakis. At about that time, Niki Kretsovali, a colleague in the mammalian group, had isolated a partial cDNA clone from a commercially available mouse library that exhibited similarities with the prokaryotic *dnaA* gene. This gene encodes the initiation factor of DNA replication in bacteria. Everybody was very excited about this finding. Babis and Professor Fotis Kafatos (director of IMBB) suggested that I clone and characterize the *Drosophila melanogaster dnaA* ortholog. Using the mouse clone as a probe, I was able to detect the gene(s) in *Drosophila* by Southern blot analysis. Then, using the same mouse clone, I started screening every available genomic and cDNA *Drosophila* library to isolate a full length or even a partial clone of this gene.

I did innumerable screenings with no success. I did hundreds of RNA *in situ*s and *in situ* hybridizations on polytene chromosomes. Despite my efforts, I could not detect the gene; there was no signal at all. The ad hoc explanation we gave was that this gene might be under-replicated, weakly expressed, or perhaps located in some "non-easily clonable" genomic region. Then, I thought that it might be better to do the screenings using a more specific probe. Working in the PCR era, I designed degenerate primers based on the multiple alignment of all available bacterial *dnaA* genes (there were not many known at that time, only five to six) and the mouse one. PCR worked nicely, and I was able to clone and sequence a 558-bp fragment of the *Drosophila* cognate *dnaA* gene. I was very excited and began a new library screening and *in situ*s. But again, no signal could be detected. I was very disappointed. I had been working for more than 18 months on this project and had the impression that I was chasing a ghost gene.

In July 1992 Babis suggested that I attend the EMBO *Drosophila* meeting in Kolymbari, Crete and talk to *Drosophila* people about my project. I was really enjoying the meeting until I heard Tim Karr's presentation. Tim presented his group's exciting findings about the presence of an intracellular bacterium, namely *Wolbachia*, in *Drosophila simulans* responsible for the induction of a kind of male sterility, cytoplasmic incompatibility. I was really shocked, thinking that "my *Drosophila dnaA* cognate" might be the *Wolbachia* ortholog. I talked with Tim about my project, and he was kind enough to accept my invitation to visit our lab. There he did some DAPI staining that indeed suggested the presence of *Wolbachia* in our *Drosophila* stocks. My further molecular work confirmed my fears.

It was really a huge shock for me. Babis suggested that I take a few days off from the lab and make up my mind about how I wanted to proceed. I told him that I did not want to hear about *Wolbachia* again, and that I wanted to work on something entirely different. Indeed, I stopped the project and became involved in the characterization of the medfly *Adh* locus. After a few months, Babis suggested that I should perhaps reconsider the *Wolbachia* project. At first, I thought that he was not serious, but he was. Ultimately, he convinced me to continue my work on this — as I discovered later — fascinating bacterium.

Babis was my mentor and my "good angel." Not only did he introduce me to the field of insect molecular biology and molecular genetics and convince me to continue my studies on *Wolbachia* but he has also supported me throughout my scientific career. If I am now considered a "Wolbachialogist," I owe this to Babis. I also owe much to Professor Scott O'Neill (University of Queensland, Australia), who greatly influenced my work on *Wolbachia*. My two postdoc years in his lab at Yale University were a really great experience. I had the unique opportunity to belong to a lab where everyone was talking about *Wolbachia* and intracellular bacteria. In addition to Scott, Henk Braig, Steve Dobson, Francois Rousset, Melinda Pettigrew, and Weiguo Zhou were in the same lab, and next door were Frank Richards, Serap Aksoy, Ravi Durvasula, and many other colleagues.

And last, but certainly not least, during my countless library screenings to clone the "*Drosophila dnaA* gene," I met and fell in love with my wife, Argyro. I would like to thank her, and my children, Nikolas and Ioanna, for sharing the struggle as well as the joys and pleasures of my everyday life. My life certainly would be miserable without them. They embody the true meaning of symbiosis.

Kostas Bourtzis

The Editors

Kostas Bourtzis, Ph.D., is Assistant Professor of Molecular Biology and Biochemistry in the Department of Environmental and Natural Resources Management, University of Ioannina, Greece. He also has a longstanding research collaboration with the Institute of Molecular Biology and Biotechnology, Heraklion, Crete, Greece. His research interests include *Wolbachia*-mediated cytoplasmic incompatibility in *Drosophila* and agricultural insect pests; genetic manipulation of *Wolbachia*; molecular mechanisms of cytoplasmic incompatibility; *Wolbachia* genomics; and the use of endosymbiotic bacteria including *Wolbachia* as a tool for the development of new environmentally friendly approaches for the control of arthropods of medical and agricultural importance.

Kostas Bourtzis (left) and Thomas A. Miller (right). Photo by Soo-ok.

Thomas A. Miller, Ph.D., is Professor of Entomology at the University of California, Riverside. His early research investigated the control of circulation in insects, mode of action of insecticides, and resistance of pest insects to insecticides. More recent projects have centered on the development of genetic control strategies to protect crops from pests and disease spread. In particular, he is part of an effort to develop a conditional lethal strain of pink bollworm to improve the radiation-based sterile insect technique for population suppression and eradication and is working on the paratransgenesis technique introduced by Frank Richards of Yale University to control the transmission of Pierce's disease to grapes by sharpshooter insects.

Contributors

Philip Agnew
Centre d'Etudes sur le Polymorphisme des Microorganisms
Centre National de la Recherche Scientifique
Institut de Recherche pour le Développement
Montpellier, France

Serap Aksoy
Department of Epidemiology and Public Health
Yale University School of Medicine
New Haven, Connecticut, U.S.A.

Siv G.E. Andersson
Department of Molecular Evolution
Evolutionary Biology Center
Uppsala University
Uppsala, Sweden

Claudio Bandi
Section of General Pathology and Parasitology
University of Milan
Milan, Italy

James J. Becnel
U.S. Department of Agriculture
Agriculture Research Service
Center for Medical Agricultural and Veterinary Entomology
Gainesville, Florida, U.S.A.

C. Ben Beard
Division of Parasitic Diseases
Centers for Disease Control and Prevention
Chamblee, Georgia, U.S.A.

Seth R. Bordenstein
The Josephine Bay Paul Center for Comparative Molecular Biology and Evolution
The Marine Biological Laboratory
Woods Hole, Massachusetts, U.S.A.

Michel Boulétreau
Laboratoire Biométrie et Biologie Evolutive
CNRS
UMR 5558
Université Lyon 1
Villeurbanne, France

Kostas Bourtzis
Department of Environmental and Natural Resources Management
University of Ioannina
Agrinio, Greece

Henk R. Braig
School of Biological Sciences
University of Wales, Bangor
Bangor, U.K.

Johannes A.J. Breeuwer
Institute of Biodiversity and Ecosystem Dynamics
University of Amsterdam
Amsterdam, the Netherlands

Celia Cordon-Rosales
Center for Health Studies
Universitad del Valle de Guatemala
Guatemala City, Guatemala

Franck Dedeine
Laboratoire Biométrie et Biologie Evolutive
CNRS
UMR 5558
Université Lyon 1
Villeurbanne, France

Stephen L. Dobson
Department of Entomology
University of Kentucky
Lexington, Kentucky, U.S.A.

Angela E. Douglas
Department of Biology
University of York
York, U.K.

Ravi V. Durvasula
Department of Internal Medicine
Yale University School of Medicine
New Haven, Connecticut, U.S.A.

Dieter Ebert
Department of Biology, Ecology and Evolution
University of Fribourg
Fribourg, Switzerland

Takema Fukatsu
National Institute of Advanced Industrial
Science and Technology
Tsukuba, Ibaraki, Japan

Abdelaziz Heddi
Biologie Fonctionnelle, Insectes et Interactions
UMR INRA/INSA de Lyon
Villeurbanne, France

Martinus E. Huigens
Laboratory of Entomology
Wageningen Agricultural University
Wageningen, the Netherlands

Gregory D.D. Hurst
Department of Biology
University College London
London, U.K.

Nobuyuki Ijichi
Department of Biological Science
University of Tokyo
Meguro, Tokyo, Japan

Hajime Ishikawa
The University of the Air
Chiba, Japan

Francis M. Jiggins
Department of Genetics
University of Cambridge
Cambridge, U.K.

Timothy L. Karr
Department of Biology and Biochemistry
University of Bath
Bath, U.K.

Natsuko Kondo
Department of Systems Science (Biology)
University of Tokyo
Meguro, Tokyo, Japan

Laura H. Kramer
Department of Animal Production
University of Parma
Parma, Italy

Carol R. Lauzon
Department of Biological Sciences
California State University, Hayward
Hayward, California, U.S.A.

Michael E.N. Majerus
Department of Genetics
University of Cambridge
Cambridge, U.K.

Kenji Matsuura
Department of Organismic
and Evolutionary Biology
MCZ Laboratories
Harvard University
Cambridge, Massachusetts, U.S.A.

Yannis Michalakis
Centre d'Etudes sur le Polymorphisme
des Microorganisms
Centre National de la Recherche Scientifique
Institut de Recherche pour le Développement
Montpellier, France

Thomas A. Miller
Department of Entomology
University of California, Riverside
Riverside, California, U.S.A.

Naruo Nikoh
Department of Genetics
North Carolina State University
Raleigh, North Carolina, U.S.A.

Pamela Pennington
Center for Health Studies
Universitad del Valle de Guatemala
Guatemala City, Guatemala

Diana L. Six
School of Forestry
University of Montana
Missoula, Montana, U.S.A.

Richard Stouthamer
Department of Entomology
University of California, Riverside
Riverside, California, U.S.A.

Ranjini K. Sundaram
Department of Internal Medicine
Yale University School of Medicine
New Haven, Connecticut, U.S.A.

Ivica Tamas
Department of Molecular Evolution
Uppsala University
Uppsala, Sweden

Andrew R. Weeks
Department of Entomology
University of California, Riverside
Riverside, California, U.S.A.

Contents

1 Insect Symbiosis: An Introduction

Hajime Ishikawa

CONTENTS

INTRODUCTION

Over the course of 4 billion years biological evolution has created millions of species on this planet. Although this is mostly due to repeated and incessant differentiation of preexisting species, species differentiation was not the only process that produced today's biodiversity. The tendency of nature to combine two species also promotes biodiversity. Combinations include predator–prey relationships, parasitism, and symbiosis. Among these, symbiosis has had major consequences for evolution. The magnitude of these consequences is immediately evident in light of the origin of eukaryotic cells, which are chimeras of several prokaryotes (Margulis, 1970; Gray and Doolittle, 1982). As multicellular organisms, insects as a group seem to be most tolerant of foreign organisms and live together with many different microorganisms, both inside and outside their bodies, in a variety of ways (Buchner, 1965). In this sense, insects may provide the best material for examining the evolutionary significance of interspecific symbioses.

0-8493-1286-8/03/$0.00+$1.50

Insects have one of the most successful lifestyles on earth. One important factor in their success is that insects have adapted to a wide variety of diets. Such flexible feeding habits have been brought to insects, at least partially, by the endosymbionts they harbor. Actually, endosymbionts are frequently observed near an insect's digestive tract, and it is widely accepted that endosymbionts play important roles in the nutrition of host insects. More recently, however, it has become increasingly apparent that symbionts' contributions to their host are not limited to nutrition. In many cases, insect endosymbionts, especially intracellular symbionts, are significant for their hosts not merely as nutritional supplements but also as DNA-containing genetic elements (Ishikawa, 1989).

Insects display the full scope of endosymbiosis. Some cases seem to be very new in an evolutionary sense because the association between host insect and symbiont is so loose and casual, whereas others are so intimate that symbionts seem to be tightly integrated into the physiology of the host. In the latter cases, endosymbionts appear to be cell organelles, and their significance for the host is no less than that of mitochondria for a eukaryotic cell.

One important point to keep in mind is that symbiosis is not necessarily an equal association in which both partners benefit equally. Mutualism, not to mention parasitism, is not always maintained on a 50–50 basis. One associate usually takes more, sometimes much more, than the other, or at least so it appears in the light of present knowledge. This is observed typically in the endosymbioses between insects and microorganisms. For example, aphids appear to profit disproportionately from the symbiosis with *Buchnera* symbionts, yet the symbiosis has lasted for some 200 million years (Moran et al., 1993). On the other hand, *Wolbachia* symbionts take full advantage of insect hosts so as to propagate their progeny, whereas the hosts do not seem to receive any reward, and yet they have harbored these selfish passengers for an evolutionarily significant length of time (Werren and O'Neill, 1997). Does such apparent unfairness really exist in symbiosis, or is it a one-sided view of symbiosis because of our incomplete understanding of this biological phenomenon?

This chapter outlines various types of interactions between insects and microorganisms, both prokaryotic and eukaryotic, mainly from physiological and evolutionary points of view. Detailed studies and recent advances in each of these interactions will be fully described in subsequent chapters.

TYPES OF ASSOCIATIONS BETWEEN INSECTS AND MICROORGANISMS

Although microbes are ubiquitous both inside and outside insect bodies, the present discussion is limited to interactions inside insect bodies. *Inside* does not necessarily refer to the inside of a cell but may mean the gut lumen, which, strictly speaking, is the outside, or the intercellular or intracellular space of the host tissue. Of course, the deeper inside the microbe resides, the more intimate the interaction with the host insect.

GUT MICROBES

Since insects are surrounded by a great variety of microbes, they should harbor some of them in the gut. It is likely that some of these gut microbes are enclosed by cells of the epithelial tissue and eventually become mutualistic symbionts. Gut microbes of insects are composed of a wide variety of species, including bacteria, archaea, and eukaryea, but their habitat in the insect body is usually restricted to one region, most commonly the hindgut. In many phytophagous insects, this region forms a large anaerobic chamber for the fermentation of cellulose and other ingested plant polysaccharides. Well-known examples include termites (Breznak, 1984) and cockroaches (Bracke et al., 1979). As in vertebrates, gut microbes of most insects are extracellular, either lying free in the lumen or adhering to the gut wall. In this sense, the yeast *Taphrina* is an exception; it lives in blind-ended caeca of the anobiid beetles that inhabit timber or other plant products (Jurzitza, 1979).

Some *Taphrina* cells are extracellular, while others are enclosed within hypertrophied epithelial cells at the distal end of the caeca.

Among all the gut microbes of insects, only those in termites and cockroaches have been studied in some detail. Gut bacteria of aphids that were later identified maintain a subtle balance with one another as well as with mutualistic symbionts inherent in the host.

The gut microbes of termites are housed in an enlarged portion of the hindgut called the paunch. This region is anaerobic, and all the microbial habitants are obligate or facultative anaerobes. The paunch of lower termites contains protozoa that are predominantly flagellate species (Honigberg, 1970). Many of these flagellates are particular to the hindgut of lower termites. There are also prokaryotes in the paunch of lower termites that include spirocetes and methanogens (Breznak, 1984). The densities of these microbes are surprisingly high, up to 10^7 protozoa and 10^9 to 10^{10} bacteria per milliliter of gut volume. Most of the microbes lie free in the lumen, but some species adhere to the gut wall. Most remarkably, some spirocetes are associated with the surface of protozoa on which they confer motility (Smith and Douglas, 1987). Higher termites are distinguished from lower ones in that their gut microbes are only prokaryotes. It is known that the gut microbes of termites are discarded at ecdysis of the host insects. However, this is of little consequence to the individual termite because these insects can easily acquire fresh microbes from the other individuals in the colony.

Cockroaches contain a complex hindgut microflora, predominantly of obligate anaerobes, including cellulolytic and methanogenic bacteria (Bracke et al., 1979). Whereas most cockroaches are omnivorous, some species, often called woodroaches, live exclusively on a diet of wood. The hindgut microbes of woodroaches are very similar to those of termites in that they contain protozoa. Many lines of evidence suggest that woodroaches and termites share a common ancestor. It is likely that the association with protozoa had already been established in the ancestor (Smith and Douglas, 1987).

By culturing honeydew, the excreta of aphids, spread over nutritional agar, Harada demonstrated that pea aphids, *Acyrthosiphon pisum*, contained several species of gut microbes (Harada and Ishikawa, 1993). These were Gram-negative, oxidase-negative, facultative anaerobic, fermentative, motile, and rod-shaped bacteria with the general characteristics of the family Enterobacteriaceae. Biochemical tests indicated that these bacteria were related most to *Erwinia herbicola* and *Pantoea agglomerans*, which are ectoparasites of many plants (Harada et al., 1997). Keeping aphids under aseptic conditions for several generations produced aseptic insects that were completely free of the gut microbes. Such aseptic aphids exhibited somewhat better performance than those with the gut microbes in terms of both growth rate and fecundity, suggesting that the gut microbes were not beneficial to the host. The result is interesting when taken together with the finding that one of the gut microbes can be monophyletic with *Buchnera*, a mutualistic, intracellular symbiont of aphids (see later). It is probable that since aphids have already established a mutualistic association with *Buchnera*, they no longer need its close relatives in the gut. In this context, it is worth noting that aposymbiotic aphids, which are artificially deprived of *Buchnera*, tend to accumulate many microbes, including both bacteria and fungi, that otherwise are not detected in the gut (A. Nakabachi, personal communications). Although it is unknown whether or not some of these microbes substitute for the mutualistic symbiont, the aposymbiotic aphids, harboring these microbes, live as long as, or even longer than, symbiotic ones. The tripartite relationship of host insect, mutualistic symbionts, and gut microbes is a potentially interesting subject of study.

Dynamics among microbes in the gut flora provide another interesting theme from a microecological point of view. Each species or strain of microbe from the aphid's gut can be cultured separately and returned to the host by mixing the cells with a synthetic diet that the host ingests. Interestingly, every species of the gut microbes, when put back into aseptic aphids by itself, tended to proliferate excessively and eventually kill the hosts (Harada and Ishikawa, 1997). This suggests that the microbes are potentially harmful to the host and can be compatible with it only when they form a multispecies community. In other words, it is likely that in the gut flora microbes keep each other in check.

The overwhelming majority of endosymbionts of insects indisputably have been acquired with food through the mouthparts. However, evidence suggests that there may be additional routes of acquisition. The tracheae of insects often contain bacteria generally related to those of the milieu. Also, the hemolymph may contain microbes without injury to the insect body. Interestingly, however, these two flora are not related to the intestinal flora of the same insect, suggesting that the spiracle can provide a route of acquisition. Honeydew, the sweet excrement of *Homoptera*, may also have led occasionally to anal acquisition of microbes. Since *Rickettsia* species are known to use this route at times, it is worth considering that *Wolbachia*, a group of so-called guest microbes, may use the same route in their horizontal transmission (see below).

Endoparasitism

Parasitism is a one-sided symbiosis in which one of the associants benefits largely at the expense of the other. As noted by de Bary (1879) and Starr (1975), however, in some cases it is difficult to draw a clear line between parasitism and symbiosis. For the sake of convenience, I distinguish between the two based on the mode of transmission. If an organism is transmitted from one host to another only horizontally, I regard it here as a parasite.

It is worth noting that mutualistic symbionts of insects are always unicellular, whether prokaryotic or eukaryotic, whereas parasites may be unicellular or multicellular. Generally speaking, insect hosts suffer more from multicellular endoparasites. Endoparasitic associations involving insects can be categorized into four types. The first type is one in which insects serve as vectors of pathogenic protozoans harmful to vertebrates. A noticeable characteristic of this association is that insect hosts suffer very little, if at all, from their unicellular parasites, from which vertebrate hosts usually suffer disastrously. As far as insects are concerned, it may be more accurate to refer to these pathogenic protozoans as commensals rather than parasites. One typical example of these parasites is hemoflagellates, which inhabit the gut of some insects. The hemoflagellates that invade the reticuloendothelial tissue of vertebrates include *Trypanosoma brucei*, *T. cruzi*, and *Leishmania donovani*, which cause African sleeping sickness, Chagas' disease, and Oriental sore, respectively. Little is known about how insect hosts are affected by these parasites. Another well-known example of parasites of this type is the *Plasmodium* species, malaria parasites that are transmitted by female *Anopheles*. Although mosquito genes that are responsible for allowing the infection of *Plasmodium* have been identified, little is known about how the insects are affected by these parasites (Ito et al., 2002).

The second type of insect endoparasitism involves nematodes. Among more than 3000 nematode–insect associations, a remarkable example involves the nematode *Neoaplectana*. The nematode carries bacterial symbionts and releases them when inside the insect's alimentary canal. The symbionts invade the insect's body cavity, rapidly multiply there, and cause the death of the insect host. Subsequently, nematodes develop rapidly in the dead host and become sexually mature (Nickle, 1984). Although endoparasitisms with nematodes that modify the insect's morphology, physiology, and behavior are considered to be a potentially important tool of biological control over noxious insects, their biochemical and molecular mechanisms are largely unknown.

The third type of endoparasitism is characterized by the fact that both the host and parasite are insects. In this association, parasitic insects are usually called parasitoids, which are either hymenopterous or dipterous insects, because their behavior is somewhat similar to that of predators, and the association necessarily results in the host's death. Although the interaction between hosts and parasitoids includes many interesting aspects from physiological and biochemical points of view, a detailed description of this association is beyond the scope of this paper.

In addition to the parasitoids mentioned above, probably the most aggravating parasites to insects are fungi. These organisms comprise the fourth type of endoparasitic association with insects. In almost every conceivable ecological niche of insects, fungi have developed associations with insects that range from casual to intimate (Anderson et al., 1984). This is, at least, partly because the two organisms share an important substance. Both contain chitin as a structural component in the cell wall of fungi and in the exoskeleton of insects. Species of *Beauveria*, *Cordyceps*, *Coelomomyces*, *Entomorphthora*, and *Metarrhizium* are well-known examples of fungi that are pathogens of many insects. These parasitic fungi infect insect hosts, assimilate their hemolymph, digest soft tissue, and produce mycotoxins, which kill the host. A typical example of this association is the parasitism by *Cordyceps*. *Cordyceps* attack many different types of insects and spiders. The infected hosts shrivel and dry after death but resist decay because of an antibiotic compound, cordycepin, produced by the fungus. After the host dies, the fungus produces a cylindrical structure that grows out from the host and produces many fruiting bodies, which contain spores. Dried bodies of insects with the outgrowth of fruiting bodies of *Cordyceps* are called "plant worm" and are believed to contain physiologically active substances good for our health (Anderson et al., 1984).

Not all parasitic fungi are harmful to insects. There are many fungus–insect associations in which the hosts are not harmed by or even benefit from fungi. A typical example of these associations is fungus gardens in which some species of termites and ants cultivate fungus for food. Termites use fungi as a supplemental source of vitamins, whereas ants rely entirely on the fungi for their nutrients. The fungi are grown in specially prepared beds that contain a mixture of plant materials and insect excrement. Unlike pathogenic fungi, all these beneficial fungi stay outside the host's body. Once taken inside, these fungi may become mutualistic symbionts that will be vertically transmitted through the host's generations (Ahmadjian and Paracer, 1986). Indeed, molecular phylogenetic evidence suggests that intercellular symbionts found in planthoppers and some aphid species are close relatives of an entomogenous fungus (see below).

Extracellular Symbiosis

All mutualistic endosymbionts that are vertically transmitted through generations of host insects are invariably unicellular organisms, either prokaryotic or eukaryotic, and either extracellular or intracellular. Although most of the gut microbes mentioned above are extracellular symbionts in a broader sense, their association with insects must be renewed whenever the host's generation is renewed. Other microbes penetrate the insect body more deeply to maintain a somewhat more prolonged and intimate association with the host. However, the location of these microbes is quite variable, and the same species are quite often found in the gut lumen and even inside certain cells, probably reflecting the fact that these microbes are at transient stages from mutualistic gut inhabitants to intracellular symbionts. Some homopteran insects, including scale insects, cicadas, and aphids, harbor yeast-like microbes in the space between hypodermis and the intestinal tract. Some aphid species belonging to the tribe Cerataphidini have been shown to harbor yeast-like extracellular symbionts in the hemocoel and fat body, instead of *Buchnera*, which are intracellular bacterial symbionts common to the rest of the family Aphididae (see below).

Intracellular Symbiosis

Many insect species harbor intracellular symbionts that are vertically transmitted through their generations. If the transient stages mentioned above are included, the percentage of insect species that contain intracellular symbionts would be as high as 70% or more. Intracellular symbiosis is the most intimate association between two different organisms, and it is generally reasoned that the association is maintained through the host's generations because the host and symbiont equally benefit from the association. In reality, however, it does not seem to apply in intracellular symbiosis between insects and microorganisms.

Intracellular symbioses of insects with microorganisms are categorized into two types, which are (or at least appear to be) quite different from each other. One is the so-called mycetocyte symbiosis, in which symbiotic microbes are harbored by the host mycetocyte, or bacteriocyte, a special cell differentiated for this purpose (Buchner, 1965; Ishikawa, 1989). In this association, the benefits to the host insects are often clear and evidenced by experimental elimination of the symbiont. By contrast, it is much less clear how symbionts truly benefit from the association. These mycetocyte symbionts seem to be domesticated by the host insects. In this respect, their relation to insects somewhat resembles that of livestock to us.

In a similar analogy, the other type of intracellular symbionts of insects may be compared to cockroaches, house rats, and mice, which sometimes share habitats with us and unilaterally exploit us. These symbionts are collectively called "guest microbes" (Ebbert, 1993). Guest microbes also differ from mycetocyte symbionts in that they are not restricted to a particular cell type, such as the mycetocyte, but are present in almost all cell types of the host insect. In this respect, again, they resemble cockroaches rather than livestock.

MUTUALISTIC SYMBIONTS IN INSECTS

F. Blochmann first discovered the rod-shaped and bacteria-like symbionts in insect cells. These symbionts were later named Blochmann bodies in honor of the discoverer (Lanham, 1968). The Blochmann body corresponds to the mycetocyte symbiont in today's terminology. Guest microbes are a very different type of inhabitant of the insect cell and were discovered comparatively recently. In view of their important influence on the insect, these microbes, despite the analogy mentioned above, may mean much more to insects than do cockroaches or house rats to human beings. In this section and the next, these two symbiont types will be described in greater detail.

MYCETOCYTE SYMBIONTS

Mycetocyte symbionts are especially characteristic of three insect groups: the order Blattaria (cockroaches), the order Homoptera (cicadas, leafhoppers, psyllids, aphids, coccids, etc.), and the family Curculionidae (weevils) of the order Coleoptera (Dasch et al., 1984). These symbionts are housed in the highly specialized somatic cells of the host insect, termed mycetocytes. The term was coined based on an early observation that the symbionts were fungi. The insect cell that harbors bacterial symbionts is more accurately called "bacteriocyte." Frequently, however, "mycetocyte" is still used regardless of which symbionts are harbored by the cell.

Mycetocytes, or bacteriocytes, are sometimes assembled into a discrete organ called a mycetome, or bacteriome (Buchner, 1965). The location of mycetocytes and mycetomes varies depending on the host species. Mycetomes of weevils and other insects form part of the cellular lining of the midgut so that the symbionts have ready access to the lumen of the alimentary canal and often are found there extracellularly (Brooks, 1963). When insects have mycetomes of this type, the association between host and symbiont is generally not very intimate, and these symbionts are possibly cultivable extracellularly on common culture media. In cockroaches and homopterans, mycetocytes are separated from the digestive tract, lying either in the body cavity or embedded in the fat body. In the insect with mycetocytes of this type, host and symbiont are, as a rule, closely interdependent, and thus the symbionts are not generally cultivable extracellularly. In proportion to the large cell size, mycetocyte nuclei are generally enlarged, lobed, and polyploid. It has been estimated that mycetocytes of cockroaches range from tetraploid in the embryo to 512-ploid for some of the largest cells in adults (Richards and Brooks, 1958).

In most cases, mycetocyte symbionts demonstrate morphological features characteristic of prokaryotes, usually bacteria, although a great variation in shape is observed from rods to lobed vesicular bodies. In many cases, the symbionts are surrounded by two endogenous membranes and an outer, third membrane derived from the host. Whereas the innermost membrane, which encloses the symbiont's cytoplasm, shows little specialization, the second membrane generally forms the lipopolysaccharide–lipoprotein layer, which is characteristic of Gram-negative bacteria, or proteobacteria. The structure is thinner and more elastic than in free-living bacteria, showing considerable adaptation to intracellular life (Houk and Griffiths, 1980).

Morphological differences among prokaryotic symbionts in mycetocytes form different insect taxa, and differences in location of their mycetocytes suggest that these symbionts are probably not monophyletic. Indeed, comparative sequence analysis of 16S rDNA has revealed that mycetocyte bacteria appear to be specialized with respect to their hosts (Baumann et al., 1993). For example, among sternorrhynchous homopterans, aphids, whiteflies, and mealybugs, each harbors its own lineage of mycetocyte bacteria, suggesting either that symbiosis took place multiple times in these insect lineages or that symbionts were replaced in certain lineages. Mycetocyte symbionts from aphids and whiteflies are members of the γ-subdivision of the proteobacteria, whereas those from mealybugs are affiliated with the β-subdivision (Munson et al., 1992; Clark et al., 1992). As for nonhomopteran insects, tsetse flies (Aksoy et al., 1995), carpenter ants (Schroeder et al., 1996), and weevils (Campbell et al., 1992) also share members of the γ-subdivision of the proteobacteria with the mycetocyte symbionts, and cockroaches (Bandi et al., 1994) and termites (Bandi et al., 1995) specifically harbor bacteria of the Flavobacterium-Bacteroides group in the mycetocyte.

That some lineages of homopteran insects acquired symbionts independently of one another is best illustrated by the fact that a few groups contain fungi instead of bacteria as mycetocyte symbionts. A well-studied example is planthoppers, which harbor yeast-like symbionts in the mycetocyte (Noda, 1974). Molecular studies have placed the symbionts in the class Pyrenomycetes in the subphylum Ascomycotina (Noda et al., 1995).

Buchnera

Of the mycetocyte symbioses, the best-studied example is that of aphids. To date, the 16S rDNA sequences of mycetocyte symbionts of numerous aphid species have been obtained and analyzed phylogenetically together with sequences of other representative prokaryotes. Results indicate that all aphid primary symbionts, the inhabitants of the large mycetocytes, belong to a single, well-supported clade within the γ-3 subdivision of the proteobacteria and have *Escherichia coli* and related bacteria as their closest relatives (Moran et al., 1993). The formal designation, *Buchnera aphidicola*, was claimed to apply to this symbiont clade (Munson et al., 1991), whereas the species name, *aphidicola*, has for the most part fallen into disuse. This is because these symbionts are too diversified to be regarded as a single species.

It turned out that the sequence-based phylogeny of *Buchnera* was completely concordant with the morphology-based phylogeny of the corresponding aphid hosts. Since the probability that such a concordance would occur by chance is infinitesimally small, this finding implies a single original infection in the common ancestor of aphid species, followed by cospeciation of aphids and *Buchnera*. The distribution of modern symbionts among their host aphids thus reflects parallel cladogenesis through consistent and long-term transmission from mother to daughter. Horizontal transfer of these symbionts apparently did not take place or was very rare (Moran et al., 1993; Moran and Baumann, 1994). The aphid fossil record implies that the symbiotic association between aphid and *Buchnera* is ancient. Several aphid lineages date from an 80-million-year-old amber deposit, implying that this is the minimum age of the most recent common ancestor of this set of aphids as well as of the original infection with *Buchnera* (Heie, 1987). Comparative sequence analyses of 16S rDNA of *Buchnera* from a dozen aphid species suggest that the original infection dates to 200 to 250 million years ago (Moran and Baumann, 1994).

The radiation of homopteran insects into the sap-feeding niches provided by vascular plants was probably dependent on the early acquisition of bacterial mutualists. This possibility is supported by the fact that most plant saps lack nutrients essential for insects and that almost all modern Homopterae harbor symbionts (Buchner, 1965; Douglas, 1989). These considerations suggest that the infection that led to modern *Buchnera* might have occurred in the ancestor common to aphids and related insects. However, sequence analyses have indicated that symbionts have been acquired more than once within the sternorrhynchous Homopterae.

The nutritional contribution of *Buchnera* to the aphid's physiology has been studied extensively, mainly using chemically defined diets containing labeled compounds and aposymbiotic insects produced by injection of antibiotics, such as rifampicin and tetracyclin, as well as heat treatment. Among the findings from these studies, the most conspicuous is amino acid metabolism in aphids, in which *Buchnera* plays a crucial role. Phloem sap, the sole diet of aphids, is rich in carbohydrates but notoriously poor in nitrogenous compounds (Douglas, 1993). Thus, it had long been an enigma that aphids were so prosperous with extraordinarily high fecundity, notwithstanding their poor diet. Sasaki et al. (1990) showed that, unlike many other insects, aphids do not excrete nitrogenous wastes as uric acid but recycle the amino group by producing the amino acid glutamine, which is taken up by *Buchnera*. It is now evident that in aphids most, if not all (Douglas, 1998), essential amino acids are synthesized by *Buchnera* using glutamine, both ingested and recycled, as a substrate (Sasaki and Ishikawa, 1995). Also, many lines of evidence strongly suggest that the vitamin riboflavin is synthesized and supplied to the host insect by *Buchnera* (Nakabachi and Ishikawa, 1999).

Studies of the primary endosymbiont of aphids later designated *Buchnera* from molecular points of view were initiated in the early 1980s. An epoch-making finding in earlier studies was that *Buchnera*, when housed in the bacteriocyte, selectively synthesized a single protein species, symbionin, in large quantities (Ishikawa, 1982), which was later demonstrated to be a homolog of GroEL, a stress protein of *Escherichia coli* that functions as a molecular chaperone (Hara et al., 1990; Kakeda and Ishikawa, 1991; Ohtaka et al., 1992). These studies led to the finding that intracellular bacterial symbionts in both insects and eukaryotes at large tended to produce GroEL homologs selectively (Ahn et al., 1994; Aksoy, 1995; Baumann et al., 1996; Charles et al., 1997a). It is likely that in addition to chaperone activity, the GroEL homologs produced by bacterial symbionts have evolutionarily acquired new functions (Morioka et al., 1994; Yoshida et al., 2001). Recently, Fares et al. (2002) suggested that the GroEL homolog produced by *Buchnera* could be essential for buffering against deleterious mutations inherent in the endosymbiont, which may also account for the rampant production of essential amino acids by *Buchnera*. The finding that the genomic DNA of *Buchnera* is very rich in A and T (Ishikawa, 1987) is now applicable to almost all known bacterial endosymbionts (Clark et al., 1999).

Recent analysis of the *Buchnera* genome confirmed the results of the nutritional studies. Phylogenetically, *Buchnera* is a close relative of *E. coli* but differs from the latter in that it contains more than 100 genomic copies per cell (Komaki and Ishikawa, 1999); in addition, its genome size is only one seventh that of *E. coli* (Charles and Ishikawa, 1999). There are genes that are used in the biosynthesis of amino acids essential for the hosts in the *Buchnera* genome, but those for nonessential amino acids are largely absent, indicating complementarity and syntrophy between the host and *Buchnera* (Shigenobu et al., 2000). The *Buchnera* genome contains a gene for the riboflavin synthase b chain, which is actively expressed, confirming the contribution of *Buchnera* with respect to the provision of this vitamin (Nakabachi and Ishikawa, 1999). By contrast, it was shown that the *Buchnera* genome did not contain genes for the sterol synthesis, excluding the long-suspected possibility that this symbiont should provide aphids with cholesterol, a precursor to the molting hormone. In addition, *Buchnera* lacks genes for the biosynthesis of cell-surface components, including lipopolysaccharides and phospholipids, regulator genes, and genes involved in the defense of the cell (Shigenobu et al., 2000). Knowing how *Buchnera* makes up for the lack of these gene products to retain its identity as a cell will be of paramount importance (Shigenobu et al., 2001; Shimomura et al., 2002).

For all the contributions of *Buchnera* to the host, it is this symbiotic system that is most readily sacrificed for the host's survival. While aphids are kept starved, their total bacteriocyte volume consistently decreases at a much higher rate than that of the decline of their body weight. Upon resumption of the feeding, their total volume of bacteriocytes immediately restores the original value. These results strongly suggest that aphids will surmount the nutritionally adverse conditions by consuming their endosymbionts as nutritional reserves (Kobayashi and Ishikawa, 1993). The fact that the reverse will never happen seems to manifest rightly the nature of this mutualism between aphids and *Buchnera*. Apparently, the insect host profits much more from this association.

Psyllids, or jumping plant lice, differ from aphids in many aspects of life history and biogeography but are similar in that they feed on phloem sap and harbor maternally transmitted endosymbionts (Fukatsu and Nikoh, 1998; Spaulding and von Dohlen, 2001). The primary endosymbionts of psyllids are housed in bacteriocytes, where they are enclosed by host-derived membrane vesicles. The phylogenetic tree derived from the 16S-23S rDNA of these symbionts agrees with the tree derived from a host gene, a result consistent with a single infection of the psyllid ancestor and subsequent cospeciation of endosymbionts and hosts (Thao et al., 2000). Thus, these symbionts are quite similar to *Buchnera* in many respects but differ from the latter in that they constitute a unique lineage within the γ-3 subdivision of the proteobacteria (Spaulding and von Dohlen, 1998). According to a recent report on 37 kb of its DNA sequenced so far, the genome of the endosymbionts of psyllids has been undergoing an evolution that is reminiscent of *Buchnera*, but its degeneration seems even more extreme than in the *Buchnera* genome. Among the unusual properties are an exceptionally low G+C content (19.9%), almost complete absence of intergenic spaces and operon fusion, and a lack of usual promoter sequence upstream of 16S rDNA (Clark et al., 2001). To further validate these findings and to indicate which genes are missing from the genome, analysis of its entire sequence is prerequisite.

Secondary Symbionts

Many insects, homopterans in particular, do not seem so ascetic as to restrict themselves to mutualistic association with only one class of endosymbiont. They harbor additional inhabitants that are collectively called secondary symbionts. Many aphids also harbor secondary symbionts (Buchner, 1965; Fukatsu and Ishikawa, 1993). Whereas *Buchnera*, the primary symbionts, are usually globular in shape with a diameter of 2 to 4 μm, secondary symbionts vary in shape. Secondary symbionts in aphids have usually been located in syncytia adjacent to the bacteriome housing *Buchnera*, but they may occupy other locations (Buchner, 1965; Hinde, 1971; Fukatsu et al., 2000). Occurrence of the secondary symbionts has no clear correlation to the phylogeny of aphids, suggesting that they were acquired many times in various lineages independently. Even within a single aphid species the presence of secondary symbionts is not consistent and varies due to its biotype or habitat. These facts suggest that the association of aphids with these symbionts is labile and that their acquisitions by aphids are much more recent than that of *Buchnera*. The secondary symbionts also vary phylogenetically. Whereas most of them are, like *Buchnera*, members of the γ-proteobacteria (Munson et al., 1991), some may be a *Rickettsia* (Chen et al., 1996) or a *Spiroplasma* (Fukatsu et al., 2001). Roles played by secondary symbionts in aphid biology are largely unknown, although a recent study demonstrated condition-dependent effects on host growth and reproduction (Chen et al., 2000). Some secondary symbionts of the pea aphid, *Acyrthosiphon pisum*, are known to harbor a bacteriophage referred to as APSE-1 (after *A. pisum* secondary endosymbiont), which belong to the Podoviridae and are related to phage P22 of *Salmonella enterica* (van der Wilk et al., 1999).

Yeast-Like Symbionts

Planthoppers differ from aphids in that they harbor eukaryotic yeast-like symbionts (YLSs) instead of *Buchnera* in the mycetocyte. YLSs are transmitted directly from mothers to their offspring by transovarial infection and are found in the host at every developmental stage (Chen et al., 1981). Many lines of evidence suggest that YLSs are essential for normal development and growth of the host. Molecular phylogenetic analysis placed YLSs in the class Pyrenomycetes in the subphylum Ascomycotina (Noda et al., 1995). Nutritional roles played by YLSs were studied extensively using the Asian rice brown planthopper *Nilaparvata lugens*. It turned out that, unlike aphids, planthoppers did produce uric acid as a nitrogenous waste product but, unlike many other insects, did not excrete it. It has been shown that uric acid synthesized by the insect is stored in insect tissues rather than excreted and that uric acid is converted into compounds of nutritional value with the aid of uricase of YLSs (Sasaki et al., 1996). Evidence suggests that planthoppers synthesize uric acid both as a waste product and as a nitrogenous storage product when they ingest nitrogen-rich diets. Under conditions of nitrogen deficiency the stored uric acid is recycled by YLSs to sustain the growth of the host insect (Hongoh and Ishikawa, 1997; Wilkinson and Ishikawa, 2001). This mode of nitrogen recycling is reminiscent of the nutritional role played by the intracellular symbiont in cockroaches (see below).

Similar eukaryotic endosymbionts are present in some aphid species. Whereas the great majority of aphid species harbor the bacterial symbiont *Buchnera* intracellularly, a few groups of the tribe Cerataphidini are exceptional in that they have extracellular, eukaryotic symbionts instead of *Buchnera* that resemble YLSs of planthoppers described above (Buchner, 1965; Fukatsu and Ishikawa, 1992). For convenience, the eukaryotic symbiont of aphids is here tentatively referred to as YLSa. Molecular phylogenetic analysis showed that the YLSa also belongs to the class Pyrenomycetes (Fukatsu and Ishikawa, 1996). Within the class, it has been suggested that YLSa are particularly close to *Cordyceps* and its relatives, which raises the interesting possibility that these endosymbionts stem from an entomogenous fungus. Since the YLSa-containing aphid species are nested within the large clade of *Buchnera*-harboring species, they appear to form a clade descended from the common ancestor of aphids in which YLSa replaced *Buchnera*. If this is the case, it follows that YLSa must be functionally homologous to *Buchnera*. However, evidence is needed for this proposition, and a function has not yet been assigned to YLSa. The cause of this takeover is also not understood. Although YLSa-containing aphids do not store uric acid, and YLSa themselves do not have uricase, the genome of YLSa contains pseudogenes that retain traces of the genes for uricases, suggesting a close relationship between YLSa and the YLSs of planthoppers. A phylogenetic tree constructed based on sequences of these genes strongly suggests that the YLSs were horizontally transferred from the aphids' lineage to the planthoppers' (Hongoh and Ishikawa, 2000).

Intracellular Symbionts of Cockroaches and Termites

It was the blattid species *Blattella germanica* in which Blochmann discovered bacterial symbionts for the first time in the late 19th century. In this regard, this group was the object of classic studies on insect symbiosis (Buchner, 1965). All the known species of cockroach harbor bacterial symbionts in the bacteriocytes embedded in the fat body. Unlike many other symbionts housed in the bactriocyte, which are members of the γ-subdivision of the proteobacteria, those of cockroaches, together with those of primitive termites, form a clade within the Bacteroides-Favobacteria group and are, thus, distantly related to other known bacteriocyte associates (Bandi et al., 1994, 1995). Despite the difference in their phylogenetic origin, the roles of these symbionts in the host's physiology do not appear to be much different from those of the others. In particular, their nutritional role in securing organic nitrogens is quite similar to that played by YLSs of planthoppers mentioned above. There is convincing evidence for nitrogen recycling in cockroaches in which intracellular symbionts play an important role. Cockroaches in general store large amounts of uric acid in specialized cells, called urate cells or urocytes, which are located adjacent to the bacteriocytes in the fat body. According to studies with *Periplaneta americana*, these insects never excrete uric acid, and the uric-acid content per unit of body weight varies with the nitrogen content of the diet they ingest (Mullins and Cochran, 1975). In addition, aposymbiotic insects accumulate more uric acid than symbiotic ones. Indeed, there is in general an inverse relation between the number of the symbionts in the fat body and the quantity of uric acid. These observations led to the conclusion that the bacteriocyte symbionts mobilized the uric-acid reserves in the urocyte under conditions of nitrogen shortage and released nitrogenous compounds, perhaps as amino acids, to the host.

As for the cockroaches rampant in cities, such a safeguard using endosymbionts seems unnecessary, because they always gain access to nitrogen-rich leftovers. The fact that they keep an apparently unnecessary safeguard reminds us that their ancestor used to live in the tropical rain forests, where nitrogenous compounds were not easily available, and that many of their remote relatives still live there. In this context, for primitive termites that feed exclusively on decayed or rotten wood a nitrogen-recycling system, such as that found in cockroaches, would be very desirable. In fact, though both termites and their intracellular symbionts are phylogenetic relatives of cockroaches and their symbionts described above, respectively, there is no evidence that the termites' symbionts mobilize uric acid. No other functions have been suggested for the intracellular symbionts of termites, either. It is possible that many microbes with various abilities, including nitrogen fixation, living in the termites' gut did not necessitate the specific contribution of intracellular symbionts. There is circumstantial evidence suggesting that some gut microbes play a role in mobilizing uric acid unaerobically without the aid of uricases (M. Ohkuma, personal communications). It is likely that the bacterium that associated with the common ancestor of cockroaches and termites has differentiated extensively since the two hosts' lineages diverged.

Intracellular Symbionts of Coleoptera

Intracellular symbiosis of microorganisms with Coleoptera has been studied in some detail using the weevil *Sitophilus oryzae*. The weevil, which is devoid of gut microbes, harbors intracellular bacterial symbionts in the bacteriocyte. These bacteriocytes assemble to form the bacteriome, which is arranged somewhat like a packsaddle around the midgut of the larvae. In adults, the bacteriomes are located at the apexes of the mesenteric caeca and ovaries. The symbionts lie free in the cytosol of bacteriocytes and are transmitted to the host's progeny through oocytes. These symbionts are intimately involved in the host's physiology and have been suggested to provide the host with several vitamins, such as riboflavin, pantothenic acid, and biotin (Wicker, 1983), and to take part in amino-acid metabolism in the host (Nardon and Grenier, 1988). It has also been suggested that these symbionts stimulate oxidative phosphorylation in mitochondria in the host cell (Heddi et al., 1993).

The symbiont of *S. oryzae* belongs to the Enterobacteriaceae of the γ-proteobacteria (Campbell et al., 1992), and its genome size was estimated to be about 3 megabases (Mb) (Charles et al., 1997b). It is likely that, as a result of their association with the host for a prolonged time, these symbionts have completely lost their ability to grow outside the host cell. However, this symbiosis is distinct from that between the aphid and *Buchnera* in that the hosts deprived of the symbionts not only are viable but also retain fecundity. A nucleocytoplasmic incompatibility between the symbiotic and aposymbiotic strain of *S. oryzae* has been reported (Nardon and Grenier, 1988).

Intracellular Symbionts of Blood Suckers

It has been pointed out that regardless of their taxonomic position, those insects that are completely dependent on mammalian blood for food throughout their life cycle harbor symbiotic bacteria, whereas those that feed on blood during only part of the life cycle do not (Trager, 1986). Thus, fleas, mosquitoes, sand flies, black flies, and house flies, which all feed as larvae on diets rich in microorganisms, do not have symbiotic bacteria, whereas the blood-sucking Hemiptera and Anoplura feed only on blood and do have symbiotic bacteria. A remarkable contrast is provided by the two closely related muscid flies, the stable fly *Stomoxys*, and the tsetse fly *Glossina*. The larvae of *Stomoxys* are free living and feed on microorganism-rich manure, whereas those of *Glossina* develop within the mother fly, where they are nourished by the milk gland, an organ used specially for this purpose. In accordance with these different feeding habits, only *Glossina* have symbiotic bacteria, which occur in bacteriomes forming a portion of the midgut of the fly. These symbionts, which are designated *Wigglesworthia glossinidae*, are transmitted transovarially, via the nurse cells, and probably also by means of the secretion of the milk gland to the feeding larvae (Aksoy et al., 1995). Treatment of female tsetse flies with bactericidal agents destroys *Wigglesworthia* and at the same time interferes with the flies' ability to reproduce, suggesting that these symbionts supply essential nutrients. Recently, the genome of *Wigglesworthia* was successfully sequenced and annotated. In addition to *Wigglesworthia,* the tsetse fly harbors a facultative, secondary symbiont called *Sodalis* and a guest microbe, *Wolbachia* (see below) (S. Aksoy, personal communications).

Likewise, the louse *Pediculus capitis* harbors obligately symbiotic bacteria. If a female louse is surgically deprived of its bacteriomes before the symbionts have migrated to the oviduct, it fails to reproduce, whereas it reproduces in a normal way if the bacteriomes are extirpated after infection of the oviducts has occurred. Larval lice deprived of their bacteriomes would die before long, but their life could be prolonged by intrarectal feeding of yeast extract, again suggesting that the symbionts supply essential nutrients to the host (Trager, 1986).

GUEST MICROBES IN INSECTS

It is generally held that once a microorganism has strict or nearly strict vertical transmission it will inevitably evolve to a mutualistic association with its host (Lipsitch et al., 1995). Indeed, all the mutualists mentioned above are heritable symbionts. However, the reverse is not necessarily true. Not all heritable symbionts in insects are necessarily mutualists, or they do not seem to be in light of our present knowledge. Several microorganisms have effects that enhance their own spread but apparently have zero or negative effect on host fitness. They are effectively parasitic, although the severity of their effects varies. In these instances, in which effects of an association are not known, microbial associates are referred to simply as maternally or cytoplasmically inherited microorganisms, which are so-called guest microbes (Ebbert, 1993). Guest microbes are apparently not mutualistic with their hosts but exploit the hosts by interfering with their sexuality and reproduction (Hoffman et al., 1986; Breeuwer and Werren, 1990).

SEX-RATIO DISTORTERS

It is now known that guest microbes inhabit many arthropod species including insects and crustaceans. In general, they were first identified as agents that interfere with the sex ratio (SR) or reproduction of their host animals rather than through microscopic observations that detect mycetocyte symbionts. It has long been known that the SR of animals is sometimes drastically shifted from the normal value of 1. When the SR distortion is inherited through generations, it is called the SR trait, a genetic trait, irrespective of its cause (Ebbert, 1993). The SR trait is caused mainly by abnormal behaviors of the animals' chromosomes or respective nuclear genes during meiosis. When elements other than nuclear genes, such as cytoplasmic factors, are suspected as the cause of the SR trait, reciprocal and backcrosses of animals are coventionally performed. To exclude the involvement of mitochondria, the trait is tested for horizontal transmission and effectively eliminated by heat shock or antibiotics. These examinations have traced some of the SR traits to guest microbes that inhabit the cytoplasm of animals and are transmitted through host generations. When the guest microbes serve as the SR distorters, they almost always somehow bias the SR toward females. This is not surprising because cytoplasmic inhabitants, including mitochondria, are necessarily transmitted to the next generation via eggs but not sperms. Thus, producing the female-biased SR is a selfish strategy inherent in the guest microbes. They produce a female bias by various means, including male killing, induction of parthenogenesis, and feminization.

SPIROPLASMA, AN EARLY MALE-KILLER

Among the extrachromosomal genetic elements that cause SR traits, the first described is *Spiroplasma*. Sakaguchi and Poulson (1961) first described the presence of a "sex-ratio organism (SRO)" in the cytoplasm of neotropical *Drosophila* species. They showed that in some *Drosophila* populations, females infected with SRO do not give birth to males by killing their sons selectively at an early stage of development, which leads eventually to an all-female population. They also showed that the SR trait could be transferred interspecifically by microinjection of hemolymph to other individuals (Sakaguchi and Poulson, 1963). Later, the SRO was identified as a spiral-shaped bacterium that lacked a cell wall and was thus designated *Spiroplasma*. Because of its spiral shape and mobility, *Spiroplasma* was once mistakenly thought to be a member of spirocetes. Also, because of its membrane filterability, it was sometimes mistaken for a virus.

Spiroplasma species belong to the order Mycoplasmatales, and more than 50 strains have been identified to date from more than 30 species (Williamson, 1998). Whereas the known genome sizes of *Mycoplasma* species are less than 1 Mb, those of *Spiroplasma* have been estimated to be around 1.5 Mb (Dybvig, 1990). Unlike mutualistic symbionts, which are usually localized in specific cells such as bacteriocytes, *Spiroplasma* are almost ubiquitous, both extracellular and intracellular, in the tissues, including hemolymph, of an infected host.

Male killing caused by *Spiroplasma* is an early killing, so called because male mortality occurs typically during embryogenesis or the first larval instar. The early male killing has commonly been studied using *Drosophila* species infected with *Spiroplasma*. Among the progeny produced by the infected female flies, males are selectively killed early in embryogenesis. The mechanisms underlying this son killing are largely unknown, whereas some observations may provide clues to understand them. For instance, in some *Drosophila* species and lineages, it is known that early and later broods will scarcely receive the bacteria from their mother. However, this incomplete transmission is not associated with the appearance of sons. The daughters that are scored as uninfected may in fact be infected with very low numbers of *Spiroplasma*, which may be sufficient, however, to kill their brothers (Ebbert, 1993). However, an alternative explanation for this phenomenon suggests that male-lethal spiroplasmas produce a diffusible toxin or virus that actually kills males (Williamson, 1965). Indeed, there are some suggestive molecular differences in the number and type of virus they harbor (Cohen et al., 1987).

It is known that *Spiroplasma* species harbor arrays of extrachromosomal genetic elements. Among them, SVC3, a polyhedrosis virus with a short tail, is shared by all the known species of *Spiroplasma*. Interestingly, it has been demonstrated that SVC3 isolated from a strain induces cell lysis of other strains of *Spiroplasma* (Whitcomb, 1980). Since every strain harbors its own SVC3, the infection of the virus itself obviously cannot be responsible for the cell lysis. It has long been known that there is a competition between strains of *Spiroplasma* that cause the SR trait and those that do not with respect to the infection in the *Drosophila* hosts. It is highly likely that this phenomenon is due to the lytic action of the SVC3 they harbor. It is a matter of conjecture whether or not the lytic action of SVC3 is somehow implicated in the male killing by spiroplasmas.

In addition to viruses such as SVC3, spiroplasmas harbor various sized, closed-circular DNA whose functions have yet to be discovered. These are collectively known as latent plasmids. Some of them are on occasion integrated into the chromosomal DNA of spiroplasmas (Whitcomb, 1980).

Microsporidia, a Late Male-Killer

Microsporidia are eukaryotes that are classified as either protists or fungi. They are well known as one of the few eukaryotes that lack mitochondria. Because of this, microsporidia were once considered to be direct descendants of a primitive eukaryote before capturing an ancestral bacterium of mitochondria. Now, however, many lines of evidence suggest that they have secondarily lost the cell organelle because of a prolonged life in anaerobic milieus.

Microsporidia are also known as male-killing agents of certain insects. More precisely, they are late killers, so called because male mortality occurs late in development, typically in late larval instars. Whereas microsporidia can be transmitted both vertically and horizontally in many insects, late male killing has so far only been recorded in mosquitoes. When microsporidia are transmitted horizontally, the first stage of infection is represented by their invasion into the epithelium of the gastric caeca, from which they invade oenocytes. In newly infected mosquitoes, the microsporidians typically develop in a benign manner in both sexes and are vertically transmitted to the next generation through the eggs. Only in this next generation will male killing arise (Bechnel and Sweeney, 1990).

Here the death of the male is associated with the rupturing of the cuticle during the release of infective spores in its later instars. The potential for horizontal transmission may explain why males are killed late in development. After the fourth instar, mosquitoes pupate in the water, after which adult mosquitoes are only rarely associated with the water. In addition, the hard pupal case can be a barrier for the guest microbes to escape from their host to water to seek a new host. Thus, the fourth instar would be the last opportunity for them to escape from the male host, through which no vertical transmission is possible. The question then arises of why they do not escape from males earlier. This is probably because of an optimizing strategy. Given that vertical transmission is not possible in males, the guest microbes wait until the host is as large as possible before entering the full-blown infectious state, thereby maximizing spore number (Hurst, 1991). This strategy somewhat resembles that of certain parasitoid wasps in which wasps manipulate the development of lepidopteran larvae and escape from them before pupation after fully exploiting the constituents of larval bodies.

Wolbachia

Of the several known guest microbes, the one that now commands the attention in almost all areas of the life sciences, from sociobiology to molecular biology, is *Wolbachia*. *Wolbachia* are strictly intracellular bacteria infecting a number of invertebrates including mites, crustaceans, filarial nematodes, and especially insects (Werren and O'Neill, 1997). A recent survey showed that about 76% of arbitrarily chosen insect species were infected with these guest microbes, making this group one of the most ubiquitous endosymbionts described to date (Jeyaprakash and Hoy, 2000). Maternally transmitted through the cytoplasm of eggs, these endosymbionts form a monophyletic group relative to other α-proteobacteria, particularly to other *Rickettsia*, causing human diseases such as typhus, Rocky Mountain spotted fever, and Q fever. In arthropods, they are distinguished by their ability to modify host reproduction in a variety of ways, such as cytoplasmic incompatibility (CI) in most species (Hoffmann and Turelli, 1997), thelytokous parthenogenesis in haplodiploid species (Stouthamer et al., 1990), male killing in several insects (Hurst et al., 1999), and feminization of genetic males in isopod crustaceans (Rigaud, 1997). All these effects are advantageous to *Wolbachia* and allow them to persist and spread their share in host populations (Hoffmann and Turelli, 1997; Rigaud, 1997). The present discussion is limited to a brief introduction of CI and the evolution and molecular biology of *Wolbachia*; other interesting phenomena caused by *Wolbachia* are described in detail in later chapters of this book.

CI occurs when crosses between *Wolbachia*-infected males and uninfected females fail to produce progeny, whereas both crosses between uninfected males and infected females and those between infected males and females are fertile (Yen and Barr, 1971). This suggests that the *Wolbachia* infection interferes with the function of sperm or the male chromosomes and that an infected oocyte is essential to rescuing proper function of the male gamete. In some insect species, chromosomes from the sperm of *Wolbachia*-infected males fail to condense in the cytoplasm of an uninfected egg but do condense in that of an infected egg (Breeuwer and Werren, 1990; O'Neill and Karr, 1990; Montchamp-Moreau et al., 1991). In infected insects, *Wolbachia* are detected in nearly all the tissues but not in the sperm. The molecular mechanisms by which *Wolbachia* modify chromosomes of the sperm of the infected males and how the modified chromosomes are rescued in the egg cytoplasm harboring *Wolbachia* are still largely unknown.

From an evolutionary point of view, incompatibility infections of *Wolbachia* in *Culex* mosquitoes are much more interesting. In *Culex* mosquitoes, no naturally occurring uninfected individuals have been reported. Notwithstanding this, it is only the matings between individuals from the same population that are always successful. Matings between individuals from distinct populations tend to fail, even if both the males and females are infected with *Wolbachia*. These observations can be explained by assuming that the genetic trait of *Wolbachia* is variable due to the host population and that only the egg cytoplasm infected with the same variant is able to rescue the chromosomes of sperm from the infected males (Ebbert, 1993). If this is the case, interpopulation incompatibility may occur between two populations that are infected with different variants of *Wolbachia*, which may eventually lead to sympatric speciation of species.

As in bacteriocyte symbionts, *Wolbachia* infections are maternally inherited through transmission from females to their eggs before oviposition. Regular maternal inheritance would seem to create the potential for cospeciation of *Wolbachia* and its hosts. However, the phylogenetic results based on the sequences of 16S rDNAs from various *Wolbachia* species provide an interesting contrast to those from *Buchnera*. Results indicate that *Wolbachia* lineages have sometimes been transmitted among members of different insect orders (Moran and Baumann, 1994). For this reason, *Wolbachia* are sometimes called evolutionary hitchhikers. The extent of divergence among the most distant of the examined *Wolbachia* 16S sequences is small (<3%) (O'Neill et al., 1992) relative to that among distantly related *Buchnera* (>8%) (Moran et al., 1993). Assuming substitution rates in *Wolbachia* are not much different from those estimated for *Buchnera* and other prokaryotes, the common ancestor of examined *Wolbachia* lived about 25 to 100 million years ago. This is much younger than any common ancestor of the hosts (>300 million years), lending further evidence that the distribution of *Wolbachia* is the result, in part, of occasional horizontal transfer of bacterial lineages among lineages of arthropods (Moran and Baumann, 1994). Evidence for more than one *Wolbachia* strain within individual insects (Rousset et al., 1992) is consistent with the view that horizontal transfer has been an important determinant of the modern distribution of *Wolbachia* among hosts.

Phylogenetic trees of *Wolbachia* strains have been constructed based on the sequences of 16S rDNA and parts of the protein-coding region of *ftsZ* (O'Neill et al., 1992; Werren et al., 1995; Tsagkarakou et al., 1996). The *ftsZ* tree had a finer resolution and divided *Wolbachia* into two major groups, designated A and B. Evidence suggests that the two diverged from each other 58 to 67 million years ago (Werren et al., 1995). The tree also implied frequent horizontal transfer of *Wolbachia*, in A group in particular, between distantly related insect orders. To further define the relationship among A group *Wolbachia* strains, a phylogenetic tree was constructed based on the sequences of parts of the *groE* operons, which suggested unexpected patterns of horizontal transfer of these bacteria (Masui et al., 1997).

It is assumed that in CI, *Wolbachia* modify host sperm during maturation, possibly by secreting some proteinaceous factors. Based on this assumption, attention was directed toward the macromolecule secretion system in *Wolbachia*. Recently, it was found that *Wolbachia* contain the genes encoding homologs to the type IV secretion system by which many pathogenic bacteria, such as *Agrobacterium tumefaciens* (Zambryski, 1988) and *R. prowazekii* (Andersson et al., 1998), secrete macromolecules. The genes identified encode most of the essential components of the secretion system and are cotranscibed as an operon (Masui et al., 2000a).

Recent studies have shown that *Wolbachia* are a playground for various mobile and extrachromosomal genetic elements. This may be one of the reasons why an earlier completion of the sequencing of the whole genome of *Wolbachia* has been prevented, despite its small size (1.2 Mb). An insertion sequence, designated ISW*1*, was identified in the genome of a *Wolbachia* strain, *w*Tai infecting the Taiwan cricket, *Teleogryllus taiwanemma* (Masui et al., 1999). ISW*1* displays a significant similarity to IS*200*, an IS element first identified in *Salmonella* that lacks the terminal invert repeat (Lam and Roth, 1986). There were at least 20 copies of ISW*1* on the chromosome of *w*Tai. Sequence analysis of 9 ISW*1* copies and their flanking regions showed that the copies were identical and suggested that ISW*1* had no preference for its insertion sites. It is hoped that this mobile genetic element, if used in combination with a strong promoter such as that of *wsp* (Braig et al., 1998), could be a useful vector for transforming *Wolbachia.*

A prophage-like genetic element was also detected in the *Wolbachia* genome and named WO (Masui et al., 2000b). All the *Wolbachia* strains examined were shown to contain the phage WO. The phylogenetic tree based on phage WO genes of several *Wolbachia* strains was not congruent with that based on chromosomal genes of the same strains, suggesting that phage WO is active and horizontally transmitted among various *Wolbachia* strains. Although the phage genome contains genes of diverse origins, their average G+C content and codon usage are quite similar to those of chromosomal genes. These results raised the possibility that phage WO has been associated with *Wolbachia* for a very long time, conferring some benefit to its host bacteria. The phage WO genome contains a gene derived from a plasmid and genes for several ankyrin-like proteins similar to those of mammalians and plants. Since it has been suggested that ankyrin-like repeats are a motif for protein–protein interaction (Sedgwick and Smerdon, 1999), it is possible that the proteins containing eukaryotic ankyrin-like repeats encoded by the phage WO genes play a role in the reproductive alteration of insect hosts by *Wolbachia* through their ability to interact with other proteins. Recently, virus-like particles that probably represent the phage WO were observed in *Wolbachia* by electron microscopy (Masui et al., 2001).

CONCLUSION

The effects of cytoplasmically inherited microbes range from those of obligate mutualists for host survival to those of selfish parasites that may cause all of a host's progeny to die prematurely. The great diversity of insects is partly due to their frequent associations with mutualistic endosymbionts, such as *Buchnera*, which enable hosts to exploit niches that would otherwise be nutritionally unsuitable. In other instances, bacterial associates may promote speciation through direct effects on reproductive systems. For example, induction of incompatibility or parthenogenesis by *Wolbachia* could effect reproductive isolation of host population (Stouthamer et al., 1990). The extent of mutualism vs. antagonism underlying a particular association will determine the degree of conflict between selection pressure on host and symbiont, with implications for the evolution of both participating lineages (Moran and Baumann, 1994).

To date, many studies have suggested how symbiotic microbes have undergone evolutionary changes in the course of association with insect hosts. To further understand the evolutionary significance of interspecific association in nature, more studies on the side of hosts are desirable. A key issue is how insects have been changed, from molecular and cellular points of view, by acquiring microbial partners.

REFERENCES

Ahmadjian, V. and Paracer, S. (1986). *Symbiosis: An Introduction to Biological Associations.* University Press of New England, Hanover, NH.

Ahn, T.I., Lim, S.T., Leeu, H.K., Lee, J.E., and Jeon, K.W. (1994). A novel strong promoter of the *groEx* operon of symbiotic bacteria in *Amoeba proteus. Gene* **148:** 43–49.

Aksoy, S. (1995). Molecular analysis of the endosymbionts of tsetse flies: 16S rDNA locus and over-expression of a chaperonin. *Insect Mol. Biol.* **4:** 23–29.

Aksoy, S., Pourhosseini, A.A., and Chow, A. (1995). Mycetome endosymbionts of tsetse flies constitute a distinct lineage related to Enterobacteriaceae. *Insect Mol. Biol.* **4:** 15–22.

Anderson, J.M., Rayer, A.D.M., and Walton, D.W.H. (1984). *Invertebrate–Microbial Interactions.* Cambridge University Press, Cambridge, U.K.

Andersson, S.G., Zomorodipour, A., Andersson, J.O., Sicheritz-Ponten, T., Alsmark, U.C., Podowski, R.M., Naslund, A.K., Eriksson, A.S., Winkler, H.H., and Kurland, C.G. (1998). The genome sequence of *Rickettsia prowazekii* and the origin of mitochondria. *Nature* **396:** 133–140.

Bandi, C., Damiani, G., Magrassi, L., Grigolo, A., Fani, R., and Sacchi, L. (1994). Flavobacteria as intracellular symbionts in cockroaches. *Proc. R. Soc. London (B)* **257:** 43–48.

Bandi, C., Sironi, M., Damiani, G., Magrassi, L., Nalepa, C.A., Laudani, U., and Sacchi, L. (1995). The establishment of intracellular symbiosis in an ancestor of cockroaches and termites. *Proc. R. Soc. London (B)* **259:** 293–299.

Baumann, P., Munson, M.A., Lai, C-Y., Clark, M.A., Baumann, L., Moran, N.A., and Campbell, B.C. (1993). Origin and properties of bacterial endosymbionts of aphids, whiteflies, and mealybugs. *ASM News*, **59:** 21–24.

Baumann, P., Baumann, L., and Clark, M.A. (1996). Levels of *Buchnera aphidicola* chaperonin GroEL during growth of the aphid *Schizaphis graminum. Curr. Microbiol.* **32:** 279–285.

Bechnel, J.J. and Sweeney, A.W. (1990). *Amblyospora trinus* n. sp. (Microsporida: Amblyosporidae) in the Australia mosquito *Culex halifaxi* (Diptera: Cullicidae). *J. Protozool.* **37:** 584–592.

Bracke, J.W., Cruden, D.L., and Markovetz, A.J. (1979). Intestinal microbial flora of the American cockroache *Periplaneta americana* L. *Appl. Environ. Microbiol.* **38:** 945–955.

Braig, H.R., Zhou, W., Doubson, S.L., and O'Neill, S.L. (1998). Cloning and characterization of a gene encoding the major surface protein of the bacterial endosymbiont *Wolbachia pipientis. J. Bacteriol.* **180:** 2373–2378.

Breznak, J.A. (1982). Intestinal microbiota of termites and other xylophagous insects. *Annu. Rev. Microbiol.* **36:** 323–343.

Breznak, J.A. (1984). Biochemical aspects of symbiosis between termites and their intestinal microbiota. In *Invertebrate–Microbial Interactions* (J.M. Anderson, A.D.M. Rayer, and D.W.H. Walton, Eds.), pp. 173–204. Cambridge University Press, Cambridge, U.K.

Breeuwer, J.A.J. and Werren, J.H. (1990). Microorganisms associated with chromosome destruction and reproductive isolation between two insect species. *Nature* **346:** 558–560.

Brooks, M.A. (1963). Symbiosis and aposymbiosis in arthropods. *Symp. Soc. Gen. Microbiol.* **13:** 200–231.

Buchner, P. (1965). *Endosymbiosis of Animals with Plant Microorganisms.* Interscience Publishers, New York.

Campbell, B.C., Bragg, T.S., and Turner C.E. (1992). Phylogeny of symbiotic bacteria of four weevil species (Coleoptera: Curculionidae) based on analysis of 16S ribosomal RNA. *Insect Biochem. Mol. Biol.* **22:** 415–421.

Charles, H. and Ishikawa, H. (1999). Physical and genetic map of the genome of *Buchnera*, the primary endosymbiont of the pea aphid *Acyrthosiphon pisum. J. Mol. Evol.* **48:** 142–150.

Charles, H., Heddi, A., Guillaud, J., Nardon, C., and Nardon, P. (1997a). A molecular aspect of symbiotic interactions between the weevil *Sitophilus oryzae* and its endosymbiotic bacteria: overexpression of a chaperonin. *Biochem. Biophys. Res. Commun.* **239:** 769–774.

Charles, H., Condemine, G., Nardon, C., and Nardon, P. (1997b). Genome size characterization of the principal endocellular symbiotic bacteria of the weevil *Sitophilus oryzae*, using pulse field gel electrophoresis. *Insect Biochem. Mol. Biol.* **27:** 345–350.

Chen, C.C., Cheng, L.-L., and Hou, R.F. (1981). Studies on the intracellular yeast-like symbionts in the brown planthopper, *Nilaparvata lugens* Stal. II. Effects of antibiotics and elevated temperature on the symbiotes and their host. *Z. Angew. Entomol.* **92:** 440–449.

Chen, D.Q., Campbell, B.C., and Purcell, A.H. (1996). A new *Rickettsia* from a herbivorous insect, the pea aphid *Acyrthosiphon pisum* (Harris). *Curr. Microbiol.* **33:** 123–128.

Chen, D.Q., Montllor, C.B., and Purcell, A.H. (2000). Fitness effects of two endosymbiotic bacteria on the pea aphid, *Acyrthosiphon pisum*, and the blue alfalfa aphid *A. kondoi. Entomol. Exp. Appl.* **95:** 315–323.

Clark, M.A., Baumann, P., Munson, M.A., Baumann, P., Campbell, B.C., Duffus, J.E., Osborne, J.S., and Moran, N.A. (1992). The eubacterial endosymbionts of whiteflies (Homoptera: Aleyrodoidea) constitute a lineage separate from the endosymbionts of aphids and mealybugs. *Curr. Microbiol.* **25:** 119–123.

Clark, M.A., Moran, N.A., and Baumann, P. (1999). Sequence evolution in bacterial endosymbionts having extreme base compositions. *Mol. Biol. Evol.* **16:** 1586–1598.

Clark, M.A., Baumann, L. Thao, M.L., Moran, N.A., and Baumann, P. (2001). Degenerative minimalism in the genome of a psyllid endosymbiont. *J. Bacteriol.* **183:** 1853–1861.

Cohen, A.J., Williamson, D.L., and Oishi, K. (1987). SpV3 viruses of *Drosophila* spiroplasmas. *Isr. J. Med. Sci.* **23:** 429–433.

Dasch, G.A., Weiss, E., and Chang, K-P. (1984). Endosymbionts of insects. In *Bergey's Manual of Systematic Bacteriology* (N.R. Krieg, Ed.), pp. 811–833. Williams & Wilkins, Baltimore.

De Bary, A. (1879). Die Erscheinung der Symbiose. *Naturforsch. Versammlung Cassel*, LI, Tagebl. p. 121.

Douglas, A.E. (1989). Mycetocyte symbiosis in insects. *Biol. Rev.* **64:** 409–434.

Douglas, A.E. (1993). The nutritional qaulity of phloem sap utilized by natural aphid populations. *Ecol. Entomol.* **18:** 31–38.

Douglas, A.E. (1998). Nutritional interactions in insect-microbial symbioses: aphids and their symbiotic bacteria *Buchnera*. *Annu. Rev. Entomol.* **43:** 17–37.

Dybvig, K. (1990). Mycoplasmal genetics. *Annu. Rev. Microbiol.* **44:** 81–104.

Ebbert, M.A. (1993). Endosymbiotic sex ratio distorters in insects and mites. In *Evolution and Diversity of Sex Ratio in Insects and Mites* (D.L. Wrensch and M.A. Ebbert, Eds.), pp. 150–191. Chapman & Hall, New York.

Fares, M.A., Ruiz-Gonzalez, M.X., Moya, A., Elena, S.F., and Barrio, E. (2002). GroEL buffers against deleterious mutations. *Nature* **417:** 398.

Fukatsu, T. and Ishikawa, H. (1992). Soldier and male of a eusocial aphid *Colophina arma* lack endosymbiont: implication for physiological and evolutionary interaction between host and symbiont. *J. Insect Physiol.* **38:** 1033–1042.

Fukatsu, T. and Ishikawa, H. (1993). Occurrence of chaperonin 60 and chaperonin 10 in primary and secondary bacterial symbionts of aphids: Implications for the evolution of an endosymbiotic system in aphids. *J. Mol. Evol.* **36:** 568–577.

Fukatsu, T. and Ishikawa, H. (1996). Phylogenetic position of yeast-like symbiont of *Hamiltonaphis styraci* (Homoptera, Aphididae) based on 18S rDNA sequence. *Insect Biochem. Mol. Biol.* **26:** 383–388.

Fukatsu, T. and Nikoh, N. (1998). Two intracellular symbiotic bacteria from the mulberry psyllid *Anomoneura mori* (Insecta, Homoptera). *Appl. Environ. Microbiol.* **64:** 3599–3606.

Fukatsu, T., Nikoh, N., Kawai, R., and Koga, R. (2000). The secondary endosymbiotic bacterium of the pea aphid, *Acyrthosiphon pisum* (Insecta: Homoptera). *Appl. Environ. Microbiol.* **66:** 2748–2758.

Fukatsu, T., Tsuchida, T., Nikoh, N., and Koga, R. (2001). *Spiroplasma* symbiont of the pea aphid, *Acyrthosiphon pisum* (Insecta: Homoptera). *Appl. Environ. Microbiol.* **67:** 1284–1291.

Gray, M.W. and Doolittle, W.F. (1982). Has the endosymbiont hypothesis been proven? *Microbiol. Rev.* **46:** 1–42.

Hara, E., Fukatsu, T., Kakeda, K., Kengaku, M., Ohtaka, C., and Ishikawa, H. (1990). The predominant protein in an aphid endosymbiont is homologous to an *E. coli* heat shock protein. *Symbiosis* **8:** 271–283.

Harada, H. and Ishikawa, H. (1993). Gut microbe of aphid closely related to its intracellular symbiont. *BioSystems* **31:** 185–191.

Harada, H. and Ishikawa, H. (1997). Experimental pathogenicity of *Erwinia aphidicola* to pea aphid, *Acyrthosiphon pisum*. *J. Gen. Appl. Microbiol.* **43:** 363–367.

Harada, H., Oyaizu, H., Kosako, Y., and Ishikawa, H. (1997). *Erwinia aphidicola*, a new species isolated from pea aphid, *Acyrthosiphon pisum*. *J. Gen. Appl. Microbiol.* **43:** 349–354.

Heddi, A., Lefebvre, F., and Nardon, P. (1993). Effect of endocytobiotic bacteria on mitochondrial enzymatic activities in the weevil *Sitophilus oryzae* (Coleoptera: Cuculionidae). *Insect Biochem. Mol. Biol.* **23:** 403–411.

Heie, O.E. (1987). Paleontology and phylogeny. In *Aphids: Their Biology, Natural Enemies and Control,* Vol. 2A (A.K. Minks and P. Harrewijn, Eds.), pp. 367–391. Elsevier, Amsterdam.

Hinde, R. (1971). The fine structure of the mycetome symbiotes of the aphids *Brevicoryne brassicae*, *Myzus persicae*, and *Macrosiphum rosae*. *J. Insect Physiol.* **17:** 2035–2050.

Hoffmann, A.A. and Turelli, M. (1997). Cytoplasmic incompatibility in insects. In *Influential Passengers: Inherited Microorganisms and Arthropod Reproduction* (S.L. O'Neill, A.A. Hoffmann, and J.H. Werren, Eds.), pp. 42–80. Oxford University Press, Oxford, U.K.

Hoffmann, A.A., Turelli, M., and Simmons, G.M. (1986). Unidirectional incompatibility between populations of *Drosophila simulans*. *Evolution* **40:** 692–701.

Hongoh, Y. and Ishikawa, H. (1997). Uric acid as a nitrogen resource for the brown planthopper, *Nilaparvata lugens*: studies with synthetic diets and aposymbiotic insects. *Zool. Sci.* **14:** 581–586.

Hongoh, Y. and Ishikawa, H. (2000). Evolutionary studies on uricases of fungal endosymbionts of aphids and planthoppers. *J. Mol. Evol.* **51:** 265–277.

Honigberg, B.M. (1970). Protozoa associated with termites and their role in digestion. In *Biology of Termites II* (K. Krishna, and F.M. Weesner, Eds.), pp. 1–36. Academic Press, London.

Houk, E.J. and Griffithis, G.W. (1980). Intracellular symbiosis of the Homoptera. *Annu. Rev. Entomol.* **25:** 161–187.

Hurst, G.D.D., Jiggins, F.M., von der Schulenburg, J.H.G., Bertrand, D., West, S.A., Goriacheva, I.I., Zakharov, I.A., Werren, J.H., Stouthamer, R., and Majerus, M.E.N. (1999). Male-killing *Wolbachia* in two species of insect. *Proc. R. Soc. London (B)* **266:** 735–740.

Hurst, L.D. (1991). The incidences and evolution of cytoplasmic male-killers. *Proc. R. Soc. London (B)* **244:** 91–99.

Ishikawa, H. (1982). Host-symbiont interactions in the protein synthesis in the pea aphid, *Acyrthosiphon pisum. Insect Biochem.* **12:** 613–622.

Ishikawa, H. (1987). Nucleotide composition and kinetic complexity of the genomic DNA of an intracellular symbiont in the pea aphid *Acyrthosiphon pisum. J. Mol. Evol.* **24:** 205–211.

Ishikawa, H. (1989). Biochemical and molecular aspects of endosymbiosis in insects. *Int. Rev. Cytol.* **116:** 1–45.

Ito, J., Ghosh, A., Moreira, L.A., Wimmer, E.A., and Jacobs-Lorena, M. (2002). Transgenic anopheline mosquitoes impaired in transmission of a malaria parasite. *Nature* **417:** 452–455.

Jeyaprakash, A. and Hoy, M.A. (2000). Long PCR improves *Wolbachia* DNA amplification: *wsp* sequences found in 76% of sixty-three arthropod species. *Insect Mol. Biol.* **9:** 393–405.

Jurzitza, G. (1979). The fungi symbiotic with anobiid beetles. In *Insect-Fungus Symbiosis* (L.R. Batra, Ed.), pp. 65–76. John Wiley & Sons, New York.

Kakeda, K. and Ishikawa, H. (1991). Molecular chaperone produced by an intracellular symbiont. *J. Biochem.* **110:** 583–587.

Kobayashi, M. and Ishikawa, H. (1993). Breakdown of indirect flight muscles of alate aphids (*Acyrthosiphon pisum*) in relation to their flight, feeding and reproductive behavior. *J. Insect Physiol.* **39:** 549–554.

Komaki, K. and Ishikawa, H. (1999). Intracellular bacterial symbionts of aphids possess many genomic copies per bacterium. *J. Mol. Evol.* **48:** 717–722.

Lam, S. and Roth, J.R. (1986). Structural and functional studies of insertion element IS*200*. *J. Mol. Biol.* **187:** 157–167.

Lanham, U.N. (1968). The Blochmann body. *Biol. Rev.* **43:** 269–286.

Lipsitch, M., Nowak, M.A., Ebert, D., and May, R.M. (1995). The population dynamics of vertically and horizontally transmitted parasites. *Proc. R. Soc. London (B)* **260:** 321–327.

Margulis, L. (1970). *Origin of Eukaryotic Cells*, Yale University Press, New Haven, CT.

Masui, S., Sasaki, T., and Ishikawa, H. (1997). *groE*-Homologous operon of *Wolbachia*, an intracellular symbiont of arthropods: a new approach for their phylogeny. *Zool. Sci.* **14:** 701–706.

Masui, S., Kamoda, S., Sasaki, T., and Ishikawa, H. (1999). The first detection of the insertion sequence ISW*1* in the intracellular reproductive parasite *Wolbachia. Plasmid* **42:** 13–19.

Masui, S., Sasaki, T., and Ishikawa, H. (2000a). Genes for the type IV secretion system in an intracellular symbiont, *Wolbachia*, a causative agent of various sexual alterations in arthropods. *J. Bacteriol.* **182:** 6529–6531.

Masui, S., Kamoda, S., Sasaki, T., and Ishikawa, H. (2000b). Distribution and evolution of bacteriophage WO in *Wolbachia*, the endosymbiont causing sexual alterations in arthropods. *J. Mol. Evol.* **51:** 491–497.

Masui, S., Kuroiwa, H., Sasaki, T., Inui, M., Kuroiwa, T., and Ishikawa, H. (2001). Bacteriophage WO and virus-like particles in *Wolbachia*, an endosymbiont of arthropods. *Biochem. Biophys. Res. Commn.* **283:** 1099–1104.

Montchamp-Moreau, C., Ferveur, J., and Jacques, M. (1991). Geographic distribution and inheritance of three cytoplasmic incompatibility types in *Drosophila simulans. Genetics* **129:** 399–347.

Moran, N.A. and Baumann, P. (1994). Phylogenetics of cytoplasmically inherited microorganisms of arthropods. *Trends Ecol. Evol.* **9:** 15–20.

Moran, N.A., Munson, M.A., Baumann, P., and Ishikawa, H. (1993). A molecular clock in endosymbiotic bacteria is calibrated using insect host. *Proc. R. Soc. London (B)* **253:** 167–171.

Morioka, M., Muraoka, H., Yamamoto, K., and Ishikawa, H. (1994). An endosymbiont chaperonin is a novel type of histidine protein kinase. *J. Biochem.* **116:** 1075–1081.

Mullins, D.E. and Cochran, D.G. (1975). Nitrogen metabolism in the American cockroach. *Comp. Biochem. Physiol.* **50A:** 489–510.

Munson, M.A., Baumann, P., and Kinsey, M.G. (1991). *Buchnera* gen. nov. and *Buchnera aphidicola* sp. nov., a taxon consisting of the mycetocyte-associated, primary endosymbionts of aphids. *Int. J. Syst. Bacteriol.* **41:** 566–568.

Munson, M.A., Baumann, P., and Moran, N.A. (1992). Phylogenetic relationships of the endosymbionts of mealybugs (Homoptera: Pseudococcidae) based on 16S rDNA sequences. *Mol. Phylogenet. Evol.* **1:** 26–30.

Nakabachi, A. and Ishikawa, H. (1999). Provision of riboflavin to the host aphid, *Acyrthosiphon pisum*, by endosymbiotic bacteria, *Buchnera*. *J. Insect Physiol.* **45:** 1–6.

Nardon, P. and Grenier, A.M. (1988). Genetical and biochemical interactions between the host and its endocytobiotes in the weevil *Sitophilus* (Coleoptera, Curculionidae) and other related species. In *Cell to Cell Signals in Plant, Animal and Microbial Symbiosis* (S. Scannerini, Ed.), pp. 255–270. Springer-Verlag, Berlin.

Nickle, W.R. (1984). *Plant and Insect Nematodes*. Marcel Dekker, New York.

Noda, H. (1974). Preliminary histological observation and population dynamics of intracellular yeast-like symbiotes in the smaller brown planthopper, *Laodelphax striatellus* (Homoptera: Dalphacidae). *Appl. Entomol. Zool.* **9:** 275–277.

Noda, H., Nakashima, N., and Koizumi, M. (1995). Phylogenetic position of yeast-like symbiotes of rice planthoppers based on partial 18S rDNA sequences. *Insect Biochem. Mol. Biol.* **25:** 639–646.

Ohtaka, C., Nakamura, H., and Ishikawa, H. (1992). Structure of chaperonins from an intracellular symbiont and their functional expression in *E. coli groE* mutants. *J. Bacteriol.* **174:** 1869–1874.

O'Neill, S.L. and Karr, T.L. (1990). Bidirectional incompatibility between conspecific populations of *Drosophila simulans*. *Nature* **348:** 178–180.

O'Neill, S.L., Giordano, R., Colbert, A.M.E., Karr, T.L., and Robertson, H.M. (1992). 16S rRNA phylogenetic analysis of the bacterial endosymbionts associated with the cytoplasmic incompatibility in insects. *Proc. Natl. Acad. Sci. U.S.A.* **89:** 2699–2702.

Richards, A.G. and Brooks, M.A. (1958). Internal symbiosis in insects. *Annu. Rev. Entomol.* **3:** 37–56.

Rigaud, T. (1997). Inherited microorganisms and sex determination of arthropod hosts. In *Influential Passengers: Inherited Microorganisms and Arthropod Reproduction* (S.L. O'Neill, J.H. Werren, and A.A. Hoffmann, Eds.), pp. 81–101. Oxford University Press, Oxford, U.K.

Rousset, F., Bouchon, D., Pintureau, B., Juchault, J., and Solignac, M. (1992). *Wolbachia* endosymbionts responsible for various alterations of sexuality in arthropods. *Proc. R. Soc. London (B)* **250:** 91–98.

Sakaguchi, B. and Poulson, D.F. (1961). Distribution of "sex-ratio" agent in tissues of *Drosophila willistoni*. *Genetics* **46:** 1665–1676.

Sakaguchi, B. and Poulson, D.F. (1963). Interspecific transfer of the "sex-ratio" condition from *Drosophila willistoni* to *D. melanogaster*. *Genetics* **48:** 841–861.

Sasaki, T. and Ishikawa, H. (1995). Production of essential amino acids from glutamate by mycetocyte symbionts of the pea aphid, *Acyrthosiphon pisum*. *J. Insect Physiol.* **41:** 41–46.

Sasaki, T., Aoki, T., Hayashi, H., and Ishikawa, H. (1990). Amino acid composition of the honeydew of symbiotic and aposymbiotic pea aphids, *Acyrthosiphon pisum*. *J. Insect Physiol.* **36:** 35–40.

Sasaki, T., Kawamura, M., and Ishikawa, H. (1996). Nitrogen recycling in the brown planthopper, *Nilaparvata lugens*: Involvement of yeast-like endosymbionts in uric acid metabolism. *J. Insect Physiol.* **42:** 125–129.

Schroeder, D., Deppisch, H., Obermeyer, M., Krohne, G., Stackebrandt, E., Holldobler, B., Goebel, W., and Gross, R. (1996). Intracellular endosymbiotic bacteria of *Camponotus* species (carpenter ants): systematics, evolution and ultrastructural characterization. *Mol. Microbiol.* **21:** 479–490.

Sedgwick, S.G. and Smerdon, S.J. (1999). The ankyrin repeat: a diversity of interactions on a common structural framework. *Trends Biochem. Sci.* **24:** 311–316.

Shigenobu, S., Watanabe, H., Hattori, M., Sakaki, Y., and Ishikawa, H. (2000). Genome sequence of the endocellular symbiont of aphids *Buchnera* sp. *Nature* **407:** 81–86.

Shigenobu, S., Watanabe, H., Sakaki, Y., and Ishikawa, H. (2001). Accumulation of species-specific amino acid replacements that cause loss of particular protein functions in *Buchnera*, an endocellular bacterial symbiont. *J. Mol. Evol.* **53:** 377–386.

Shimomura, S., Shigenobu, S., Morioka, M., and Ishikawa, H. (2002). An experimental validation of orphan genes of *Buchnera*, a symbiont of aphids. *Biochem. Biophys. Res. Commun.* **292:** 263–267.

Smith, D.C. and Douglas, A.E. (1987). *The Biology of Symbiosis*. Edward Arnold, London.

Spaulding, A.W. and von Dohlen, C.D. (1998). Phylogenetic characterization and molecular evolution of bacterial endosymbionts in psyllids (Hemiptera: Sternorrhyncha). *Mol. Biol. Evol.* **15:** 510–524.

Spaulding, A.W. and von Dohlen, C.D. (2001). Psyllid endosymbionts exhibit patterns of co-speciation with hosts and destabilizing substitutions in ribosomal RNA. *Insect Mol. Biol.* **10:** 57–67.

Starr, M.B. (1975). A general scheme for classifying organismic associations. *Symp. Soc. Exp. Biol.* **29:** 1–20.

Stouthamer, R., Luck, R.F., and Hamilton, W.D. (1990). Antibiotics cause parthenogenetic *Trichogramma* to revert to sex. *Proc. Natl. Acad. Sci. U.S.A.* **87:** 2424–2427.

Thao, M.L., Moran, N.A., Abbot, P., Brennan, E.B., Burckhardt, D.H., and Baumann, P. (2000). Cospeciation of psyllids and their prokaryotic endosymbionts. *Appl. Environ. Microbiol.* **66:** 2898–2905.

Trager, W. (1986). *Living Together: The Biology of Animal Parasitism.* Plenum Press, New York.

Tsagkarakou, A., Guillemaud, T., Rousset, F., and Navajas, M. (1996). Molecular identification of a *Wolbachia* endosymbiont in a *Tetranychus urticae* strain (Acari: Tetranychidae). *Insect Mol. Biol.* **5:** 217–221.

van der Wilk, F., Dullemans, A.M., Verbeek, M., and van den Heuvel, J.F. (1999). Isolation and characterization of APSE-1, a bacteriophage infecting the secondary endosymbiont of *Acyrthosiphon pisum. Virology* **262:** 104–113.

Werren, J.H. and O'Neill, S.L. (1997). The evolution of heritable symbionts. In *Influential Passengers: Inherited Microorganisms and Arthropod Reproduction* (S.L. O'Neill, A.A. Hoffmann, and J.H. Werren, Eds.), pp. 1–41. Oxford University Press, Oxford, U.K.

Werren, J.H., Zhang, W., and Guo, L.R. (1995). Evolution and phylogeny of *Wolbachia*: reproductive parasites of arthropods. *Proc. R. Soc. London (B)* **261:** 55–71.

Whitcomb, R.F. (1980). The genus *Spiroplasma. Annu. Rev. Microbiol.* **34:** 677–709.

Wicker, C. (1983). Differential vitamin and choline requirements of symbiotic and aposymbiotic *S. oryzae* (Coleoptera: Curculionidae). *Comp. Biochem. Physiol.* **76A:** 177–182.

Wilkinson, T.L. and Ishikawa, H. (2001). On the functional significance of symbiotic microorganisms in the Homoptera: a comparative study of *Acyrthosiphon pisum* and *Nilaparvata lugens. Physiol. Entomol.* **26:** 86–93.

Williamson, D.L. (1965). Kinetic studies of "sex-ratio" spirochetes in *Drosophila melanogaster* Meigen females. *J. Invertebrate Pathol.* **7:** 493–501.

Williamson, D.L. (1998). Revised group classification of the genus *Spiroplasma. Int. J. Syst. Bacteriol.* **48:** 1–12.

Yen, J.H. and Barr, A.R. (1971). New hypothesis of the cause of cytoplasmic incompatibility in *Culex pipiens* L. *Nature* **232:** 657–658.

Yoshida, N., Oeda, K., Watanabe, E., Mikami, T., Fukita, Y., Nishimura, K., Komai, K., and Matsuda, K. (2001). Chaperonin turned insect toxin. *Nature* **411:** 44.

Zambryski, P. (1988). Basic processes underlying *Agrobacterium*-mediated DNA transfer to plant cells. *Annu. Rev. Genet.* **22:** 1–30.

2 *Buchnera* Bacteria and Other Symbionts of Aphids

Angela E. Douglas

CONTENTS

A HISTORICAL OVERVIEW OF SYMBIOTIC BACTERIA IN APHIDS

The bacterial symbiosis in aphids was first detected, but not recognized, by the eminent 19th-century evolutionary biologist. T. H. Huxley. Huxley noticed cells with "granular protoplasm" in the body cavity of aphids. He mistakenly identified these cells as accessory fat body, which he called the "pseudovitellus," and the granular inclusions as "vitelline spheres." But perhaps this error should not affect Huxley's wider reputation, for, as Huxley's biographer Adrian Desmond writes, "aphids might not amount to much on the cosmic scale" (Desmond, 1997)!

The true nature of the granular protoplasm in the aphid "pseudovitellus" — bacteria tightly packed into the cytoplasm of insect cells — was first appreciated by Sulc and Pierantoni in 1910 (Pierantoni, 1910; Sulc, 1910). Microscopic analyses conducted during the first half of the last century (summarized in Buchner, 1966) revealed that virtually all aphids apart from the Phylloxeridae and Adelgidae bore coccoid bacteria in large cells, termed mycetocytes. Furthermore, these bacteria were transmitted from mother aphid to offspring via the maternal ovaries, a process called transovarial transmission, such that each egg produced by sexual oviparous females and each larva deposited by asexual parthenogenetic females bore bacteria derived from the mother. Some, but not all, aphid species also bore rod-shaped bacteria, readily distinguishable from the coccoid bacteria and transmitted transovarially with the cocci. Buchner (1966) termed the universal cocci "primary symbionts" and the other bacteria "accessory symbionts"; the latter are also often known as "secondary symbionts."

The remarkable bacterial associations in aphids were deemed intractable to study. In particular, all attempts to isolate the primary symbionts into axenic culture failed; consequently, the bacteria

0-8493-1286-8/03/$0.00+$1.50

could not be identified and the relationship could not be manipulated. Until about 10 years ago, the primary symbionts were described as "mycoplasma-like" or "rickettsia-like" (e.g., Hinde, 1971a) simply because of their shared trait of living within animal cells. In 1991, Baumann et al. identified the primary symbionts from their 16S ribosomal RNA gene sequence. The bacteria were not mycoplasmas or rickettsias but γ-proteobacteria allied to *E. coli* (Munson et al., 1991a). Baumann et al. assigned the bacteria to a novel genus, *Buchnera*, in recognition of the remarkable achievements of Paul Buchner (Munson et al., 1991b). Subsequent research has revealed that the accessory symbionts of Buchner include five taxa, none of which has been brought into axenic culture. None has been described formally, and most are known by acronyms. These five bacteria are PASS (= R type), PABS (= T type) and U type in γ-proteobacteria, PAR (= S type) in α-proteobacteria, and a *Spiroplasma* species (Chen et al., 1996; Chen and Purcell, 1997; Fukatsu et al., 2000, 2001; Darby et al., 2001; Sandström et al., 2001).

Buchner and colleagues believed that the aphids required their complement of primary symbionts. The evidence was slight. Some aphid species included morphs that lacked primary symbionts, and these were invariably small, e.g., the dwarf males of *Stomaphis* spp. The obligate nature of the symbiosis was not demonstrated definitively until 1971 when Mittler showed that aphids treated with the antibiotic auromycin (= tetracycline) grew very poorly and produced no offspring (Mittler, 1971).

It has also been accepted widely that the accessory/secondary symbionts are not essential to the aphid. This issue is currently being researched actively and is reviewed below. However, the term "symbiont" has led to the unfortunate assumption that aphids "should" benefit from all the bacteria they contain. To avoid any preconceived notions as to the significance of these bacteria to the aphid, I prefer to call them "accessory bacteria."

This article reviews our current understanding of *Buchnera* and other bacteria associated with aphids. It records the dramatic increase in knowledge and understanding of the symbiosis over the last 10 years, achieved largely through the application of molecular approaches including the public availability of the complete genome sequence of *Buchnera* in the pea aphid, *Acyrthosiphon pisum* (Shigenobu et al., 2000). Some important areas of current ignorance are also identified.

LOCALIZATION OF BACTERIA

Buchnera

Through most of the aphid lifespan, the *Buchnera* cells are restricted to mycetocytes (also known as bacteriocytes), cells whose sole function appears to be to house the bacteria. In most aphid species studied, the mycetocytes do not divide after the birth of the aphid and accommodate the proliferating *Buchnera* population entirely by increasing in size. Mycetocytes are often very large cells. For example, a young adult pea aphid, *Acyrthosiphon pisum*, weighing 4 mg may bear ca. 100 mycetocytes, of mean diameter 85 μm, each of which contains 23,500 *Buchnera* cells (calculated from data in Wilkinson and Douglas, 1998).

Within the mycetocytes, each *Buchnera* cell is separated from the cytoplasmic contents by a membrane of insect origin, known as the symbiosomal membrane. Two or more *Buchnera* cells are virtually never observed in a single symbiosome, suggesting that when each *Buchnera* cell divides and its daughter cells separate, the surrounding insect membrane pinches off to form two separate symbiosomes. The insect cell cytoskeleton probably mediates and controls both these processes and the regular distribution of *Buchnera* cells across the cell cytoplasm, but this has not been studied. Also unknown is whether the conditions experienced by *Buchnera* cells vary between the periphery and central regions of each mycetocyte.

The mycetocytes of aphids lie in the hemocoel (body cavity) of the insect. In embryos and young larvae, they are aggregated together as a coherent V-shaped organ, lying dorsal to the gut, with the base of the V to the posterior and bounded by a single layer of cells known as sheath cells. As the aphids develop and the ovarioles increase in volume through the production of eggs or embryos, the mycetocytes separate and are distributed to the anterior and posterior ends of the haemocoel, as either isolated cells or small groups of cells. This breakup of the cells is usually attributed to the reduced space in the hemocoel, e.g., Buchner (1966), but the consequences for mycetocyte and *Buchnera* function have not been considered.

Why are *Buchnera* cells restricted to mycetocytes? Both limited access to other cells and poor survival external to mycetocytes may be involved. *Buchnera* apparently lacks the general capacity to invade insect cells because the gene inventory of *Buchnera* from *A. pisum*, as obtained from the complete genome sequence (Shigenobu et al., 2000), has revealed none of the suite of genes implicated in adherence to and invasion of animal cells by other bacteria including related members of the γ-proteobacteria (Hentschel et al., 2000; Goebel and Gross, 2001). *Buchnera* cells injected into the hemocoel of aphids are lysed rapidly (Douglas, unpublished data); it is not known whether this reflects insect defense against bacteria or poor *Buchnera* tolerance of the conditions (ionic composition, osmotic pressure, etc.) in the hemolymph. These data, although fragmentary, raise the possibility that *Buchnera* have trophic or other requirements met only within the symbiosomal membrane in mycetocytes. Pertinent to this possibility, the genomic data also indicate that *Buchnera* cells have very few membrane transporters to mediate nutrient exchange and maintain the bacterial metabolic pools. (*Buchnera* cells of necessity derive all their nutrients from the surrounding mycetocyte cytoplasm.) Perhaps metabolic and signaling interactions between *Buchnera* and the insect are mediated primarily by the transport and receptor functions of the symbiosomal membrane. If this is correct, then the symbiosomal membrane is a crucial, but completely unstudied, element in our understanding of the function of *Buchnera* cells and their interactions with aphids.

Accessory Bacteria

Generally, the accessory bacteria are restricted to the sheath cells bounding mycetocytes. They have also been described free in the hemolymph, where they may attain high densities, and occasionally in cells of the fat body, in hemocytes, and in mycetocytes that lack *Buchnera* cells (Douglas, 1998; Fukatsu et al., 2000). One accessory bacterium, PABS, has additionally been reported as associated with the gut (Darby et al., 2001). Taken together, these data suggest that the accessory bacteria are less fastidious and more invasive than *Buchnera*.

The density of *Buchnera* cells in aphids is ca. 1 to 3×10^7 cells mg^{-1} aphid fresh weight, and the density of accessory bacteria is about tenfold lower (Humphreys and Douglas, 1997; Wilkinson et al., 2001a). These values are equivalent to *Buchnera* cells and accessory bacteria, accounting for 8 to 24% and 0.8 to 2.4% of aphid volume, respectively. Limited data suggest that bacterial density varies with various factors, including temperature, host plant, aphid morph, and developmental age (e.g., Baumann and Baumann, 1994; Humphreys and Douglas, 1997; Wilkinson et al., 2001a), but the impact of environmental factors on the symbiosis has not been studied systematically.

MODES OF TRANSMISSION OF BACTERIA

Buchnera

The aphid–*Buchnera* symbiosis is a "closed system," i.e., the *Buchnera* cells are obligately vertically transmitted, such that all the descendants of each *Buchnera* cell are either descendants of the aphid bearing that *Buchnera* cell or dead; the descendants have no access to either the external environment or other aphid lineages.

The chief evidence is molecular: the phylogenies of *Buchnera* and aphids are strictly congruent, i.e., the phylogenetic trees of the two partners map onto each other. This was first demonstrated by Moran et al. (1993) for trees constructed for 16S rRNA genes of *Buchnera* and the morphological characteristics of aphids across all subfamilies of the Aphididae (Figure 2.1) and has since been

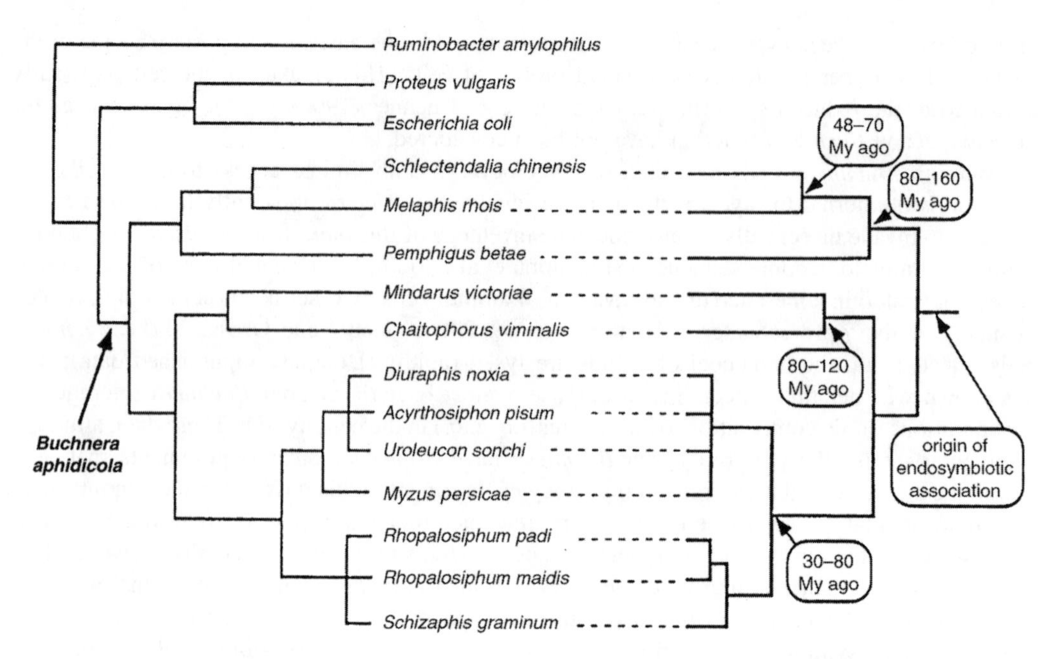

FIGURE 2.1 Phylogeny of *Buchnera* bacteria and corresponding aphid hosts, as determined from 16S rRNA sequence and morphology, respectively. [From Moran, N.A. and Baumann, P. (1994). *Trends Ecol. Evol.* **9:** 15–20: Fig. 1. With permission.]

confirmed using sequences from an array of *Buchnera* and aphid genes at phylogenetic scales ranging from the family Aphididae to intraspecies variation (e.g., Clark et al., 2000; Funk et al., 2000 and 2001).

The processes underpinning the obligate vertical transmission and congruent phylogenies have been described at the electron microscopic level, especially by Hinde (1971b) and Brough and Dixon (1990). Mycetocytes become associated with the germarium of the aphid ovaries. As the parthenogenetic embryos in viviparous aphids develop to the blastoderm stage and eggs in oviparous aphids initiate vitellogenesis, *Buchnera* cells are expelled from mycetocytes by exocytosis, pass between follicle cells of the ovaries, and become incorporated into the embryo or unfertilized egg. Thereafter, they become incorporated into the mycetocytes as these cells differentiate.

An added level of complexity is generated by the telescoping of generations in parthenogenetic aphids. The capacity of aphids to initiate larviposition within a few hours of reaching adulthood derives from their initiation of embryogenesis as larvae or even as embryos. Because transmission of *Buchnera* occurs early in embryo development, an adult aphid may contain multiple symbioses at different developmental stages: one in her hemocoel and one in each of her daughter embryos and some of her granddaughter embryos.

At present, essentially no information is available on the cellular and molecular interactions underpinning vertical transmission of *Buchnera*. For example, what is the nature of the signal exchange triggering exocytosis of *Buchnera* from mycetocytes and their passage to embryos and eggs? Do *Buchnera* cells participate in this intercell communication, and is there any "quality control" by which the condition of the transferred *Buchnera* is tested? In connection with this, are the transmitted *Buchnera* cells functionally equivalent to other *Buchnera* cells, or do they display modified surface features or metabolic traits linked to their brief extracellular status? With the increasing availability of specific probes to interrogate gene expression, signaling events, and metabolites in single cells, these questions can, in principle, be answered.

A further point of ignorance is the number of *Buchnera* cells acquired by each embryo/egg from its mother, which is equivalent to the effective size (N_e) of the *Buchnera* population in aphids.

One might expect the number to vary between species with the size and level of nutrient provisioning of the egg (Krakauer and Mira, 1999), but this has not been investigated, and informal estimates of N_e range from 50 to 500 *Buchnera* cells (Douglas, 1998). These values can be translated to the proportion of *Buchnera* cells in the mother's mycetocytes available for transmission. Let us consider the pea aphid *Acyrthosiphon pisum*, which on reaching adulthood bears 100 mycetocytes with a total volume of 0.03 mm^3 and has 100 offspring (Wilkinson and Douglas, 1998). *Buchnera* cells have an average diameter of 2.5 μm and occupy 60% of the mycetocyte volume (Whitehead and Douglas, 1993a), and there are therefore 2.2×10^6 *Buchnera* cells available for transmission. The 50 to 500 cells transmitted per offspring account for 2 to 22% of the *Buchnera* cells available, leaving 78 to 98% of the cells to die with the mother. (This calculation is a simplification because transmission is initiated in larvae, but it illustrates the point.) Do *Buchnera* cells compete for access to offspring? The highly ordered process of transmission suggests not; *Buchnera* cells do not proliferate outside of mycetocytes around the ovaries or invade embryos/eggs beyond a very restricted developmental stage. The very small N_e may provide the explanation for the altruism of the majority. If few *Buchnera* cells are transmitted and all derived from one mycetocyte (i.e., within a few cell divisions of a common ancestor), then most of their progeny in the offspring aphid will be clonemates, closely related to each other by descent. These clonemates are predicted to cooperate, just as the cells in the aphid body cooperate and do not compete to be a reproductive cell (Maynard Smith and Szathmary, 1995).

Depressed conflict among *Buchnera* cells is not the only predicted evolutionary consequence of the low N_e. The second consequence is genomic deterioration of *Buchnera*. If the complement of *Buchnera* acquired by an aphid from its mother uniformly bears a slightly deleterious mutation, then there is no way to replace these slightly inferior *Buchnera* cells by superior forms. The resultant gradual genetic decay in the vertically transmitted cells is expected to be exacerbated if genes mediating DNA repair and recombination accumulate mutations, reducing the capacity to correct replication errors and purge mutations by sexual reproduction. There is now overwhelming evidence that genomic decay has been a major factor in the evolution of *Buchnera* genomes (Moran, 1996; Rispe and Moran, 2000). Protein-coding genes have a much higher incidence of nonsynonymous substitutions (i.e., ones that alter the amino acid in the protein product) (Moran, 1996), and the substitutions in the 16S rRNA gene tend to destabilize the secondary structure of the rRNA molecule (Lambert and Moran, 1998) (Figure 2.2). As Figure 2.2B illustrates, these effects are not unique to *Buchnera* but are common to microorganisms that are persistently transmitted vertically (Moran and Wernegreen, 2000).

One would expect pseudogenes to accumulate in the *Buchnera* genome, as has been described, for example, in *Mycobacterium leprae* (Cole et al., 2001). Certainly, pseudogenes have been described in certain *Buchnera* genes (Lai et al., 1996; Baumann et al., 1997; van Ham et al., 1999), but the strong deletional bias in bacterial genomes presumably results in the total elimination of most genes that have become nonfunctional (Lawrence et al., 2001). The predicted consequence is genome reduction; only those genes whose function is under strong selection pressure are retained. Consistent with this prediction, the genome size of *Buchnera* is small at 0.45 to 0.64 megabases (Mb) (Charles and Ishikawa, 1999; Wernegreen et al., 2000; Gil et al., 2002). Many genes that would be essential for independent existence, including almost all genes for phospholipid synthesis and the TCA cycle, are absent, and many genes for DNA repair, including *recA* and *uvrABC*, are missing (Shigenobu et al., 2000).

Accessory Bacteria

The vertical transmission of accessory bacteria has been demonstrated for multiple aphid species and bacterial taxa by both microscopic and molecular methods (e.g., Darby et al., 2001; Fukatsu et al., 2000 and 2001; Sandström et al., 2001). Buchner (1966, p. 310) comments that the coccoid primary symbionts (*Buchnera*) and rod-shaped accessory bacteria transmitted to the winter eggs of *Drepanosiphum* "appear

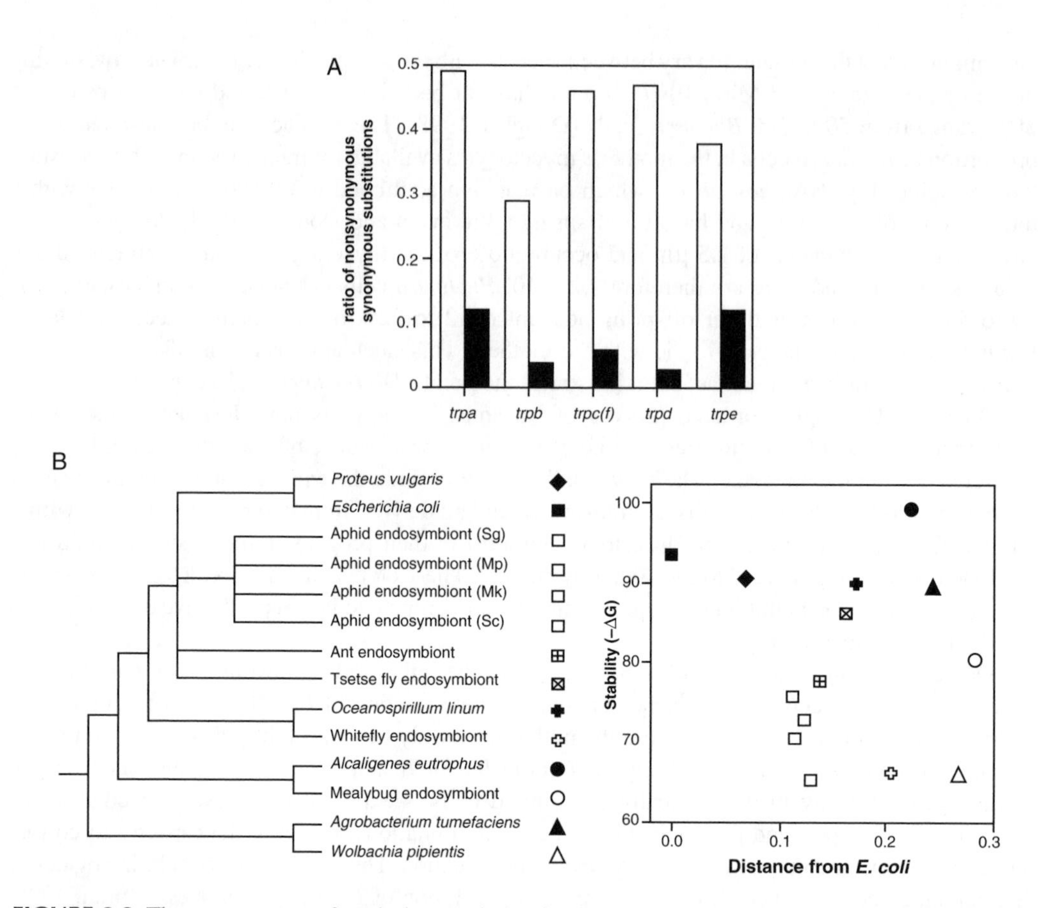

FIGURE 2.2 The consequences of vertical transmission for the molecular evolution of *Buchnera.* (A) Nonsynonymous substitutions in five protein-coding *trp* genes of *Buchnera.* The number of substitutions between *Buchnera* in the aphids *Schizaphis graminum* and *Schlechtendalia chinensis* (open bars) and between the two enteric bacteria *E. coli* and *Salmonella typhimurium* (closed bars) is expressed as the ratio of nonsynonymous:synonymous substitutions per nucleotide site. [From Hurst and McVean (1996). *Nature* **381:** 650–651: Fig. 1]. (B) The stability of Domain I of the 16S rRNA in symbiotic bacteria, including *Buchnera*, and free-living bacteria, showing the relationships of bacteria and stabilities (–ΔG) summed over Domain I for each organism. [From Lambert, J.D. and Moran, N.A. (1998). *Proc. Natl. Acad. Sci. U.S.A.* **95:** 4458–4462: Fig. 2.]

at approximately the same time at the follice zone [of the ovaries] and are transmitted together to the oocyte through the gaps," but modern studies are generally lacking. In one recent study on *Aphis fabae*, the abundance of accessory bacteria relative to *Buchnera* was lower in embryos than in the maternal tissues (Wilkinson et al., 2001a), suggesting that either very few accessory bacteria are transmitted or that embryos provide a poor environment for their proliferation.

The dynamics of accessory bacterial populations are, however, fundamentally different from *Buchnera* because the accessory bacteria are not universally present in aphids. For example, the pea aphid, *A. pisum*, may bear PASS, PABS, PAR, and U-type, but their incidence varies between different clones (Sandström et al., 2001) (Figure 2.3). These data suggest strongly that accessory bacteria can be lost through failure of vertical transmission and gained horizontally by individual aphids. The frequencies of such gains and losses are unknown and may vary with environmental conditions.

Horizontal transmission can be demonstrated experimentally in the laboratory by the injection of isolated accessory bacteria from a "donor" aphid to a "recipient" that lacks the accessory bacteria; the injected accessory bacteria are stably inherited through multiple generations (e.g., Chen et al., 2000; Fukatsu et al., 2001). In natural populations, an equivalent process may be mediated by a

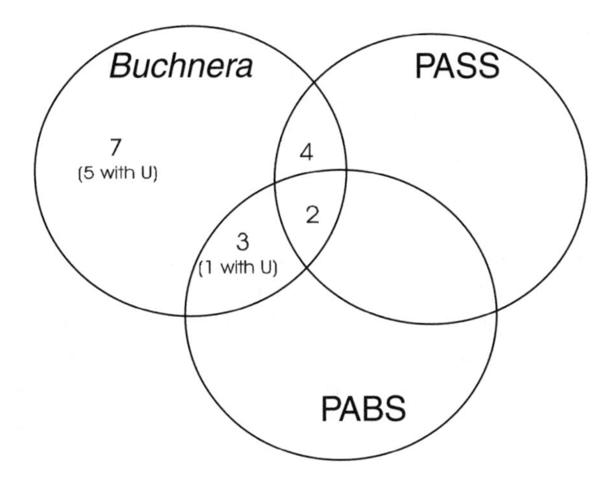

FIGURE 2.3 Bacteria detected in pea aphids *Acyrthosiphon pisum*. US clones (n = 16) from various sources. [Redrawn from Sandström, J.P., Russell, J.A., White, J.P., and Moran, N.A. (2001). *Mol. Ecol.* **10:** 217–228: Table 2.]

parasitoid whose ovipositor may become contaminated with accessory bacteria at a previous oviposition event. Since the successful transmission of accessory bacteria requires the parasitoid infection to abort (essentially all successfully parasitized aphids produce no offspring), accessory bacteria may be predicted to accumulate in aphids with physiological resistance to parasitoids. This expectation, however, remains to be tested.

An alternative transmission route for accessory bacteria is oral. Any accessory bacterial cells released into the environment via aphid secretions (honeydew, cornicle secretions, saliva) are potentially available for ingestion by other aphids. However, the incidence of accessory bacteria in these aphid secretions, bacterial persistence in the free-living condition, and the capacity of the bacteria to survive in aphid guts after ingestion all remain unknown.

The expected evolutionary consequences of mixed vertical/horizontal transmission are that the phylogenies of accessory bacteria and aphids will not be congruent and that their genomes will not be subject to decay. Consistent with the first expectation, the 16S rRNA sequence of PABS (= T type) has >98% similarity to that of bacteria described in the whitefly *Bemisia tabaci* (Darby et al., 2001; Sandström et al., 2001). Using published substitution rates for 16S rRNA genes of bacteria, PABS and the bacteria in *B. tabaci* are estimated to have diverged 17 to 34 million years ago, considerably more recently than the likely common ancestor of aphids and whitefly (Darby et al., 2001), indicating that these two bacteria have not evolved from a common bacterial ancestor in the common ancestor of aphids and whitefly. It is unknown whether the bacteria have switched directly between these two insect families or whether their common ancestor was a free-living form.

MUTUAL DEPENDENCE

This section considers whether bacteria and aphids are dependent on their association and the underlying causes of the observed level of dependence. *Buchnera* and accessory bacteria will be considered first and then the aphid.

BACTERIA

Buchnera cells are absolutely dependent on their intracellular habitat in aphids. As considered in the section entitled Modes of Transmission of Bacteria, their genome is very small, close to the size of the theoretical minimal genome size for any living organism (Maniloff, 1996). In addition, the amino acid and pantothenate–coenzyme A biosynthetic pathways of *Buchnera* are inextricably linked to aphid metabolism (Shigenobu et al., 2000). As a consequence, *Buchnera* cannot be isolated into long-term axenic culture, even though viable and metabolically active preparations of *Buchnera* can be maintained for several hours *in vitro* (Whitehead and Douglas, 1993b). To quote Shigenobu et al. (2000), "the gene repertoire of the *Buchnera* genome is so specialized to intracellular life that it cannot survive outside the eukaryotic cell."

No accessory bacteria have been brought into axenic culture. There could be several reasons for this. First, the correct conditions (especially media) may not have been tested yet. Second, the accessory bacteria may be among the >90% of all bacteria that apparently are not amenable to current culture methodologies. Finally, these bacteria, like *Buchnera*, may have an absolute requirement for conditions or resources provided by aphids. To date, virtually all research on the relationship between aphids and their accessory bacteria has been conducted from the perspective of the aphid (see below), and as a consequence we are ignorant of the importance of aphids as a habitat for these taxa. We lack the information to answer such simple questions as: What proportion of the global population of accessory bacteria is in aphids? And if all aphids went extinct today, would their accessory bacteria go extinct too? The answers to these questions may vary among the different taxa of accessory bacteria.

The Aphid

Aphids require their association with bacteria. When young larval aphids are administered antibiotics at a dose that disrupts the bacteria but has minimal direct effects on aphid metabolism or behavior, the aphids grow very poorly and usually produce either no offspring or, less commonly, a few offspring that are dead at birth or die within a few days without growing or developing. Adults treated with antibiotics produce bacteria-free offspring that, in turn, grow slowly and are reproductively sterile. The effect of antibiotic treatment on embryo growth is particularly severe. As Figure 2.4 illustrates, the embryos in a young adult of *A. pisum* (11 days old) account for 65% of the total protein of aphids containing their bacteria but just 12% of the protein in bacteria-free aphids. There is, however, no evidence that embryogenesis is halted at a specific developmental stage (Douglas, 1996). This dependence of aphids on their bacteria has been attributed to *Buchnera* because only *Buchnera*, and none of the accessory bacteria, is universal.

It is widely accepted that the requirement of aphids for *Buchnera* has a nutritional basis. Aphids experimentally deprived of their bacteria are deficient in protein, but not lipid, and have

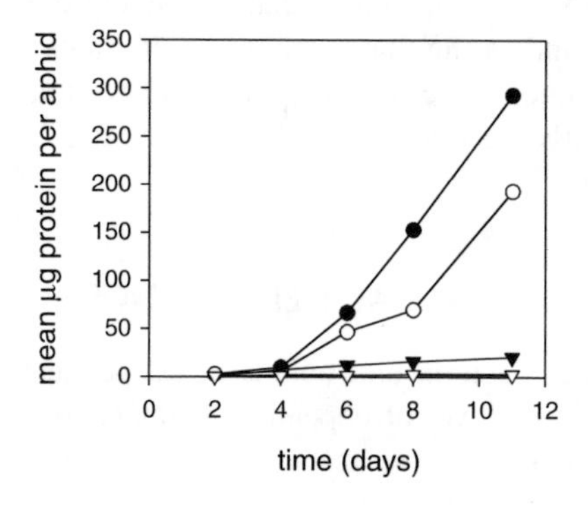

FIGURE 2.4 Protein growth of pea aphids *Acyrthosiphon pisum* and their embryos. Total protein growth (closed symbols) and embryo protein growth (open symbols) are shown for aphids containing (circles) and experimentally deprived of their symbiotic bacteria (triangles). [Redrawn from Douglas, A.E. (1996). *J. Insect Physiol.* **42:** 247–255: Fig. 2.]

elevated levels of free amino acids. For example, the *Buchnera*-free pea aphids *A. pisum* studied by Prosser and Douglas (1991) had protein content depressed by 20% and free-amino-acid content increased by 40% relative to untreated controls. These data suggest that protein synthesis by the *Buchnera*-free aphids is limited by a shortfall of certain amino acids, resulting in an accumulation of other amino acids in the free amino acid pool. The key amino acids with low concentrations in the free-amino acid pool of *Buchnera*-free aphids have been identified as phenylalanine and methionine in *A. pisum* (Prosser and Douglas 1991; Douglas, 1996). These are among the nine amino acids that aphids and other animals cannot synthesize *de novo*, known as the essential amino acids. The implication that these essential amino acids are provided to the aphid by *Buchnera* fits with the finding that the amino acid composition of plant phloem sap, the diet of aphids, is grossly deficient in these nine essential amino acids. Whereas animal protein contains on average 50% essential amino acids, plant phloem sap contains 10 to 25% essential amino acids (Douglas, 1993).

The direct evidence that *Buchnera* provides aphids with essential amino acids is threefold. The first line of evidence derives from the development of methods to raise aphids on a chemically defined liquid diet that mimics phloem sap. When individual essential amino acids are deleted from these diets, aphid growth and reproduction are generally not affected, but the growth of experimentally generated *Buchnera*-free aphids is depressed by elimination of some or all of these nutrients, varying among aphid species (e.g., Mittler, 1971; Douglas et al., 2001).

Second, the intact aphid–*Buchnera* association and isolated *Buchnera* cells can synthesize essential amino acids, but *Buchnera*-free aphids (as expected) have no essential-amino-acid biosynthetic capability (e.g., Douglas, 1988; Sasaki and Ishikawa, 1995; Febvay et al., 1999; Douglas et al., 2001). Essential amino acids synthesized by *Buchnera* are recovered from aphid proteins, which indicates that they are made available to aphid tissues. Recently, Douglas et al. (2001) quantified the rates of essential amino acid production by the *Buchnera* population in *A. fabae*. When these data are compared against the rate of increase in *Buchnera* protein content (as calculated from the estimated proliferation rate of 0.3 cells/day and amino acid composition of *Buchnera* protein), the amount of each essential amino acid made available to the aphid tissues is estimated at 50 to 75% for all essential amino acids except tryptophan (Table 2.1).

TABLE 2.1
Essential Amino Acid Production by *Buchnera* in *Aphis fabae*

Amino Acid	Production per *Buchnera* Cells per Day (fmol)[a]	Amount Released to Aphid (%)[b]
Phenylalanine	0.28	76
Isoleucine	0.19	75
Leucine	0.33	74
Lysine	0.26	74
Valine	0.22	70
Methionine	0.04	68
Threonine	0.19	56
Histidine	0.02	54
Tryptophan	0.02	19

[a] Data from Douglas, A.E. et al. (2001).

[b] Calculation assumes that *Buchnera* cells proliferate at rate of 0.3 cells/day and that their protein has the same amino acid composition as that in *Escherichia coli* (Schaechter, 1992).

The final line of evidence for *Buchnera* provisioning of essential amino acids to aphids is genomic. Despite its tiny size, the *Buchnera* genome includes virtually all the genes coding for enzymes in the essential amino acid biosynthetic pathways (Shigenobu et al., 2000). The most parsimonious explanation for this finding is that *Buchnera* is under intense selection pressure to retain these genes because *Buchnera* are consistently net providers of essential amino acids to the aphid tissues. A further line of genomic evidence comes from the discovery of two plasmids in *Buchnera*, one bearing the genes *leuA-D* coding for enzymes in the dedicated leucine biosynthetic pathway and the other bearing multiple tandem repeats of *trpEG*, key genes in the tryptophan biosynthetic pathway (Lai et al., 1994; Bracho et al., 1995). These genes are amplified relative to the chromosomal genes of *Buchnera* by up to 18 fold, varying among and within aphid species (e.g., Lai et al., 1994; Thao et al., 1998; Birkle et al., in press). It is widely accepted that this gene amplification promotes the production of leucine and tryptophan (Douglas, 1998; Moran and Baumann, 2000), although this remains to be tested definitively.

The strong evidence that essential amino acid production is the principal, and possibly sole, function of *Buchnera* generates the prediction that *Buchnera*-free aphids raised on a diet with optimally balanced amino acid content would perform as well as aphids containing *Buchnera*. This expectation is not fulfilled. For example, *Buchnera*-free *Aphis fabae* perform maximally on a chemically defined diet containing 75% essential amino acids, but these aphids attain an adult weight and fecundity that are 40% and 80% lower than aphids containing *Buchnera* (L.B. Minto and A.E. Douglas, unpublished data). Douglas (1998) has suggested that either *Buchnera* fulfill other functions — e.g., provision of a vitamin, riboflavin (Nakabachi and Ishikawa, 1999) — or aphids have a limited capacity to assimilate dietary essential amino acids, reflecting their long evolutionary history with an endogenous (bacterial) source of these nutrients. The latter possibility was tested by Douglas et al. (2001). *Buchnera*-free *A. fabae* showed poor assimilation efficiency for the essential amino acid leucine at 38%, but untreated aphids containing their normal complement of *Buchnera* had an assimilation efficiency of 98%. These data suggest that aphids have very effective essential amino acid uptake systems, but that one consequence of the depressed protein synthesis in *Buchnera*-free aphids may be reduced synthesis of these key gut transporter proteins. Thus, disruption of the bacteria eliminates an endogenous supply of essential amino acids and impairs the aphid capacity to exploit exogenous sources of these nutrients.

An alternative approach to providing aphids with exogenous essential amino acids is to inject them into the hemolymph. This treatment also fails to enhance aphid performance (Wilkinson and Ishikawa, 2000) probably because hemolymph amino acids are turned over very rapidly such that injected amino acids are cleared rapidly from the system by respiration (Wilkinson et al., 2001b).

The demonstration that *Buchnera* provide aphids with essential amino acids raises many issues that remain unresolved. Chief among them is the nature of the mobile compounds, i.e., the form in which the nutrients are transferred from *Buchnera* cells to the surrounding mycetocyte cytoplasm and then distributed to the other aphid tissues. The *Buchnera* genome sequence offers few clues because few transporter genes are present: a GlpF and OmpF-like porin that may promote passive diffusion; a few ABC transporter genes and phosphoenolpyruvate-carbohydrate phosphotransferase systems for transport of small molecules; and SecB, a protein that promotes the correct conformation in proteins destined for export (Lai and Baumann, 1992; Shigenobu et al., 2000). Perhaps both free amino acids and *Buchnera* proteins are released from *Buchnera* cells and translocated across the symbiosomal membrane into the mycetocyte cytoplasm. The recovery of one protein, GroEL, abundant in *Buchnera* cytoplasm, from aphid hemolymph (van den Heuvel et al., 1994) suggests that this protein is made available to the aphid, although it is unknown whether it is released from living *Buchnera* cells or lysed cells. The ultrastructure of mycetocytes likewise offers few clues. The mycetocyte cytoplasm is not richly endowed with ribosomes, suggesting that if *Buchnera* cells release amino acids at high rates, the mycetocytes do not have the machinery to incorporate them into aphid protein. Additionally, the cell membrane

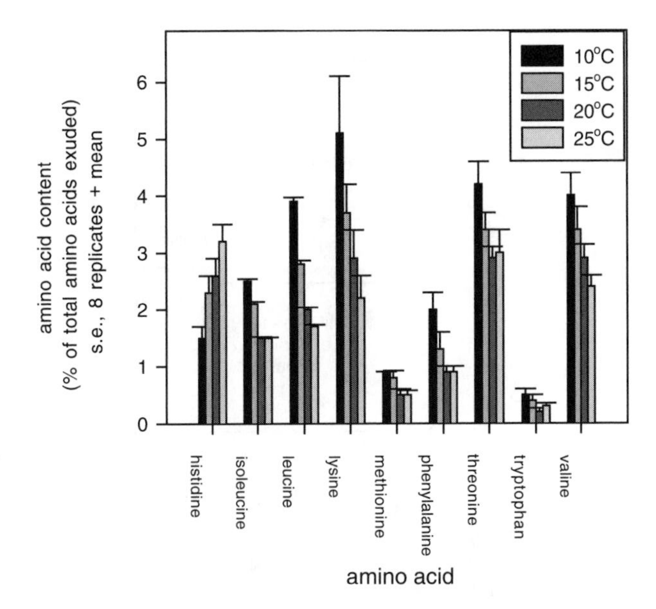

FIGURE 2.5 Essential amino acid content of phloem sap exudates from *Vicia faba* plants (cv. The Sutton) reared at different temperatures. The EDTA exudates were taken from the top fully expanded leaf of pre-flowering plants at 3 weeks post sowing by the protocol of Douglas (1993). [Previously unpublished data from C.R. Tosh and A.E. Douglas.]

of these very large cells is not thrown into extensive folds or underlain by large numbers of mitochondria as would be expected if mycetocytes had a major function in the secretion of essential amino acids.

A second issue relates to the response of the symbiosis to varying dietary amino acid supply. The composition of phloem sap varies among plant species and with developmental age and environmental conditions (Blackmer and Byrne, 1999; Karley et al., in press). This is illustrated by the data in Figure 2.5; the total essential amino cid concentration in the phloem sap of *Vicia faba* declines with temperature from 24 ± 3.5% of the total amino acid content at 10°C to 16 ± 1.2% at 25°C (mean ± s.e., 8 reps), and this general pattern is followed by all individual essential amino acids except histidine, which shows the reverse relationship. How does the symbiosis respond to such variability? Febvay et al. (1999) and Douglas et al. (2001) have shown that the incorporation of radioactivity from ^{14}C-sucrose and ^{14}C-glutamate, respectively, into essential amino acids is elevated in aphids reared on diets of low essential amino acid content and interpreted these data as elevated *Buchnera* provisioning of these nutrients under conditions of increased aphid demand. This conclusion is, however, premature because the elevated specific activity of ^{14}C in the essential amino acids probably reflected, in part or wholly, the decreased total concentration of these compounds. It will be necessary to explore the total flux of metabolites, not just radiotracer flux, to establish definitively whether *Buchnera* metabolism responds to variation in aphid nutritional demand.

A final unresolved issue relates to the developmental biology of the symbiosis. Aphid symbiosis is developmentally complex, with multiple associations of different ages (see the section entitled Modes of Transmission of Bacteria), and the embryos are particularly dependent on the *Buchnera* (Figure 2.4). Does this mean that the symbiosis in the embryos is more active than that in the maternal tissues? Or, alternatively, that the maternal symbiosis supplements the function of the symbiosis in the embryos? The finding that radioactivity from ^{14}C-labeled essential amino acids injected into aphids is recovered from embryos (Wilkinson and Ishikawa, 1999) is consistent with the notion of net transfer of *Buchnera*-derived nutrients from the maternal symbiosis to embryos but also open to the interpretation that the injected amino acids are catabolized to products

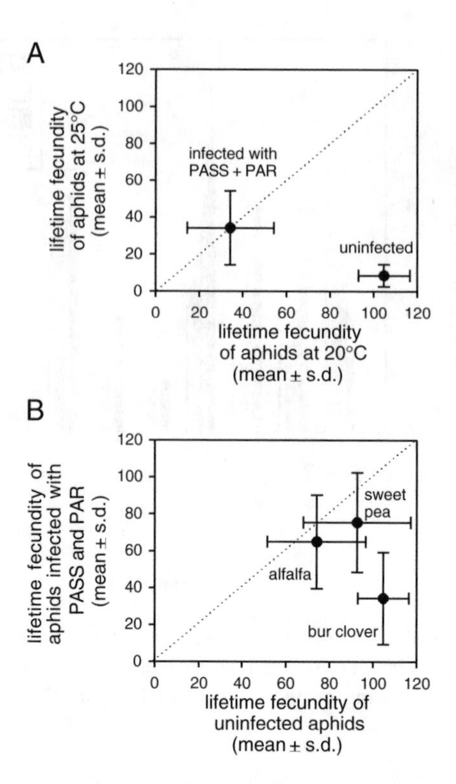

FIGURE 2.6 Mean lifetime fecundity of pea aphids *Acyrthosiphon pisum,* which naturally contain no accessory bacteria (uninfected) and have been infected experimentally with the accessory bacteria PASS and PAR. (A) Impact of temperature on aphids reared on bur clover. (B) Impact of rearing plant at 20°C. [Redrawn from Chen, D.-Q., Montllor, C.B., and Purcell, A.H. (2000). *Entomol. Exp. Appl.* **95:** 315–323: Tables 1 and 3.]

(Wilkinson et al., 2001b) subsequently taken up by embryos and used as precursors for essential amino acid synthesis by the embryo symbiosis.

In contrast to their requirement for *Buchnera*, aphids apparently do not require accessory bacteria. Aphids lacking any accessory bacteria detectable by microscopy and molecular methods can be isolated from natural populations, and they generate vigorous parthenogenetic colonies in the laboratory (unpublished results). These data beg the question whether accessory bacteria are of any significance to the insect; perhaps they just "hitch a ride."

The experiments of Chen et al. (2000) are directly relevant to this question. When a clone of *A. pisum* naturally lacking accessory bacteria was injected with the bacteria PASS and PAR, stable vertically transmitted associations were formed. The response of the resultant associations to temperature and host plants was altered relative to the uninfected controls. Aphid fecundity was depressed at the standard rearing temperature of 20°C, but the thermal range of the aphids was expanded (Figure 2.6A), and the effective plant range was increased (Figure 2.6B). The underlying mechanisms are obscure, but these results raise the possibility that accessory bacteria have a greater role than is currently appreciated in shaping ecologically important traits of aphids.

Other experiments have also implicated accessory bacteria as determinants of the plant range of aphids. *Aphis fabae* performs relatively poorly on one occasional host plant, *Lamium purpureum*. Unusually, the performance of aphids treated with antibiotic to eliminate their bacteria is not reduced (relative to bacteria-free aphids on other plant species), and the reduction in performance is restricted to aphids bearing their normal complement of bacteria (Adams and Douglas, 1997). This specific effect is correlated with elevated levels of the accessory bacteria (Wilkinson et al., 2001a), raising the possibility that these bacteria are acting as opportunistic pathogens that may tend to reduce or limit the plant range of their aphid hosts.

CONCLUDING COMMENTS

Our understanding of the relationship between bacteria and aphids has been transformed over the last decade, largely through the application of molecular biology. The various bacteria have been identified and their incidence and characteristics established, principally by molecular techniques. Molecular approaches have also been crucial to our appreciation of the importance of vertical transmission and genomic decay in shaping the gene content and capabilities of *Buchnera*.

Most recent studies, however, have concentrated on individual members of the association. A task for the future is to address interactions between partners. This requires us to view the association as an interactive relationship, not an assemblage of microbial taxa coexisting in a habitat known as an aphid. Symbiosis function and aphid traits are influenced by the total complement of microorganisms and their interrelationships. What are the nutrients and signals that mediate these diverse interactions? How should we regard the accessory bacteria — as commensals, opportunistic pathogens, or "helper bacteria" that are potential replacements for *Buchnera* ("primary symbionts in waiting"), should the latter suffer genomic meltdown? A firm understanding of these interactions will lead to a better appreciation of this astonishing symbiosis and may provide a basis for novel approaches to aphid pest management.

ACKNOWLEDGMENTS

Research leading to previously unpublished data was funded by grant GR3/10491 from the Natural Environment Research Council, U.K.

REFERENCES

Adams, D. and Douglas, A.E. (1997). How symbiotic bacteria influence plant utilisation by the polyphagous aphid, *Aphis fabae*. *Oecologica* **110:** 528–532.

Baumann, L. and Baumann, P. (1994). Growth kinetics of the endosymbiont *Buchnera aphidicola* in the aphid *Schizaphis graminum*. *Appl. Environ. Microbiol.* **60:** 3440–3443.

Baumann, L., Clark, M.A., Rouhbakhsh, D., Baumann, P., Moran, N.A., and Voegtlin, D.J. (1997). Endosymbionts (*Buchnera*) of the aphid *Uroleucon sonchi* contain plasmids with trpEG and remnants of *trpE* pseudogenes. *Curr. Microbiol.* **35:** 18–21.

Birkle, L.M., Minto, L.B., and Douglas, A.E. Relating genotype and phenotype for tryptophan synthesis in an aphid-bacterial symbiosis. *Physiol. Entomol.* In press.

Blackmer, J.L. and Byrne, D.N. (1999). Changes in the amino acids in *Cucumis melo* in relation to life-history traits and flight propensity of *Bemisia tabaci*. *Entomol. Exp. Appl.* **93:** 29–40.

Bracho, A.M., Martinez-Torres, D., Moya, A., and Latorre, A. (1995). Discovery and molecular characterisation of a plasmid localized in *Buchnera* sp. bacterial endosymbiont of the aphid *Rhopalosiphum padi*. *J. Mol. Evol.* **41:** 67–73.

Brough, C.N. and Dixon, A.F.G. (1990). Ultrastructural features of egg development in oviparae of the vetch aphid, *Megoura viciae*. *Tissue Cell* **22:** 51–63.

Buchner, P. (1966). *Endosymbiosis of Animals with Plant Micro-Organisms*. John Wiley & Sons, London.

Charles, H. and Ishikawa, H. (1999). Physical and genetic map of the genome of *Buchnera*, the primary endosymbiont of the pea aphid *Acyrthosiphon pisum*. *J. Mol. Evol.* **48:** 142–150.

Chen, D.-Q. and Purcell, A.H. (1997). Occurrence and transmission of facultative endosymbionts in aphids. *Curr. Microbiol.* **34:** 220–225.

Chen, D.-Q., Campbell, B.C., and Purcell, A.H. (1996). A new rickettsia from a herbivorous insect, the pea aphid *Acyrthosiphon pisum* (Harris). *Curr. Microbiol.* **33:** 123–128.

Chen, D.-Q., Montllor, C.B., and Purcell, A.H. (2000). Fitness effects of two facultative endosymbiotic bacteria on the pea aphid, *Acyrthosiphon pisum*, and the blue alfalfa aphid, *A. kondoi*. *Entomol. Exp. Appl.* **95:** 315–323.

Clark, M.A., Moran, N.A., Baumann, P., and Wernegreen, J.J. (2000). Cospeciation between bacterial endosymbionts (*Buchnera*) and a recent radiation of aphids (*Uroleucon*) and pitfalls of testing for phylogenetic congruence. *Evolution* **54:** 517–525.

Cole, S.T. and 43 others. (2001). Massive gene decay in the leprosy bacillus. *Nature* **409:** 1007–1011.

Darby, A.C., Birkle, L.M., Turner, S.L., and Douglas, A.E. (2001). An aphid-borne bacterium allied to the secondary symbionts of whitefly. *FEMS Microbiol. Ecol.* **36:** 43–50.

Desmond, A. (1997). *Huxley*, p. 238. Penguin Books, London.

Douglas, A.E. (1988). Sulphate utilisation in an aphid symbiosis. *Insect Biochem.* **18:** 599–605.

Douglas, A.E. (1993). The nutritional quality of phloem sap utilized by natural aphid populations. *Ecol. Entomol.* **18:** 31–38.

Douglas, A.E. (1996). Reproductive failure and the amino acid pools in pea aphids (*Acyrthosiphon pisum*) lacking symbiotic bacteria. *J. Insect Physiol.* **42:** 247–255.

Douglas, A.E. (1998). Nutritional interactions in insect-microbial symbioses. *Annu. Rev. Entomol.* **43:** 17–37.

Douglas, A.E., Minto, L.B., and Wilkinson, T.L. (2001). Quantifying nutrient production by the microbial symbiosis in an aphid. *J. Exp. Biol.* **204:** 349–358.

Febvay, G., Rahbe, Y., Rynkiewiecz, M., Guillaud, J., and Bonnot G. (1999). Fate of dietary sucrose and neosynthesis of amino acids in the pea aphid, *Acyrthosiphon pisum*, reared on different diets. *J. Exp. Biol.* **202:** 2639–2652.

Fukatsu, T., Nikoh, N., Kawai, R., and Koga, R. (2000). The secondary endosymbiotic bacterium of the pea aphid *Acyrthosiphon pisum* (Insecta: Homoptera). *Appl. Environ. Microbiol.* **66:** 2748–2758.

Fukatsu, T., Tsuchida, T., Nikoh, N., and Koga, R. (2001). *Spiroplasma* symbiont of the pea aphid, *Acyrthosiphon pisum* (Insecta: Homoptera). *Appl. Environ. Microbiol.* **67:** 1284–1291.

Funk, D.J., Helbling, L., Wernegreen, J.J., and Moran, N.A. (2000). Intraspecific phylogenetic congruence among multiple symbiont genomes. *Proc. R. Soc. London (B)* **267:** 2517–2521.

Funk, D.J., Wernegreen, J.J., and Moran, N.A. (2001). Intraspecific variation in symbiont genomes: bottlenecks and the aphid–*Buchnera* association. *Genetics* **157:** 477–489.

Gil, R., Sabater-Munoz, B., Latorre, A., Silva, F.J., and Moya, A. (2002). Extreme genome reduction in *Buchnera* spp.; toward the minimal genome needed for symbiotic life. *Proc. Natl. Acad. Sci. U.S.A.* **99:** 4454–4458.

Goebel, W. and Gross, R. (2001). Intracellular survival strategies of mutualistic and parasitic prokaryotes. *Trends Microbiol.* **9:** 267–273.

Haynes, S., van Veen, F., Daniell, T.J., Darby A.C., Godfray, H.C.J., and Douglas, A.E. The diversity of bacteria associated with natural populations of aphids. Unpublished manuscript.

Hentschel, U., Steinert, M., and Hacker, J. (2000). Common molecular mechanisms of symbiosis and pathogenesis. *Trends Microbiol.* **8:** 226–231.

Hinde, R. (1971a). Maintenance of aphid cells and the intracellular symbiotes of aphids *in vitro*. *J. Invertebrate Pathol.* **17:** 333–338.

Hinde, R. (1971b). The control of the mycetome symbiotes of the aphids *Brevicoryne brassicae, Myzus persicae,* and *Macrosiphum rosae*. *J. Insect Physiol.* **17:** 1791–1800.

Humphreys, N.J. and Douglas, A.E. (1997). The partitioning of symbiotic bacteria between generations of an insect: a quantitative study of *Buchnera* in the pea aphid (*Acyrthosiphon pisum*) reared at different temperatures. *Appl. Environ. Microbiol.* **63:** 3294–3296.

Hurst, L.D. and McVean, G.T. (1996). Evolutionary genetics ... and scandalous symbionts. *Nature* **381**: 650–651.

Karley, A.J., Douglas, A.E., and Parker, W.E. Amino acid composition and nutritional quality of potato leaf phloem sap for aphids. *J. Exp. Biol.* In press.

Krakauer, D.C. and Mira, A. (1999). Mitochondria and germ-cell death. *Nature* **400:** 125–126.

Lai, C.-Y. and Baumann, P. (1992). Sequence analysis of a DNA fragment from *Buchnera apidicola* (an endosymbiont of aphids) containing genes homologous to *dnaG*, *rpoD*, *cysE* and *secB*. *Gene* **119:** 113–118.

Lai, C.-Y., Baumann, L., and Baumann, P. (1994). Amplification of *trpEG*: adaptation of *Buchnera aphidicola* to an endosymbiotic association with aphids. *Proc. Natl. Acad. Sci. U.S.A.* **91:** 3819–3823.

Lai, C.-Y., Baumann, P., and Moran, N.A. (1996). The endosymbiont (*Buchnera* sp.) of the aphid *Diuraphis noxia* contains plasmids consisting of *trpEG* and tandem repeats of *trpEG* pseudogenes. *Appl. Environ. Microbiol.* **62:** 332–339.

Lambert, J.D. and Moran, N.A. (1998). Deleterious mutations destabilize ribosomal RNA in endosymbiotic bacteria. *Proc. Natl. Acad. Sci. U.S.A.* **95:** 4458–4462.

Lawrence, J.G., Hendrix, R.W., and Casjens, S. (2001). Where are the pseudogenes in bacterial genomes? *Trends Microbiol.* **9:** 535–540.

Maniloff, J. (1996). The minimal cell genome: "on being the right size." *Proc. Natl. Acad. Sci. U.S.A.* **93:** 10004–10006.

Maynard Smith, J. and Szathmary, E. (1995). *The Major Transitions in Evolution.* W.H. Freeman, Oxford, U.K.

Mittler, T.E. (1971). Dietary amino acid requirement of the aphid *Myzus persicae* affected by antibiotic uptake. *J. Nutr.* **101:** 1023–1028.

Moran, N.A. (1996). Accelerated evolution and Muller's rachet in endosymbiotic bacteria. *Proc. Natl. Acad. Sci. U.S.A.* **93:** 2873–2878.

Moran, N.A. and Baumann, P. (1994). Phylogenetics of cytoplastically inhererited microorganisms of arthropods. *Tr. Ecol. Evol.* **9**: 15–20.

Moran, N.A. and Baumann, P. (2000). Bacterial endosymbionts in animals. *Curr. Opinions Microbiol.* **3:** 270–275.

Moran, N.A. and Wernegreen, J.J. (2000). Lifestyle evolution in symbiotic bacteria: insights from genomics. *Trends Ecol. Evol.* **15:** 321–326.

Moran, N.A., Munson, M.A., Baumann, P., and Ishikawa, H. (1993). A molecular clock in endosymbiotic bacteria is calibrated using the insect hosts. *Proc. R. Soc. London (B)* **253:** 167–171.

Munson, M.A., Baumann, P., Clark, M.A., Baumann, L., Moran, N.A., Voegtlin, D.J., and Campbell, B.C. (1991a). Evidence for the establishment of aphid-eubacterium endosymbiosis in an ancestor of four aphid families. *J. Bacteriol.* **173:** 6321–6324.

Munson, M.A., Baumann, P., and Kinsey, M.G. (1991b). *Buchnera* gen. nov. and *Buchnera aphidicola* sp. nov., a taxon consisting of the mycetocyte-associated, primary endosymbionts of aphids. *Int. J. Syst. Bacteriol.* **41:** 566–568.

Nakabachi, A. and Ishikawa, H. (1999). Provision of riboflavin to the host aphid, *Acyrthosiphon pisum*, by endosymbiotic bacteria, *Buchnera. J. Insect Physiol.* **45:** 1–6.

Pierantoni, U. (1910). Origine e struttura del corpo ovale del Dactylopius citri e del corpo verde dell'Aphis brassicae. *Boll. Soc. Nat. Napoli* **24:** 1–43.

Prosser, W.A. and Douglas, A.E. (1991). The aposymbiotic aphid: an analysis of chlortetracycline-treated pea aphid, *Acyrthosiphon pisum. J. Insect Physiol.* **37:** 713–719.

Rispe, C. and Moran, N.A. (2000). Accumulation of deleterious mutations in endosymbionts: Muller's ratchet with two levels of selection. *Am. Nat.* **156:** 425–441.

Sandström, J.P., Russell, J.A., White, J.P., and Moran, N.A. (2001). Independent origins and horizontal transfer of bacterial symbionts of aphids. *Mol. Ecol.* **10:** 217–228.

Sasaki, T. and Ishikawa, H. (1995). Production of essential amino acids from glutamate by mycetocyte symbionts of the pea aphid *Acyrthosiphon pisum. J. Insect Physiol.* **41:** 81–86.

Schaechter, M. (1992). *Escherichia coli*, general biology. In *Encyclopedia of Microbiology* (J. Lederberg, Ed.), pp. 115–124. Academic Press, New York.

Shigenobu, S., Watanabe, H., Hattori, M., Sakaki, Y., and Ishikawa, H. (2000). Genome sequence of the endocellular bacterial symbiont of aphids *Buchnera* sp. APS. *Nature* **407:** 81–86.

Sulc, K. (1910). "Pseudovitellus" und ähnliche Gewebe der Homopteren sind Wohnstatten symbiontischer Saccharomyceten. *Sitzber. Bohm. Ges. Wiss.* **25:** 108–134.

Tamas, I., Klasson, L., Canback, B., Naslund, A.K., Eriksson, A.S., Wernegreen, J.J., Sandström, J.P., Moran, N.A., and Andersson, S.G.E. (2002). 50 million years of genomic stasis in endosymbiotic bacteria. *Science* **296**: 2376–2379.

Thao, M.L., Baumann, L., Baumann, P., and Moran, N.A. (1998). Endosymbionts (*Buchnera*) from the aphids *Schizaphis graminum* and *Diuraphis noxia* have different copy numbers of the plasmid containing the leucine biosynthetic genes. *Curr. Microbiol.* **36:** 238–240.

Van den Heuvel, J.F.J.M., Verbeek, M., and van der Wilk, F. (1994). Endosymbiotic bacteria associated with circulative transmission of potato virus by *Myzus persicae. J. Gen. Virol.* **75:** 124–142.

Van Ham, R.C.H.J, Martinez-Latorre, D., Moya, A., and Latorre, A. (1999). Plasmid-encoded anthranilate synthase (TrpEG) in *Buchnera aphidicola* from aphids of the family Pemphigidae. *Appl. Environ. Microbiol.* **65:** 117–125.

Wernegreen, J.J., Ochman, H., Jones, I.B., and Moran, N.A. (2000). Decoupling of genome size and sequence divergence in a symbiotic bacterium. *J. Bacteriol.* **182:** 3867–3869.

Whitehead, L.F. and Douglas, A.E. (1993a). Populations of symbiotic bacteria in the parthenogenetic pea aphid (*Acyrthosiphon pisum*) symbiosis. *Proc. R. Soc. London (B)* **254:** 29–32.

Whitehead, L.F. and Douglas, A.E. (1993b). A metabolic study of *Buchnera*, the intracellular bacterial symbionts of the pea aphid, *Acyrthosiphon pisum. J. Gen. Microbiol.* **139:** 821–826.

Wilkinson, T.L. and Douglas, A.E. (1998). Host cell allometry and regulation of the symbiosis between pea aphids, *Acyrthosiphon pisum*, and bacteria, *Buchnera. J. Insect Physiol.* **44:** 629–635.

Wilkinson, T.L. and Ishikawa, H. (1999). The assimilation and allocation of nutrients by symbiotic and aposymbiotic pea aphids, *Acyrthosiphon pisum. Entomol. Exp. Appl.* **91:** 195–201.

Wilkinson, T.L. and Ishikawa, H. (2000). Injection of essential amino acids substitutes for bacterial supply in aposymbiotic pea aphids (*Acyrthosiphon pisum*). *Entomol. Exp. Appl.* **94:** 85–91.

Wilkinson, T.L., Adams, D., Minto, L.B., and Douglas, A.E. (2001a). The impact of host plant on the abundance and function of symbiotic bacteria in an aphid. *J. Exp. Biol.* **204:** 3027–3038.

Wilkinson, T.L., Minto, L.B., and Douglas, A.E. (2001b). Amino acids as respiratory substrates in aphids: an analysis of *Aphis fabae* reared on plants and diets. *Physiol. Entomol.* **26:** 225–228.

3 Comparative Genomics of Insect Endosymbionts

Ivica Tamas and Siv G.E. Andersson

CONTENTS

INTRODUCTION

To date, approximately 60 microbial genomes have been sequenced, with sizes ranging from the smallest known genome of *Mycoplasma genitalium* of only 0.58 megabases (Mb) to *Myxococcus xanthus* and *Nostoc punctiforme* of 9.5 Mb and 10 Mb, respectively (www.nlm.ncbi.nih.gov/entrez/query.fcgi). The genomes of obligate host-associated bacteria tend to be smaller than the genomes of their close free-living relatives. The smallest genomes are thought to be not ancestral but the result of reductive genome evolution (Andersson and Kurland, 1998; Moran, 2002). The sequence loss has come about through a massive loss of phages, mobile elements, and repeated sequences that are present in abundance in most genomes of free-living bacteria as well as through the loss of gene sequences that are essential in most other systems.

Unlike free-living bacteria, which are capable of growth in a variety of different environments, intracellular bacteria grow optimally inside other eukaryotic cells, either freely in the cytoplasm or inside phagosomes or within specialized cells whose sole function is to contain bacteria. Some bacteria have lost their ability to grow in a free-living mode; these are obligate host-associated. Others, referred to as facultative intracellular, are capable of switching between the intracellular and the extracellular growth environments. In addition to differences in the extent of host-association, intracellular bacteria also differ with respect to the type of host associations, as summarized by the following terms:

0-8493-1286-8/03/$0.00+$1.50

Symbiosis — Association where one organism (the symbiont) lives within the body of another organism (the host), regardless of the actual effect on the host; this term is often used to describe physically associated mutualistic relationships

Mutualism — Association in which both partners benefit from their interaction; develops most often between organisms with widely different living requirements

Commensalism — Association where one of the species obtains some benefit from the interaction without either harming or benefiting the other organism; this association is most often seen between a larger host and a smaller commensal

Parasitism — Relationship between two species in which one partner benefits at the expense of the other

Most intracellular bacteria whose genomes have been sequenced are human parasites that often kill their hosts after exploiting amino acids, nucleotides, and other small molecules present in the nutrient-rich intracellular growth milieu. *Rickettsia prowazekii* is an excellent example of an obligate intracellular human pathogen with a small genome (1.1 Mb) that is transmitted by insect vectors, in this case lice (Andersson et al., 1998). Not surprisingly, the reduction of genome size has mostly been achieved at the expense of genes coding for proteins involved in metabolic pathways that are no longer required in the new, host-associated environment. Furthermore, most of the nonessential genetic material, such as repetitive elements, insertion sequences, prophages, transposons, and other kinds of foreign DNA, has been eliminated from these small genomes. Conversely, genes involved in basic information processes are less affected by the reduction of genome size, and a basic set of essential genes is present in most intracellular parasites irrespective of their different phylogenetic affiliation and the extent of damage caused to their host cells.

Other intracellular bacteria have established mutualistic interactions with their host cells. These relationships are typically based on the bacterial supply of compounds such as amino acids and other small metabolites to the host. The aphid endosymbiont *Buchnera aphidicola* is perhaps the best-known example of a bacterium that has established a long-term obligate relationship with its host (Baumann et al., 1995). However, in contrast to obligate intracellular parasites, obligate endosymbionts have often retained genes for basic biosynthetic processes and in some cases even amplified them on plasmids in response to the host demand for small molecules.

Thus, the process of reductive evolution is associated with host specialization as well as with genome deterioration and loss of mutational variability in the population. This seems to be an irreversible process in small, isolated, and asexual populations that frequently undergo bottlenecks (Moran, 1996, 2002; Andersson and Kurland, 1998) irrespective of the effects on host phenotypes. Thus, there is a theoretical danger that deleterious mutations will accumulate to such an extent that the fitness of the bacterium (and eventually its host) will start to decrease. This is because few, if any, genome variants are present in the population and because the loss of genes involved in DNA repair and recombination processes further reduces the possibility of repairing a sequence loss by recombination with a variant in which the lost fragment is still present.

Most published work on the interactions of endosymbionts with their hosts has focused on the nutritional dependencies of the host and its symbiont and on the physiological consequences of the interaction. However, recent research on the genomes of endosymbiotic bacteria has started to reveal the genetic basis for this relationship from the bacterial perspective. In this review, we will discuss the genome content and structure of bacteria that have developed mutualistic relationships with insects, based primarily on the recent publication of genome-sequence information from aphid endosymbionts, including our own genome-sequence data (Tamas et al., 2002).

APHID ENDOSYMBIONTS: *BUCHNERA*

Associations with intracellular prokaryotes are common among members of Arthropoda, which has probably contributed to their evolutionary success (Goebel and Gross, 2001). This is the most numerous animal group, with the number of described species exceeding 750,000. A majority of the endosymbionts identified within these species belong to the γ-subdivision of the proteobacteria (von Dohlen et al., 2001). Numerous obligate and facultative intracellular bacteria are found within the proteobacteria including genera such as *Buchnera, Francisella,* and *Legionella* in the γ-proteobacteria and *Rickettsia, Wolbachia, Ehrlichia,* and *Bartonella* in the α-proteobacteria.

The Homoptera (Insecta) contains about 4500 aphid species that are known for the presence of primary bacterial endosymbionts of the genus *Buchnera* (Blackman and Eastop, 1994). This particular association represents the best-studied example of animal endosymbionts. *Buchnera* are Gram-negative cocci with average cell size ranging in diameter from 2 to 5 μm (Houk and Griffiths, 1980). Ultrastructural studies revealed the presence of the bacteria surrounded by host-derived membrane in the cytoplasm of specialized polyploidic host cells termed bacteriocytes or mycetocytes (Buchner, 1965; Griffiths and Beck, 1973). These are grouped in bilobed structures (bacteriomes) within the aphid body cavity (Buchner, 1965).

In a typical case, there are 5.6 million bacterial cells per adult aphid (Baumann and Baumann, 1994). The bacteriomes may be surrounded by a sheath of cells that contains rod-shaped bacteria related to *Escherichia coli* in young individuals of some aphid species. These bacteria are referred to as secondary (S-) symbionts. In most cases, the effect of these bacteria on host phenotypes is not known (Moran and Baumann, 2000). In adult aphids, the bacteriome disintegrates into the abdomen, and S-endosymbionts can be isolated from a variety of body parts including the hemolymph (Baumann et al., 1995). Some members of Cerataphidini, a group within Aphidoidea, do not seem to contain any bacterial endosymbionts but harbor extracellular, yeast-like organisms (Buchner, 1965).

The association between *Buchnera* and its hosts is strictly mutualistic. Thus, *Buchnera* has never been cultured on artificial media outside of its aphid, emphasizing the obligate nature of the association. Likewise, it has been experimentally shown (Houk and Griffiths, 1980) that antibiotic treatment of the aphid host, which eliminates the bacteria, reduces aphid growth and often induces sterility, stressing the need for a bacterial partner. Therefore, aphids are best considered chimerical or composite organisms consisting of one partner that is an insect and another partner that is a bacterium.

Buchnera is transmitted vertically, from the mother aphid to the progeny via the infection of eggs or embryos (Baumann et al., 1995). It has been reported that *Buchnera* cells while transferred to the progeny are not surrounded by the host-derived membrane (Baumann et al., 1995). This can be viewed as their last connection to the outer world, perhaps imposing certain limits on the extent to which the bacterial membrane structure can be degraded and eliminated. Interestingly, *Blochmannia* species, the endosymbionts of ants closely related to Enterobacteriaceae, are found free in the cytoplasm of bacteriocytes (Schröder et al., 1996).

ENDOSYMBIOSIS: METABOLIC CONSIDERATIONS

From the host point of view, the aphid–*Buchnera* association is nutritionally based, while the host in return provides a stable, nutrient-rich environment to the bacterium. The primary function of the endosymbionts appears to be the synthesis of essential amino acids that are strictly required by the aphids. The phloem sap provides a diet rich in carbohydrates but very poor in amino acids (Baumann et al., 1995; Douglas, 1992, 1998). The concentration of essential amino acids in the sap is very low, further reducing its already limited nutritional value (Sandström and Pettersson, 1994; Sandström et al., 1999). Aphids, like other animals, require ten essential amino acids. Every single amino acid represents a growth-limiting factor if not provided in adequate quantities (Sandström et al., 1999). Thus, it may not be surprising that as much as 10% of *Buchnera* genomes are dedicated to amino-acid biosynthesis (Shigenobu et al., 2000; Tamas et al., 2002).

Tryptophan and leucine are present in extremely low concentrations in the phloem sap. With respect to these amino acids, *Buchnera* plays a particularly important role for the aphid in that the leucine and tryptophan biosynthetic genes have amplified on plasmids, leading to an overproduction of these amino acids (Bracho et al., 1995; Baumann et al., 1997; Baumann et al., 1999). The importance of tryptophan overproduction by *Buchnera* has been experimentally demonstrated (Douglas and Prosser, 1992). In this study, pea aphids were cured of the bacteria by chlortetracycline treatment and fed a diet deficient in tryptophan. Under these circumstances the pea aphids rarely reached adulthood.

The key enzyme in the metabolic pathway leading to tryptophan biosynthesis is anthranilate synthase. The two subunits of the enzyme are encoded by the *trpEG* genes that are amplified in the form of tandem repeats amounting to 3.6 kb of the *Buchnera* genome (Sg) (Lai et al., 1994). The tandem repeats account per se for a 16-fold amplification, not including additional effects caused by high plasmid copy numbers. The remaining genes dedicated to tryptophan biosynthesis (*trpDC(F)BA*) are located on the main chromosome (Munson and Baumann, 1993). The same pattern is seen in several *Buchnera* species from the Aphididae, indicating that the amplification occurred prior to the divergence of this group of aphid endosymbionts (Lai et al., 1996). However, this pattern is not universal among the aphid endosymbionts because the *trpEG* genes are located on the chromosome in at least one representative of the Pemphigidae (van Ham et al., 1999). The most likely explanation is that a single ancestral transfer of *trpEG* to a *RepA/C*-like replicon took place, followed by a reverse transfer to the chromosome in some *Buchnera* lineages (Roeland et al., 1999; van Ham et al., 2000).

Genes involved in the biosynthesis of leucine are also located on plasmids (*leu*ABCD) in *Buchnera* residing in *Schizaphis graminum*, *Diuraphis noxia*, and *Rhopalosiphum padi* (Baumann et al., 1999). These plasmids vary in size, from 3.0 to 12.8 kb, and contain two additional genes, *repA*1 and *repA*2, involved in plasmid replication (Lai et al., 1994). There are 24 copies of the leucine genes in *Buchnera* (Sg) when compared to the single copy identified in *Buchnera* of *D. noxia*.

The presence or absence of plasmid amplification in particular lineages is perhaps connected to different needs for tryptophan/leucine, which is related to differences in the overall life cycle such as developmental time and reproduction rates (Lai et al., 1995). The origin of the tryptophan and leucine biosynthetic plasmids in *Buchnera* is not clear, as they are not universally present in all *Buchnera* strains of the Aphidae families. In addition, different gene contents and orders have been reported in different strains of *Buchnera* (Soler et al., 2000). The process of relocation and amplification of these amino-acid biosynthetic genes on plasmids is most likely relatively novel in evolutionary terms and at least not older than the establishment of the endosymbiotic relationship.

BUCHNERA AND THEIR FREE-LIVING RELATIVES

Members of the Enterobacteriaceae, such as *E. coli* and *Haemophilus influenzae,* are the closest modern free-living relatives of *Buchnera*. Unlike *Buchnera*, members of Enterobacteriaceae are commensals, which is reflected in genome sizes, contents, and architectures. For example, there is a considerable genome size variation among *E. coli* strains. Currently, there are three sequenced *E. coli* genomes: the K-12 strain with a genome size of 4.60 Mb, the 0157:H7EDL strain with a size of 5.50 Mb, and the 0157:H7 strain totaling 5.60 Mb (www.ncbi.nlm.nih.gov/entrez/query.fcgi). These differences appear to be largely explained by the presence or absence of bacteriophages, prophages, and virulence genes in the genome (Ohnishi et al., 2001). In between the variable parts is a well-conserved core of the *E. coli* chromosome with no large rearrangements. This "backbone" of the *E. coli* genome, which is estimated to be around 4.1 Mb, is topologically fragmented by strain-specific pieces distributed around the chromosome (Hayashi et al., 2001).

The *Buchnera* genomes are less than one eighth of the *E. coli* 0157 genome (Shigenobu et al., 2000; Tamas et al., 2002). There are no examples of horizontally transferred DNA or any obvious virulence-associated genes, suggesting that the *Buchnera* genome is essentially a small subset of the *E. coli* genome. The loss of genes is evident in all categories but particularly so in categories such as gene regulation. This extreme genome reduction raises questions about the rate and process of genome deterioration in aphid endosymbionts. The acquisition of complete genome-sequence data from aphid endosymbionts now provides an opportunity to study these processes at the whole-genome level.

DIVERGENCE DATES FOR *BUCHNERA*

The *Buchnera*–Aphidae association extends back to the origin of Aphidoidea approximately 150 to 200 million years ago (Moran et al., 1993), and the origin of the aphidiform ancestor can be traced back to Jurassic or earlier, about 250 million years ago (Heie, 1987). Aphids are a morphologically defined group of insects with a rich fossil record, making it possible to assign divergence dates for the separation of the different species. Since the aphid endosymbionts have been vertically transmitted, with no evidence of horizontal transmission (Simon et al., 1996), it is in principle possible to infer divergence dates for the aphid endosymbionts based on dates estimated for their aphid hosts. Indeed, a comparison of phylogenies based on genes from *Buchnera* and their aphid hosts confirms that the tree topologies are identical, indicative of synchronous divergences (Moran et al., 1993; Moran and Baumann, 1994). These findings support an exclusively vertical transmission of *Buchnera* via maternal inheritance.

Other cases of bacteriocyte-associated bacteria showing a phylogenetic congruence with hosts have also been identified in tsetse flies, carpenter ants, and members of Blattaria (Bandi et al., 1995; Chen et al., 1999; Sauer et al., 2000). A well-studied example of an association between arthropods and bacteria is *Wolbachia* spp. that are maintained in a wide range of hosts (Breeuwer et al., 1992; Vavre et al., 1999). However, unlike *Buchnera*, the obtained phylogenies for *Wolbachia* and its host are not congruent. This implies a loss of the ancestral bacterium in one or more lineages, followed by multiple subsequent infections (O'Neill et al., 1992).

COMPARATIVE GENOMICS OF *BUCHNERA*

The aphid *Schizaphis graminum* (Sg) and its relative *Acyrtosyphon pisum* (Ap) are thought to have diverged about 50 to 70 million years ago. Thanks to the availability of genome-sequence data for *Buchnera aphidicola* (Ap) (Shigenobu et al., 2000) and *Buchnera* (Sg) (Tamas et al., 2002), it is now possible to quantify the changes introduced during the past 50 million years. The two genomes are of approximately the same size — 641,454 bp for *Buchnera* (Sg) and 640,681 bp for *Buchnera* (Ap) (Table 3.1). The similarity of genome sizes is in accordance with highly conserved genome sizes among aphid endosymbionts, showing less than a 5% difference even among lineages separated by 100 to 200 million years of independent evolution (Wernegreen et al., 2000).

There are 545 identified genes in *Buchnera* (Sg), compared to 564 genes in *Buchnera* (Ap) (Tamas et al., 2002). A total of 526 orthologous genes were identified, while pseudogenes and

TABLE 3.1
Comparison of Genome Features for *B. aphidicola* (Sg) and *B. aphidicola* (Ap)

Feature	*B. aphidicola* (Sg)	*B. aphidicola* (Ap)
Genome size (bp)	641,454	640,681
Genic G + C content (%)	26.2	26.3
Intergenic G + C content (%)	14.8	16.1
Protein coding genes (no.)	545	564
Pseudogenes (no.)	38	13
Orthologous genes (no.)	526	526

noncoding intergenic regions account for the difference between the two genomes (Table 3.1). One of the most significant findings obtained from the *Buchnera* (Sg) genome sequence and its subsequent comparison to the *Buchnera* (Ap) genome is the fact that the two genomes show perfect synteny with no detected inversions, translocations, or duplications, despite 50 to 70 million years of independent evolution. There are also no indications of genes horizontally transferred into these genomes subsequent to their divergence (Tamas et al., 2002).

Unlike the remarkable stability observed in gene content and gene order, nucleotide substitutions at synonymous sites (Ks) are surprisingly high, close to saturation (Tamas et al., 2002). Since little variation is observed among genes, these rates are expected to reflect the intrinsic mutation rate. Based on the assigned dates for divergences of strains of *Buchnera*, it has been possible to estimate substitution rates among lineages. The genomic substitution rate has been estimated at 9.0×10^{-9} synonymous and 1.6×10^{-9} nonsynonymous substitutions per site per year for chromosomal genes in *Buchnera* (Sg) and *Buchnera* (Ap), averaged over all genes (Tamas et al., 2002).

The genomic stasis observed in *Buchnera* can most likely be attributed to its specific ecological niche. Locked in their intracellular environment with few or no possibilities to take up external DNA or a need for doing so, these genomes have remained remarkably conserved. In a global comparison of the relative frequency of rearrangements and insertions/deletions compared to nucleotide substitutions, *Buchnera* were found to represent the least variable genomes, whereas *E. coli* and *Salmonella* were found to have the most variable genomes among all the analyzed genome pairs (Tamas et al., 2002).

This extreme stasis may be related to the fact that the *Buchnera* genomes are the only bacterial genomes known so far in which a *recA* gene is not present (Tamas et al., 2002). The product of this gene, recombinase A, catalyzes the hydrolysis of ATP in the presence of single-stranded DNA, the ATP-dependent uptake of single-stranded DNA by duplex DNA, and the ATP-dependent hybridization of homologous single-stranded DNAs (Dale, 1998; Belotserkovski and Zarling, 2002). The loss of this gene is expected to lower the incidence of recombination events, although it cannot be excluded that *recA*-independent recombination pathways are present and may partially compensate for this loss.

The *recF* gene involved in DNA replication and normal SOS induction has also been eliminated in these genomes. In *Buchnera* (Sg), additional genes involved in base-excision repair, such as *endA*, *lig*, *ung*, *mfd*, and *phrB*, have started to accumulate mutations. In addition to impaired capabilities for DNA recombination, this is expected to influence the process of DNA repair in *Buchnera* (Sg). The presence of as many as 120 genomic copies per *Buchnera* cell (Komaki and Ishikawa, 1999) may be a creative way of compensating for the loss of *recA* and DNA-repair-associated genes: once severe damage to the active genomic copy occurs, another copy takes on the leading role.

DETERIORATION OF THE *BUCHNERA* GENOMES

The *Buchnera* genomes are two of the most highly reduced genomes described to date. A comparison of the two genomes confirms that they are still in the process of erasing gene sequences, albeit at a low rate. Detailed analyses of pseudogenes and intergenic DNA have provided clues about the patterns and processes of genome deterioration. Interestingly, the genes targeted by mutations are not a random subset of the *Buchnera* genomes; almost half of the identified pseudogenes (18 out of 38) in *Buchnera* (Sg) belong to the three blocks of genes coding for functionally related proteins (Table 3.2).

CYSTEINE BIOSYNTHETIC GENES

A total of 6 of the 46 genes in *Buchnera* (Sg) that are related to amino-acid biosynthesis appear to be mutationally inactivated and are all involved in cysteine biosynthesis (*cysDGHINQ*). The

TABLE 3.2
Examples of *Buchnera* Pseudogenes in Three Functional Categories

Function	*Buchnera* (Sg)	*Buchnera* (Ap)
Amino acid metabolism		
Cysteine biosynthesis	*cysNDGHIQ*	
Asparaginase		*asnA*
Methionine regulation		*metR*
Restriction-repair systems		
Base excision repair	*endA, lig, ung*	
Uvr excision repair	*mfd*	
Direct damage reversal	*phrB*	
Membrane components		
Lipoproteins	*lgt, nlpD, mrcB*	
Phospholipids	*fabZ*	*fabD*
Peptidoglycan	*murCEF, mraY, mltE*	*ddlB*

Description of pseudogenes: In addition to pseudogenes that show clear homology to functional genes in other species, some sequences show signs of more extensive degradation and are recognizable as gene remnants only through short stretches of homology plus a conserved position and orientation relative to flanking, intact genes. Among these are five genes (*ansA, hemD, apbE, cmk,* and *cvpA*) containing mutations in both *Buchnera* genomes with more extensive degradation in one genome. [Data from Tamas, I., Klasson, L., Canback, B., Naslund, A.K., Eriksson, A.-S., Wernegreen, J.J., Sandström, J.P., Moran, N.A., and Andersson, S.G.E. (2002). *Science* **296:** 2376–2379 and Shigenobu, S., Watanabe, H., Hattori, M., Sakaki, Y., and Ishikawa, H. (2000). *Nature* **407:** 81–86.]

homologous genes in *Buchnera* (Ap) are not affected. This may reflect the particular interaction of *Buchnera* (Sg) with its host. As reported elsewhere (Sandström et al., 1999), some aphid species inflict distinctive types of chlorotic lesions on their host plants, elevating the concentration of amino acids such as cysteine in the phloem sap. This phenomenon may be related to the observed silencing of genes for tryptophan biosynthesis in some lineages. On the assumption that these are novel behaviors in evolutionary terms, it has been interpreted as a decay of the mutualistic potential (Wernegreen and Moran, 2000).

Cell-Envelope Genes

The accumulation of mutations in cell-envelope genes *(murCEF, mraY, mltE, ddlB, nlpD,* and *mrcB*) is most likely related to the status of intracellular symbionts residing in the host-derived vesicles. The mutations identified in these genes are mainly single-nucleotide deletions and insertions in homopolymeric tracts. Their effects on cell membrane structures remain to be determined.

DNA Repair–Associated Genes

Genes coding for proteins involved in recombination/repair seem to have been targeted by mutations to a larger extent than other genes. These mutated genes include *endA*, *lig*, *ung*, *mfd,* and *phrB*. The functional inactivation of these genes may be related to the presence of, on average, 120 genomic copies per *Buchnera* cell.

Genome Degradation: A Gradual Process

Currently, there are two different models that attempt to explain the extensive genome reduction observed in *Buchnera*. On the one hand, it has been suggested that genome reduction has resulted from the removal of large pieces of DNA mediated by repeated sequences (Moran and Mira, 2001; Tamas et al., 2002). On the other hand, genome reduction has also been explained by the gradual accumulation of deletions of small sizes (Silva et al., 2001). As we will argue below, these scenarios are not mutually exclusive.

Thus, it seems likely that the process of genome reduction was initially obtained by large deletion events leading to a relatively rapid disappearance of large fragments of DNA. We can call this stage phase I; it is characterized by the early stages of the integrative process, soon after acquisition of the obligate host-associated lifestyle. Unfortunately, it is difficult to confirm that large fragments of DNA were once eliminated from the ancestral genome because these processes happened hundreds of millions of years ago and left few traces in modern *Buchnera* genomes. However, the rate of sequence evolution during the past 50 million years is too low to account for the extensive sequence loss that has occurred since the divergence of *Buchnera* from its free-living relatives, providing indirect support for large, repeat-mediated deletions at an earlier stage.

Once recombination frequencies at repeated sites were reduced due to the loss of repeated sequences and recombination genes, deletions involving single to several nucleotides started to dominate the reductive processes. We can call this stage phase II; the rate of gene removal witnessed today (14 genes/50 million years per two genomes) (Tamas et al., 2002) is an example of the slow rate at which sequences are lost during this second phase.

Several examples of weakly as well as heavily mutated genes have been identified in *B. aphidicola* (Sg). In addition to genes like the *cysNDHGIQ*, *murCEF*, and *ddlB* homologs, which are only weakly affected by mutations, we have identified noncoding DNA that shows little or no similarity with orthologs in the other genome (*cmk, ycfM, asnA, bioH, folE*) (Tamas et al., 2002). In some cases, an existing functional gene has been identified in one genome, whereas there is a long spacer at the same position in the other genome showing no sequence similarity to the identified gene. This may be explained by an early inactivation followed by a rapid rate of nucleotide substitutions, eroding all traces of the ancestral gene. Examples of such extreme cases are *cspC, hns,* and *ycfM,* which display no deducible sequence similarity to the corresponding locus in the other genome (Tamas et al., 2002). Thus, gene elimination in *Buchnera* has been shown to occur in a slow step-by-step manner, also observed in other obligate host-associated genomes such as the *Rickettsia* genomes (Andersson et al., 1998; Andersson and Andersson, 1999a,b, 2001).

Genes in *Buchnera* tend to be slightly shorter than their orthologs in *E. coli* (Charles et al., 1999). The reduction of gene length is mainly seen at the 3′ ends of *Buchnera*, which may be due to a higher frequency of termination codons in A+T-rich genomes (Oliver and Marin, 1996). Perhaps the most drastic example of gene shortening described so far is found in the psyllid endosymbiont *Carsonella ruddii,* yet another example of parallel evolution of a prokaryote and its host (Clark et al., 2001). *C. rudii* proteins appear to be on average 9% shorter than their *E. coli* homologs. The reduction of intergenic regions is in some cases so extreme that it results in operon fusions. In addition, they have extremely low G+C content values, sometimes as low as 10%. These genomic features suggest that this could be one of the most highly derived endosymbiont genomes identified so far.

Despite the conservation of the gene order in *Buchnera* (Sg)–*Buchnera* (Ap) (Tamas et al., 2002) and genome size in *Buchnera* of *R. padi* and *Melaphis rhois* (Wernegreen et al., 2000), a further reduction of *Buchnera* genomes down to 450 kb has been observed (Gil et al., 2002). These, supposedly the smallest bacterial genomes, have been identified in *Buchnera* present in Lachninae, a subfamily distant to Aphidinae and considered the most ancestral group within the Aphidoidea (Martinez-Torres et al., 2001). These genomes may contain as few as 396 protein-coding genes. The variation in genome size of the primary aphid endosymbionts may be attributed to differences in life style, time since the first infection, and order of gene loss. For example, the *Buchnera* lineages of Lachninae may have deleted a more substantial fraction of their genomes prior to the loss of the *recA* gene. Or, alternatively, their evolution may have been faster than that of the other genomes.

INTRACELLULAR MUTUALISM VS. PARASITISM

It is conceivable that the colonization of an intracellular environment, whether leading to a mutualistic or a parasitic relationship, is based on a common strategy. This is reflected in a number of shared genomic features that appear to be characteristic of bacteria that reproduce in a host-associated manner: small genome sizes when compared to their free-living relatives, a reduced gene repertoire, absence of extensive repetitive elements, and a tendency toward AT richness. Possible explanations for the increase in AT content are a reduced capacity to repair misinsertions of thymidine and adenine into DNA via replication errors and C to U deaminations. The use of similar invasion strategies is also reflected by a certain amount of damage, which in some cases is inflicted on a host not only by pathogens but also at early stages of colonization by the intracellular mutualists. In addition, elevated production of stress-related proteins in both intracellular parasites and mutualists has been observed (Goebel and Gross, 2001).

Nevertheless, the genomes of mutualists and parasites, though both well suited for intracellular existence, do show characteristic differences that arise from the particular lifestyles of the organisms. Although there is a general loss of genes involved in regulation, biosynthesis, and repair and recombination processes (and to a lesser extent in energy metabolism and basic cellular processes) in both organisms, specific parts of the metabolic systems are often preserved or even amplified in the endosymbionts.

This may not be surprising because the eukaryote provides a stable and isolated environment that is also rich in nutrients, while the bacteria are expected in return to produce nutrients desired by the host. The effect is that genes associated with these desired metabolic pathways are under particularly high selective pressure. As a result, in addition to their being maintained in the genome, their functions may even be improved or altered according to the demands of the host. This is observed in *Buchnera* for genes associated with the biosynthesis of amino acids. For example, the *trpEG* genes have been relocated and subsequently amplified on plasmids to escape feedback inhibition. In contrast, intracellular pathogens such as *Rickettsia* have lost most genes dedicated to biosynthetic functions because the corresponding metabolites are present and easily accessible in the host cell. Other differences observed between *Rickettsia* and *Buchnera* include the lack of a glycolytic system in *R. prowazekii* (Andersson et al., 1998) and of the TCA cycle in *Buchnera* (Shigenobu et al., 2000; Tamas et al., 2002).

In parasites, the loss of basic biosynthetic capabilities is often complemented by elevated export–import capabilities. Moreover, the presence of pathogenicity-related genes in parasites is necessary for ensuring the invasion of the host cell. This functional category is largely eliminated from genomes of obligate intracellular mutualists, such as *Buchnera* spp. Those few *Buchnera* genes, which according to homology searches should be classified as virulence factors, are probably either functionally altered or simply required for entry into the host cell. For example, genes encoding flagellar structures, typically used for bacterial motility, have atypically high substitution rates and may have evolved a novel function (transport) in *Buchnera* due to the mutualistic lifestyle (Tamas et al., 2002).

Also in contrast to parasites, a reduced repertoire of genes encoding cell-surface structures has been observed in *Buchnera*. Such a loss is not an option for parasites, which utilize cell-surface proteins for continued virulence. Intracellular pathogens like *R. prowazekii* must penetrate host defense mechanisms to gain access to the intracellular environment. Here, cell-surface structures are assumed to play a prominent role.

Are Switches between Parasitism and Mutualism Possible?

The assumption that entry into the intracellular environment is based on the same or related genetic adaptations has produced some speculation about whether a switch between these two lifestyles is possible (Moran and Wernegreen, 2000; Goebel and Gross, 2001). Thus, it has been suggested that invertebrates and their endosymbionts act as a "playground" for the evolution of virulence factors in bacteria, which are later used to invade mammals (Goebel and Gross, 2001). This line of thinking is supported by a substantial overlap in taxonomic designation (α-, β-, or γ-proteobacteria) of both intracellular mutualists found in invertebrates and many pathogenic bacteria in mammals.

For example, several associations involving members of Enterobacteriaceae in tsetse flies, psyllids, and ants have been reported (Schröder et al., 1996; Chen et al., 1999; Spaulding and von Dohlen, 2001). In addition, in some cases the presence of common virulence factors has also been observed: the maternally transmitted secondary symbiont of tsetse flies *Sodalis glossinidius* requires a type III secretion system to invade the host. The same type of secretion system has been identified in pathogenic Enterobacteriaceae, including *S. typhimurium* (Dale et al., 2001). Another common feature is tissue tropism, which is found in many pathogenic bacteria. Examples among bacterial mutualists are bacteriomes in insect species and the root nodules of leguminous plants or the crown gall tumors (Hentschel et al., 2000). Finally, the identification of a "symbiosis island" in *Mesorhizobium loti* much resembling in its nature "pathogenicity islands" commonly found in pathogenic bacteria has been described (Sullivan et al., 2002).

However, in stable and well-established associations based on mutualism, a switch toward pathogenicity is not likely to occur. This is because the amount of genetic change required to reach this point would probably deter such a redirection. The highly reduced genomes of obligate host-associated bacteria mean that they have in a sense been "painted into a (metabolic) corner" (Tamas et al., 2001). Thus, once amino acid biosynthetic genes have been lost, such as in *Rickettsia*, it would be impossible for these bacteria to develop a symbiotic relationship based on the supply of amino acids to a presumptive host. Another major limitation is the impaired ability to import foreign DNA (Tamas et al., 2002), which could prevent the acquisition of novel virulence factors required for a pathogenic lifestyle.

A more likely scenario is that intracellular mutualists may in some species be "attenuated pathogens" while retaining their pathogenic status in other species (Corsaro et al., 1999). For example, *Bartonella henselae* causes no observable disease to the cat that represents its animal reservoir, while it causes a variety of diseases in humans, depending on the activity of the immune system. Numerous such cases have been described for a variety of bacterial and viral systems, including HIV, which is pathogenic for humans, whereas its close relative SIV causes no observable harm to its animal host. Thus, although adaptations to unusually stable intracellular environments impose drastic, largely irreversible limits on the direction for further evolution, the borderline between mutualists, parasites, and commensals may not always be distinct and well defined.

CONCLUDING REMARKS

The majority of recently published papers in the field of bacterial genomics suggest that bacterial genomes are highly dynamic structures. Large intragenomic alterations, such as chromosomal rearrangements and insertions/deletions, occur frequently, probably mediated by insertion sequences, prophages, and repetitive sequences. As a consequence, closely related species may differ drastically in gene content and gene order. Even very closely related strains of the same species have been demonstrated to differ by up to 20% in genome size (Boucher at al., 2001). Therefore, the most provocative finding of the genomic work on aphid endosymbionts is that the *Buchnera* genomes are so remarkably conserved in both genome content and gene order. There are no signs of inversions, translocations, duplications, or horizontal gene transfer (Tamas et al., 2002).

This may be partially explained by the loss of repeated sequences and *recA*, which is likely to have had a profound effect on DNA-repair and recombination abilities. The presence of more than 100 genomic copies per cell may represent a compensatory phenomenon directly related to defective DNA repair. In contrast to the extreme stability in structure, nucleotide substitutions occur at high rates, and gene loss has been estimated to be about one eliminated gene per 5 to 10 million years (Tamas et al., 2002).

The observation that the *Buchnera*–host association is stable indicates that these endosymbionts are at an advanced, later stage of the internalization process. Most of the large-scale modifications, primarily deletions, already happened far back in evolutionary time. Although *Buchnera* are no longer a source of ecological novelties for their host (Tamas et al., 2002), the converse might still be true. Changes in the feeding behavior of aphids could induce a genetic change in the *Buchnera* genome, as seen in the case of *cys* genes. These genomes are therefore excellent model systems for tracing individual genetic changes that can be directly correlated to phenotypes of a particular host or have occurred as a response to a changing environment.

An evolutionary reconstruction of the events leading to the colonization of intracellular growth habitats that is applicable to both mutualists and parasites is as follows. The ancestral free-living bacterium first learned how to enter a host cell, then how to multiply within it, and finally how to exit again, maybe via the acquisition of virulence genes on pathogenicity islands. The initial association may have been either parasitic or mutualistic or it may have had no effect on the selected host cells. During the early stages of integration, the bacteria were most likely sporadic cell-surface-associated organisms that mostly reproduced in a free-living manner.

At a later stage, the bacteria invaded the intracellular environment regularly while still maintaining their free-living growth capabilities; they developed a facultative intracellular lifestyle. For example, *F. tularensis* is an intracellular parasite of both macrophages and parenchymal cells, although it can be still cultivated *in vitro* on complex, cysteine-supplemented media (Conlan and North, 1992; Maurin et al., 2000). However, most facultative intracellular bacteria grew substantially better within their new host-associated habitat than in a free-living mode. As these bacteria gradually deepened their relationship with the eukaryotic host cell, the obligatory dimension of the association started to dominate in such a way that it became increasingly difficult for the bacterium to grow outside of its host. Finally, the process of reductive evolution, first based on rapid, large-scale deletions mediated by repeated sequences and later on small intragenic deletion mutations, pushed the organism to the "point of no return." Genome reduction eventually became so extensive that a free-living lifestyle was no longer an option.

Thus, the cost of colonizing a host-associated environment is extreme sequence loss; the genome is in all cases studied so far reduced to a fraction of what it was at the start of the internalization process. The reward is a minimal genome stripped of everything not essential for survival — and access to a protective nutrient-rich environment. At a certain stage of this mingling process it may no longer be meaningful to speak about an insect host and a bacterial guest; the two have merged to become a new, single organism.

REFERENCES

Andersson, J.O. and Andersson, S.G.E. (1999a). Genome degradation is an ongoing process in *Rickettsia*. *Mol. Biol. Evol.* **16:** 1178–1191.

Andersson, J.O. and Andersson, S.G.E. (1999b). Insights into the evolutionary process of genome degradation. *Curr. Opinions Genet. Dev.* **9:** 664–671.

Andersson, J.O. and Andersson, S.G.E. (2001). Pseudogenes, junk DNA and the dynamics of *Rickettsia* genomes. *Mol. Biol. Evol.* **18:** 829–839.

Andersson, S.G.E. and Kurland, C.G. (1998). Reductive evolution of resident genomes. *Trends Microbiol.* **6:** 263–278.

Andersson, S.G.E., Zomorodipour, A., Andersson, J.O., Sicheritz-Ponten, T., Alsmark, U.C.M., Podowski, R.M., Naslund, A.K., Eriksson, A.-S., Winkler, H.H., and Kurland, C.G. (1998). The genome sequence of *Rickettsia prowazekii* and the origin of mitochondria. *Nature* **396:** 133–140.

Bandi, C., Sironi, M., Damiani, G., Magrassi, L., Nalepa, C.A., Laudani, U., and Sacchi, L. (1995). The establishment of intracellular symbiosis in an ancestor of cockroaches and termites. *Proc. R. Soc. London (B)* **259:** 293–299.

Baumann, P. and Baumann, L. (1994). Growth kinetics of the endosymbiont *Buchnera aphidicola* in the aphid *Schizaphis graminum. Appl. Environ. Microbiol.* **49:** 55–94.

Baumann, P., Baumann, L., Lai, C.Y., Rouhbakhsh, D., Moran, N.A., and Clark, M.A. (1995). Genetics, physiology, and evolutionary relationships of the genus *Buchnera*: intracellular symbionts of aphids. *Annu. Rev. Microbiol.* **49:** 55–94.

Baumann, L., Clark, M.A., Rouhbakhsh, D., Baumann, P., Moran, N.A., and Voegtlin, D. (1997). Endosymbionts (*Buchnera*) of the aphid *Uroleucon sonchi* contain plasmids with *trpEG* and remnants of *trpE* pseudogenes. *Curr. Microbiol.* **35:** 18–21.

Baumann, P., Baumann, L., Moran, N.A., Sandström, J., and Thao, M.L. (1999). Genetic characterization of plasmids containing genes encoding enzumes of leucine biosynthesis in endosymbionts (*Buchnera*) of aphids. *J. Mol. Evol.* **48:** 77–85.

Belotserkovski, B.P. and Zarling, D.A. (2002). Peptide nucleic acid (PNA) facilitates multistranded hybrid formation between linear double-stranded DNA targets and *recA* protein-coated complementary single-stranded DNA probes. *Biochemistry* **41:** 3686–3692.

Blackman, R.L. and Eastop, V.F. (1994). *Aphids on the World's Trees.* CAB International, Wallingford, U.K.

Boucher, Y., Nesbo, C.L., and Doolittle, W.F. (2001). Microbial genomes: dealing with diversity. *Curr. Opinions Microbiol.* **4:** 285–289.

Bracho, A.M., Martinez-Torres, D., Moya, A., and Latorre, A. (1995). Discovery and molecular characterization of a plasmid localized in *Buchnera sp.* bacterial endosymbiont of the aphid *Rhopalosiphum padi. J. Mol. Evol.* **41:** 67–73.

Breuwer, J.A., Stouthamer, R., Barnes, S.M., Pelletier, D.A., Weisburg, W.G., and Werren, J.H. (1992). Phylogeny of cytoplasmic incompatibility microorganisms in the parasitoid wasp genus *Nasonia* (Hymenoptera:Pteromalidae) based on 16S ribosomal DNA sequences. *Insect Mol. Biol.* **1:** 25–36.

Buchner, P. (1965). In *Endosymbiosis of Animals with Plant Microorganisms*, pp. 210–332. Interscience, New York.

Charles, H., Mouchiroud, D., Lobry, J., Goncalves, I., and Rahbe, Y. (1999). Gene size reduction in the bacterial aphid endosymbiont *Buchnera. Mol. Biol. Evol.* **16:** 1820–1822.

Chen, X., Li, S., and Aksoy, S. (1999). Concordant evolution of a symbiont with its host insect species: molecular phylogeny of genus *Glossina* and its bacteriome-associated endosymbiont *Wigglesworthia glossinidia. J. Mol. Evol.* **48:** 49–58.

Clark, M.A., Baumann, L., Thao, M.L., Moran, N.A., and Baumann, P. (2001). Degenerative minimalism in the genome of a psyllid endosymbiont. *J. Bacteriol.* **183:** 1853–1861.

Conlan, J.W. and North, R.J. (1992). Early pathogenesis of infection in the liver with the facultative intracellular bacteria *Listeria monocytogenes*, *Francisella tularensis*, and *Salmonella typhimurium* involves lysis of infected hepatocytes by leukocytes. *Infect. Immunol.* **60:** 5164–5171.

Corsaro, D., Venditti, D., Padula, M., and Valassina, M. (1999). Intracellular life. *Crit. Rev. Microbiol.* **25:** 39–79.

Dale, C., Young, S.A., Haydon, D.T., and Welburn, S.C. (2001). The insect endosymbiont *Sodalis glossinidius* utilizes a type III secretion system for cell invasion. *Proc. Natl. Acad. Sci. U.S.A.* **98:** 1883–1888.

Dale, J.W. (1998). Mutation and variation. In *Molecular Genetics of Bacteria*, 3rd ed., pp. 39–67. John Wiley & Sons, New York.

Douglas, A.E. (1992). Requirement of pea aphids (*Acyrthosiphon pisum*) for their symbiontic bacteria. *Entomol. Exp. Appl.* **65:** 195–198.

Douglas, A.E. (1998). Nutritional interactions in insect-microbial symbioses: aphids and their symbiotic bacteria. *Annu. Rev. Entomol.* **43:** 17–37.

Douglas, A.E. and Prosser, W.A. (1992). Synthesis of the essential amino acid tryptophan in the pea aphid (*Acyrthosiphon pisum*) symbiosis. *J. Insect Physiol.* **38:** 565–568.

Gil, R., Sabater-Munoz, B., Latorre, A., Silva, F.J., and Moya, A. (2002). Extreme genome reduction in *Buchnera spp.*: Toward the minimal genome needed for symbiotic life. *Proc. Natl. Acad. Sci. U.S.A.* **99:** 4454–4458.

Goebel, W. and Gross, R. (2001). Mutualistic and parasitic intracellular survival strategies of prokaryotes. *Trends Microbiol.* **9:** 267–273.

Griffiths, G.W. and Beck, S.D. (1973). Intracellular symbionts of the pea aphid *Acyrthosiphon pisum*. *J. Insect Physiol.* **19:** 75–84.

Hayashi, T., Makino, K., Ohnishi, M., Kurokawa, K., Ishii, K., Yokoyama, K., Han, C.G., Ohtsubo, E., Nakayama, K., Murata, T., Tanaka, M., Tobe, T., Iida, T., Takami, H., Honda, T., Sasakawa, C., Ogasawara, N., Yasunaga, T., Kuhara, S., Shiba, T., Hattori, M., and Shinagawa, H. (2001). Complete genome sequence of enterohemorrhagic *Escherichia coli* O157:H7 and genomic comparison with a laboratory strain K-12. *DNA Res.* **8:** 11–22.

Heie, O.E. (1987). Paleontology and phylogeny. In *Aphids: Their Biology, Natural Enemies and Control*, World Crop Pests, Vol. 2A (A.K. Minks and P. Harrewijn, Eds.), pp. 367–391. Elsevier, Amsterdam.

Hentschel, U., Steinert, M., and Hacker, J. (2000). Common molecular mechanisms of symbiosis and pathogenesis. *Trends Microbiol.* **8:** 2226–2231.

Houk, E.J. and Griffiths, G.W. (1980). Intracellular symbionts of the Homoptera. *Annu. Rev. Entomol.* **25:** 161–187.

Komaki, K. and Ishikawa, H. (1999). Intracelullar bacterial symbionts of aphids possess many genomic copies per bacterium. *J. Mol. Evol.* **48:** 717–722.

Lai, C.Y., Baumann, L., and Baumann, P. (1994). Amplification of *trpEG*: adaptation of *Buchnera aphidicola* to an endosymbiotic association with aphids. *Proc. Natl. Acad. Sci. U.S.A.* **91:** 3819–3823.

Lai, C.Y., Baumann, P., and Moran, N.A. (1995). Genetics of the tryptophan biosynthetic pathway of the prokaryotic endosymbiont (*Buchnera*) of the aphid *Schlechtendalia chinensis*. *Insect Mol. Biol.* **4:** 47–59.

Lai, C.Y., Baumann, P., and Moran, N. (1996). The endosymbiont (*Buchnera* sp.) of the aphid *Diuraphis noxia* contains plasmids consisting of *trpEG* and tandem repeats of *trpEG* pseudogenes. *Appl. Environ. Microbiol.* **62:** 332–339.

Martinez-Torres, D., Buades, C., Latorre, A., and Moya, A. (2001). Molecular systematics of aphids and their primary endosymbionts. *Mol. Phylogenet. Evol.* **20:** 437–449.

Maurin, M., Mersali, N.F., and Raoult, D. (2000). Bactericidal activities of antibiotics against intracellular *Francisella tularensis*. *Antimicrob. Agents Chemother.* **44:** 3428–3431.

Moran, N.A. (1996). Accelerated evolution and Muller's rachet in endosymbiotic bacteria. *Proc. Natl. Acad. Sci. U.S.A.* **93:** 2873–2878.

Moran, N.A. (2002). Microbial minimalism: genome reduction in bacterial pathogens. *Cell* **108:** 583–586.

Moran, N.A. and Baumann, P. (1994). Phylogenetics of cytoplasmically inherited microorganisms of arthropods. *Trends Ecol. Evol.* **9:** 15–20.

Moran, N.A. and Baumann, P. (2000). Bacterial endosymbionts in animals. *Curr. Opinions Microbiol.* **3:** 270–275.

Moran, N.A. and Mira, A. (2001). The process of genome shrinkage in the obligate symbiont *Buchnera aphidicola*. *Genome Biol.* **2:** 0054.1–0054.12.

Moran, N.A. and Wernegreen, J.J. (2000). Lifestyle evolution in symbiotic bacteria: insights from genomics. *Tree* **15:** 321–326.

Moran, N.A, Munson, M.A., Baumann, P., and Ishikawa, H. (1993). A molecular clock in endosymbiotic bactera is calibrated using the insect hosts. *Proc. R. Soc. London (B)* **253:** 167–171.

Munson, M.A. and Baumann, P. (1993). Molecular cloning and nucleotide sequence of a putative *trpDC(F)BA* operon in *Buchnera aphidicola* (endosymbiont of the aphid *Schizaphis graminum*). *J. Bacteriol.* **175:** 6426–6432.

Ohnishi, M., Kurokawa, K., and Hayashi, T. (2001). Diversification of *Escherichia coli* genomes: are bacteriophages the major contributors? *Trends Microbiol.* **10:** 481–485.

Oliver, J.L. and Marin, A. (1996). A relationship between GC content and coding-sequence length. *J. Mol. Evol.* **43:** 216–223.

O'Neill, S.L., Giordano, R., Colbert, A.M., Karr, T.L., and Robertson, H.M. (1992). 16S rRNA phylogenetic analysis of the bacterial endosymbionts associated with cytoplasmic incompatibility in insects. *Proc. Natl. Acad. Sci. U.S.A.* **89:** 2699–2702.

Roeland, C.H.J., van Ham, R., Martinez-Torres, D., Moya, A., and Latorre, A. (1999). Plasmid-encoded anthranilate synthase (*trpEG*) in *Buchnera aphidicola* from aphids of the family Pemphigidae. *Appl. Environ. Microbiol.* **65:** 117–125.

Sandström, J. and Pettersson, J. (1994). Amino acid composition of phloem sap and the relation to intraspecific variation in pea aphid (*Acyrthosiphon pisum*) performance. *J. Insect Physiol.* **40:** 947–955.

Sandström, J., Telang, A., and Moran, N.A. (1999). Nutritional enhancement of host plants by aphids—a comparison of three aphid species on grasses. *J. Insect Physiol.* **46:** 33–40.

Sauer, C., Stackebrandt, E., Gadau, J., Holldobler, B., and Gross, R. (2000). Systematic relationships and cospeciation of bacterial endosymbionts and their carpenter ant host species: proposal of the new taxon Candidatus *Blochmannia* gen. nov. *Int. J. Syst. Evol. Microbiol.* **5:** 1877–1886.

Schroder, D., Deppisch, H., Obermayer, M., Krohne, G., Stackebrandt, E., Holldobler, B., Goebel, W., and Gross, R. (1996). Intracellular endosymbiotic bacteria of *Camponotus* species (carpenter ants): systematics, evolution and ultrastructural characterization. *Mol. Microbiol.* **21:** 479–489.

Shigenobu, S., Watanabe, H., Hattori, M., Sakaki, Y., and Ishikawa, H. (2000). Genome sequence of the endocellular bacterial symbiont of aphids *Buchnera* sp. APS. *Nature* **407:** 81–86.

Silva, J.F., Amparo, L., and Moya, A. (2001). Genome size reduction through multiple events of gene disintegration in *Buchnera* APS. *Trends Genet.* **17:** 615–618.

Simon, J.C., Martinez-Torres, D., Latorre, A., Moya, A., and Hebert, P.D. (1996). Molecular characterization of cyclic and obligate parthenogens in the aphid *Rhopalosiphum padi* (L). *Proc. R. Soc. London (B)* **263:** 481–486.

Soler, T., Latorre, A., Sabater, B., and Silva, F.J. (2000). Molecular characterization of the leucine plasmid from *Buchnera aphidicola*, primary endosymbiont of the aphid *Acyrthosiphon pisum*. *Curr. Microbiol.* **40:** 264–268.

Spaulding, A.W. and von Dohlen, C.D. (2001). Psyllid endosymbionts exhibit patterns of co-speciation with hosts and destabilizing substitutions in ribosomal RNA. *Insect Mol. Biol.* **10:** 57–67.

Sullivan, J.T., Trzebiatowski, J.R., Cruickshank, R.W., Gouzy, J., Brown, S.D., Elliot, R.M., Fleetwood, D.J., McCallum, N.G., Rossbach, U., Stuart, G.S., Weaver, J.E., Webby, R.J., De Brujin, F.J., and Ronson, C.W. (2002). Comparative sequence analysis of the symbiosis island of *Mesorhizobium loti* strain R7A. *J. Bacteriol.* **184:** 3086–3095.

Tamas, I., Klasson, L., Sandström, J.P., and Andersson, S.G.E. (2001). Mutualists and parasites: how to paint yourself into a metabolic corner. *FEBS Lett.* **498:** 135–139.

Tamas, I., Klasson, L., Canback, B., Naslund, A.K., Eriksson, A.-S., Wernegreen, J.J., Sandström, J.P., Moran, N.A., and Andersson, S.G.E. (2002). 50 million years of genomic stasis in endosymbiotic bacteria. *Science* **296:** 2376–2379.

Van Ham, R.C., Martinez-Torres, D., Moya, A., and Latorre, A. (1999). Plasmid-encoded anthranilate synthase (*trpEG*) in *Buchnera aphidicola* from aphids of the family pemphigidae. *Appl. Environ. Microbiol.* **65:** 117–125.

Van Ham, R.C., Gonzalez-Candelas, F., Silva, F.J., Sabater, B., Moya, A., and Latorre, A. (2000). Postsymbiotic plasmid acquisition and evolution of the repA1-replicon in *Buchnera aphidicola*. *Proc. Natl. Acad. Sci. U.S.A.* **97:** 10855–10860.

Vavre, F., Fleury, F., Lepetit, D., Fouillet, P., and Boulétreau, M. (1999). Phylogenetic evidence for horizontal transmission of *Wolbachia* in host–parasitoid associations. *Mol. Biol. Evol.* **16:** 1711–1723.

Von Dohlen, C.D., Kohler, S., Alsop, S.T., and McManus, W.R. (2001). Mealybug-proteobacterial endosymbionts contain γ-proteobacterial symbionts. *Nature* **412:** 433–435.

Wernegreen, J.J. and Moran, N.A. (2000). Decay of mutualistic potential in aphid endosymbionts through silencing of biosynthetic loci: *Buchnera* of *Diuraphis*. *Proc. R. Soc. London (B)* **267:** 1423–1431.

Wernegreen, J.J., Ochman, H., Jones, I.B., and Moran, N.A. (2000). Decoupling of genome size and sequence divergence in a symbiotic bacterium. *J. Bacteriol.* **182:** 3867–3869.

4 Symbiosis in Tsetse

Serap Aksoy

CONTENTS

INTRODUCTION

Tsetse flies (Diptera: Glossinidae) are important agricultural and medical vectors that transmit the protozoan African trypanosomes, the agents of sleeping sickness disease in humans and various diseases in animals (nagana). In addition to the parasite they vector, tsetses provide a suitable niche for multiple bacterial symbionts. The evolutionary histories of the symbionts with their tsetse host vary, and their functional associations range from obligate mutuals to facultative parasites. As a well-defined microhabitat, tsetses provide an ideal system for observing the dynamics of interactions among the multiple symbionts and investigating host responses to the different associations. From a practical perspective, symbiotic organisms stand to benefit from therapeutic approaches to controlling plant and animal diseases transmitted by their arthropod hosts.

CHARACTERIZATION OF TSETSE SYMBIONTS

Insects with a single diet throughout their entire developmental cycle (such as blood, plant sap, or wood) rely on microbial symbionts for additional nutrients that are not found in their restricted diet and that they are unable to synthesize (Stuhlman, 1907; Roubaud, 1919; Wigglesworth, 1929; Pinnock and Hess, 1974; Shaw and Moloo, 1991). Microorganisms with different ultrastructural characteristics have been reported from various tissues of tsetse, which only feeds on vertebrate blood during all developmental stages (Stuhlman, 1907; Roubaud, 1919; Wigglesworth, 1929; Pinnock and Hess, 1974; Shaw and Moloo, 1991). Recent phylogenetic analyses have confirmed that these organisms represent three distinct associations (Aksoy, 2000). Two of these symbionts are members of the Enterobacteriaceae: the primary (P)-symbiont (genus *Wigglesworthia*) (Aksoy, 1995; Aksoy et al., 1995) and the secondary (S)-symbiont (genus *Sodalis*) (Aksoy et al., 1995; Cheng and Aksoy, 1999; Dale and Maudlin, 1999). *Wigglesworthia* lives intracellularly within the

0-8493-1286-8/03/$0.00+$1.50

specialized epithelial cells (bacteriocytes) in the bacteriome organ located in the anterior midgut. In contrast, *Sodalis* is harbored both inter- and intracellularly, principally in the midgut, but has also been detected in the hemolymph (Cheng and Aksoy, 1999). The third symbiont present in some tsetse species is related to *Wolbachia pipientis*, which in many insects is confined primarily to the reproductive organs. A phylogenetic analysis of the *Wolbachia* strains infecting various species of tsetse has shown that they are unique and as such represent independent acquisitions by each species (Cheng et al., 2000).

Tsetse has a viviparous reproductive biology — an adult female produces a single egg at a time that hatches and develops *in utero*. After a period of maturation and sequential molting within the mother, a third instar larva is deposited and pupates shortly thereafter. Each female can deposit five to seven offspring during her 3- to 4-month life span. During its intrauterine life, progeny receives nutrients along with the tsetse symbionts (*Sodalis* and *Wigglesworthia*) through the mother's milk gland secretions (Ma and Denlinger, 1974; Cheng and Aksoy, 1999). It is not known, however, whether whole bacteriocytes or free-living *Wigglesworthia* cells are transferred from the mother to her larva. Unlike the gut symbionts, *Wolbachia* is transovarially transmitted through maternal lineages. Given the unique reproductive biology of tsetse, all three symbionts are in essence vertically acquired by the progeny.

EVOLUTIONARY HISTORIES OF SYMBIONTS

Because it has been difficult to cultivate many of these fastidious and often intracellular insect symbionts *in vitro*, their physiological characterization and correct taxonomic positioning have been controversial. Recent advances in polymerase-chain-reaction (PCR)-based technologies as well as the use of nucleic-acid sequences in phylogenetic reconstruction have provided additional insight into the evolutionary relationships among bacteria (Woese, 1987). Based on the 16S rDNA sequence comparison, many insect symbionts (including the tsetse gut organisms) belong to the γ-subdivision of Proteobacteria in the family Enterobacteriaceae, suggesting that microorganisms originating from this group have been able to enter into and fulfill symbiotic roles in various insect hosts.

Characterization of different members of genus *Wigglesworthia* from the four species groups of tsetse — *fusca, morsitans, palpalis* and *austeni* — has shown that they form a distinct lineage (Aksoy et al., 1995). The evolutionary relationship of the various tsetse species has been independently determined based on observed variation in the internal transcribed spacer (ITS-2) regions of rDNA (Chen et al., 1999). This analysis has shown similarity between the phylogeny of genus *Glossina* and the phylogeny of its bacterial symbionts, implying that a tsetse ancestor had been infected with a bacterium some 50 to 80 million years ago and from this ancestral pair, species of tsetse and their associated *Wigglesworthia* strains radiated without horizontal transfer events among species. The bacteriome-associated symbionts of distant insect orders, such as *Buchnera* from aphids (Hemiptera) (Munson et al., 1991), whitefly (Hemiptera) (Campbell, 1993), Candidatus *Carsonella* from psyllids (Hemiptera) (Thao et al., 2000b), and Candidatus *Blochmania* from carpenter ants (Hymenoptera) (Schröder et al., 1996), have each been found to form distinct lineages closely related to *Wigglesworthia* from tsetse (Diptera) (Figure 4.1). Similar concordant evolutionary history of bacteriome symbionts with their host insect species has been reported for Candidatus *Blochmania*, *Buchnera*, whitefly symbionts, and Candidatus *Carsonella*.

In addition to the obligate P-symbionts, whose mutualistic biology is intimately linked with their host physiology, many insects harbor one or more secondary microorganisms (S-symbionts) that often display commensal relationships with their host. These organisms confer no known benefits to hosts, do not appear to influence the nutritional and reproductive biology of their host, and often infect various host tissues. Observations in psyllids have indicated independent acquisitions of closely related but different S-symbionts (Thao et al., 2000a). In contrast, members of genus *Sodalis* characterized from distant tsetse species do not exhibit any significant

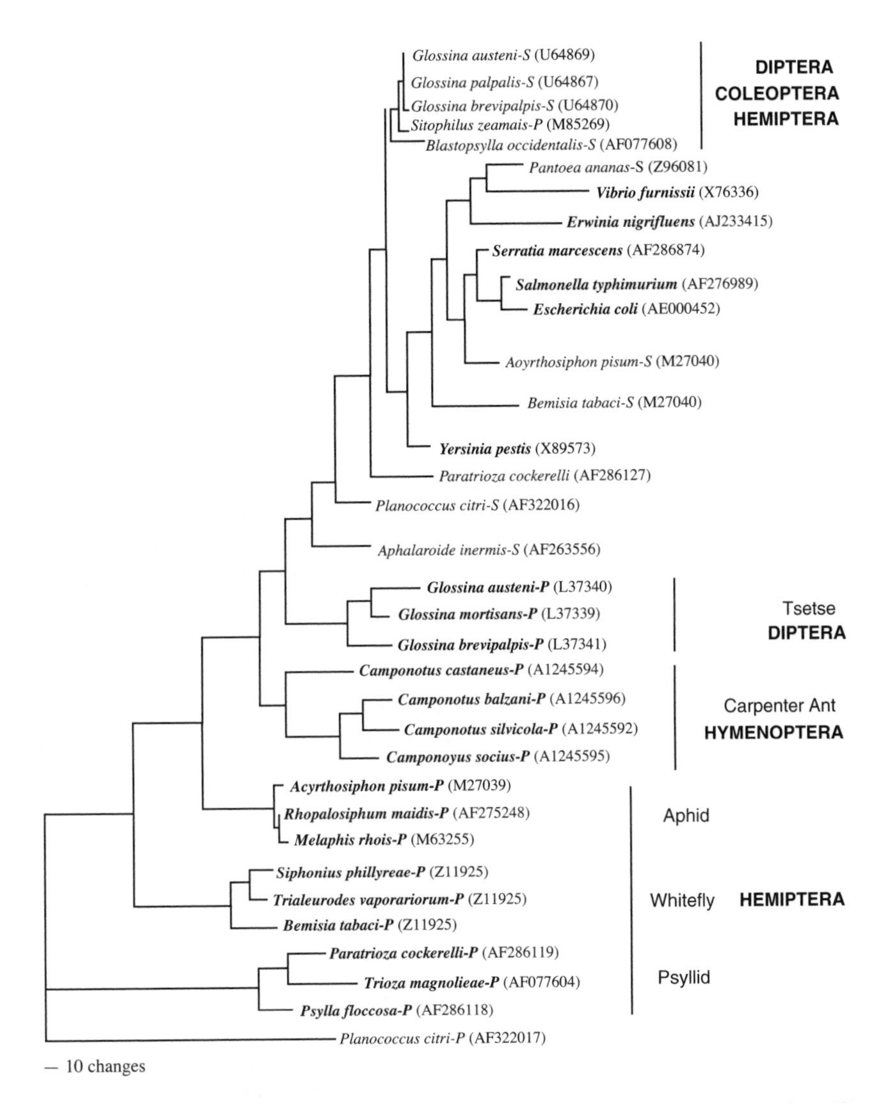

FIGURE 4.1 Phylogenetic tree showing the taxonomic position of primary (P) and secondary (S) symbionts characterized from insect species in distant orders determined by 16S rDNA sequence. Bold italicized names indicate the free-living bacteria within λ-proteobacteria. The nodes shown had bootstrap values above 50 using the neighbor-joing method with 100 replications. The P-symbionts of the mealy bugs in the β-subdivision of proteobacteria was used as the outgroup. The Genbank accession numbers for each 16S rDNA sequence are indicated in parentheses.

phylogenetic differences based on their 16S rDNA sequence comparison, implying that this symbiosis is recent in origin. Each tsetse species may have acquired the *Sodalis* symbiont independently, or horizontal transfer events, which are common among the different tsetse species, may have occurred. Interestingly, symbionts of the rice weevil (*Sitophilus)* and the psyllid (*Blastopsylla*) form a close lineage with *Sodalis*, which suggests that they share a recent common ancestor (Figure 4.1). Despite their close taxonomic relatedness, the *Sitophilus* symbiont, SOPE, appears to play an obligate role in its weevil host and is intracellular within the bacteriome organ. *Sodalis*, on the other hand, has a wide tissue prevalence in tsetse and is likely facultative since it can survive in cell-free media *in vitro*. Hence, comparative analysis of the genomes from these two closely related organisms may help elucidate loci responsible for the different functional biologies manifested in their respective tsetse and weevil hosts.

GENOMICS OF THE OBLIGATE SYMBIONT *WIGGLESWORTHIA*

Each tsetse fly harbors about 1×10^8 cells of *Wigglesworthia* in its bacteriome, with bacteria lying free in the cytoplasm of the bacteriocytes (Figure 4.2).

The genome size of *Wigglesworthia* from several distant tsetse host species had been estimated to range from 705 to 770 kilobases (kb) based on pulsed-field gel electrophoresis analysis (Akman and Aksoy, 2001). Recently, its completely sequenced genome was found to be 697,724 base pairs (bp) in *Wigglesworthia brevipalpis* (Akman et al., 2002) — about one sixth of that of the related free-living *Escherichia coli* (4.6 megabases [Mb]). Similar drastic genome size reductions have been observed for intracellular pathogens such as *Chlamydia trachomatis* (1.04 Mb), *Treponema pallidum* (1.14 Mb), *Mycoplasma genitalium* (0.58 Mb) (Fraser et al., 1995), and *Rickettsia prowazekii* (1.1 Mb), as well as for the primary endosymbiont of aphids, *Buchnera* (640 kb) (Charles and Ishikawa, 1999; Shigenobu et al., 2000). The genome sizes of the mutualists *Wigglesworthia* and *Buchnera* are apparently approaching that of *M. genitalium,* the smallest bacterial genome reported thus far. As a reflection of gene loss (and presumably loss of their associated functions), neither of the modern-day obligates can live outside its host-insect niche.

Another hallmark of intracellular bacteria is the high A+T content in their coding sequences. The *Wigglesworthia* genome has an average A+T content of 78% and the *Buchnera* genome is 74% (Shigenobu et al., 2000), while the overall A+T content of the genomes of *R. prowazekii*, *M. genitalium,* and *C. trachomatis* are 71, 68, and 59%, respectively. In contrast, the A+T content of free-living bacteria are lower, such as 45% in *E. coli* (Blattner et al., 1997). Analysis of the coding capacity of these reduced genomes has indicated that loci encoding for DNA repair and recombination functions have been lost or limited in many cases (Moran and Wernegreen, 2000). It is thought that this loss of the repair functions might have led to the A+T bias observed in these intracellular genomes (Moran, 1996).

The completely annotated sequence of *Wigglesworthia* genome has revealed the presence of 621 predicted coding sequences (CDSs) with an average length of 988 bp. It has been possible to assign biological roles to 522 (86%) of these putative proteins, while 95 proteins (14%) matched hypothetical proteins of unknown function. Comparative analysis of the CDSs indicates that the *Wigglesworthia* genome contains a subset of the genes of free-living bacteria, such as the enteric *E. coli* and *Salmonella typhimurium*, further supporting the idea that it shares an ancestor with them.

One of the surprising findings of the *Wigglesworthia* genome was the absence of the important gene coding for the DNA replication-initiation protein, DnaA — an observation unprecedented in eubacteria. As would be expected, its genome also did not exhibit any DnaA boxes or a clear GC

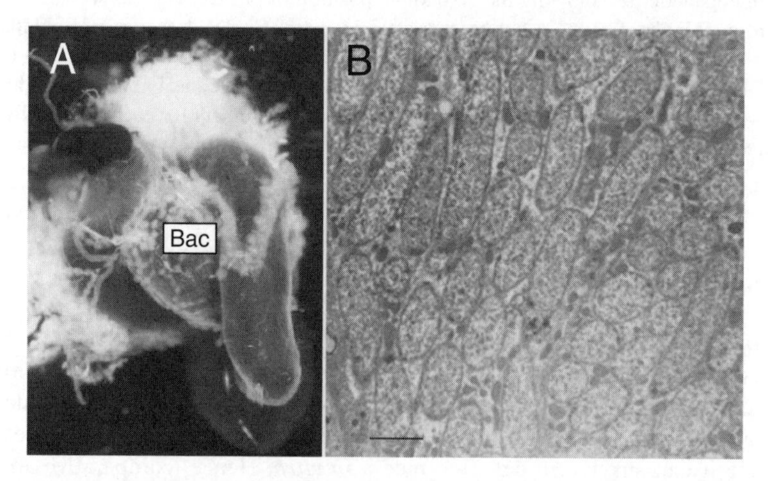

FIGURE 4.2 (A) The U-shaped bacteriome (Bac) structure in the anterior midgut. (B) *Wigglesworthia* cells lying within the bacteriocyte cytoplasm. Bar = 5 μm.

skewing. The lack of an autonomous DNA-replication machinery may reflect the dependency of *Wigglesworthia* on its tsetse host genome functions and may be one of the mechanisms by which the host bacteriocyte cell regulates symbiont numbers.

Wigglesworthia does not infect the egg tissue but is maternally transmitted to tsetse intrauterine larva via milk-gland secretions. It is not clear, however, whether whole bacteriocytes or *Wigglesworthia* cells are transferred from the mother to her larva. While the commensal symbiont *Sodalis* has been shown to utilize a Type III secretion system to invade the larval tsetse cells (Dale et al., 2001), the *Wigglesworthia* genome does not encode for a Type III secretion system that could be used for invasion. The *Wigglesworthia* genome, however, has retained the machinery for the synthesis of a complete flagellar apparatus including the hook, filament, filament cap regions, and the integral membrane proteins required for motility functions, *motA* and *motB*. While retention of the genes associated with the flagellar operons is suggestive of its functional role, neither flagellum nor motility has been observed in *Wigglesworthia* in adult bacteriocytes. It remains to be seen whether the flagellum is expressed in the different development stages of tsetse. It is possible that components of the *Wigglesworthia* flagellum may provide an alternate to the Type III secretion mechanism of cell invasion to enable entry into the host larval or pupal gut cells destined to be bacteriocytes. A mechanism utilizing the flagellar export system has been shown for *Yersinia* invasion of mammalian gut cells (Young et al., 1999). Alternatively, since the *Wigglesworthia* genome encodes for very few transporters and has only a partial sec-dependent export pathway, components of the flagellar structure may fulfill export functions, as has been suggested in the case of *Buchnera* (Shigenobu et al., 2000).

Among the physiological traits shared by obligate genomes that set them apart from small pathogens is their ability to supplement their host's genome needs. For example, *Buchnera* provide all essential amino acids that the host diet apparently lacks. Similarly, the small genome of *Wigglesworthia* has retained the ability to synthesize various vitamin metabolites including biotin, thiazole, lipoic acid, FAD (riboflavin, B2), folate, pantothenate, thiamine (B1), pyridoxine (B6), protoheme, and nicotinamide, which are known to be low in the single diet of tsetse–vertebrate blood. Hence, supplementing the eukaryotic diet with vitamin products may play a central role in the functional basis of this mutualistic relationship.

Despite the apparent functional and evolutionary similarities of their symbiotic associations, the genetic blueprints of the obligates *Buchnera* and *Wigglesworthia* are quite distinct. *Wigglesworthia* shares only 69% of its CDSs with *Buchnera*, and these represent mostly the indispensable housekeeping genes involved in transcription, translation, and cellular functions (Figure 4.3). Also, in comparison to *Buchnera*, a significant portion of the small *Wigglesworthia* genome encodes for lipopolysaccharides and phospholipids, components necessary for a Gram-negative cell-wall structure, and for a full flagellar structure. Since both of these characteristics are associated with free-living or parasitic microbes such as *Rickettsia*, either *Wigglesworthia* represents a much "younger symbiotic" association than a true obligate like *Buchnera* or its unique route of intrauterine transmission may have shaped its genome to retain functions associated with parasitic or free-living microbes.

GENOMICS OF GENUS *SODALIS*

Sodalis, the secondary symbiont of tsetse, shows similarities to the free-living enterics. Its genome size is approximately 2 Mb, which is significantly larger than those of the intracellular pathogens and obligate symbionts yet smaller than those of the closely related free-living enterics (Akman et al., 2001). In addition, *Sodalis* cells harbor an abundant large plasmid of about 135 kb. Based on the *groEL* and *ftsZ* gene sequences, the A+T content of the *Sodalis* genome is less than 45%, which is more similar to free-living organisms, unlike the A+T-rich intracellulars described earlier.

Although genome-wide sequence data are not yet available for *Sodalis*, hybridization of its genomic DNA to *E. coli* macroarrays revealed the presence of 1800 orthologs (ORFs), which

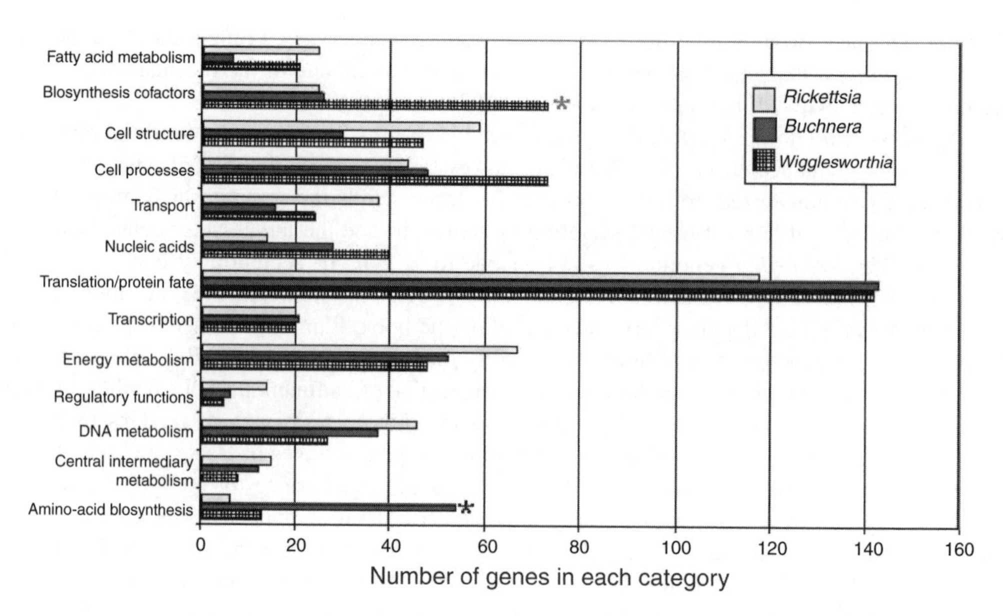

FIGURE 4.3 Comparative analysis of the number of genes present in each functional category described in the genomes of *Wigglesworthia*, *Buchnera*, and *Rickettsia*. * denotes the categories, biosynthesis of cofactors in *Wigglesworthia* genome, and amino acid biosynthesis in *Buchnera* genome that contain higher numbers of genes in comparison to the other two symbiont genomes analyzed.

represents about 85% of the genome (Akman et al., 2001), assuming an average size of 1 kb per gene (Shigenobu et al., 2000). The *E. coli* array contains 4290 ORFs corresponding to its sequenced genome, and functional roles have been assigned to 1938 of these ORFs. Of the 1800 heterologous genes detected from *Sodalis*, 1158 had functional roles assigned in *E. coli,* while the remaining 642 ORFs detected corresponded to genes with hypothetical functions. Although the *Sodalis* genome is about half the size of that of *E. coli*, comparative analysis revealed that it contained a high proportion of the genes necessary for all of the amino acid biosynthetic pathways, regulatory functions, translation, and transcription and for nucleic acid biosynthesis. Many genes involved in the biosynthesis of cofactors, replication, and transport functions were also found to be present, and most of the DNA repair and recombinase orthologs of *E. coli* involved in direct damage reversal, base-excision repair, mismatch repair, recombinase pathways, and nucleotide-excision repair were retained. Genes involved in carbon-compound catabolism, central intermediary metabolism, fatty acid phospholipid metabolism, cell processes, and cell structure, however, were fewer in number in comparison to the *E. coli* genome. Based on array-hybridization analysis, *Sodalis* appears to have respiratory oxidases, NADH dehydrogenase complex enzymes, and a complete TCA cycle. It can grow on several sugars including galactose, fructose, and raffinose, as well as the amino sugars *N*-acetyl-D-glucosamine, the methylpentoses L-fucose, L-rhamnose, L-arabinose, and xylose. In fact, *in vitro* carbon-substrate-assimilation tests suggest that *Sodalis* may be utilizing primarily *N*-acetylglucosamine and raffinose as its primary carbon sources *in vitro* (Dale and Maudlin, 1999). Apparently *Sodalis* has many of the capabilities of free-living bacteria, which is supported by *in vitro* cell-free culture for this organism (Welburn et al., 1987; Beard et al., 1993). While the array-hybridization approach has provided us with a general understanding of the genomic coding capacity of *Sodalis*, it lacks information on loci not represented in the *E. coli* genome. There are at least two such examples; the first is a *chitinase* characterized from *Sodalis* that is absent in the *E. coli* genome (Welburn et al., 1993), and the second is the pathogenicity island genes that may help *Sodalis* invade the tsetse cells via a Type III secretion system (Dale et al., 2001). Thus, the array analysis has provided a broad understanding of the general aspects of the *Sodalis* genome functions.

FUNCTIONAL ROLE OF TSETSE SYMBIONTS

It has been difficult to study the individual functions of the multiple symbionts in tsetse. Attempts to eliminate the symbionts by administration of antibiotics, lysozyme, and specific antibodies have resulted in retarded growth of the insect and a decrease in egg production, preventing the ability of the aposymbiotic host to reproduce (Nogge, 1976, 1978, 1980). The ability to reproduce, however, can be partially restored when the aposymbiotic tsetse flies receive a blood meal that is supplemented with B-complex vitamins (thiamine, pantothenic acid, pyridoxine, folic acid, and biotin), suggesting that the endosymbionts probably play a role in metabolism that involves these compounds (Nogge, 1981). The sequenced genome of *Wigglesworthia,* which has extensive vitamin-coding capabilities, indeed supports this role, as described earlier.

The functional significance of *Sodalis* for tsetse biology is unknown at present. Its genomic aspects, such as its size, A+T bias, and genome contents, are reminiscent of free-living organisms rather than mutualists. While all tsetse species harbor bacteriome-associated *Wigglesworthia*, the prevalence of *Sodalis* infections in the various tsetse species studied has been shown to vary widely (Cheng and Aksoy, 1999). In at least one study, it has been possible to preferentially eliminate *Sodalis* by using the sugar analog and antibiotic streptozotocin without drastically reducing fitness of the flies, implying that its presence may be dispensable (Dale and Welburn, 2000). The presence of *Sodalis* in tsetse has also been implicated in enhancing the susceptibility trait of tsetse for trypanosome transmission (Welburn and Maudlin, 1991). It has been shown that *Sodalis* produces at least one type of chitinase enzyme that may be responsible for the increased trypanosome susceptibility of its tsetse host (Welburn et al., 1993; Welburn and Maudlin, 1999).

The *W. pipientis*-like symbiont detected from several tsetse species has been found to infect a wide range of invertebrate hosts. In one survey in the neotropics, more than 15% of the analyzed taxa were reported to carry this group of microorganisms (Werren et al., 1995). The functional presence of *Wolbachia* has been shown to result in a variety of reproductive abnormalities in the various hosts they infect. One of these abnormalities is termed cytoplasmic incompatibility (CI) and when expressed commonly results in embryonic death due to disruptions in early fertilization events (Hoffmann and Turelli, 1997). In an incompatible cross, the sperm enters the egg but does not successfully contribute its genetic material to the potential zygote. In most species, this results in very few hatching eggs. The infected females have a reproductive advantage over their uninfected counterparts as they can produce successful progeny with both the imprinted and normal sperm. This reproductive advantage allows the infected insects to spread into populations. Most functional studies have involved curing insects of their *Wolbachia* infections by administering antibiotics in their diet. This approach, however, has not been feasible in tsetse because the antibiotic treatment of flies results in the clearing of all bacterial symbionts, including *Wigglesworthia*, which results in fly sterility. The analysis of tsetse laboratory colonies has shown that 100% of sampled individuals carry *Wolbachia* infections, making the analysis of *Wolbachia*-mediated effects impossible by traditional mating experiments. The infection prevalence in field populations, however, has shown significant polymorphism (Cheng et al., 2000). Perhaps *Wolbachia*-infected and uninfected lines can be developed from these polymorphic field populations and used to elucidate the functional role of this organism in tsetse biology.

SYMBIONT–HOST INTERACTIONS

In addition to differences in genomic characteristics and evolutionary history with the tsetse host, symbionts also display differences in tissue tropism as shown by a PCR-based assay. While *Wigglesworthia* appears to be strictly associated with the bacteriome, *Sodalis* has been detected in midgut, muscle, fat body, hemolymph, milk gland, and salivary glands of certain tsetse species (Cheng and Aksoy, 1999). Furthermore, the density of infections with *Sodalis* appears to vary widely in the different species analyzed. Infections in *G. morsitans* and *G. palpalis* midgut tissues are maintained at high density, while infections in *G. austeni* and *G. brevipalpis* are significantly less, as observed by both PCR (Cheng and Aksoy, 1999) and microscopic analysis (Moloo and Shaw, 1989). The factors that control the tissue tropism and density of symbionts in different host species are not known but may be attributable to both symbiont- and host-controlled functions.

Although *Wolbachia* infections were initially thought to be largely restricted to the germ-line tissue of their insect hosts, they recently have been observed in a variety of somatic tissues from several different insects (Min and Benzer, 1997; Dobson et al., 1999). Analysis of tissue tropism of *Wolbachia* in tsetse indicates that while in *G. morsitans* and *G. brevipalpis* the infections are restricted to gonads, in *G. austeni* they are detected from various somatic tissues, even in 2-d-old teneral flies. The extensive tissue infections observed in *G. austeni* may reflect the particular strain harbored in this species, similar to what has been reported with the *W. popcorn* strain characterized from *D. melanogaster* (Min and Benzer, 1997).

Given that most insects have robust immune mechanisms to combat foreign pathogens, it is of interest how the commensal symbionts are maintained in their hosts. Recent studies with the tsetse immunity genes showed in midgut and fat body tissues constitutive expression of *diptericin*, an antimicrobial peptide gene specific for Gram-negative bacteria (Hao et al., 2001). Constitutive expression of *diptericin* in insects is unprecedented and might reflect natural tsetse immune responses to the presence of its symbionts. Our analysis with synthetic *diptericin* has shown that *Sodalis* is more resistant to its antibacterial action than the related bacterium *E. coli in vitro* (Hao et al., 2001) and hence may escape its intended harmful effects. In the presence of trypanosome infections in tsetse, expression of several immunity genes *(attacin, diptericin, defensin)* has been found to be induced. It remains to be seen if the expression of these antimicrobial products affects the viability of the tsetse symbionts — in particular *Wigglesworthia*, which might lead to reduced fecundity of parasite-infected flies.

SYMBIONTS AS GENE-EXPRESSION VEHICLES

From an applied perspective, beneficial microbes can be used for drug and vaccine delivery (Hooper and Gordon, 2001) or for the expression of foreign genes designed to block the development of pathogens (Aksoy et al., 2001). Since transgenic approaches involving egg manipulation are difficult due to the viviparous reproductive biology of tsetse, transgenic symbionts provide an alternative method for examining gene functions *in vivo*.

The availability of an *in vitro* culture of *Sodalis* has allowed for the development of a genetic transformation system to introduce and express foreign gene products in these cells (Welburn et al., 1987; Beard et al., 1993). It has been found that the *in vitro* manipulated recombinant *Sodalis* cells are successfully acquired by the intrauterine progeny when microinjected into the female parent hemolymph. The adults that have acquired the recombinant symbionts as larvae have been shown to continue to express the marker gene product green fluorescent protein (GFP) (Cheng and Aksoy, 1999). Using a similar microinjection approach, the facultative S-symbiont (PASS) in aphids has also been successfully introduced from *Acyrthosiphon pisum* into *A. kondoi* Shinji (blue alphalfa aphid) as well as into *A. pisum* symbiont-free clones, where it has been found to be maintained in the offspring of the injected mother aphids with a high rate of maternal transmission (Chen and Purcell, 1997).

Since the symbionts live in close proximity to the developing trypanosomes in midgut, antipathogenic products expressed and secreted from these cells could adversely affect parasite transmission. The identification of monoclonal antibodies (mABs) with parasite-transmission-blocking characteristics and their expression as single-chain antibody gene fragments in the symbionts provides a vast array of potential antipathogenic products. To this end, several transmission-blocking antibodies targeting the major surface protein of the insect-stage procyclic trypanosomes have already been reported (Nantulya and Moloo, 1988). Recently it has been possible to express and secrete a single-chain antibody gene product in the transformed symbionts of reduviid bugs *in vivo* (Durvasula et al., 1999). Hence, expression of single-chain antibody gene fragments specific for these monoclonals in *Sodalis* is now desirable. The multitude of potential antiparasitic targets that can be expressed in bacteria makes this a desirable system for transgenic approaches. Should resistance develop in pathogens against any of the expressed foreign gene products, bacteria could be transfected with new and different genes. Alternatively, several target genes could potentially be expressed simultaneously in the symbionts to prevent the development of resistance against any one individual target. Most insects mount significant immune responses to the presence of pathogens and can effectively clear the majority of these infections, making these immunity genes good candidates for further exploration. Constitutive and abundant expressions of these immunity molecules in transgenic symbionts in the relevant compartments in the fly can enhance the ability of insects to clear pathogens *in vivo*.

The relative ease of DNA transformation systems in bacteria can facilitate the development of similar disease-intervention strategies in other medically and agriculturally important vector/symbiont systems such as in ticks, mites, bed bugs, lice, some species of fleas, planthoppers, whiteflies, and termites.

TRYPANOSOMIASIS CONTROL AND THE IMPACT OF PARASITE REFRACTORY TSETSE ON DISEASE TRANSMISSION

Human sleeping sickness has been on the rise and is believed to claim over 50,000 lives per year, with epidemics that can devastate whole communities. The impact of the animal disease nagana is equally devastating, depriving much of Africa of livestock — milk, meat, and draught oxen for plowing and transport. There is little doubt that elimination of the burden of tsetse-transmitted trypanosomiasis would be of major benefit for the social and economic development of sub-Saharan Africa. Because of the antigenic variation trypanosomes display in their mammalian host, there are no candidate mammalian vaccines in sight, and most disease-control programs rely on active surveillance and chemotherapy and tsetse-control approaches. Among the techniques used for eliminating the tsetse populations are odor-baited traps and targets, pour-ons, ultra-low-volume aerial spraying, and the sterile insect technique (SIT), which can each be used, singly or in combination, to maintain cleared barriers between treated and untreated areas.

SIT is a genetic population-suppression approach and involves sustained, systematic releases of irradiated sterile male insects among the wild population. Releasing sterile males in high numbers over a period of three to four generations, after population density has been reduced by other techniques (trapping, insecticide spraying, etc.), the target population can be eradicated (Politzar and Cuisance, 1984; Vreysen et al., 2000). The recent successful eradication of *G. austeni* from the island of Zanzibar using such an integrated approach of population suppression followed by SIT has demonstrated that SIT can be a valuable vector-management tool for selective foci (Vreysen et al., 2000). There has been an ongoing debate on the effectiveness of SIT strategy for continent-wide tsetse control given the high cost and time associated with mass rearing the large numbers of sterile males needed and the potential for reinvasion of areas cleared of tsetse (Molyneux, 2001). While it may indeed be difficult to employ SIT in all tsetse habitats, it may be a very efficient and important tool within the context of an integrated tsetse-management program.

Improvements in two aspects of the current SIT technology have the potential to enhance its efficacy for future programs (Aksoy et al., 2001). The first is the development of parasite refractory strains. Because the large numbers of male flies released can potentially contribute to a temporary increase in disease transmission, the incorporation of refractory traits into the SIT release strains will greatly enhance the efficacy of this approach, especially in human disease endemic foci. During the current field SIT programs male tsetse flies are provided with a blood meal containing a trypanocide before release. The second is the use of *Wolbachia*-mediated cytoplasmic incompatibility

(CI) as a method of inducing sterility as an alternative to irradiation. With CI, the release strain of tsetse would carry a *Wolbachia* infection that would induce CI when males are mated with wild females. The competitiveness of these males would be expected to be much higher than irradiated males and as a result fewer insects would need to be released to achieve the same level of sterility in the wild population, significantly reducing the cost of the approach. This strategy depends on the use of a very efficient sexing system. If *Wolbachia*-infected females are released in sufficient quantities, then *Wolbachia* would have the opportunity to invade the target population, which would render subsequent releases ineffective. If it were impossible to guarantee extremely low quantities of released females, then it would be possible to incorporate low levels of irradiation with *Wolbachia*-induced sterility to prevent released females from successfully reproducing.

An alternative control strategy involves replacing the susceptible natural tsetse populations with their engineered parasite refractory counterparts. However, for the symbiont-based transformation approach to be successful, a significant proportion of the natural symbiont population of the gut will need to be reconstituted by its recombinant counterpart so that foreign expressed products can accumulate in the gut to levels where they can interfere with trypanosome biology. The eventual replacement of parasite-susceptible vector populations with engineered refractory flies could provide an additional strategy to reduce disease in the field (Aksoy et al., 2001). If *Wolbachia* infections in tsetse do express strong CI phenotypes, the two symbiotic systems could be coupled so as to drive the phenotypes conferred by the engineered gut-symbionts into the field. This can be achieved because as *Wolbachia* infections rapidly invade populations by virtue of the CI phenomenon they confer, they can drive other maternally inherited elements, such as mitochondria (Turelli et al., 1992) or the engineered gut-symbionts, into that same population (Sinkins et al., 1997). While no naturally occurring infectious transfer of *Wolbachia* has been observed, it has become increasingly common to experimentally transfer *Wolbachia* among different hosts and even into insects with no prior infection history, making it an attractive gene-expression system with a naturally associated driving mechanism (Boyle et al., 1993; Chang and Wade, 1994; Rousset and de Stordeur, 1994; Poinsot et al., 1998).

The success of symbiont-based transgenic strategies in insects relies on a solid understanding of the molecular and developmental biology of the symbionts as well as the pathogens transmitted by each system so that genes with transmission-blocking activities can be identified and efficiently expressed in the correct tissues to adversely affect pathogen viability. In addition to questions regarding technical success and efficacy, much current debate focuses on the safety and regulatory concerns for release of genetically modified insects, especially human-biting vectors. Before any release studies can be entertained with recombinant animals, information on environmental and ecological hazards associated with the releases and potential public health risks will need to be deliberated.

ACKNOWLEDGMENTS

Many colleagues have contributed to the development of this body of knowledge in tsetse. I am grateful to the past and present members of my laboratory for their continued interest and hard work, in particular Leyla Akman, Xiao-ai Chen, Song Li, Quiying Cheng, Yian Jian, Rita V.M. Rio, Patricia Strickler, Irene Kasumba, Dana Nayduch, Youjia Hu, and Zhengrong Hao.

REFERENCES

Akman, L. and Aksoy, S. (2001). A novel application of gene arrays: *Escherichia coli* array provides insight into the biology of the obligate endosymbiont of tsetse flies. *Proc. Natl. Acad. Sci. U.S.A.* **98:** 7546–7551.

Akman, L., Rio, R.V.M., Beard, C.B., and Aksoy, S. (2001). Genome size determination and coding capacity of *Sodalis glossinidius*, an enteric symbiont of tsetse flies, as revealed by hybridization to *Escherichia coli* gene arrays. *J. Bacteriol.* **183:** 4517–4525.

Akman, L., Yamashita, A., Watanabe, H., Oshima, K., Shiba, T., Hattori, M., and Aksoy, S. (2002). Genome sequence of the endocellular obligate symbiont of tsetse, *Wigglesworthia glossinidia*. *Nat. Genet.* **32(3):** 402–407.

Aksoy, S. (1995). *Wigglesworthia* gen. nov. and *Wigglesworthia glossinidia* sp. nov., taxa consisting of the mycetocyte-associated, primary endosymbionts of tsetse flies. *Int. J. Syst. Bacteriol.* **45:** 848–851.

Aksoy, S. (2000). Tsetse: a haven for microorganisms. *Parasitol. Today* ***16:*** 114–119.

Aksoy, S., Maudlin, I., Dale, C., Robinson, A., and O'Neill, S. (2001). Prospects for control of African trypanosomiasis by tsetse vector manipulation. *Trends Parasitol.* **17:** 29–35.

Aksoy, S., Pourhosseini, A.A., and Chow, A. (1995). Mycetome endosymbionts of tsetse flies constitute a distinct lineage related to *Enterobacteriaceae*. *Insect Mol. Biol.* **4:** 15–22.

Andersson, S.G. and Kurland, C.G. (1998). Reductive evolution of resident genomes. *Trends Microbiol.* **6:** 263–268.

Beard, C.B., O'Neill, S.L., Mason, P., Mandelco, L., Woese, C.R., Tesh, R.B., Richards, F.F., and Aksoy, S. (1993). Genetic transformation and phylogeny of bacterial symbionts from tsetse. *Insect Mol. Biol.* **1:** 123–131.

Blattner, F.R., Plunkett, G., III, Bloch, C.A., Perna, N.T., Burland, V., Riley, M., Collado-Vides, J., Glasner, J.D., Rode, C.K., Mayhew, G.F., Gregor, J., Davis, N.W., Kirkpatrick, H.A., Goeden, M.A., Rose, D.J., Mau, B., and Shao, Y. (1997). The complete genome sequence of *Escherichia coli* K-12. *Science* **277:** 1453–1474.

Boyle, L., O'Neill, S.L., Robertson, H.M., and Karr, T.L. (1993). Interspecific and intraspecific horizontal transfer of *Wolbachia* in Drosophila. *Science* **260:** 1796–1799.

Campbell, B.C. (1993). Congruent evolution between whiteflies (Homoptera: Aleyrodidae) and their bacterial endosymbionts based on respective 18S and 16S rDNAs. *Curr. Microbiol.* **26:** 129–132.

Chang, N. and Wade, M. (1994). The transfer of *Wolbachia pipientis* and reproductive incompatibility between infected and uninfected strains of the flour beetle, *Tribolium confusum,* by microinjection. *Can. J. Microbiol.* **40:** 978–981.

Charles, H. and Ishikawa, H. (1999). Physical and genetic map of the genome of *Buchnera*, the primary endosymbiont of the pea aphid *Acrythosiphon pisum*. *J. Mol. Evol.* **48:** 142–150.

Chen, D. and Purcell, A. (1997). Occurrence and transmission of facultative endosymbionts in aphids. *Curr. Microbiol.* **34:** 220–225.

Chen, X., Song, L., and Aksoy, S. (1999). Concordant evolution of a symbiont with its host insect species: molecular phylogeny of genus *Glossina* and its bacteriome-associated endosymbiont, *Wigglesworthia glossinidia*. *J. Mol. Evol.* **48:** 49–58.

Cheng, Q. and Aksoy, S. (1999). Tissue tropism, transmission and expression of foreign genes *in vivo* in midgut symbionts of tsetse flies. *Insect Mol. Biol.* **8:** 125–132.

Cheng, Q., Ruel, T., Zhou, W., Moloo, S., Majiwa, P., O'Neill, S., and Aksoy, S. (2000). Tissue distribution and prevalence of *Wolbachia* infections in tsetse flies, *Glossina* spp. *Med. Vet. Entomol.* **14:** 51–55.

Dale, C. and Maudlin, I. (1999). *Sodalis* gen. nov. and *Sodalis glossinidius* sp. nov., a microaerophilic secondary endosymbiont of the tsetse fly *Glossina morsitans morsitans*. *Int. J. Syst. Bacteriol.* **49:** 267–275.

Dale, C. and Welburn, S. (2000). The endosymbionts of tsetse flies: manipulating host-parasite interactions. *Int. J. Parasitol.* **31:** 628–631.

Dale, C., Young, S.A., Haydon, D.T., and Welburn, S.C. (2001). The insect endosymbiont *Sodalis glossinidius* utilizes a type III secretion system for cell invasion. *Proc. Natl. Acad. Sci. U.S.A.* **98:** 1883–1888.

Dobson, S., Bourtzis, K., Braig, H., Jones, B., Zhou, W., Rousset, F., and O'Neill, S. (1999). *Wolbachia* infections are distributed throughout insect somatic and germ line tissues. *Insect Biochem. Mol. Biol.* **29:** 153–160.

Durvasula, R., Gumbs, A., Panackal, A., Kruglov, O., Aksoy, S., Merrifield, R., Richards, F., and Beard, C. (1997). Prevention of insect borne disease: an approach using transgenic symbiotic bacteria. *Proc. Natl. Acad. Sci. U.S.A.* **94:** 3274–3278.

Durvasula, R., Gumbs, A., Panackal, A., Kruglov, O., Taneja, J., Kang, A., Cordon-Rosales, C., Richards, F., Whitham, R., and Beard, C. (1999). Expression of a functional antibody fragment in the gut of *Rhodnius prolixus* via transgenic bacterial symbiont *Rhodococcus rhodnii*. *Med. Vet. Entomol.* **13:** 115–119.

Fraser, C., Gocayne, J., and 27 others. (1995). The minimal gene complement of *Mycoplasma genitalium*. *Science* **270:** 397–403.

Hao, Z., Kasumba, I., Lehane, M.J., Gibson, W.C., Kwon, J., and Aksoy, S. (2001). Tsetse immune responses and trypanosome transmission: implications for the development of tsetse-based strategies to reduce trypanosomiasis. *Proc. Natl. Acad. Sci. U.S.A.* **98:** 12648–12653.

Hoffmann, A.A. and Turelli, M. (1997). Cytoplasmic incompatibility in insects. In *Influential Passengers* (S.L. O'Neill, A.A. Hoffmann, and J.H. Werren, Eds.), pp. 42–80. Oxford University Press, Oxford, U.K.

Hooper, L.V. and Gordon, J.I. (2001). Commensal host-bacterial relationships in the gut. *Science* **292:** 1115–1118.

Ma, W.-C. and Denlinger, D.L. (1974). Secretory discharge and microflora of milk gland in tsetse flies. *Nature* **247**: 301–303.

Min, K.-T. and Benzer, S. (1997). *Wolbachia*, normally a symbiont of *Drosophila*, can be virulent, causing degeneration and death. *Proc. Natl. Acad. Sci. U.S.A.* **94:** 10792–10796.

Moloo, S.K. and Shaw, M.K. (1989). Rickettsial infections of midgut cells are not associated with susceptibility of *Glossina morsitans centralis* to *Trypanosoma congolense* infection. *Acta Trop.* **46:** 223–227.

Molyneux, D.H. (2001). Sterile insect release and trypanosomiasis control: a plea for realism. *Trends Parasitol.* **17:** 413–414.

Moran, N.A. (1996). Accelerated evolution and Muller's ratchet in endosymbiotic bacteria. *Proc. Natl. Acad. Sci. U.S.A.* **93:** 2873–2878.

Moran, N. and Wernegreen, J. (2000). Lifestyle evolution in symbiotic baceria: insights from genomics. *Trends Ecol. Evol.* **15:** 321–326.

Munson, M., Baumann, P., and Kinsey, M. (1991). *Buchnera* gen. nov. and *Buchnera aphidicola* sp. nov., a taxon consisting of the mycetocyte-associated, primary endosymbionts of aphids. *Int. J. Syst. Bacteriol.* **41:** 566–568.

Nantulya, V.M. and Moloo, S.K. (1988). Suppression of cyclical development of *Trypanosoma brucei brucei* in *Glossina morsitans centralis* by an anti-procyclics monoclonal antibody. *Acta Trop.* **45:** 137–144.

Nogge, G. (1976). Sterility in tsetse flies (*Glossina morsitans* Westwood) caused by loss of symbionts. *Experientia* **32:** 995–996.

Nogge, G. (1978). Aposymbiotic tsetse flies, *Glossina morsitans morsitans* obtained by feeding on rabbits immunized specifically with symbionts. *J. Insect Physiol.* **24:** 299–304.

Nogge, G. (1980). Elimination of symbionts of tsetse flies (*Glossina m. morsitans* Westw.) by help of specific antibodies. In *Endocytobiology* (W. Schwemmler and H. Schenk, Eds.), pp. 445–452. W. de Gruyter, Berlin.

Nogge, G. (1981). Significance of symbionts for the maintenance of an optional nutritional state for successful reproduction in hematophagous arthropods. *Parasitology* **82:** 101–104.

Pinnock, D.E. and Hess, R.T. (1974). The occurrence of intracellular rickettsia-like organisms in the tsetse flies, *Glossina morsitans, G. fuscipes, G. brevipalpis* and *G. pallidipes*. *Acta Trop.* **31:** 70–79.

Poinsot, D., Bourtzis, K., Markaris, G., Savakis, C., and Mercot, H. (1998). *Wolbachia* transfer from *Drosophila melanogaster* into *D. simulans*: host effect and cytoplasmic incompatibility relationships. *Genetics* **150:** 227–237.

Politzar, H. and Cuisance, D. (1984). An integrated campaign against riverine tsetse *Glossina palpalis gambiensis* and *G. tachinoides* by trapping and the release of sterile males. *Insect Sci. Appl.* **5:** 439–442.

Roubáud, B. (1919). Les particularités de la nutrition et de la vie symbiotique chez les mouches tsetse. *Ann. Inst. Pasteur* **33:** 489–537.

Rousset, F. and de Stordeur, E. (1994). Properties of *Drosophila simulans* strains experimentally infected by different clones of the bacterium *Wolbachia*. *Heredity* **72:** 325–331.

Schröder, D., Deppisch, H., Obermayer, M., Krohne, G., Stackebrandt, E., Holldobler, B., Goebel, W., and Gross, R. (1996). Intracellular endosymbiotic bacteria of Camponotus species (carpenter ants): systematics, evolution and ultrastructural characterization. *Mol. Microbiol.* **21:** 479–489.

Shaw, M.K. and Moloo, S.K. (1991). Comparative study on Rickettsia-like organisms in the midgut epithelial cells of different Glossina species. *Parasitology* **102:** 193–199.

Shigenobu, S., Watanabe, H., Hattori, M., Sakaki, Y., and Ishikawa, H. (2000). Genome sequence of the endocellular bacterial symbiont of aphids *Buchnera* sp. APS. *Nature* **407:** 81–86.

Sinkins, S.P., Curtis, C.F., and O'Neill, S.L. (1997). The potential application of inherited symbiont systems to pest control. In *Influential Passengers* (S.L. O'Neill, A.A. Hoffmann, and J.H. Werren, Eds.), pp. 155–175. Oxford University Press, Oxford, U.K.

Stuhlman, F. (1907). Beiträge zur Kenntnis der Tsetse Fliege. *Arb. Gesundh. Amte (Berlin)* **26:** 301–308.

Thao, M.L., Clark, M., Baumann, L., Brennan, E., Moran, N., and Baumann, P. (2000a). Secondary endosymbionts of psyllids have been acquired multiple times. *Curr. Microbiol.* **41:** 300–304.

Thao, M.L., Moran, N.A., Abbot, P., Brennan, E.B., Burckhardt, D.H., and Baumann, P. (2000b). Cospeciation of psyllids and their primary prokaryotic endosymbionts. *Appl. Environ. Microbiol.* **66:** 2898–2905.

Turelli, M., Hoffmann, A.A., and McKechnie, S.W. (1992). Dynamics of cytoplasmic incompatibility and mtDNA variation in natural *Drosophila simulans* populations. *Genetics* **132:** 713–723.

Vreysen, M.J., Saleh, K.M., Ali, M.Y., Abdulla, A.M., Zhu, Z., Juma, K.G., Dyck, A., Msangi, A.R., Mkonyi, P.A., and Feldmann, A.M. (2000). *Glossina austeni* (Diptera: Glossinidae) eradicated on the Island of Unguga, Zanzibar, using the sterile insect technique. *J. Econ. Entomol.* **93:** 123–135.

Welburn, S.C. and Maudlin, I. (1991). Rickettsia-like organisms, puparial temperature and susceptibility to trypanosome infection in *Glossina morsitans*. *Parasitology* **102:** 201–206.

Welburn, S.C. and Maudlin, I. (1999). Tsetse-trypanosome interactions: rites of passage. *Parasitol. Today* **15:** 399–403.

Welburn, S.C., Maudlin, I., and Ellis, D.S. (1987). *In vitro* cultivation of rickettsia-like-organisms from *Glossina* spp. *Ann. Trop. Med. Parasitol.* **81:** 331–335.

Welburn, S.C., Arnold, K., Maudlin, I., and Gooday, G.W. (1993). Rickettsia-like organisms and chitinase production in relation to transmission of trypanosomes by tsetse flies. *Parasitology* **107:** 141–145.

Werren, J.H., Windsor, D., and Guo, L.R. (1995). Distribution of *Wolbachia* among neotropical arthropods. *Proc. R. Soc. London (B)* **262:** 197–204.

Wigglesworth, V.B. (1929). Digestion in the tsetse fly: a study of the structure and function. *Parasitology* **21:** 288–321.

Woese, C.R. (1987). Bacterial evolution. *Microbiol. Rev.* **51:** 221–271.

Young, G.M., Schmiel, D.H., and Miller, V.L. (1999). A new pathway for the secretion of virulence factors by bacteria: the flagellar export apparatus functions as a protein-secretion system. *Proc. Natl. Acad. Sci. U.S.A.* **96:** 6456–6461.

5 Endosymbiosis in the Weevil of the Genus *Sitophilus*: Genetic, Physiological, and Molecular Interactions among Associated Genomes

Abdelaziz Heddi

CONTENTS

INTRODUCTION

Interspecific associations are currently believed to take part in evolution either by improving a partner's fitness through integrated endosymbioses or by causing reproductive isolation and subsequent host speciation, such as in the *Wolbachia* endosymbioses (Nardon and Grenier, 1991; Margulis, 1993a; Bordenstein et al., 2001). Most often, associations involve very distant species like bacteria with invertebrates, but several examples are known where eukaryotes associate together to form lichen or coral holobiont units (Rowan, 1997). Recently, an unusual symbiosis has been discovered within the mealy bugs wherein a γ-proteobacterium lives inside a β-proteobacterium (von Dohlen et al., 2001), which demonstrates that even among prokaryotes a tendency exists to form tight associations among the genomes (Heddi et al., 2001).

In insects of the order Coleoptera, a large array of symbioses occurs with three principal types of association: ectosymbiosis with fungi in *Ambrosia* beetles, intestinal lumen endosymbiosis in Scarabaeidae, and intracellular endosymbiosis in Chrysomelidae, Scolytidae, and Dryophthoridae (for review see Nardon and Grenier, 1989). The Dryophthoridae family (or

0-8493-1286-8/03/$0.00+$1.50

Rhynchophoridae) comprises as many as 145 genera that feed on plant stems, roots, and seeds. Among them, *Sitophilus* spp. are the most characterized in terms of symbiosis because of the damage they cause on cereals. Three cereal-feeding species (*S. oryzae*, *S. zeamais*, and *S. granarius*) destroy up to 40% of stored cereals, particularly in tropical regions. All are naturally symbiotic and transmit maternally to the offspring an intracellular γ-proteobacterium. Females lay eggs inside the seeds, and larvae develop within the grain and emerge from the grain as adults. During this period, larvae feed on seed albumen only, which is starch rich but with low amino acid, lipid, and vitamin content. A strong database has demonstrated that intracellular bacteria help the host to balance these nutritional deficiencies (Wicker, 1983; Wicker et al., 1985; Nardon and Grenier, 1988).

This chapter describes intracellular symbioses in the weevils of the genus *Sitophilus*, reviews the recent knowledge on genetic, physiological, and molecular interactions between the host and the symbiont, and explores the role of intracellular symbiosis as a driving force in evolution.

THE BIOLOGY OF *SITOPHILUS* SSP. SYMBIOSIS

Symbiosis in members of the genus *Sitophilus* was first described by Pierantoni (1927), then by Mansour (1930, 1934, 1935). The best-understood symbiont is the intracellular bacterium, recently called SOPE (*Sitophilus oryzae* principal endosymbiont) by Heddi et al. (1998), which is a Gram-negative bacterium located in the ovaries and in the larval bacteriome formed by polyploid cells called bacteriocytes (Figure 5.1). It occurs naturally in all populations of the three *Sitophilus* species analyzed (*S. oryzae*, *S. zeamais,* and *S. granarius*) and always lives free in the cytosol of oocytes and bacteriocytes without any surrounding membrane. This feature suggests that either SOPE does not behave (or behaves only slightly) as a pathogenic cell or that host cells regulate their immune system to control and tolerate the bacterium. Nevertheless, the tissue specificity of SOPE, which is restricted to two kinds of tissues (i.e., bacteriocytes and oocytes), argues in favor of molecular signals or "dialogue" between partners leading to the control of the bacterial population.

In the young nymph, the larval bacteriome dissociates and bacteriocytes migrate to reach the apex of mesenteric caeca, where they form mesenteric bacteriomes in the young adults (Figure 5.1). The density of SOPE changes with the developmental stage and with the age of the insect. Their number increases from the first instar larvae and reaches the highest level in the fourth and last instar larvae, with as many as 3 million bacteria per larval bacteriome (Nardon and Wicker, 1981). In adults, SOPE density decreases with age, and 3-week-old adults are completely devoid of symbiotic bacteria (Mansour, 1930; Schneider, 1956). However, adult females retain them in the apical bacteriome of the ovaries and in the oocytes from which the bacteria are transmitted to the progeny. Thus, from a physiological point of view, symbiosis in the weevil seems to be necessary during the larval and the young-adult developmental stages only.

In addition to SOPE, *Sitophilus* weevils harbor a second type of bacterium, which belongs to the genus *Wolbachia* (Figures 5.1 and 5.2) (Heddi et al., 1999). This rickettsia-like organism is widespread in insects and is estimated to occur in at least 20% of the insect species (Werren, 1997; Jeyaprakash and Hoy, 2000). It has also been detected in crustaceans, nematodes, and arachnids. Unlike integrated endosymbionts, *Wolbachia* spp. show different behavior in that, in parallel with vertical inheritance throughout the egg cytoplasm, they have undergone an extensive intertaxon transmission. Moreover, *in vitro* transmission by microinjection has been accomplished between insect taxa, particularly within the *Trichogramma* genus (Grenier et al., 1998). In the weevils, several populations are infected with *Wolbachia*. Among 23 *Sitophilus* stains belonging to three species (*S. oryzae, S. zeamais, S. granarius*) collected worldwide, 13 are infected with *Wolbachia*, constituting 9 strains totally infected and 4 strains partially infected (Heddi et al., 1999). Cytologically, the distribution of *Wolbachia* exhibits different features from that of SOPE. Fluorescence *in situ* hybridization (FISH) using specific oligonucleotide probes

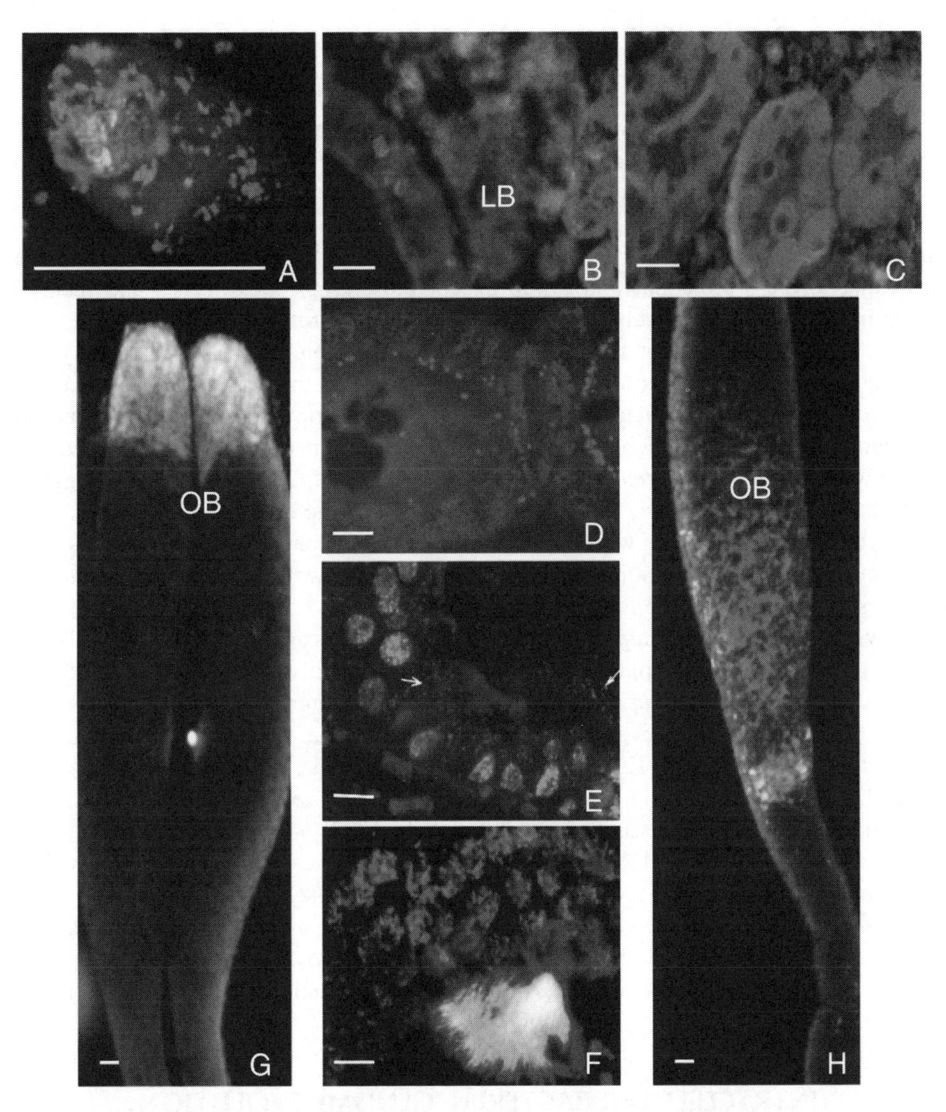

FIGURE 5.1 (Color figure follows p. 206.) Fluorescent *in situ* hybridization (FISH) of *Sitophilus oryzae* intracellular bacteria. Specific oligonucleotide probes were designed by sequence alignment of *Wolbachia* and SOPE 16S rDNA. Two *Wolbachia* probes (W1, W2) 5′ end labeled with rhodamine were used to increase the signals. The SOPE probe (S) was 5′ end labeled with rhodamine except in panel B (with fluorescein). Hybridization was performed as described by Heddi et al. (1999). Slides were mounted in Vectashield medium containing DAPI. (A) Bacteriocyte labeled with W1W2. (B) Larval bacteriome (LB) labeled with W1W2 and S (fluorescein). (C, D, E, F, and H) Adult mesenteric caeca bacteriomes, oocyte, follicular cells, testis, and ovary, respectively, labeled with W1W2. (G) Ovary labeled with S. Scale bar = 10 μm.

designed for SOPE and *Wolbachia* 16S rDNA showed that although SOPE was limited to the bacteriocytes, *Wolbachia* were disseminated throughout all insect tissues even within the bacteriocytes, coexisting with SOPE (Figure 5.1). *Wolbachia* spp. are present in low density in muscles, adipocytes, and the intestine, whereas they are highly abundant in the male and female germ cells, located around the periplasmic membrane of the oocyte and mixed with many spermatid nuclei in the testis (Figure 5.1). Here also *Wolbachia* do not interact with the host in the same way as integrated endosymbionts do because they are not exclusive to specific host cells (such as the bacteriocytes), but rather germ cells are the privileged site for the development of these bacterial cells.

PHYLOGENETIC CHARACTERIZATION OF SOPE AND *WOLBACHIA*

The molecular phylogenetic positioning of insect intracellular bacteria began in 1989 with the sequence of the 16S rDNA of *Buchnera aphidicola*, the endosymbionts of aphids (Untermann et al., 1989; Munson et al., 1991). Currently, the bacterial symbionts of more than ten insect models are described, including those of aphids, psyllids, cicadas, planthoppers, mealy bugs, glossines, weevils, cockroaches, termites, ants, and bugs. Most of the insect-integrated endosymbionts belong to the γ-proteobacteria group except those from cockroaches and mealy bugs, which are included in, respectively, the Flavobacteria and β-proteobacteria groups (Figure 5.2) (Bandi et al., 1994; Kantheti et al., 1996; Fukatsu and Nikoh, 2000). Using a phylogenetic analysis based on a heterogeneous model of DNA evolution that takes into account the GC-content heterogeneity among bacterial sequences, we have shown recently that the majority of the γ-proteobacteria endosymbionts are closely related within the Enterobacteriaceae family, located between the genera *Proteus* and *Yersinia* (Charles et al., 2001). This result suggests that the Enterobacteriaceae group may possess features allowing the establishment and maintenance of stable symbiotic relationships with insects. The nature of such features and traits is unknown currently, but endosymbiont-genome-sequencing projects, as well as host-symbiont molecular-interaction studies, may provide insights into this aspect.

In the weevils of the genus *Sitophilus*, two types of symbionts are present, SOPE and *Wolbachia* (Figure 5.2), and both are transmitted to the offspring maternally. The 16S rDNA sequences of *Wolbachia* spp. that occur in *Sitophilus* spp. fall into the B-group of *Wolbachia* with which they form a monophyletic group relative to the other α-proteobacteria. However, the SOPE 16S rDNA sequence was placed within the Enterobacteriaceae (Heddi et al., 1998). It shares 95% sequence identity with *Escherichia coli* and 87% sequence identity with *B. aphidicola*. SOPE is also affiliated with the endosymbionts of the tsetse fly (Aksoy et al., 1997) and those of the carpenter ants, *Camponotus* spp. (Schröder et al., 1996). Interestingly, the closest intracellular endosymbiont to SOPE is *Sodalis glossinidius*, the secondary endosymbiont of tsetse. Recent data suggest that SOPE and *S. glossinidius* may have evolved from the same ancestor or that horizontal gene transfer may have occurred between tsetse and weevil over the course of evolution (unpublished data). These hypotheses are under investigation in conjunction with phylogenetic studies of the Dryophthoridae family endosymbionts.

INTRACELLULAR BACTERIAL GENOME EVOLUTION

It is now well documented that intracellular bacteria, as well as other intracellular pathogenic organisms, undergo particular evolutionary constraints with respect to their DNA G+C content and their genome sizes (Moran, 1996; Heddi et al., 1998; Sun et al., 2001). In general, and regardless of the bacterial physiology and behavior, genome sizes are reduced and genomes themselves are biased toward A+T, as compared with their closest free-living bacterial relatives. Presently, five insect endosymbiont genomes have been analyzed for genome size and G+C content, four belonging to the Enterobacteriaceae family and the fifth one from the *Wolbachia* group. As shown in Table 5.1, all intracellular bacterial genomes that have been studied so far have reduced genome sizes and are A+T-biased with regard to the free-living bacterium *E. coli*. The reasons for these genetic changes are not precisely understood, but they could be related to the vertical mode of bacterial transmission. Indeed, most intracellular bacteria are transmitted to the progeny through the oocytes, which generates a bottleneck effect and prevents endosymbiont recombination with free-living organisms, thereby facilitating accumulation of deleterious mutations and compositional bias due to directional mutation pressure (Moran, 1996; Heddi et al., 1998). Moreover, genome-size reduction could be interpreted as being the result of serial deletions of gene fragments, which thus are no longer necessary for the new host–symbiont unit. Likewise, gene transfer to the nucleus could

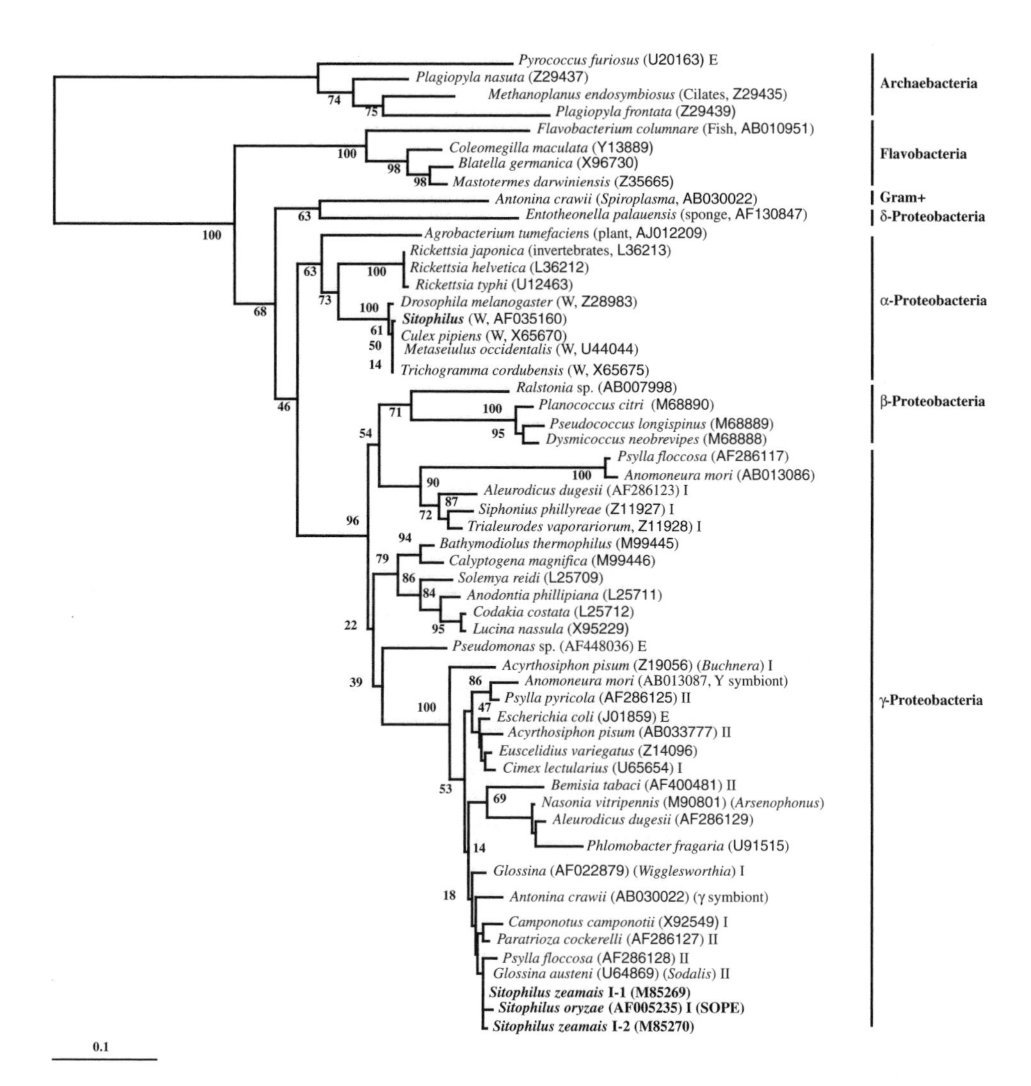

FIGURE 5.2 The phylogenetic tree of the bacterial endosymbionts based on 16S rDNA genes. Alignments were performed and the phylogenetic tree was constructed as described previously (Heddi et al., 1998). Briefly, DNA sequence database searches were performed at the National Center for Biotechnology Information by using the BLAST network service. Alignments were generated by Clustal W software and modified by visual inspection. The phylogenetic tree was generated using the neighbor-joining program of Clustal W software. Genetic distance (percentage divergence) was calculated excluding positions with gaps or unknown residues. Confidence limits on grouping were determined by the Clustal W bootstrapping technique (1000 repeats). Bacteria are indicated by their host names or by their names when they infect several hosts. In this case, the bacterial names are indicated in italics. Species are followed by GenBank accession number. W: *Wolbachia,* I: principal endosymbiont, II: secondary endosymbiont, E: extracellular bacteria (in italics). *Sitophilus zeamais* harbors two principal endosymbionts (1 and 2).

be an alternative explanation for this phenomenon. In agreement with these findings is the genome sequence of *B. aphidicola*. This genome encodes for 583 ORFs only and, by comparison with *E. coli* ORFs, it may have lost most genes encoding for nonessential amino acids and the Krebs cycle (Shigenobu et al., 2000). Nevertheless, it is noteworthy that SOPE and *S. glossinidius* genomes are neither very A+T-rich nor of a very size-reduced genome (Table 5.1). One interpretation suggests that this could be linked to the fact that both associations (SOPE–weevil and *Sodalis*–tsetse) are relatively recent and that their DNA composition has not diverged much from those of other free-

TABLE 5.1
Genome Sizes and A+T Content of Insect Intracellular Bacteria

	Insect Host	Genome Size	A+T Content	Ref.
Buchnera spp.	Aphid PE	From 450 kb to 670 kb	74%	Charles and Ishikawa, (1999); Shigenobu et al. (2000); Gil et al. (2002)
Wigglesworthia spp.	Tsetse PE	From 705 kb to 770 kb	60%[a]	Akman and Aksoy (2001)
Sodalis glossinidius	Tsetse SE	2.2 Mb	45%[a]	Akman et al. (2001)
SOPE	Weevil PE	3 Mb	46%	Charles et al. (1997); Heddi et al. (1998)
Wolbachia	Invertebrate endosymbiont	From 0.95 Mb to 1.66 Mb	60%[a]	Sun et al. (2001)
Escherichia coli	Free-living bacteria	4.6 Mb	47%	Smith et al. (1987)

[a] Values determined from gene sequences only. PE = principal endosymbiont; SE = secondary endosymbiont.

living Enterobacteriaceae. This may also partly explain why the weevil association could be artificially dissociated (Nardon, 1973) or why *Sodalis* can grow outside the host (Beard et al., 1993), whereas the other endosymbionts (*B. aphidicola* and *Wigglesworthia* spp.) are strictly dependent on their hosts.

INSECT–BACTERIA INTERACTIONS IN THE WEEVIL SYMBIOTIC ASSOCIATION

Living in association implies interaction, which starts from the first contact between partners and continues during the process of permanent mutual establishment. Today, most biologists admit that the evolution of eukaryotes resulted from symbiotic interaction between several independent ancestors such as ancestors of mitochondria and plastids (Sogin, 1997), although the origin of other cellular organelles, such as undulipodia, microtubules, and nucleus, is controversial (Maynard Smith, 1989; Margulis, 1993b). However, what remains unresolved is an understanding of the behavior of bacteria at the early phases of host infection as well as the mechanism leading to stable unit formation. This section describes how SOPE interacts with the weevil at three levels: genetic, showing how the insect controls the SOPE population; physiological, by which the association improves its *fitness*; and molecular interactions that define the molecular "conversation" between partners.

Genetic Interactions between Endosymbionts and Weevils

As reported above, SOPE is located specifically in the bacteriocytes that form the bacteriome; no bacterial proliferation has been observed elsewhere. Moreover, symbiont density is different among *Sitophilus* strains, but it is consistent within a given population. Finally, symbiont numbers from the egg (20,000 cells/embryo) to the last larval stage (3×10^6 cells/larvae) indicate that SOPE may divide seven times over 21 d (Nardon et al., 1998). Taken together, these results support the existence of a host genetic system controlling both the location and the population size of bacteria within the insect. To shed light on this aspect, genetic experiments were conducted on strains selected on the basis of larval development times (Figure 5.3). One Low (LL) and one Rapid (RR) strain were obtained from the SFr wild-type strain, and reciprocal crosses and backcrosses were conducted between them (Nardon et al., 1998). Interestingly, the number of symbionts is identical in both

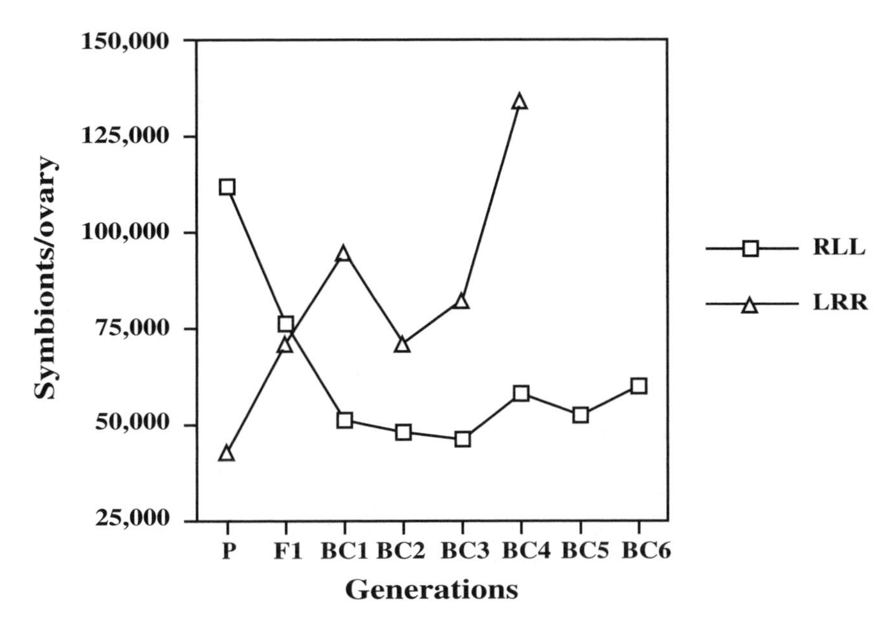

FIGURE 5.3 The evolution of ovarian symbiont number in successive reciprocal backcrosses. LRR: introduction of RR genome in LL cytoplasm, RLL: introduction of LL genome in RR cytoplasm. [Modified from Nardon, P., Grenier, A.M., and Heddi, A. (1998). *Symbiosis* **25:** 237–250.]

reciprocal F1 crosses and represents more or less the mean value of bacterial density in RR and LL strains (Figure 5.3). Moreover, reciprocal backcrosses indicate that the symbiont population decreases in LRR crosses when the RR genome is introduced into the LL cytoplasm, whereas it increases in RLL lines where the LL genome is replacing the RR genome in the RR cytoplasm. This demonstrates clearly that symbiont numbers are being controlled by a host genetic mechanism. The molecular mechanism is currently under investigation using the cDNA subtraction approach, which has revealed the existence of several bacteriocyte-specific genes that may regulate bacterial populations in symbiotic systems (A. Heddi et al., unpublished data).

For *Wolbachia* endosymbioses, no extensive work has been carried out on this aspect, so it remains unclear whether or not the host genetic system controls bacterial populations. However, we know from previous work that not all weevil populations are infected with *Wolbachia* and not all individuals are *Wolbachia* symbiotic when a given population is infected. Bourtzis et al. (1996) surveyed 41 stocks from the *Drosophila* genus and showed, with dot blot assays, that the infection levels were not equal among populations. These results together provide support for bacterial control by the host genetic system. However, *Wolbachia* genome variability could also explain why some insect strains are infected while others are not at all or only partly infected.

Wolbachia Induce Cytoplasmic Incompatibility in Weevils

In insects, *Wolbachia* do not seem to improve the host *fitness* significantly except in one case analyzed so far, where *Wolbachia* were shown to be necessary for the oogenesis of the parasitic wasp *Asobara tabida* (Dedeine et al., 2001). Otherwise, *Wolbachia* are known to alter host reproduction mainly in four ways: cytoplasmic incompatibility (Breeuwer and Werren, 1990; O'Neill and Karr, 1990), parthenogenesis (Stouthamer et al., 1993), feminization of genetic males (Rigaud et al., 1991), and male killing (Jiggins et al., 2000). In weevils, to investigate separately and to distinguish the roles of SOPE and *Wolbachia* in host-symbiont biology, genetic analyses were conducted with different crosses between individuals treated or not treated with heat or tetracycline. Antibiotic treatment resulted in the elimination of both SOPE and *Wolbachia*, while heat treatment disrupted SOPE only. From these experiments we have determined that *Wolbachia* are not highly involved in weevil physiology but rather in the reproduction of the insect by causing unidirectional cytoplasmic incompatibility (Heddi et al., 1999). The molecular mechanism that triggers the host-genome deregulation was not studied. Nevertheless, the intimate contact between *Wolbachia* and the germ-cell nucleus (Figure 5.1) argues in favor of protein–protein or DNA–protein interaction processes.

Weevil–SOPE Physiological Interactions

In the last decade, studies of the weevil symbiosis were mainly focused on how integrated bacteria influence the behavior and physiology of the insect. Nardon provided the first data when he succeeded in obtaining aposymbiotic insects from a wild-type symbiotic strain named SFr (Nardon, 1973). This system is unique in insect symbioses as a heat treatment (35°C) and 90% relative humidity for 1 month resulted in bacterial elimination from both somatic and germ cells. However, the resulting aposymbiotic strain is less fertile than the wild type; it develops slowly during the larval stages and is unable to fly as an adult. These perturbations have led to the conclusion that SOPE influences the physiology and behavior of the insect (Nardon and Grenier, 1988, 1989), and these impact host *fitness*. These data provided evidence that intracellular symbiosis has an impact on the evolution of insect characters.

Several biochemical aspects were tested to understand how SOPE enhanced the host *fitness*. First, SOPE provides the insect with many components that are poorly represented in wheat grains or completely absent from the albumen part of the grain on which *Sitophilus* spp. larvae feed. These include pantothenic acid, biotin, riboflavin (Wicker, 1983), and amino acids, particularly the aromatic amino acids phenylalanine and tyrosine (Wicker and Nardon, 1982). Second, SOPE interferes with insect metabolism either directly through the modification of some products or indirectly by increasing some insect enzymatic pathways (Gasnier-Fauchet et al., 1986; Heddi et al., 1993).

Investigating the metabolic pathways of amino acids, Gasnier-Fauchet and Nardon (1987) have noticed that methionine, an amino acid in excess in wheat grains, is not metabolized the same way in the aposymbiotic insect as in the symbiotic. Methionine sulfoxide is always at high levels in symbiotic insects, while sarcosine levels increase regularly during the last instar larvae of aposymbiotic insects (Gasnier-Fauchet and Nardon, 1986), reaching the highest value at the end of this instar (43.5 nmol/insect). In symbiotic insects, sarcosine levels never exceed 3.8 nmol/insect. Hence, it was demonstrated that SOPE helped the insect to catabolize methionine into methionine sulfoxide by a nonenergy-consuming and reversible reaction. In contrast, aposymbiotic insects fail to carry out this pathway, and methionine is preferably demethylated via a glycine *N*-methyltransferase-like activity leading to the accumulation of sarcosine, which is neither incorporated into proteins nor excreted in the feces (Gasnier-Fauchet et al., 1986).

An additional indirect metabolic effect was shown on mitochondrial energy metabolism. Mitochondria isolated from symbiotic insects exhibited high levels of mitochondrial respiratory control (RCR) when compared with the aposymbiotic insect, indicating that ATP production is more efficient in insects harboring intracellular bacteria (Heddi and Lefebvre, 1990a; Heddi et al., 1991). This metabolic aspect was confirmed later by measuring six mitochondrial enzymatic activities, three belonging to the oxidative phosphorylation chain (cytochrome *c* oxidase, succinate cytochrome *c* reductase, and glycerol-3-phosphate cytochrome *c* reductase) and three other enzymes from the Krebs cycle (isocitrate dehydrogenase, pyruvate dehydrogenase, and α-ketoglutarate dehydrogenase) (Heddi et al., 1993). As a result, symbiotic insects always show higher specific enzymatic activities regardless of the insect stage. However, the differences between symbiotic and aposymbiotic insects are attenuated in the adult stage, where symbiont density decreases with age until the third week of adult development, at which time the mesenteric caeca are completely devoid of symbionts.

To determine how specific enzymatic activities are enhanced in the symbiotic strain, two hypotheses were tested: (1) that aposymbiotic insect genetic drift and mitochondrial expression failure may occur after the heat treatment of the symbiotic strain while obtaining the aposymbiotic insects and (2) that nutritional interaction would occur between the host and the symbiont. The former hypothesis was ruled out by performing backcrosses between symbiotic and aposymbiotic insects and measuring mitochondrial enzymatic activities (Heddi and Lefebvre, 1990b; Heddi et al., 1993) and by studying mitochondrial protein neosynthesis (Heddi and Nardon, 1993). The nutritional hypothesis was tested with vitamin (pantothenic acid and riboflavin)-supplementation experiments conducted on wheat-flour artificial pellets (Heddi et al., 1999). These experiments showed clearly that differences between symbiotic and aposymbiotic strains were attenuated or became nonsignificant when insects were reared on vitamin-supplemented pellets (Figure 5.4). It was therefore concluded that SOPE improved mitochondrial oxidative phosphorylation in a nutritional way (Heddi et al., 1999). Indeed, by supplying the host with vitamins that are in low quantities in

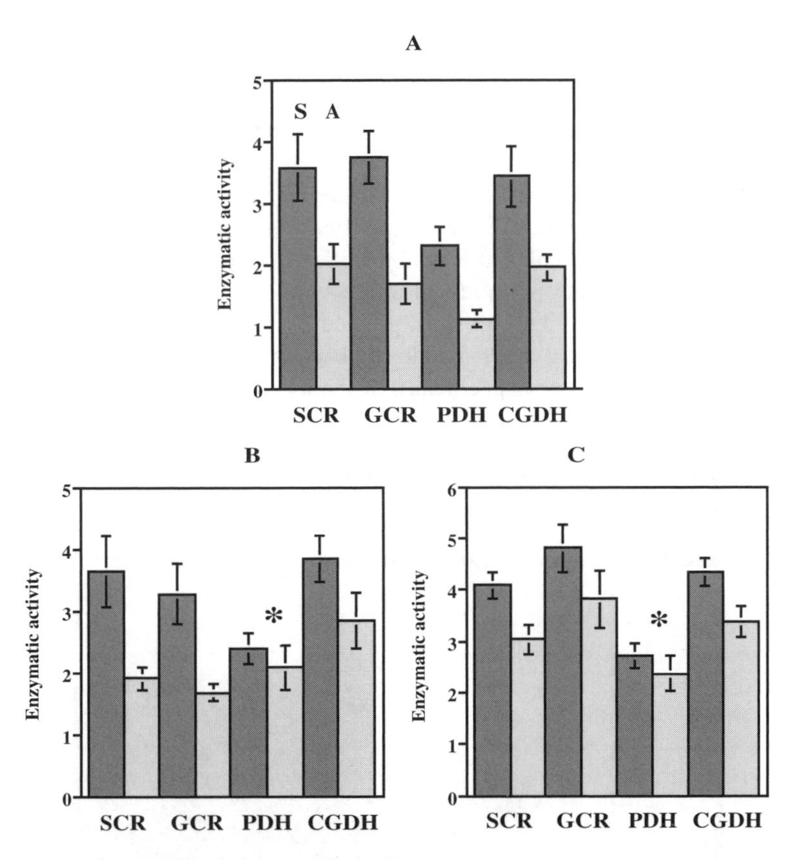

FIGURE 5.4 Specific enzymatic activities of mitochondrial suspensions isolated from symbiotic (S) and aposymbiotic (A) adults of the weevil *S. oryzae*. Insects were reared for one generation on wheat flour artificial pellets supplemented or not with vitamins. (A) Control pellets without any supplementation. (B) Pellets supplemented with pantothenic acid. (C) Pellets supplemented with riboflavin. SCR = succinate cytochrome *c* reductase (y unit = μmol mg^{-1}s^{-1}), GCR = glycerol-3-phosphate cytochrome *c* reductase (y = 0.5 μmol mg^{-1}s^{-1}), PDH: pyruvate dehydrogenase (y = 0.1 μmol mg^{-1}s^{-1}), KGDH = α-ketoglutarate dehydrogenase (y = 0.2 μmol mg^{-1}s^{-1}). Data are expressed as the mean of ten measurements ± tsvn. * = No significant difference between symbiotic and aposymbiotic insects (*t* test, α = 0.05) [Modified from Heddi, A., Lefebvre, F., and Nardon, P. (1993). *Insect Biochem. Mol. Biol.* **23:** 403–411 and Heddi, A., Grenier, A.M., Khatchadourian, C., Charles, H., and Nardon, P. (1999). *Proc. Natl. Acad. Sci. U.S.A.* **96:** 6814–6819.]

wheat, SOPE interacts with the insect metabolic pathways involved in the biosynthesis of coenzyme A and NAD^+-FAD, which are the required cofactors for mitochondrial enzyme function. Moreover, similar experiments had shown previously that adult aposymbiotic flight ability was partly restored when insects were supplemented with vitamins (Grenier et al., 1994). This finding links mitochondrial oxidative phosphorylation with flight, an ATP-consuming phenomenon, and indicates that SOPE interacts with both physiological traits and the behavior of weevils through flight.

Molecular Interaction between SOPE and Weevils

Intracellular bacterial integration within the bacteriocytes has certainly evolved from prokaryotic and eukaryotic gene "conversation." The nature of these genetic communications is currently under investigation in the weevil association, but several genes from other models have been studied in terms of host–symbiont interactions. Excluding microorganism–plant associations, the marine squid–*Vibrio* association is the best-studied animal model. The development of the squid light organ requires the presence of the species-specific bioluminescent bacterial symbiont, *Vibrio fischeri* (McFall-Ngai, 1998). This extracellular symbiont is recruited from the surrounding seawater by the newly hatched squid (Ruby, 1996; Nyholm et al., 2000). The establishment of the light organ involves prokaryotic–eukaryotic gene signaling that triggers many developmental changes such as apoptosis, cell swelling, and oxidative stress (Foster et al., 2000; Lemus and McFall-Ngai, 2000; Visick et al., 2000).

In the weevil symbiosis, the first evidence of gene interaction was provided by two-dimensional electrophoresis experiments (Charles et al., 1997). Comparison between symbiotic and aposymbiotic protein profiles revealed at least four differential proteins — α, β, γ, and δ (Figure 5.5). The protein α is 30 kD and is expressed in the symbiotic strain only. Since it is not seen on SOPE protein profiles, it is likely to be expressed by the insect in the presence of SOPE unless the protein is expressed by SOPE within the bacteriocyte and exported to the host cells. The protein β (33 kD) is visualized specifically in the aposymbiotic panel only. Its expression could be related to either the absence of bacteria in aposymbiotic insects or the inhibition by

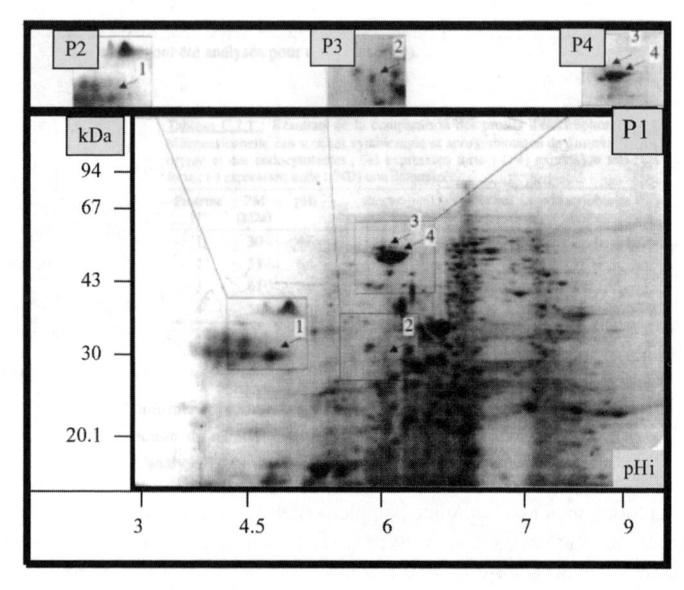

FIGURE 5.5 Two-dimensional electrophoresis protein pattern of *S. oryzae* symbiotic strain (P1). Differences observed in the aposymbiotic strain are reported in the panels P2, P3, and P4. α, β, γ, and δ proteins are represented by 1, 2, 3, and 4, respectively. KDa = kilo Dalton; pHi = isoelectric pH. [Modified from Charles, H., Heddi, A., Guillaud, J., Nardon, C., and Nardon, P. (1997). *Biochem. Biophys. Res. Commun.* **239:** 769–774.]

bacteria in symbiotic insects. The two other proteins (γ and δ) are from the bacteria; γ has not yet been characterized, and δ was shown to be a chaperonin (GroEL) (Charles et al., 1995). *Ex vivo* protein labeling showed that the GroEL protein was prominent among the SOPE-produced proteins and could represent up to 40% of the total neosynthesized symbiont proteins. In contrast, GroEL expression diminishes when SOPE is incubated *in vitro*, indicating that the *GroEL* gene may play an important role in the association, and its regulation may be under the control of bacteriocyte cytosolic parameters.

Currently, although many symbiotic teams have been studying the *GroEL* gene (Ishikawa, 1985; Aksoy, 1995; Jeon, 1995), the gene function has not been clearly elucidated in relation to intracellular symbiotic conditions. Nevertheless, considerating that most endosymbionts have reduced their genome considerably in the course of their intracellular life history and consequently lack several ORFs, one possible *GroEL* gene function could be that of protein folding and unfolding during the shuttling of protein between bacteriocytes and bacteria. Regardless of the exact function of these proteins in symbiotic systems, their differential expression profiles suggest the existence of host–symbiont molecular interactions.

CONCLUSION

Symbiotic associations are widespread through the biotic world, and they occur at different levels of organismal complexity, ranging from bacterial associations to animal and plant symbioses. The newly formed unit involves at least two separate species and may include several organisms belonging to distant phyla (Fukatsu and Ishikawa, 1996; Fukatsu, 2001).

In insects, symbionts can occur as fungi, protists, or bacteria that exhibit extraorganismal relationships with ectosymbiosis or intraorganismal associations with endosymbiosis. The latter could be extracellular, such as in the gut–lumen endosymbioses, or intracellular, such as in insects that feed permanently on a nutritionally poor and unbalanced diet. The early steps in symbiogenesis are often believed to be harmful (or mildly harmful), during which time the partners exchange virulence gene products, leading to population selection (Paillot, 1933; Jeon, 1983). When mutual genome complementarity occurs establishing new adaptive traits on the interactive unit, and when constraint pressures exerted on the association do not completely eliminate the populations, virulence genes can be silenced (through expression or gene deletion) or at least controlled by the association favoring host–symbiont coadaptation. Insects may have selected the bacteriome organ for the control of the symbiont population.

A successful coadaptation often leads to symbiont–host coevolution (Baumann et al., 1995), where the symbiont finds a way to transmit to the offspring permanently. In insect bacteriocyte-inducing symbioses, symbionts are always transmitted through the female germ cells, where they persist, as in the weevils, or they infect the oocytes during the vitellogenesis process, as in cockroaches, bugs, and aphids. This restricted way of symbiont transmission, along with particular selection pressures exerted by the intracellular environment, appears to result in different selection pressures on microorganism evolution. Symbionts are no longer able to recombine (or can recombine only indirectly) with extracellular microorganisms; their DNA undergoes directional mutation bias toward A+T, and their genomes reduce in size either through gene-fragment loss or gene transfer to the host chromosomes. Early DNA rearrangements lead to bacterial domestication by the host. From this symbiogenesis phase, intracellular bacteria start losing the ability to grow and divide outside the association until their genomes reach a critical gene composition where the bacteria become dependent on the host. The ultimate case of this progression is the organogenetic symbioses of mitochondria, plastids, and hydrogenosomes (Gupta et al., 1989; Palmer, 1997; Sogin, 1997; Cavalier-Smith, 1999; Gray, 1999; Doolittle, 2000), where not only are genome sizes drastically reduced but also gene composition for a given organelle exhibits functional specialization. However, it should be emphasized that not all symbioses end in organogenetic relationships. Secondarily integrated symbioses occur very frequently in insects (Fukatsu et al., 2000), and it is possible that secondary endosymbionts may succeed the one formerly established as a consequence of a switch in the insect environment.

Symbiogenesis is a fundamental biological process that has accompanied cell evolution from the beginning of life and continues today. Indeed, similar genome reduction and specialization are found in bacteriocyte-associated symbionts of insects. *B. aphidicola* (the endosymbiont of aphids) has reduced its genome seven- to tenfold during 250 million years of coevolution with aphids (Charles and Ishikawa, 1999; Gil et al., 2002), and *Wigglesworthia* spp. (the endosymbionts of the tsetse fly) have had almost as much genome reduction during their 80 million years of evolution (Akman and Aksoy, 2001). In both, bacterial genome specialization seems to correspond with the composition of the insect diet. This phenomenon is strongly suggested with *B. aphidicola*, where the genome has lost most genes encoding for nonessential amino acids and the Krebs cycle (Shigenobu et al., 2000). However, *B. aphidicola* has kept genes for the biosyntheses of amino acids essential for the aphid hosts, indicating that host–symbiont complementarity and synthrophy have helped selection to remove redundant functions from the prokaryotic genome and may favor the maintenance of the functions adaptive for the unit.

Hence, regardless of the scenario by which serial genomes have become established in the association, symbiogenesis creates the possibility for life to diversify through the colonization of new and empty biotopes. For instance, aerobic constraints selected for the acquisition of α-Proteobacteria by early eukaryotic host cells around 2 billion years ago, permitting the diversification of Eukarya. One billion years later, animals and plants diverged when cyanobacteria invaded the ancestor protist (Doolittle, 1997).

In *Sitophilus* weevils, the SOPE did not reduce its genome size as much, perhaps because of the relative recent age of the symbiosis in this model. However, from a symbiogenesis point of view, at least four genomes have integrated serially into the weevil cells in the course of evolution: the nucleus, mitochondria, SOPE, and *Wolbachia* (Heddi et al., 1999). Each of them through the interaction with the others has separate functions with regard to the biology of the association. The most striking conclusion is that insect evolution should be considered not only the consequence of the classical "eukaryotic system" balance between mutation and selection but also the result of selective constraints that have been exerted on the whole association.

ACKNOWLEDGMENTS

I thank M. McFall-Ngai, C. Vieira, and P. Nardon for critical comments on the manuscript; H. Charles for the phylogenetic tree construction; and C. Khatchadourian for assistance.

REFERENCES

Akman, L. and Aksoy, S. (2001). A novel application of gene arrays: *Escherichia coli* array provides insight into the biology of the obligate endosymbiont of tsetse flies. *Proc. Natl. Acad. Sci. U.S.A.* **98:** 7546–7551.

Akman, L., Rio, R.V.M., Beard, C.B., and Aksoy, S. (2001). Genome size determination and coding capacity of *Sodalis glossinidius*, an enteric symbiont of tsetse flies, as revealed by hybridization to *Escherichia coli* gene arrays. *J. Bacteriol.* **183:** 4517–4525.

Aksoy, S. (1995). Molecular characteristics of the endosymbionts of tsetse flies: 16S rDNA locus and over-expression of chaperonins. *Insect Mol. Biol.* **4:** 23–29.

Aksoy, S., Chen, X., and Hypsa, V. (1997). Phylogeny and potential transmission routes of midgut-associated endosymbionts of tsetse (Diptera: Glossinidae). *Insect Mol. Biol.* **6:** 183–190.

Bandi, C., Damiani, G., Magrassi, L., Grigolo, A., Fani, R., and Sacchi, L. (1994). Flavobacteria as intracellular symbionts in cockroaches. *Proc. R. Soc. London (B)* **257:** 43–48.

Baumann, P., Baumann, L., Lai, C.-Y., Rouhbakhsh, D., Moran, N.A., and Clark, M.A. (1995). Genetics, physiology, and evolutionary relationships of the genus *Buchnera*: intracellular symbionts of aphids. *Annu. Rev. Microbiol.* **49:** 55–94.

Beard, C.B., O'Neill, S.L., Mason, P., Mandelco, L., Woese, C.R., Tesh, R.B., Richards, F.F., and Aksoy, S. (1993). Genetic transformation and phylogeny of bacterial symbionts from tsetse. *Insect Mol. Biol.* **1:** 123–131.

Bordenstein, S.R., O'Hara, F.P., and Werren, J.H. (2001). *Wolbachia*-induced incompatibility precedes other hybrid incompatibilities in *Nasonia*. *Nature* **409:** 707–710.

Bourtzis, K., Nirgianaki, A., Markakis, G., and Savakis, C. (1996). *Wolbachia* infection and cytoplasmic incompatibility in *Drosophila* species. *Genetics* **144:** 1063–1073.

Breeuwer, J.A.J. and Werren, J.H. (1990). Microorganisms associated with chromosome destruction and reproductive isolation between two insect species. *Nature* **346:** 558–560.

Cavalier-Smith, T. (1999). Principles of protein and lipid targeting in secondary symbiogenesis: Euglenoid, Dinoflagellate, and Sporozoan plastid origins and the Eukaryote family tree. *J. Eukaryotic Microbiol.* **46:** 347–366.

Charles, H. and Ishikawa, H. (1999). Physical and genetic map of the genome of *Buchnera*, the primary endosymbiont of the pea aphid *Acyrthosiphon pisum*. *J. Mol. Evol.* **48:** 142–150.

Charles, H., Ishikawa, H., and Nardon, P. (1995). Presence of a protein specific of endocytobiosis (symbionin) in the weevil *Sitophilus*. *C. R. Acad. Sci. Paris* **318:** 35–41.

Charles, H., Heddi, A., Guillaud, J., Nardon, C., and Nardon, P. (1997). A molecular aspect of symbiotic interactions between the weevil *Sitophilus oryzae* and its endosymbiotic bacteria: overexpression of a chaperonin. *Biochem. Biophys. Res. Commun.* **239:** 769–774.

Charles, H., Heddi, A., and Rahbé, Y. (2001). A putative insect intracellular endosymbiont stem clade, within the Enterobacteriaceae, inferred from phylogenetic analysis based on a heterogeneous model of DNA evolution. *C. R. Acad. Sci. Paris* **324:** 489–494.

Dedeine, F., Vavre, F., Fleury, F., Loppin, B., Hochberg, E., and Boulétreau, M. (2001). Removing symbiotic Wolbachia bacteria specifically inhibits oogenesis in a parasitic wasp. *Proc. Natl. Acad. Sci. U.S.A.* **98:** 6247–6252.

Doolittle, R.F. (2000). Searching for the common ancestor. *Res. Microbiol.* **151:** 85–89.

Doolittle, W.F. (1997). Fun with genealogy. *Proc. Natl. Acad. Sci. U.S.A.* **94:** 12751–12753.

Foster, J.S., Apicella, M.A., and McFall-Ngai, M. (2000). *Vibrio fischeri* lipopolysaccharide induces developmental apoptosis, but not complete morphogenesis, of the *Euprymna scolopes* symbiotic light organ. *Dev. Biol.* **226:** 242–254.

Fukatsu, T. (2001). Secondary intracellular symbiotic bacteria in aphids of the genus *Yamatocallis* (Homoptera: Aphididae: Drepanosiphinae). *Appl. Environ. Microbiol.* **67:** 5315–5320.

Fukatsu, T. and Ishikawa, H. (1996). Phylogenetic position of yeast-like symbiont of *Hamiltonaphis styraci* (Homoptera, Aphididae) based on 18S rDNA sequence. *Insect Biochem. Mol. Biol.* **26:** 383–388.

Fukatsu, T. and Nikoh, N. (2000). Endosymbiotic microbia of the bamboo pseudococcid *Antonina crawii* (Insecta, Homoptera). *Appl. Environ. Microbiol.* **66:** 643–650.

Fukatsu, T., Nikoh, N., Kawai, R., and Koga, R. (2000). The secondary endosymbiotic bacterium of the pea aphid *Acyrthosiphon pisum* (Insecta: Homoptera). *Appl. Environ. Microbiol.* **66:** 2748–2758.

Gasnier-Fauchet, F. and Nardon, P. (1986). Comparison of methionine metabolism in symbiotic and aposymbiotic larvae of *Sitophilus oryzae* L. (Coleoptera: Curculionidae) 2. Involvement of the symbiotic bacteria in the oxidation of methionine. *Comp. Biochem. Physiol.* **85B:** 251–254.

Gasnier-Fauchet, F. and Nardon, P. (1987). Comparison of sarcosine and methionine sulfoxide levels in symbiotic larvae of two sibling species, *Sitophilus oryzae* L. and *S. zeamais* mots. (Coleoptera: Curculionidae). *Insect Biochem.* **17:** 17–20.

Gasnier-Fauchet, F., Gharib, A., and Nardon, P. (1986). Comparison of methionine metabolism in symbiotic and aposymbiotic larvae of *Sitophilus oryzae* L. (Coleoptera:Curculionidae) 1. Evidence for a glycine *N*-methyltransferase-like activity in the aposymbiotic larvae. *Comp. Biochem. Physiol.* **85B:** 245–250.

Gil, R., Sabater-Muñoz, B., Latorre, A., Silva, F.J., and Moya, A. (2002). Extreme genome reduction in *Buchnera* spp.: toward the minimal genome needed for symbiotic life. *Proc. Natl. Acad. Sci. U.S.A.* **99:** 4454–4458.

Gray, M.W. (1999). Evolution of organellar genomes. *Curr. Opinion Genet. Dev.* **9:** 678–687.

Grenier, A.M., Nardon, C., and Nardon, P. (1994). The rôle of symbiotes in flight activity of *Sitophilus* weevils. *Entomol. Exp. Appl.* **70:** 201–208.

Grenier, S., Pintureau, B., Heddi, A., Lassablière, F., Jager, C., Louis, C., and Khatchadourian, C. (1998). Successful horizontal transfer of *Wolbachia* symbionts between *Trichogramma* wasps. *Proc. R. Soc. London (B)* **265:** 1441–1445.

Gupta, R.S., Picketts, D.J., and Ahmad, S. (1989). A novel ubiquitous protein chaperonin supports the endosymbiotic origin of mitochondrion and plant chloroplast. *Biochem. Biophys. Res. Commun.* **163:** 780–787.

Heddi, A. and Lefebvre, F. (1990a). Comparative study of the respiratory control (RCR) and ADP/O ratio in symbiotic and aposymbiotic adults of *Sitophilus oryzae* (Coleoptera, Curculionidae). In *Microbiology in Poecilothermes* (R. Lesel, Ed.), pp. 29–32. Elsevier, Paris.

Heddi, A. and Lefebvre, F. (1990b). Energetic metabolism of mitochondria in hybrids of symbiotic and aposymbiotic larvae of *Sitophilus oryzae* (Coleoptera: Curculionidae). In *Endocytobiology IV* (P. Nardon et al., Eds.), pp. 497–500. INRA Press, Paris.

Heddi, A. and Nardon, P. (1993). Mitochondrial DNA expression in symbiotic and aposymbiotic strains of *Sitophilus oryzae*. *J. Stored Prod. Res.* **29:** 243–252.

Heddi, A., Lefebvre, F., and Nardon, P. (1991). The influence of symbiosis on the respiratory control ratio (RCR) and the ADP/O ratio in the adult weevil *Sitophilus oryzae* (Coleoptera, Curculionidae). *Endocytobiosis Cell Res.* **8:** 61–73.

Heddi, A., Lefebvre, F., and Nardon, P. (1993). Effect of endocytobiotic bacteria on mitochondrial enzymatic activities in the weevil *Sitophilus oryzae* (Coleoptera, Curculionidae). *Insect Biochem. Mol. Biol.* **23:** 403–411.

Heddi, A., Charles, H., Khatchadourian, C., Bonnot, G., and Nardon, P. (1998). Molecular characterization of the principal symbiotic bacteria of the weevil *Sitophilus oryzae*: a peculiar G–C content of an endocytobiotic DNA. *J. Mol. Evol.* **47:** 52–61.

Heddi, A., Grenier, A.M., Khatchadourian, C., Charles, H., and Nardon, P. (1999). Four intracellular genomes direct weevil biology: nuclear, mitochondrial, principal endosymbionts, and *Wolbachia*. *Proc. Natl. Acad. Sci. U.S.A.* **96:** 6814–6819.

Heddi, A., Charles, H., and Khatchadourian, C. (2001). Intracellular bacterial symbiosis in the genus *Sitophilus:* the "biological individual" concept revisited. *Res. Microbiol.* **152:** 431–437.

Ishikawa, H. (1985). Symbionin, an aphid endosymbiot-specific protein. II. Diminution of symbionin during post embryonic development of aposymbiotic insects. *Insect Biochem.* **15:** 165–174.

Jeon, K.W. (1983). Integration of bacterial endosymbionts in Amoebae. *Int. Rev. Cytol.* **14:** 29–47.

Jeon, K.W. (1995). Bacterial endosymbiosis in amoebae. *Tr. Cell Biol.* **5:** 137–140.

Jeyaprakash, A. and Hoy, A. (2000). Long PCR improves *Wolbachia* DNA amplification: *wsp* sequences found in 76% of sixty-three arthropod species. *Insect Mol. Biol.* **9:** 393–405.

Jiggins, F.M., Hurst, G.D.D., Dolman, C.E., and Majerus, M.E.N. (2000). High-prevalence male-killing *Wolbachia* in the butterfly *Acraea encedana*. *J. Evol. Biol.* **13:** 495–501.

Kantheti, P., Jayarama, K.S., and Chandra, H.S. (1996). Developmental analysis of a female-specific 16S rRNA gene from mycetome-associated endosymbionts of a mealybug, *Planococcus lilacinus*. *Insect Biochem. Mol. Biol.* **26:** 997–1009.

Lemus, J.D. and McFall-Ngai, M. (2000). Alterations in the proteome of the *Euprymna scolopes* light organ in response to symbiotic *Vibrio fischeri*. *Appl. Environ. Microbiol.* **66:** 4091–4097.

Mansour, K. (1930). Preliminary studies on the bacterial cell mass (accessory cell mass) of *Calandra oryzae*: the rice weevil. *Q. J. Microsc. Sci.* **73:** 421–436.

Mansour, K. (1934). On the so-called symbiotic relationship between Coleopterous insects and intracellular micro-organisms. *Q. J. Microsc. Sci.* **77:** 255–272.

Mansour, K. (1935). On the microorganisms free and the infected *Calandra granaria*. *Bull. Soc. R. Entomol. Egypt* **19:** 290–306.

Margulis, L. (1993a). Origins of species: acquired genomes and individuality. *Biosystems* **31:** 121–125.

Margulis, L. (1993b). *Symbiosis in Cell Evolution,* 2nd ed. Freeman, New York.

Maynard Smith, J. (1989). Generating novelty by symbiosis. *Nature* **341:** 224–225.

McFall-Ngai, M.J. (1998). The development of cooperative associations between animals and bacteria: establishment détente between domains. *Am. Zool.* **38:** 3–18.

Moran, N.A. (1996). Accelerated evolution and Muller's rachet in endosymbiotic bacteria. *Proc. Natl. Acad. Sci. U.S.A.* **93:** 2873–2878.

Munson, M.A., Baumann, P., Clark, M.A., Baumann, L., Moran, N., Voegtlin, D.J., and Campbell, B.C. (1991). Evidence for the establishment of aphid eubacterium endosymbiosis in an ancestor of four aphid families. *J. Bacteriol.* **173:** 6321–6324.

Nardon, P. (1973). Obtention d'une souche asymbiotique chez le charançon *Sitophilus sasakii* Tak: différentes méthodes d'obtention et comparaison avec la souche symbiotique d'origine. *C. R. Acad. Sci. Paris* **277D:** 981–984.

Nardon, P. and Wicker, C. (1981). La symbiose chez le genre *Sitophilus* (Coleoptère Curculionidae). Principaux aspects morphologiques, physiologiques et génétiques. *Ann. Biol.* **4:** 329–373.

Nardon, P. and Grenier, A.M. (1988). Genetical and biochemical interactions between the host and its endosymbiotes in the weevil *Sitophilus* (Coleoptere Curculionidae) and other related species. In *Cell to Cell Signals in Plant, Animal and Microbial Symbiosis* (S. Scannerini et al., Eds.), pp. 255–270. Springer-Verlag, Berlin.

Nardon, P. and Grenier, A.M. (1989). Endocytobiosis in coleoptera: biological, biochemical and genetic aspects. In *Insect Endocytobiosis: Morphology, Physiology, Genetics, Evolution* (W. Schwemmler and G. Gassner, Eds.), pp. 175–215. CRC Press, Boca Raton, FL.

Nardon, P. and Grenier, A. (1991). Serial endosymbiosis theory and weevil evolution: the role of symbiosis. In *Symbiosis As a Source of Evolutionary Innovation* (L. Margulis and R. Fester, Eds.), pp. 153–169. MIT Press, Cambridge, MA.

Nardon, P., Grenier, A.M., and Heddi, A. (1998). Endocytobiote control by the host in the weevil *Sitophilus oryzae*, Coleoptera, Curculionidae. *Symbiosis* **25:** 237–250.

Nyholm, S.V., Stabb, E.V., Ruby, E.G., and McFall-Ngai, M.J. (2000). Establishment of an animal-bacterial association: recruiting symbiotic vibrios from the environment. *Proc. Natl. Acad. Sci. U.S.A.* **97:** 10231–10235.

O'Neill, S.L. and Karr, T.L. (1990). Bidirectional incompatibility between conspecific populations of *Drosophila simulans*. *Nature* **348:** 178–180.

Paillot, A. (1933). L'infection chez les insectes — immunité et symbiose. Imprimerie de Trévoux, Trévoux, France.

Palmer, J.D. (1997). Organelle genomes: going, going, gone! *Science* **275:** 790–791.

Pierantoni, U. (1927). L'organo simbiotico nello sviluppo di *Calandra oryzae*. *Rend Reale Acad. Sci. Fis. Mat. Napoli* **35:** 244–250.

Rigaud, T., Souty-Grosset, C., Raimond, R., Mocquard, J.P., and Juchault, P. (1991). Feminizing endocytobiosis in the terrestrial crustacean *Armadilidium vulgare* Latv: recent acquisition. *Endocytobiosis Cell Res.* **7:** 259–273.

Rowan, R. (1997). Diversity and ecology of zooxanthellae on coral reefs. *J. Phycol.* **34:** 407–417.

Ruby, E.G. (1996). Lessons from cooperative, bacterial-animal associations: the *Vibrio fischeri–Euprymna scolopes* light organ symbiosis. *Annu. Rev. Microbiol.* **50:** 591–624.

Schneider, H. (1956). Morphologische und experimentelle Untersuchungen über die Endosymbiose der Korn und Reiskäfer (*Calandra granaria* und *C. oryzae*). *Z. Morphol. Oekol. Tiere* **44:** 555–625.

Schroder, D., Deppisch, H., Obermayer, M., Krohne, G., Stackebrandt, E., Holldobler, B., Goebel, W., and Gross, R. (1996). Intracellular endosymbiotic bacteria of *Camponotus* species (carpenter ants): systematics, evolution and ultrastructural characterization. *Mol. Microbiol.* **21:** 479–489.

Shigenobu, S., Watanabe, H., Hattori, M., Sakaki, Y., and Ishikawa, H. (2000). Genome sequence of the endocellular bacterial symbiont of aphids *Buchnera* sp. APS. *Nature* **407**: 81–86.

Smith, C.L., Econome, J.G., Schutt, A., Klco, S., and Cantor, C.R. (1987). A physical map of the *Escherichia coli* K12 genome. *Science* **236:** 1448–1453.

Sogin, M.L. (1997). Organelle origins: energy-producing symbionts in early eukaryotes? *Curr. Biol.* **7:** R315–R317.

Stouthamer, R., Breeuwers, J.A.J., Luck, R.F., and Werren, J.H. (1993). Molecular identification of microorganisms associated with parthenogenesis. *Nature* **361:** 247–252.

Sun, L.V., Foster, J.M., Tzertzinis, G., Ono, M., Bandi, C., Slatko, B.E., and O'Neill, S.L. (2001). Determination of *Wolbachia* genome size by pulsed-field gel electrophoresis. *J. Bacteriol.* **183:** 2219–2225.

Untermann, B.M., Baumann, P., and McLean, D.L. (1989). Pea aphid symbiont relationships established by analysis of 16S rRNA. *J. Bacteriol.* **171:** 2970–2974.

Visick, K.L., Foster, J., Doino, J., McFall-Ngai, M., and Ruby, E. (2000). *Vibrio fischeri lux* genes play an important role in colonization and development of the host light organ. *J. Bacteriol.* **182:** 4578–4586.

von Dohlen, C.D., Kohler, S., Alsop, S.T., and McManus, W.R. (2001). Mealybug β-proteobacterial endosymbionts contain γ-proteobacterial symbionts. *Nature* **412:** 433–436.

Werren, J.H. (1997). Biology of *Wolbachia. Annu. Rev. Entomol.* **42:** 587–609.

Wicker, C. (1983). Differential vitamin and choline requirements of symbiotic and aposymbiotic *S. oryzae* (Coleoptera: Curculionidae). *Comp. Biochem. Physiol.* **76A:** 177–182.

Wicker, C. and Nardon, P. (1982). Development responses of symbiotic and aposymbiotic weevil *Sitophilus oryzae* L. (Coleoptera, Curculionidae) to a diet supplemented with aromatic amino acids. *J. Insect Physiol.* **28:** 1021–1024.

Wicker, C., Guillaud, J., and Bonnot, G. (1985). Comparative composition of free, peptide and protein amino acids in symbiotic and aposymbiotic *Sitophilus oryzae* (Coleoptera, Curculionidae). *Insect Biochem.* **15:** 537–541.

6 *Rhodnius prolixus* and Its Symbiont, *Rhodococcus rhodnii*: A Model for Paratransgenic Control of Disease Transmission

Ravi V. Durvasula, Ranjini K. Sundaram, Celia Cordon-Rosales, Pamela Pennington, and C. Ben Beard

CONTENTS

INTRODUCTION

Despite many advances in vaccines and public-health measures, insect-borne diseases remain a leading cause of human illness throughout the world. It is estimated that between 300 and 500 million cases of malaria alone occur annually, with 2.3 billion humans living in areas of malaria risk (Gratz, 1999). Visceral leishmaniasis, transmitted by phlebotomine sandflies, has undergone resurgence in eastern regions of India, with spread into neighboring areas of Nepal. Epidemics of dengue and dengue hemorrhagic fever continue in Latin America and Southeast Asia, linked closely to the introduction and spread of the mosquito vector, *Aedes aegypti*. Whereas many of these established diseases remain largely unchecked around the world, novel or emerging arthropod-borne diseases have further complicated the situation. Tick-borne diseases such as Lyme disease and Ehrlichiosis and mosquito-borne illnesses such as West Nile disease have emerged in recent years in the United States, illustrating the global impact of vector-borne diseases.

0-8493-1286-8/03/$0.00+$1.50

Despite extensive study of disease-transmitting arthropods and pathogens, few methods are available for preventing most vector-borne diseases. Pending development of effective vaccines and improvement in socioeconomic conditions for the majority of the populations affected by these diseases, strategies for disease control rely principally on chemical insecticides that eliminate or reduce numbers of vector insects. Historically, campaigns such as the DDT spraying efforts in the Indian subcontinent have yielded spectacular short-term results. Several ongoing vector-control programs, such as the Southern Cone Initiative for control of Chagas' disease, have been quite successful in decreasing transmission of arthropod-borne diseases to humans (Schofield and Dujardin, 1997; Schofield and Dias, 1999). However, long-term use of chemical insecticides can be problematic. Insecticides cause environmental toxicity, and many classes of these chemicals harm humans. Insects have evolved resistance to many of these agents (World Health Organization, 1992), and insecticide failure is common. Furthermore, insecticide programs are expensive and difficult to sustain over prolonged periods of time.

TRANSGENIC MODIFICATION

Genetic manipulation of disease-transmitting insects is a potential alternative to strategies aimed at elimination of vector populations. Expression products of foreign genes that block or eliminate the ability of the arthropod to transmit pathogens could provide a valuable tool in the control of several vector-borne diseases. In broad terms, there are two approaches to genetic transformation of arthropod vectors. The first involves direct genome transformation via inserted genetic material. This is accomplished via a variety of mobile elements that are reviewed elsewhere (Coates et al., 1998; Jasinskiene et al., 1998; Catteruccia et al., 2000). The second approach involves expression of foreign genes using engineered symbiotic or commensal bacteria that reside within the disease-transmitting insect. This "Trojan Horse" method is termed paratransgenesis and will be discussed in detail in this chapter.

PARATRANSGENESIS

Paratransgenesis is a novel approach to controlling arthropod-borne diseases and is derived from naturally occurring interactions between vectors, pathogens, and populations of symbiotic or commensal bacteria that reside within the vectors. Many disease-transmitting insects harbor populations of bacteria that serve important physiologic roles. Obligate blood-sucking insects, such as bed bugs, sucking lice, and triatomines, employ symbiotic bacteria to provide nutrients that may not be found in restricted diets. A wide variety of insects harbor specialized bacteria of the genus *Wolbachia* that mediate reproductive fitness. Whereas the presence of such microbes within the insect vector is a foundation for paratransgenic transformation, the following requirements exist for a successful paratransgenic strategy.

1. A population of microbes that is amenable to culture and genetic manipulation *in vitro* must exist within a disease-transmitting vector.
2. Facile methods for isolating and transforming the symbiotic bacteria must be present.
3. Transformation of the symbiotic bacteria must result in stable mutants, without loss of reproductive fitness of the symbiont.
4. Genetic manipulation of the bacteria should not affect their symbiotic functions in the host vector. Paratransgenic vectors that harbor altered symbionts should maintain growth and reproductive rates comparable to wild-type vectors.
5. Expression products of the genetically altered bacteria must target pathogens and disrupt transmission cycles in the host vector. Therefore, localization of transmission-blocking molecules to appropriate sites of pathogen development is required.

6. Genetic manipulation of symbiotic bacteria should not render them virulent, either to the target vector or other organisms in the environment. Furthermore, bacteria chosen as gene-delivery vehicles must not be pathogens themselves.
7. A robust method must exist for delivery of genetically altered symbionts within field populations of the disease-transmitting arthropods. Ideally, this method would mimic naturally occurring mechanisms for symbiont dispersal and would not result in increased populations of the disease-transmitting vectors. Strategies of foreign gene dispersal should target appropriate vectors selectively and minimize nontarget uptake and retention of recombinant genetic material. Methods for assessing environmental spread of foreign DNA and prediction of limits of spread must exist as part of a paratransgenic approach.

In summary, paratransgenic manipulation of disease-transmitting vectors requires critical evaluation of the natural biology of vector populations. Careful assessment of patterns of symbiont transmission, population dynamics, reproductive strategies, and vector interactions with the environment is the cornerstone for the development of a paratransgenic approach that mimics natural phenomena.

THE TRIATOMINE BUG *RHODNIUS PROLIXUS*: A PARADIGM FOR PARATRANSGENESIS

Though the paratransgenic strategy may be applied to a variety of disease-transmitting arthropods, it has been most successfully developed in the reduviid bug, *Rhodnius prolixus*. A member of the Triatominae, this bug is widely distributed in Central America and northern regions of South America. It is an important vector of *Trypanosoma cruzi*, the causative agent of Chagas' disease. Over 50,000 people die annually of this disease, and nearly 90 million individuals are at risk for this disease due to exposure to the reduviid vectors. Neither a cure nor a vaccine exists for Chagas' disease. Control of transmission has hinged largely on insecticide campaigns aimed at elimination of reduviid vectors. These have been quite successful in areas of the Southern Cone countries, but issues of cost, maintenance, environmental toxicity, and vector resistance remain.

The symbiotic relationships between triatomine bugs and soil-associated actinomycete bacteria are well described. *R. prolixus* maintains a close relationship with the actinomycete *Rhodococcus rhodnii*. The bacterium aids in the processing of B-complex vitamins found in the diet of *Rhodnius prolixus* and aids in sexual maturation of the bug. *R. prolixus*, which are reared from surface-sterilized eggs in sterile laboratory chambers, fail to develop beyond the second molt. Delivery of the bacteria to aposymbiotic nymphs via blood meals across a membrane restores normal growth and fecundity of the nymphs.

Rhodococcus rhodnii exists as an extracellular, intraluminal symbiont in the hindgut of *Rhodnius prolixus* where it may reach concentrations of 10^8 colony-forming units (CFU)/ml. It is juxtaposed in this environment with the infective trypomastigote form of *T. cruzi* (Figure 6.1). The amenability of *Rhodococcus rhodnii* to isolation and genetic manipulation and its proximity to the pathogen *T. cruzi* in the bug gut satisfies several requirements for paratransgenic manipulation of the host reduviid bug.

PARATRANSGENIC MANIPULATION OF *RHODNIUS PROLIXUS*

Isolation and culture of *Rhodococcus rhodnii* from both laboratory and field-caught *Rhodnius prolixus* is quite simple. Fecal extracts of bug hindgut contents are suspended in physiologic saline and plated aerobically on brain heart infusion agar at 28°C. *Rhodococcus rhodnii* appear within 3 to 5 days as discrete, rough colonies of a pink-white hue. Older colonies exhibit heaped or "volcanic" growth patterns characteristic of several actinomycetes. In liquid BHI medium under conditions of vigorous agitation (200 to 225 rpm), *R. rhodnii* reach concentrations of 10^8 to 10^9 CFU/ml with a generation time of approximately 18 h.

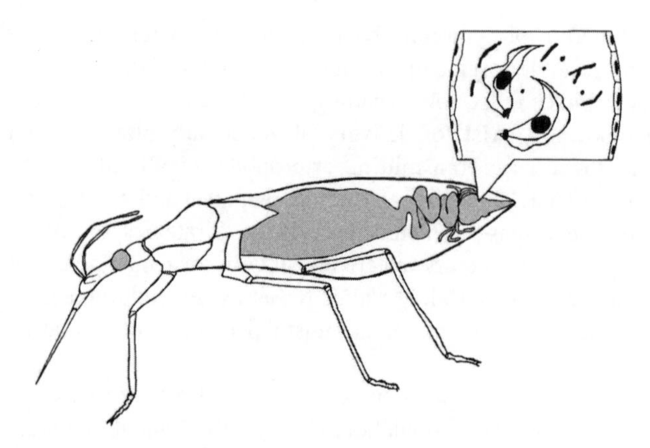

FIGURE 6.1 The actinomycete symbiont *Rhodoccocus rhodnii* lives in the gut of *Rhodnius prolixus*, in direct proximity to the Chagas' disease agent *Trypanosoma cruzi*. [From Beard, C.B., Durvasula, R.V., and Richards, F.F. (2000). In *Insect Transgenesis: Methods and Applications* (A.M. Handler and A.A. James, Eds.), pp. 289–303. CRC Press, Boca Raton, FL. With permission.]

Methods for genetic manipulation of *R. rhodnii* are well described (Beard et al., 1993; Beard and Aksoy, 1997). We have developed a series of shuttle plasmids that maintain low copy numbers in *R. rhodnii*. In initial studies, *R. rhodnii* transformed to express resistance to the antibiotic thiostrepton were introduced to first-instar aposymbiotic *R. prolixus*. These bacteria were maintained in the host bugs throughout sexual maturation for a period of 6.5 months (Beard et al., 1992). Assay of the host *R. prolixus* revealed thiostrepton-resistant *Rhodococcus rhodnii* at all developmental stages. No adverse effects of the genetically altered bacteria on growth and fecundity of the bugs were noted. This study established the principle of paratransgenic expression of foreign genetic material in an arthropod vector.

CECROPIN A EXPRESSION IN PARATRANSGENIC *RHODNIUS PROLIXUS*

The search for antitrypanosomal molecules that could be used in a paratransgenic strategy led to studies involving the immune peptide L-cecropin A. Cecropin A, isolated from the moth *Hyalophora cecropia*, is a member of a family of nonspecific peptides present in insects and certain vertebrates. It is a pore-forming peptide that has a broad spectrum of antibacterial activity and plays an important role in insect innate immunity (Boman, 1991). We demonstrated paracidal activity of cecropin A against several strains of *T. cruzi*, with virtually no activity against *Rhodococcus rhodnii* (Table 6.1).

We then cloned cDNA for cecropin A in the shuttle plasmid pRrThioCec (Figure 6.2) and established a line of cecropin A–producing *R. rhodnii* (Durvasula et al., 1997).

An experimental colony of aposymbiotic *Rhodnius prolixus* was fed cecropin-producing *Rhodococcus rhodnii* and challenged with strain DM28 *T. cruzi*. A control group of *Rhodnius prolixus* was fed wild-type *Rhodococcus rhodnii* and subjected to the same *T. cruzi* challenge. These studies revealed clearance of *T. cruzi* in 65% of experimental group *Rhodnius prolixus*, with a 2 to 3 log reduction of parasite count in the remaining 35%. Control group *R. prolixus* supported *T. cruzi* at concentrations of 10^5 to 10^6 parasites/bug. Results shown in Figure 6.3 are from the first experiment; several repeat trials involving 100 insects in each group confirmed our initial findings.

The transformed *Rhodococcus rhodnii* in these studies served normal symbiotic functions in the host bugs. Growth rates and fecundity of paratransgenic *Rhodnius prolixus* equaled those of wild-type bugs. Furthermore, expression of cecropin A and resistance to thiostrepton persisted throughout the 7-month duration of the study in the absence of antibiotic selection. The pRrThioCec

TABLE 6.1
Concentrations of Synthetic Cecropin A Lethal for *Escherichia coli*, the *Rhodococcus rhodnii* Symbiont, and Three Strains of *Trypanosoma cruzi*

Strain′	Concentration
E. coli	23 μ*M*[a]
DH5αMCR	
R. rhodnii	500 μ*M*[a]
ATCC 35071	
T. cruzi	240 μ*M*[b]
Gainesville	
DM28	140 μ*M*[b]
"Y"	150 μ*M*[b]

Note: The DM28 strain used in these *in vivo* experiments is a common human Chagas pathogen in Brazil.

[a] Values for *E. coli* and *R. rhodnii* are minimum bactericidal concentrations (MBC).
[b] The values for *T. cruzi* (epimastigote forms derived from cell culture) are concentrations needed to kill 100% of parasites at 24 h of incubation (LC).

Source: Data from Durvasula, R.V., Gumbs, A., Panackal, A., Kruglov, O., Aksoy, S., Merrifield, R.B., Richards, F.F., and Beard, C.B. (1997). *Proc. Natl. Acad. Sci. U.S.A.* **94:** 3274–3278.

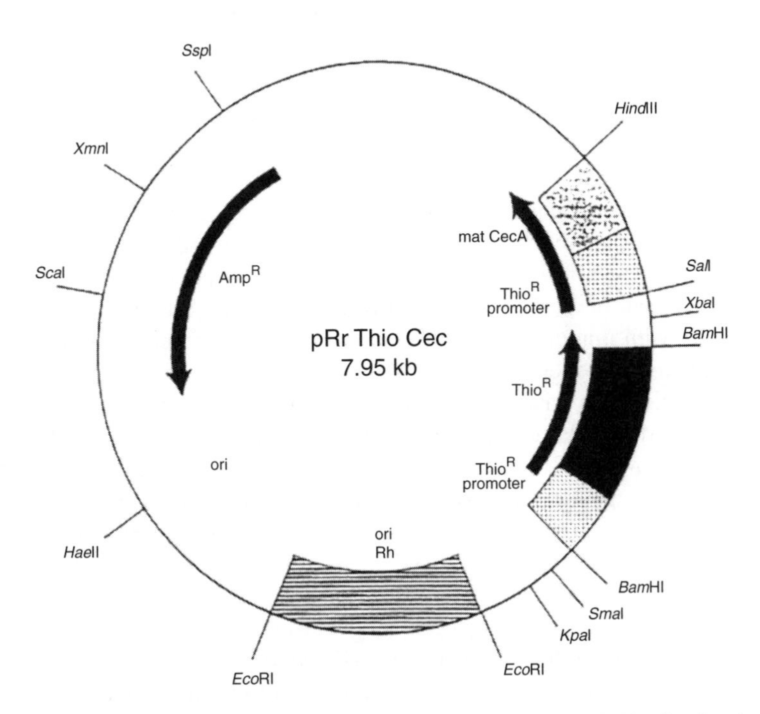

FIGURE 6.2 The shuttle plasmid pRrThioCec. [Adapted from Durvasula, R.V., Gumbs, A., Panackal, A., Kruglov, O., Aksoy, S., Merrifield, R.B., Richards, F.F., and Beard, C.B. (1997). *Proc. Natl. Acad. Sci. U.S.A.* **94:** 3274–3278.]

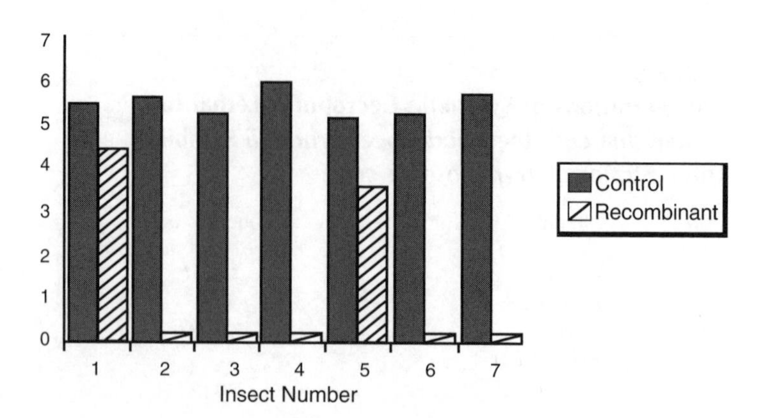

FIGURE 6.3 Number of *T. cruzi* in the hind gut of *Rhodnius prolixus* in control (black bars), *R. prolixus* carrying native symbiotic *Rhodococcus rhodnii* in the hind gut, and (hatched bars) recombinant group where *Rhodnius prolixus* carry a genetically modified *Rhodococcus rhodnii,* which expresses the gene for cecropin A peptide. Unstained metacyclic trypomastigotes were counted using a Neubauer hemocytometer, and the count number is expressed as $\log_{10}$. All values are the mean of four measurements. No trypanasomes were seen in recombinant insects 2, 3, 4, 6, and 7, and the hatched bars in these columns indicate that counts were performed. [Adapted from Durvasula, R.V., Gumbs, A., Panackal, A., Kruglov, O., Aksoy, S., Merrifield, R.B., Richards, F.F., and Beard, C.B. (1997). *Proc. Natl. Acad. Sci. U.S.A.* **94:** 3274–3278.]

plasmid was quite stable in the absence of antibiotic selection, with a 0.5% loss per generation of *Rhodococcus rhodnii*, under *in vitro* continuous culture conditions. In the gut of *Rhodnius prolixus*, where concentrations of the bacteria increase rapidly only after blood meals and are stable during interim periods, plasmid stability appears to be much greater.

These studies were the first demonstration of paratransgenic expression of foreign genes in a disease-transmitting arthropod with reduction in vector competence.

ANTIBODY EXPRESSION

Cecropin A and related immune peptides hold promise as effector molecules in a paratransgenic strategy. However, these compounds lack specificity for *T. cruzi* and exhibit bactericidal activity against a variety of Gram-negative bacteria that may be found in reduviid bugs under field conditions. Furthermore, resistance to such peptides among target populations of *T. cruzi* is likely to occur over time. The need for another class of effector molecules that could be expressed concurrently or sequentially with immune peptides led to work with single-chain antibodies.

The use of antibodies for therapeutic and diagnostic purposes has gained prominence in the past decade. Immunoglobulins have excellent target specificity, but their clinical applications are limited because of short circulating half-lives and their need for associated effector functions. Biomolecular engineering technology has provided the tools to design high-affinity-based reagents for immunotherapeutic applications. Several diseases, such as rheumatoid arthritis, Crohn's disease, and cancer, are now being approached by antibody treatments. In recent years, the Food and Drug Administration has approved the first engineered antibodies for cancer therapy (Rituxan for non-Hodgkins lymphoma, Herceptin for breast cancer). These immunotherapeutic antibodies are designed to include the Fc domain for prolonged serum half-life and complement mediated effects. Engineered antibodies may be designed to have various structural changes that include reduction in size to single-chain Fvs, dissection into minimal binding fragments such as Vh domains, and rebuilding of scFvs into multivalent, high-avidity oligomeric scFvs (Huse et al., 1989; Holliger et al., 1993). In recent years, through phage-display technology, it has become possible to clone single-chain antibody-encoding genes that can be expressed in a variety of

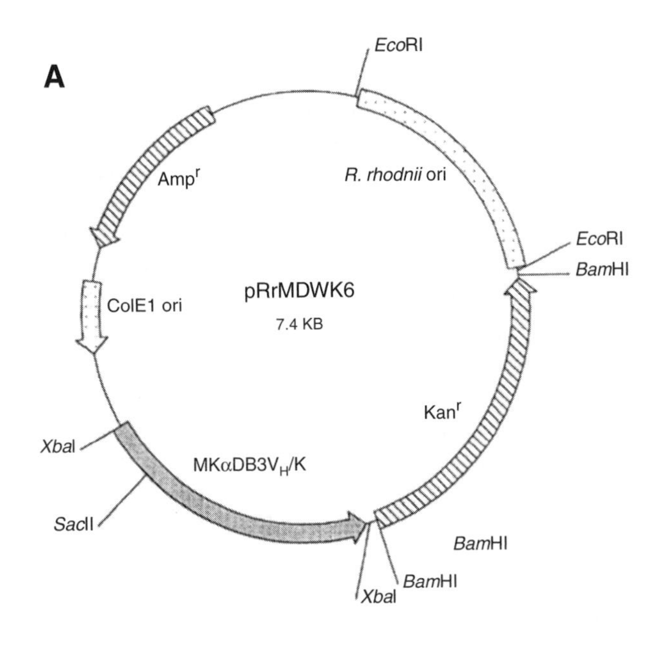

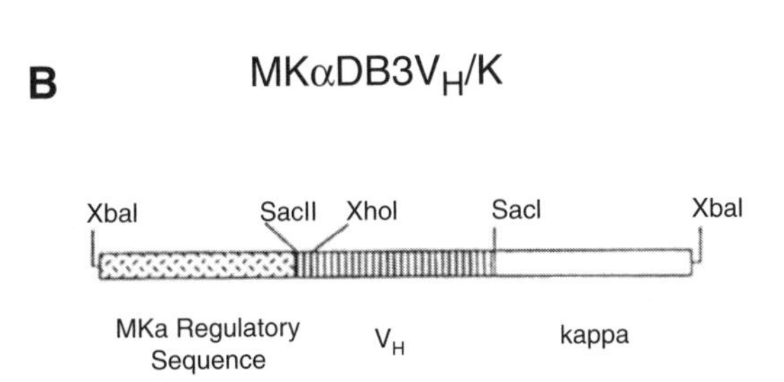

FIGURE 6.4 (A) The shuttle plasmid pRrMDWK6. (B) MKαDB3WV$_H$K: rDB3 expression/secretion cassette. [From Durvasula, R.V., Gumbs, A., Panackal, A., Kruglov, O., Taneja, J., Kang, A.S., Cordon-Rosales, C., Richards, F.F., Whitham, R.G., and Beard, C.B. (1999a). *Med. Vet. Entomol.* **13:** 115–119. With permission.]

systems. These antibody fragments are composed of variable regions of heavy chains linked to kappa light chains (VH-Kappa) and are selected from combinatorial libraries of heavy- and light-chain genes. Selective panning of VH-Kappa fragments in a phage-display system yields molecules with target specificity and affinity that approximate parent IgG.

Insects lack immunoglobulin-mediated defenses and secondary cascades such as complements. Hence, a strategy to express recombinant single-chain antibody fragments that directly disrupt transmission of a pathogen by an arthropod is highly desirable.

In 1999, we first described the expression of a functional single-chain antibody in an insect (Durvasula et al., 1999a). *Rhodoccus rhodnii* was transformed using the expression plasmid pRrMDWK6 to secrete the single-chain antibody rDB3 (Figure 6.4), a murine VH-Kappa fragment that binds progesterone and served as a marker antibody in our system.

Expression and secretion of rDB3 was under the control of the heterologous element Mkα derived from *Mycobacterium kansasii*. *R. rhodnii* transformed with pRrMDWK6 were introduced via blood meal to aposymbiotic first-instar nymphs of *Rhodnius prolixus*. The recombinant DB3 antibody fragment was synthesized and secreted into the insect gut lumen by the engineered symbiont in a stable fashion for the 6-month duration of the study. Gut contents of the experimental-group bugs tested positive by ELISA for progesterone binding activity. This study established that paratransgenic transformation of *R. prolixus* could result in constitutive and stable expression of a functional single-chain antibody. It provided the basis for expression of single-chain antibodies that target surface antigens of *T. cruzi* to inactivate the trypanosome within the gut of the reduviid bug.

T. cruzi has a life cycle consisting of four major stages. The parasite multiplies as an epimastigote within the gut of the insect host and differentiates into the trypomastigote in the hindgut. The trypomastigotes are metacyclic, nondividing invasive forms that are transmitted to the mammalian host by fecal droplets released by the bug during a blood meal. In the mammalian host, *T. cruzi* invades the macrophages and multiplies as an amastigote that later differentiates into the trypomastigote form. Many surface-specific antigens of *T. cruzi* have been cloned and sequenced. Gp72 is a 72-kd glycoprotein that was first identified on epimastigote cells using a specific monoclonal antibody WIC29.26 (Haynes et al., 1996). The WIC29.26 epitope was characterized on epimastigotes and insect-derived metacyclic trypomastigotes and to a lesser extent on trypomastigotes that were derived from a culture of epimastigotes. The WIC 29.26 epitope is not present in the amastigote form of *T. cruzi*.

Gp72 null mutant strains of *T. cruzi* are morphologically distinct from wild-type parasites, predominantly due to detachment of the flagellum from the flagellar pocket. The exact function of Gp72 is unclear, but it has been implicated to have a role in maintenance of morphology, to function as an acceptor for complement factor C3, and to help in differentiation of *T. cruzi*. Gp72 null mutant strains of *T. cruzi* are grossly misshapen, have impaired motility, and, in a *Triatoma infestans* model, are incapable of maturation in the bug gut (Nozaki and Cross, 1994).

Studies involving the monoclonal anti-Gp72 antibody WIC29.26 are currently under way. *Trypanosoma cruzi* strains “Y” and “Telhuen” that have been coated with WIC29.26 (Figure 6.5) have been introduced into *R. prolixus*. Survival and maturation of the antibody-coated trypanosomes will be determined. Ultimately, Vh-Kappa fragments that target the Gp72 epitope and, potentially, other key surface epitopes of *T. cruzi* will be expressed in the gut of *R. prolixus* by engineered *Rhodococcus rhodnii*. Antibody-mediated disruption of the *T. cruzi* life cycle coupled with lytic functions of immune peptides such as cecropin A may provide the necessary molecular armamentarium for a successful paratransgenic strategy.

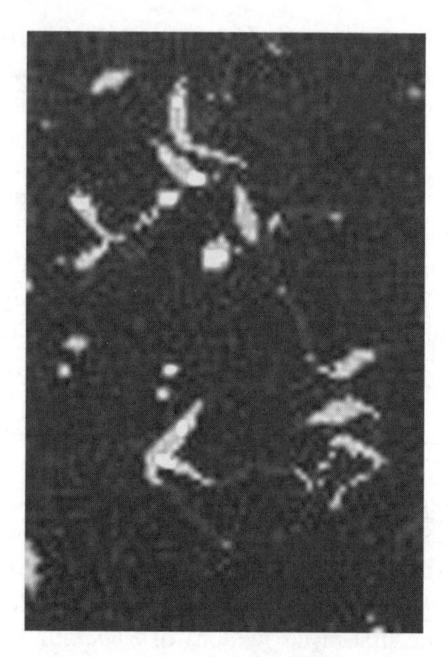

FIGURE 6.5 (Color figure follows p. 206.) Immunofluorescence stain of “Y” strain *Trypanosoma cruzi* coated with the anti-Gp72 antibody, WIC 29.26 at 400× magnification.

A SPREADING STRATEGY FOR POTENTIAL FIELD APPLICATION

Field use of the paratransgenic approach to control of Chagas' disease requires a strategy for delivery and spread of transgenic *R. rhodnii* among natural populations of *Rhodnius prolixus*. Modes of delivery of engineered bacteria to vector populations must simulate naturally occurring methods of symbiont spread. An optimal spreading strategy would involve minimum numbers of genetically altered organisms with maximum uptake and activity of the transgenic material. The transgenic symbionts should be competitive with wild-type flora to establish predominant infections in the target vector populations. Furthermore, spread of foreign genes via engineered bacteria should occur only in targeted populations of vectors, with limited uptake by nontarget arthropods or other environmental bacteria. In the following sections we review our ongoing and proposed studies aimed at development of a robust spreading mechanism for engineered *Rhodococcus rhodnii*.

CRUZIGARD: A SUBSTRATE FOR GENE DELIVERY

Spread of *R. rhodnii* within populations of *Rhodnius prolixus* occurs via coprophagy, the ingestion of fecal deposits. Newly emerging first-instar nymphs are transiently aposymbiotic (devoid of symbiotic bacteria). Probing of fecal droplets deposited by adult bugs either on the eggshell or in the immediate environment permits first instar nymphs to ingest *R. rhodnii* and establish gut infection. Nymphs reared in sterile chambers from surface-sterilized eggs fail to mature beyond the second instar stage. Delivery of *R. rhodnii* via membrane feeder to aposymbiotic nymphs by the second molt permits normal development and sexual maturation.

We have developed a simulated fecal paste — termed CRUZIGARD — that permits delivery of *R. rhodnii* to colonies of *Rhodnius prolixus*. CRUZIGARD is comprised of an inert guar gum matrix impregnated with 10^8 CFU/ml of *Rhodococcus rhodnii*. India ink is added to achieve the black color of natural *Rhodnius* feces. Survival of *R. rhodnii* in this form is between 6 and 8 weeks.

Initial studies of CRUZIGARD involved aposymbiotic colonies of *Rhodnius prolixus*. Aposymbiotic first-instar nymphs exposed to CRUZIGARD containing genetically altered *Rhodococcus rhodnii* reached sexual maturity at a rate comparable to similar nymphs exposed to natural feces. Sampling of adult bugs in this study revealed antibiotic resistance markers indicative of genetically altered *R. rhodnii*. These initial studies confirmed that CRUZIGARD could approximate natural mechanisms for delivery of genetically altered bacteria without causing excess toxicity to the bugs. However, these studies were performed in very confined spaces under sterile conditions. Field populations of *Rhodnius prolixus* and other reduviid bugs may harbor over 20 different types of bacteria and fungi (P. Pennington, unpublished data). Competition from these microbes is an important determinant of the ultimate success of the CRUZIGARD strategy.

To address issues of bacterial competition, a study was conducted at the Medical Entomology Research and Training Unit/Guatemala (MERTU/G) (Durvasula et al., 1999b). In this trial, our goal was to assess the efficacy of CRUZIGARD under simulated field conditions. Insects and cage materials for this study were collected from the Chagas' disease–endemic region of Olopa in Eastern Guatemala. Lucite cages ($0.6 \times 0.6 \times 0.6$ m) were constructed and lined with dirt. Panels of thatch and adobe measuring 20×20 cm were made and impregnated with CRUZIGARD to cover 50% of the surface area (Figure 6.6). *Rhodococcus rhodnii* transformed to express DB3 and kanamycin resistance were used in the experimental cage; wild-type *R. rhodnii* were applied to the control cage. Eight adult male and female *Rhodnius prolixus* were placed in each cage until egg laying was complete. Fecal droplets deposited by these adults remained in the cages, providing bacterial competition to the genetically altered bacteria in CRUZIGARD.

Sampling of instar nymphs at the fourth, fifth, and adult stages revealed that approximately 50% of bugs carried kanamycin-resistant *Rhodococcus rhodnii*, confirmed by both culture and polymerase chain reaction. Other environmental microbes, such as *Staphylococcus* spp., *Pseudomonas* spp., and Candida, were isolated from the bugs. Bugs that harbored recombinant *R. rhodnii*

FIGURE 6.6 Thatch panels used in the stimulated field study at MERTU/G. Guatemala. Thatch panels constructed of materials from the Chagas' disease–endemic region of Olopa, Guatemala. CRUZIGARD droplets are visible as black spots. [Adapted from Durvasula, R.V., Kroger, A., Goodwin, M., Panackal, A., Kruglov, O., Taneja, J., Gumbs, A., Richards, F.F., Beard, C.B., and Cordon-Rosales, C. (1999b). *Ann. Entomol. Soc. Am.* **92:** 937–943.]

were further tested by replica plating of gut contents. Between 89 and 96% of total CFUs of *R. rhodnii* in these bugs were comprised of genetically altered bacteria.

This study demonstrated the efficacy of CRUZIGARD under conditions of microbial competition. Though genetically altered *R. rhodnii* were present in only 50% of the target bugs, they comprised nearly 100% of the total CFUs in the bug, when infection was successful. Mixed infections, involving transformed and wild-type *R. rhodnii* or other environmental microbes, were not found. Previously, in experiments involving sequential infection of aposymbiotic first-instar nymphs of *Rhodnius prolixus* with wild-type and recombinant bacteria, we determined that the initial infection established as the predominant gut organism throughout maturation of the bug. Data from the Guatemala trial suggest that in 50% of the target bugs, recombinant *Rhodococcus rhodnii* were successfully established as the predominant gut bacteria. Perhaps a larger number of first-instar larvae were exposed to the transgenic bacteria. This, however, cannot be confirmed with this study. Nevertheless, this study provided evidence that the CRUZIGARD mechanism of gene delivery could be used under conditions that simulated the field environment.

FOREIGN GENE SPREAD UNDER SPATIALLY ACCURATE CONDITIONS

Initial studies with CRUZIGARD were conducted in small laboratory chambers that maximized the likelihood of contact between emerging nymphs of *Rhodnius prolixus* and recombinant bacteria. To test efficacy of CRUZIGARD under conditions that approximate the dimensions of a real hut, we are currently conducting trials in a greenhouse on the Centers for Disease Control campus in Chamblee, Georgia. An environmentally controlled greenhouse has been fitted for arthropod containment with multiple screen barriers (Figure 6.7). A plywood "hut" measuring $5' \times 5' \times 8'$ has been constructed. The "thatch" roof of the hut has been treated with CRUZIGARD impregnated with *Rhodococcus rhodnii* expressing the foreign gene β-galactosidase. Adult *Rhodnius prolixus* carrying wild-type *Rhodococcus rhodnii* have been introduced to the hut, and we are assaying F1 progeny for spread of recombinant *R. rhodnii.* These studies will provide valuable information about the efficacy of CRUZIGARD under spatially accurate conditions, ultimate dosing of transgenic bacteria in the CRUZIGARD formulation, and the possible role of CRUZIGARD baited with attractants in a gene-delivery system.

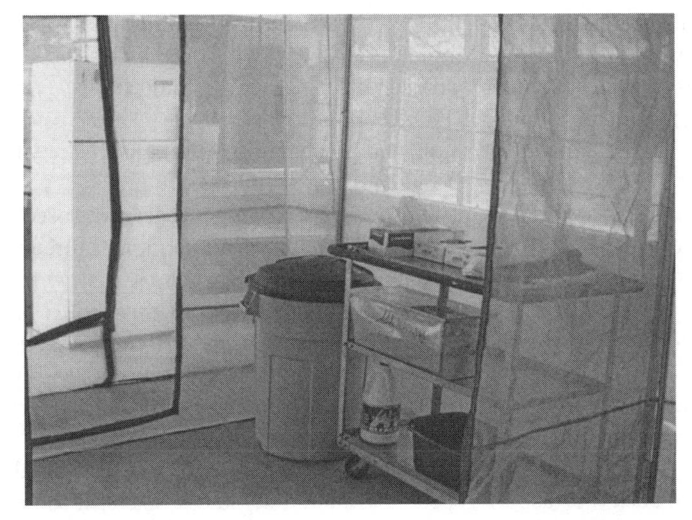

FIGURE 6.7 Containment chamber in greenhouse facility, Centers for Disease Control and Prevention, Chamblee campus, Georgia.

ENVIRONMENTAL IMPLICATIONS OF A PARATRANSGENIC STRATEGY

The symbiont/vector association of *R. rhodnii* and *Rhodnius prolixus* satisfies several requirements of a paratransgenic strategy:

1. Symbiotic bacteria that are closely associated with a disease-transmitting vector can be readily isolated *in vitro.*
2. Methods exist for genetic transformation of the symbiont.
3. The genetically transformed symbiont is stable, with minimal loss of the foreign genes both *in vitro* and *in vivo.*
4. Genetic transformation of the symbiont does not attenuate its role in the host arthropod. Sexual maturation and fecundity of the host is not affected by genetic manipulation of the symbiont.
5. Delivery of antiparasite molecules by the genetically altered symbiont *in vivo* results in reduction or clearance of the parasite. Appropriate targeting of the parasite effects a significant reduction in host vector competence.
6. A method exists for field delivery of genetically altered symbionts. This method mimics naturally occurring modes of symbiont dispersal in the target vector.
7. The approach may be customized to other symbiont/reduviid vector relationships. Application of this strategy to the Chagas' disease vectors *Triatoma infestans* and *T. dimidata* is under way.

Field use of the paratransgenic approach, with environmental release of genetically altered bacteria, remains a future prospect. The environmental impact of such a release must be considered carefully, with attention to limits of foreign-gene spread. Issues of regulatory control and policy pertaining to release of genetically modified organisms are reviewed elsewhere (Beard et al., 2002). Ongoing studies address issues of nontarget spread of foreign genetic material introduced via engineered bacterial symbionts.

The paratransgenic approach is predicated on the specific relationship between the target arthropod and its bacterial symbiont. Methods of dispersal of foreign genetic material to field vectors must ensure that nontarget uptake of recombinant DNA is minimal. We have designed the model of *R. prolixus* and its proven symbiont, *Rhodococcus rhodnii,* with these caveats in mind.

Two scenarios in which nontarget spread of foreign DNA could be hastened are (1) uptake, retention, and propagation of foreign DNA by arthropods that have faster generation times than *Rhodnius prolixus* and (2) uptake, retention, and propagation of foreign DNA by bacteria that have faster generation times than *Rhodococcus rhodnii.*

We are currently assessing uptake of transgenes by nontarget species exposed to either CRUZIGARD or genetically altered *R. rhodnii.* Testing for uptake and retention of foreign genetic material by grazing insects, such as cockroaches, ants, and crickets, which coexist with reduviid bugs, is under way in Guatemala. Laboratory tests of gene transfer from *R. rhodnii* to other environmental microbes, including nonactinomycetes such as coliforms, *Staphylococcus* spp., and *Candida* spp., are ongoing. Furthermore, a predictive mathematical model that incorporates several elements of gene spread, including stability, vector migration, physical dispersal, and nontarget uptake, is being constructed (R.V. Durvasula and S. Matthews, unpublished data).

These studies will provide information that is required for possible field applications of paratransgenesis. Indeed, the use of genetically altered symbiotic bacteria to battle vector-borne diseases is a potentially powerful tool. We are exploring application of this approach to the Old World vectors of leishmaniasis, *Phlebotomus papatasi*, and *P. argentipes* (R.V. Durvasula and K. Ghosh, unpublished data). We believe that this approach may someday be used as part of an integrated strategy for control of Chagas' disease.

SUMMARY

Paratransgenesis is a novel approach to controlling vector-borne infectious diseases. It involves genetic manipulation of commensal or symbiotic bacteria that reside within certain arthropod hosts. The bacteria are transformed to export molecules that interrupt transmission of a target pathogen. *Rhodnius prolixus*, an important reduviid bug vector of Chagas' disease, harbors the actinomycete symbiont *Rhodococcus rhodnii*. This symbiont has been transformed to export Cecropin A, which is lethal to the Chagas' disease parasite *Trypanosoma cruzi*, and marker single-chain antibodies that form the basis of antibody-mediated disruption of *T. cruzi* transmission. A spreading strategy for delivery of genetically altered *R. rhodnii* to field populations of *Rhodnius prolixus* has been devised. Safety and efficacy of this approach are being tested under simulated field conditions.

REFERENCES

Beard, C.B. and Aksoy, S. (1997). Genetic manipulation of insect symbionts. In *Molecular Biology of Insect Disease Vectors: A Methods Manual* (J.M. Crampton, C.B. Beard, and C. Louis, Eds.), pp. 555–560. Chapman & Hall, London.

Beard, C.B., Mason, P.W., Aksoy, S., Tesh, R.B., and Richards, F.F. (1992). Transformation of an insect symbiont and expression of a foreign gene in the Chagas' disease vector *Rhodnius prolixus*. *Am. J. Trop. Med. Hyg.* **46:** 195–200.

Beard, C.B., O'Neill, S.L., Tesh, R.B., Richards, F.F., and Aksoy, S. (1993). Modification of arthropod vector competence via symbiotic bacteria. *Parasitol. Today* **9:** 179–183.

Beard, C.B., Durvasula, R.V., and Richards, F.R. (2000). Bacterial symbiont transformation in Chagas disease vectors. In *Insect Transgenesis: Methods and Applications* (A.M. Handler and A.A. James, Eds.), pp. 289–303. CRC Press, Boca Raton, FL.

Beard, C.B., Cordon-Rosales, C., and Durvasula, R.V. (2002). Bacterial symbionts of the triatominae and their potential use in control of Chagas disease transmission. *Annu. Rev. Entomol.* **47:** 123–141.

Boman, H.G. (1991). Antibacterial peptides: key components needed in immunity. *Cell* **65:** 205–207.

Catteruccia, F., Nolan, T., Loukeris, T.G., Blass, C., Savakis, C., Kafatos, F.C., and Crisanti, A. (2000). Stable germline transformation of the malaria mosquito *Anopheles stephensi*. *Nature* **405:** 959–962.

Coates, C.J., Jasinskiene, N., Miyashiro, L., and James, A.A. (1998). Mariner transposition and transformation of the yellow fever mosquito, *Aedes aegypti*. *Proc. Natl. Acad. Sci. U.S.A.* **95:** 3748–3751.

Durvasula, R.V., Gumbs, A., Panackal, A., Kruglov, O., Aksoy, S., Merrifield, R.B., Richards, F.F., and Beard, C.B. (1997). Prevention of insect-borne disease: an approach using transgenic symbiotic bacteria. *Proc. Natl. Acad. Sci. U.S.A.* **94:** 3274–3278.

Durvasula, R.V., Gumbs, A., Panackal, A., Kruglov, O., Taneja, J., Kang, A.S., Cordon-Rosales, C., Richards, F.F., Whitham, R.G., and Beard, C.B. (1999a). Expression of a functional antibody fragment in the gut of *Rhodnius prolixus* via transgenic bacterial symbiont *Rhodococcus rhodnii*. *Med. Vet. Entomol.* **13:** 115–119.

Durvasula, R.V., Kroger, A., Goodwin, M., Panackal, A., Kruglov, O., Taneja, J., Gumbs, A., Richards, F.F., Beard, C.B., and Cordon-Rosales, C. (1999b). Strategy for introduction foreign genes into field populations of Chagas disease vectors. *Ann. Entomol. Soc. Am.* **92:** 937–943.

Gratz, N.G. (1999). Emerging and resurging vector-borne diseases. *Annu. Rev. Entomol.* **44:** 51–75.

Haynes, P.A., Russell, D.G., and Cross, G.A. (1996). Subcellular localization of *Trypanasoma cruzi* glycoprotein Gp72. *J. Cell Sci.* **109:** 2979–2988.

Holliger, P., Prospero, T., and Winter, G. (1993). "Diabodies": small bivalent and bispecific antibody fragments. *Proc. Natl. Acad. Sci. U.S.A.* **90:** 6444–6448.

Huse, W.D., Sastry, L., Iverson, S.A., Kang, A.S., Alting-Mees, M., Burton, D.R., Benkovic, S.J., and Lerner, R.A. (1989). Generation of a large combinatorial library of the immunoglobulin repertoire in phage lambda. *Science* **246:** 1275–1281.

Jasinskiene, N., Coates, C.J., Benedict, M.Q., Cornel, A.J., Rafferty, C.S., James, A.A., and Collins, F.H. (1998). Transformation of the yellow fever mosquito, *Aedes aegypti*, with the *Hermes* element from the housefly. *Proc. Natl. Acad. Sci. U.S.A.* **95:** 3743–3747.

Nozaki, T. and Cross, G.A.M. (1994). Functional complementation of Glycoprotein 72 in a *Trypanasoma cruzi* glycoprotein 72 null mutant. *Mol. Biochem. Parasitol.* **67:** 91–102.

Schofield, C.J. and Dias, J.C. (1999). The Southern Cone Initiative against Chagas disease. *Adv. Parasitol.* **42:** 1–27.

Schofield, C.J. and Dujardin, J.P. (1997). Chagas disease vector control in Central America. *Parasitol. Today* **13:** 141–144.

World Health Organization. (1992). Resistance to pesticides. Fifteenth report of the WHO Expert Committee on Vector Biology and Control. *WHO Tech. Rep.* **818:** 62.

7 Bark Beetle–Fungus Symbioses

Diana L. Six

CONTENTS

INTRODUCTION

Fungi are ubiquitous associates of bark beetles, and while not all form close associations with their hosts, many depend on their hosts for dissemination. In turn, many beetles exhibit at least some degree of dependence on the fungi. Recognition of these reciprocal effects led to a long-standing view by many entomologists that these associations are primarily mutualistic. However, closer investigation has revealed a diverse array of interactions including antagonism and commensalism. Indeed, considering the extensive variation in life histories of the host beetles and their associated fungi and the range of taxonomic variation in fungal associates, the discovery of a wide range of complex interactions is not surprising.

Our current understanding of bark beetle–fungal associations has been limited by a tendency to disregard the vastly more numerous and ecologically diverse nonaggressive species of bark beetles in favor of the economically important tree-killing species. Further, ecological theory on the development and maintenance of symbiotic associations has only been spottily applied to the study of these systems. Past bark beetle–fungal research has resulted in the development of a

0-8493-1286-8/03/$0.00+$1.50

comprehensive body of literature on host-tree defense reactions to invasion by aggressive beetles and their associated fungi as well as a more limited literature characterizing interactions between beetles and their associates. Ultimately, however, if we are to gain a fuller understanding of these systems, it will be necessary to develop a broader conceptual framework that considers the full range of diverse bark beetle hosts and their associated fungi.

BARK BEETLES

Bark beetles are in the Scolytidae (alt. Scolytinae), a family that includes the cone beetles and many ambrosia beetles. Bark beetles typically invade the bark and phloem of plants but not the woody tissues (Wood, 1982). These beetles exhibit great variation in life histories, ranging from highly aggressive (tree-killing) to facultative (colonizing weak or recently killed trees) to parasitic (using living trees) to saprophagous (using dead hosts) (Paine et al., 1997). Aggressive beetles have attracted the greatest interest due to their eruptive population dynamics and subsequent economic impact. However, relative to the large number of bark-beetle species that have been described, the number of aggressive species is very small. For example, less than 2% of the nearly 500 species of bark beetles described in North America commonly kill trees.

Scolytids typically feed in nutritionally poor substrates. They colonize a great variety of plant tissues including woody tissues, bark and phloem, fruits, and the pith of twigs (Wood, 1982). Only the ambrosia beetles do not feed on plant tissues, feeding solely on their associated fungi. Other scolytids feed on plant tissues or, more often, on a combination of plant and fungal tissues. Berryman (1989) grouped the Scolytidae into three broad categories based on feeding strategy: saprophages, phytophages, and mycetophages. Saprophages include scolytids that feed exclusively on dead or decaying tree tissues. While numerous, this group of beetles is not well studied, and little information exists on their associations with fungi. Because these insects arrive late in the sequence of colonization of plant host material, they are more likely to encounter competition from other insect groups (Wood, 1982). Likewise, any fungi carried by these late-arriving beetles would face resource depletion or strong competition with earlier arriving saprophytic species and, in some cases, with fungi introduced by earlier arriving scolytids. Berryman (1989) predicted that strict associations with specific fungi were unlikely in the saprophagous beetles. However, it is highly plausible that these beetles ingest fungal tissues along with wood. Rather than transporting specific fungi into wood, these beetles may instead rely upon fungi already present, although the species composition of the fungal community encountered is likely to vary.

Some insects that feed on dead wood show marked attraction to decayed wood and, in particular, prefer decay fungi (French et al., 1981). Further, feeding rate and survivorship increase on wood containing some fungi but decrease on wood containing others (Moein and Rust, 1992). Similar relationships may also exist with saprophagous bark beetles and decay fungi and may even account for the differential attraction of beetles to various substrates. Alternatively, these beetles may have no attraction to, or dependence on, decay fungi and may merely feed on fungi incidentally as an inevitable and unavoidable component of their food.

The phytophages feed on living or recently killed plant tissues. This category includes bark beetles that feed primarily on the phloem and cambial tissues of trees, although some species that feed on these tissues may spend a portion of their developmental period in the outer bark (Wood, 1982). Associations with fungi are apparently universal within this group of scolytids, and many possess specialized structures, termed mycangia (treated below), for the transport of fungi. These beetles use living or freshly killed plant tissues and consequently experience less interspecific competition for resources. The fungi they carry are able to exploit uncolonized tissues well in advance of highly competitive saprophytes. Indeed, Harrington (1993) hypothesized that the weak pathogenicity exhibited by some bark beetle–associated fungi may allow these fungi a competitive advantage over strictly saprophytic species.

Many phytophages feed on fungi and plant tissues as both larvae and adults. The degree of dependence of beetles on fungi ranges from obligate to facultative to opportunistic. Associations further range from highly species specific to merely incidental. Unfortunately, for most species virtually nothing is known regarding fungal associates, specificity, or degree of obligacy of association.

The mycetophage category includes the ambrosia beetles that live in wood or other plant tissues but cultivate and feed strictly on associated fungi. These associations are obligate, and, predictably, mycetophagous beetles possess mycangia, ensuring the continuity of insect–fungus association from generation to generation. Most of these beetles colonize nutritionally poor and mostly indigestible substrates including xylem. The fungal associates allow the exploitation of these otherwise marginal habitats by concentrating nitrogen and providing essential nutrients, including sterols and vitamins, to the beetle host (Beaver, 1989). Despite the complete dependence of mycetophagous beetles on their fungal associates, these associations are seldom monophilic (one symbiotic partner per beetle species) but are more often oligophilic (two to several symbiotic partners per beetle species) (Batra, 1966).

BARK BEETLE–ASSOCIATED FUNGI

Mycelial Fungi

Bark beetles are associated with filamentous fungi in the Ascomycotina and Basidiomycotina. Most ascomycete associates are in three genera — *Ophiostoma* Syd., *Ceratocystiopsis* Upadhyay and Kendrick, and *Ceratocystis* Ell. and Halst., included in a group of morphologically similar fungi collectively referred to as the ophiostomatoid fungi. *Ceratocystis* possess *Chalara* Corda anamorphs (asexual states), while *Ophiostoma* possess several anamorph states including *Leptographium* Lagerb. and Melin, *Graphium* Corda, *Hyalorhinocladiella* Upadhyay and Kendrick, and *Sporothrix* Hektoen and Perkins. *Ceratocystiopsis* anamorphs include *Hyalorhinocladiella* and *Sporothrix*.

Despite many morphological similarities, *Ophiostoma* and *Ceratocystiopsis* are phylogenetically distinct from *Ceratocystis* (Hausner et al., 1993a,b). *Ophiostoma* and *Ceratocystiopsis* form a highly variable but monophyletic group in the Diaporthiales, while *Ceratocystis* are most similar to the Microascales (Hausner et al., 1993b; Spatafora and Blackwell, 1994).

Ophiostoma and *Ceratocystiopsis* species are vectored primarily by subcortical insects, including bark beetles, and are often found in temperate forest trees (Harrington, 1993). Most are competitive saprophytes, and only a few are known plant pathogens (Harrington, 1993). *Ophiostoma* species associated with bark beetles are strongly pleomorphic. The mycelial state is produced when the fungi grow in phloem and wood; however, when the fungi are growing in mycangia, or adjacent to beetles in pupal chambers (Figure 7.1), growth can become highly concentrated and sporogenous (Tsuneda, 1988) or even yeast-like (Barras and Perry, 1972; Happ et al., 1976).

Ceratocystis species, on the other hand, grow on a wide variety of herbaceous as well as woody plants in temperate and tropical zones. While two bark-beetle species vector *Ceratocystis* species at least occasionally (Furniss et al. 1990; Solheim and Safranyik, 1997; Viiri, 1997), most vectors are sap-feeding insects (Harrington, 1987). In contrast to *Ophiostoma*, many *Ceratocystis* are capable of causing plant disease.

Other ascomycetes have on occasion been isolated from bark beetles or from their galleries; however, effects of these apparently incidental fungi on the beetles remain unknown. Some, such as *Trichoderma* and *Penicillium* species, are common in older galleries (Whitney, 1971). These fungi may be present as ubiquitous saprophytes, but due to their antagonistic mycoparasitic nature they may interact in significant, but unknown, ways with bark beetle–fungal associates.

While most associations among bark beetles and fungi involve ascomycetes, a few associations with basidiomycetes have been characterized. *Gloeocystidium ipidophilum* Siemasko was described from galleries of *Ips typographus* L. (Siemasko, 1939), and a *Sebacina*-like basidiomycete was

FIGURE 7.1 (Color figure follows p. 206.) Pupae of *Dendroctonus ponderosae* in fungal-lined pupal chambers.

isolated from pupal chambers of *I. avulsus* (Eichoff) (Moser and Roton, 1971). Tsuneda et al. (1992) isolated an arthroconidial basidiomycete from pupal chambers of *Dendroctonus ponderosae* Hopkins. *Entomocorticium* species appear to be the most common and widespread basidomycete associates of bark beetles. *I. avulsus*, a nonmycangial beetle, is associated with an *Entomocorticium* species that acts as a larval nutritional symbiont (B.T. Sullivan, personal communication cited in Klepzig et al., 2001). *Entomocorticium dendroctoni* Whitney, Bandoni, and Oberwinkler can sometimes be found lining the pupal chambers of *D. ponderosae*, either alone or in combination with ascomycete mycangial fungi, and some evidence suggests that this fungus may contribute nutritionally to the beetle (Whitney et al., 1987). A closely related *Entomocorticium* can also sometimes be found in pupal chambers of *D. jeffreyi* (Hsiau, 1996). *Entomocorticium* sp. A (previously called SJB 122) was the first basidiomycete to be isolated from a bark-beetle mycangium and is commonly found in the mycangia of *D. frontalis* (Barras and Perry, 1972; Hsiau, 1996). *Dendroctonus brevicomis* was subsequently found to carry a related *Entomocorticium* in its mycangium (Whitney and Cobb, 1972; Paine and Birch, 1983; Hsiau, 1996).

Yeasts

Although the occurrence of ascomycete yeasts with bark beetles has been common knowledge for many years (Whitney, 1982), relatively little is known of their taxonomy or effects on the beetle host. Yeasts are associated with all developmental stages of the beetle. They are common on the outer surface of eggs, larvae, pupae, and adults, in the intestinal tracts of larvae and adults (Grosmann, 1930; Leach et al., 1934; Shifrine and Phaff, 1956; Lu et al., 1957), and occasionally in the alimentary canal of pupae (Shifrine and Phaff, 1956). They are commonly carried in mycangia and in pits of the exoskeleton of adult beetles (Whitney and Farris, 1970; Whitney and Cobb, 1972; Furniss et al., 1990, 1995; Lewinsohn et al., 1994), and they densely colonize walls of beetle galleries and pupal chambers (Bridges et al., 1984) as well as xylem tissues of the host tree (Caird, 1935). Individual adult beetles often carry two or more yeast species, and, unlike the filamentous fungi, which tend to have species-specific associations with beetles, several yeast species are commonly carried by several species of beetles. Two species, *Pichia capsulata* (Wickerham) Kurtzmann and *P. pini* (Holst) Phaff, are associated with most bark beetles thus far surveyed, and for two of three beetle species where yeast associations have been quantified they are also the most prevalent associates (Shifrine and Phaff, 1956; Bridges et al., 1984; Leufven and Nehls, 1986).

The yeasts associated with bark beetles do not exhibit pathogenicity to host trees (Callaham and Shifrine, 1960). Like the ophiostomatoid fungi, they are not highly competitive and are limited to colonizing tree tissues relatively early in the colonization process (Bridges et al., 1984).

MAINTENANCE OF ASSOCIATIONS

Beetle Adaptations

Many bark beetles possess structures of the integument that function in the transport of fungi and contribute to the maintenance of bark beetle–fungal symbioses. Some of these structures have been termed mycangia (Batra, 1963). Use of the term mycangium can vary from highly restrictive to very broad. In the most restrictive sense, a mycangium is defined as an invagination of the integument lined with glands or secretory cells that is specialized for the acquisition and transport of fungi (Batra, 1963; Levieux et al., 1991). More loosely defined, the term mycangium has been applied to any structure that consistently transports fungi regardless of form or presence of secretory cells (Farris and Funk, 1965; Livingston and Berryman, 1972; Nakashima, 1975; Beaver, 1986; Furniss et al., 1987). This broader definition allows the inclusion of shallow pits and setae along with deeper pockets that act as fungal repositories but are not known to be associated with glands.

At present, strict categorization of beetles as mycangial or nonmycangial is difficult due to the great variation in structures involved and the current ambiguity in the literature of what constitutes a "true" mycangium. Given that there is a wide variety of structures, including pits, punctures, setal brushes, and highly developed sac-like structures, that function in a biologically similar manner, it may be appropriate to consider any structure that consistently functions to transport specific fungi as mycangia. Further subdivisions of mycangia could then be delineated on the basis of coarse and fine structure, depending on what is known about the repository. Categories of coarse structure would include pit, sac, and setal-brush mycangia. Pit mycangia include all fungal repositories formed by shallow depressions of the exoskeleton. These pits may or may not be associated with one or several setae. Sac mycangia consist of more complex invaginations forming deep pockets, tubes, or cavities in the exoskeleton, while setal-brush mycangia consist of dense brushes of setae that may or may not arise from depressions in the exoskeleton. Upon investigation of fine structure, further subdivision could be made depending on the presence or absence of glands. In the future, as more is learned about gland types or secretory cells associated with various bark beetle mycangia, further subdivision by gland type may also prove useful.

The proposed two-tier system of classification of scolytid mycangia is presented in Figure 7.2. This system avoids the confusing and potentially misleading use of the term nonmycangial to refer to beetles with fungal repositories and close associations with fungi but whose structures lack glands, or for which the presence of glands is uninvestigated. For example, under the previous strict definition of what constitutes a mycangium, *Dendroctonus ponderosae* would be considered nonmycangial because the sac-like structures on the maxillary cardines described by Whitney and Farris (1970) have not been investigated for the presence of glands, despite the fact that they are involved in maintaining a highly consistent association with two specific fungi and function biologically as true mycangia. Additionally, beetles with pits that consistently transport specific fungi have been traditionally considered nonmycangial, even though at least some of those pits are associated with glands. The proposed system allows initial classification by biological function followed by subsequent classification by structure, rather than the reverse.

The presence of glands is uninvestigated for many mycangia. For some sac and pit mycangia, the structures have been observed to be associated with waxy or oily secretions, indicating the presence of glands (Livingston and Berryman, 1972; Furniss et al., 1987, 1995). For others, glands

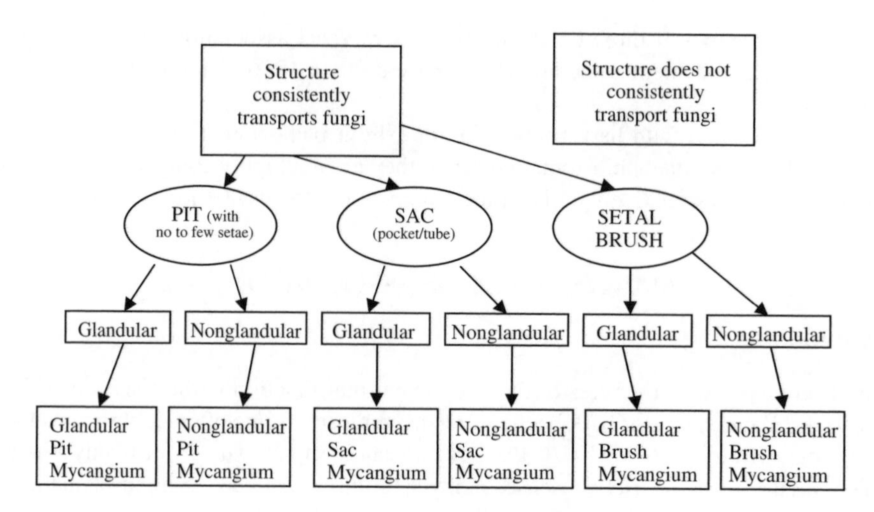

FIGURE 7.2 Schematic for classification of mycangia by coarse structure and presence of glands.

have been observed, but their fine structure has not been fully described (Farris and Funk, 1965; Livingston and Berryman, 1972). A number of glandular types are associated with insect exoskeletons (Noirot and Quennedey, 1974). Bark and ambrosia beetle mycangial glands thus far described are most commonly class III glands (glands with direct ducts to the surface of the exoskeleton) (Farris and Funk, 1965; Schneider and Rudinsky, 1969; Barras and Perry, 1971; Levieux et al., 1991), although some mycangia may be associated with more than one gland type (Happ et al., 1971; Cassier et al., 1996).

The secretions produced by glands associated with bark beetle mycangia are believed to support the growth of fungal propagules, protect spores and mycelium from desiccation, and act selectively against fungi not symbiotic with the beetle (Schneider and Rudinsky, 1969; Happ et al., 1971; Barras and Perry, 1971; Barras and Perry, 1972; Paine and Birch, 1983). Glandular secretions of several ambrosia beetles contain fatty acids, phospholipids, sterols, and amino acids and influence the growth of ambrosial fungi while in the mycangium (Norris, 1979). Glands associated with mycangia of the related bark beetles may produce similar secretions and function in a similar manner, but there have been few if any studies to document the nature and function of gland products in these insects.

Mycangia have arisen independently several times in the Scolytidae, underscoring the importance of fungi to this group of beetles. Although type of mycangium is often genus specific, the presence or type of mycangium can vary even within a single genus. For example, within *Dendroctonus*, five species possess complex glandular sac mycangia on the pronotum functional only in females (Barras and Perry, 1971, 1972), while two species possess sac mycangia (glandular status unknown) on the maxillary cardines that are functional in both sexes (Whitney and Farris, 1970; Six and Paine, 1997). Still other *Dendroctonus* species possess pit mycangia consisting of shallow pits in combination with sparse setae (Lewinsohn et al., 1994). Some *Dendroctonus* species, including *D. valens* LeConte, *D. micans* (Kugelann), and *D. terebrans* (Olivier), remain uninvestigated for the presence of mycangia; however, as these beetles show little consistency in fungal associates (Lieutier et al., 1992; Klepzig et al., 1995; D.L. Six, unpublished data), they may be nonmycangial.

Fungal Adaptations

Ophiostomatoid fungi are well adapted to dispersal by arthropods (Malloch and Blackwell, 1993). Most produce ascomata with ostiolate necks that extrude ascospores (sexual spores) at heights where they are most likely to be encountered by insects and other arthropods. Ascospores are

FIGURE 7.3 (Color figure follows p. 206.) Conidiophores (anamorph) of *Ophiostoma clavigerum* lining a pupal chamber of *Dendroctonus ponderosae.*

probably important for some fungi in long-distance dispersal between host trees (Malloch and Blackwell, 1993). The ascospores produced by *Ophiostoma* and *Ceratocystiopsis* are coated with an adhesive material and are of various shapes that allow for multiple contact points with the vector, ensuring that they are not easily removed in transit. The adhesive coats of some *Ophiostoma* ascospores disperse in resin but not in water, which may provide a mechanism for removal of the sticky spores only when an appropriate substrate (i.e., the new host tree) is encountered (Whitney and Blauel, 1972).

The ophiostomatoid fungi also produce a wide range of anamorphs that produce conidia (asexual spores) in slimy masses that readily adhere to the insect cuticle (Tsuneda and Hiratsuka, 1984; Tsuneda, 1988; Malloch and Blackwell, 1993). The conidia are often round, cylindrical, or oval, permitting only one contact point with the vector arthropod, which may allow the conidia to be easily dislodged and dispersed within beetle galleries (Malloch and Blackwell, 1993). The conidia are also often found in pits on the beetle exoskeleton or in mycangia, indicating they also play a critical role in long-distance dispersal for some species.

The basidiomycetes (*Entomocorticium* spp.) associated with bark beetles produce both chlamydospores and conidia. However, in mycangia, only chlamydospores or a yeast-like form of the fungus is present (Barras and Perry, 1972; Happ et al., 1976; Goldhammer et al., 1990).

For beetles possessing sac mycangia, apparently only the anamorph is acquired and disseminated in the structure, although sexual reproduction of the fungi may also occur (Paine and Birch, 1983; Moser et al., 1995). For these beetles often only the anamorph is produced in the pupal chambers (Figure 7.3) where mycangia of teneral adults are charged with fungal propagules prior to dispersal (Whitney, 1971). In contrast, ascomata often form in old galleries distant to where teneral adult beetles develop (Figure 7.4), and therefore contact between the sexual stage and new adults prior to emergence is unlikely. If and how the sexual stages of these fungi are disseminated has posed a difficult problem for investigators. The answer to this conundrum for at least one bark beetle system may be found by looking at phoretic mites associated with the insect (Klepzig et al., 2001). The ascospores of a mycangial fungus of *D. frontalis*, *C. ranaculosus,* are transported in sporothecae of phoretic mites while the conidia are carried in the mycangium of the host beetle (Moser et al., 1995). The mites colonize the insect host within the pupal chambers prior to beetle emergence and dispersal. It is not known if mites are important vectors of ascospores of mycangial fungi in other bark-beetle systems.

For other fungi associated with sac mycangia, the production of the sexual state may be rare or even lacking. The sibling species, *D. ponderosae* and *D. jeffreyi*, both carry *O. clavigerum* in sac mycangia. Laboratory pairings of this fungus have not produced ascomata, and there is some question as to whether a sexual state exists for this fungus (D.L. Six, T.C. Harrington, D. McNew,

FIGURE 7.4 (Color figure follows p. 206.) Ascomata (teleomorph) of *Ophiostoma montium* in parental galleries of *Dendroctonus ponderosae*.

J. Steimel, and T.D. Paine, unpublished observations). The most common mycangial associates of *D. adjunctus* and *D. approximatus* are *Leptographium* species with no known sexual states (Six and Paine, 1996 and 1998). Furthermore, *Entomocorticium* spp. associated with bark beetles are not known to produce sexual states (Goldhammer et al., 1989; Hsiau, 1996). The lack or rarity of sexual recombination in some bark beetle–associated fungi may be an adaptation to mutualism. If a fungus possesses a genotype that confers high fitness to both itself and to the host beetle, sexual reproduction may be disadvantageous by continually breaking up successful gene combinations (Wulff, 1985).

Production and dispersal of sexual and asexual spores by fungi associated with pit mycangia appear variable. *D. pseudotsugae* carries both conidia and ascospores of *O. pseudotsugae* in pit mycangia (Furniss et al., 1990, 1995; Lewinsohn, 1994), indicating that this fungus produces ascomata in the host's pupal chambers. *O. ips* is known to produce ascomata in pupal chambers of *I. pini* where the beetle acquires both ascospores and conidia in pit mycangia prior to emergence (Furniss et al., 1995). Spore types have not been determined for fungi associated with beetles possessing setal-brush mycangia (Furniss et al., 1987).

BARK BEETLE–FUNGUS SYMBIOSES

Many scolytid–fungus symbioses are considered mutualisms; however, many other types of associations also exist, including antagonism, predation, and commensalism. Some saprophage associations may be commensal if the beetles benefit from the presence of a fungus without adversely affecting the fungus, or antagonistic if the fungus or effects of the fungus on the wood substrate are detrimental to the beetle, or they may involve predation if the beetle benefits by feeding on fungal tissues but does not disseminate or otherwise benefit the fungus in return.

In contrast to saprophagous beetles, mutualism is probably widespread, although not universal, in other scolytid groups. As with most mutualistic associations among insects and microorganisms, benefits gained by either host or symbiont in these associations can be placed into three categories: nutrition, transport, and direct protection.

Nutrition

Nutrition is the major driving force in mycetophage scolytid associations (Norris and Baker, 1967; Norris et al., 1969; Abrahamson and Norris, 1970; Kok, 1979) and in many other obligate insect–fungal symbioses (Wetzel et al., 1992). It is also likely to be an important factor driving many associations of fungi with phytophage bark beetles. Bark beetle larvae feed on mycelial fungi as larvae and teneral adults and probably on yeasts throughout development. In many cases, mycelium, yeast cells, and conidia are ingested; however, *I. avulsus* and *I. calligraphus* have been observed to seek out and ingest entire *O. ips* perithecia during maturation feeding as young adults (Yearian et al., 1972).

The two mycangial fungi of *D. frontalis* impart positive impacts on brood development and survival that likely occur through nutrition gained during feeding on fungi by larvae and adult beetles (Barras, 1973; Bridges, 1983; Goldhammer et al., 1990; Ayres et al., 2000). However, the two fungi are not equal in their effects (Bridges, 1983; Goldhammer et al., 1990). Female *D. frontalis* carrying *Entomocorticium* sp. A or *C. ranaculosus* and *Entomocorticium* sp. A are larger and possess higher lipid content than those carrying only *C. ranaculosus* (Coppedge et al., 1995). Such differences can greatly affect the fitness of host beetles. Greater body size and weight positively affect survival (Safranyik, 1976; Botterweg, 1982, 1983; Anderbrandt, 1988), dispersal ability (Atkins, 1969; Thompson and Bennett, 1971), and reproductive capacity of female beetles (Reid, 1962; Amman, 1972; Clarke et al., 1979). Furthermore, a large proportion of *Dendroctonus* eggs is composed of lipid (Barras and Hodges, 1974), indicating that the lipid content, and consequently the nutrition, of parental female beetles is important for optimal egg production.

The two mycangial fungi of *D. ponderosae* also differentially affect the reproductive potential of their host in a way that indicates nutritional effects (Six and Paine, 1998). The production of progeny *D. ponderosae* is significantly higher, and emergence occurs significantly earlier (indicating a more nutritious food source) for brood developing with *O. clavigerum* than with *O. montium*. For this beetle, only one mycangial fungus appears to benefit larval development.

Nonmycangial beetles appear to be less specific in their associations and are likely to be less dependent on their fungal associates than mycangial beetles. For example, *I. avulsus* (Eichoff) is considered nonmycangial. It is associated with two fungi, *Entomocorticium* sp. and *O. ips* (Klepzig et al., 2001). Wild beetles associated with their full fungal complement (including *Entomocorticium* sp.) are larger, produce more brood, and exhibit higher brood survival than beetles associated only with *O. ips*, a potential antagonist, or beetles that are fungus free (Yearian et al., 1972). Association with fungi, however, does not appear to be obligate with *I. avulsus*, as the beetles can be reared without fungi, although with a reduced level of success (Yearian et al., 1972). In contrast, fungus-free *I. calligraphus* (Germar) and *I. grandicollis* (Eichoff) laid more eggs and produced more brood than when reared with fungi (Yearian et al., 1972). Fungus-free *I. paraconfusus* were also observed to develop successfully to adult, but size of brood and developmental rate of larvae were reduced in fungus-free beetles compared with beetles developing with fungi (Fox et al., 1993). Caution, however, must be exercised in interpreting effects of so-called nonmycangial fungi because many "nonmycangial" beetles have not been formally investigated for the presence of mycangia.

Many of the observed positive effects of fungal association on phytophagous bark beetles are likely to be directly related to nutrition, either through modification of nutrient form or availability, production of essential nutrients not found in phloem or bark (or present only in inadequate amounts), or concentration of nutrients by the fungal mycelium. Wood is a poor source of vitamins, sterols, and other growth factors (Norris and Baker, 1967), and fungi may provide alternative sources. Recently, evidence was found supporting the role of nitrogen concentration by fungi associated with *D. frontalis* (Ayres et al., 2000). In that study, feeding by *D. frontalis*, a mycangium-bearing species, was compared with that of *I. calligraphus*, a nonmycangium-bearing species. Results indicate that the mycangial fungi of *D. frontalis* concentrate nitrogen better than an antagonistic nonmycangial associate, *O. minus*, reducing the amount of phloem required for development. *I. calligraphus,* which is associated primarily with *O. minus*, had to consume additional phloem to obtain its nitrogen requirements. Similarly, other nonmycangial beetles may also need to feed more extensively in phloem to compensate for low nutrient content in the absence of nutritionally beneficial fungi. For example, *D. rufipennis*, an apparently nonmycangial beetle, feeds extensively in the phloem and in many populations has a 2-year life cycle (Furniss and Carolin, 1977). This feeding pattern is in strong contrast with sympatric mycangial *Dendroctonus* species, which typically produce short feeding galleries and exhibit one or more generations per year (Furniss and Carolin, 1977).

It is likely that nutritionally beneficial fungi not only concentrate nitrogen but also provide a source of sterols to their host beetles. Insects depend on dietary sources of sterols for growth, molting, and reproduction (Clayton, 1964; Svoboda et al., 1978). Sterols are important elements of cellular structure, provide necessary precursors for hormone synthesis (including juvenile hormone), and are critical for the production of viable eggs (Clayton, 1964). Sterol concentrations in plant tissues like phloem and xylem are typically low. A lack of fungal-produced sterols may account for observations that fungus-free larvae produce longer mines, are smaller than normal, and do not pupate (Webb and Franklin, 1978; Strongman, 1982). Furthermore, beetles developing with some fungal associates pupate successfully, while those developing with other fungi either fail to pupate or have reduced rates of pupation (Strongman, 1982).

Such differential effects on beetle development may relate to the quantity and quality of sterols present in the beetle's diet. It is clear that not all sterols are equally suitable for use by insects. Ergosterol, a major, and often the only, sterol produced by many fungi, has been found to be one of the most suitable sterols for insect nutrition, whereas many plant-produced sterols are not usable (Clayton, 1964).

In several insect–fungus symbioses, the insect associate has been found to be dependent on sterols provided by the fungal associate (Norris et al., 1969; Kok et al., 1970; Norris, 1972; Maurer et al., 1992; Morales-Ramos et al., 2000; Mondy and Corio-Costet, 2000). For the ambrosia beetle, *Xyleborus ferrigineus* (Fabricius), the presence of an associated fungus in its diet is required for pupation (Kok, 1979) and oviposition of viable eggs (Norris and Baker, 1967) and for oocyte development, oviposition, larval development, and pupation in a related species, *X. dispar* (French and Roeper, 1975). For both beetles, ergosterol produced by their associated fungi is responsible for these effects (Norris et al., 1969). Similar results were also found for another scolytid, *Hypothenemus hampei* (Ferrari). This beetle bores into, and feeds on, coffee berries but cannot molt or reproduce without ergosterol from its symbiotic fungus, *Fusarium solani* (Morales-Ramos et al., 2000). For several phytophage beetles, reduced levels of pupation or lack of oviposition have been noted in fungus-free beetles (Strongman, 1982; Fox et al., 1993; Six and Paine, 1998), indicating that for at least some beetles in this group, fungus-derived compounds may also be important.

Ergosterol production varies considerably by fungal species (Kok and Norris, 1973), and this differential production may account for different fungal effects on the beetle host. Kok et al. (1970) and Kok and Norris (1973) found that the species of fungi associated with *Xyleborus* ambrosia beetles each produced different amounts of ergosterol, the only sterol detected in any of the fungi. The fungi most beneficial for beetle success were found to contain the greatest concentration of ergosterol. Similar investigations being conducted with phytophagous scolytids may help elucidate the potential roles of fungal-produced sterols with these insects (D.L. Six, B.J. Bentz, K. Wallin, and K. Bleiker, unpublished data).

Effects, if any, of yeast associates on host beetles are largely uninvestigated. Grosmann (1930), after rearing a single larva successfully to adulthood without yeasts, concluded that yeasts were not nutritional mutualists. Holst (1937) reached a similar conclusion after successfully rearing beetles to the adult stage under sterile *in vitro* conditions. However, the limited scope of these studies, combined with a failure to extend the experiment into a second generation to test whether yeast feeding is critical for reproduction, makes it difficult to accept their conclusions without reservation.

Fungi may also benefit nutritionally from their associations with bark beetles. The glands associated with mycangia may provide a source of nutrition for fungi during transit between host trees. For example, small amounts of fungi are acquired in the mycangia of teneral adult *D. brevicomis* during maturation feeding in the pupal chamber. The fungi then grow within and fill the mycangium by the time the beetle matures and colonizes a new tree (Paine and Birch, 1983). Such extensive growth within the mycangium would be unlikely in the absence of nutrients.

Transport

In bark beetle–fungal symbioses, benefit through transport is gained strictly by the fungi. Bark beetle-associated fungi are completely dependent, or nearly so, on their hosts for transport from tree to tree. Sticky spores, or spore production within plant tissues, primarily in galleries, pupal chambers, and the interface between the phloem and outer bark layers of the tree, preclude dispersal by wind or rain splash. Spores of *Ophiostoma* may also sometimes be isolated from the bodies of other insects that develop in bark beetle–colonized trees such as clerid beetles (D.L. Six, unpublished data); however, incidence and reliability of dissemination by these potential alternate vectors is unknown. In some cases, mites may be important vectors (Klepzig et al., 2001).

Protection

As with nutrition, protection is a benefit that may be gained by both beetle and fungus. Beetle-associated fungi likely gain direct physical protection from desiccation and UV light while in transit within mycangia. Beetle brood, on the other hand, may in some instances receive protection from detrimental contact with antagonistic fungi when developing in phloem colonized by competitive beneficial fungi. Competition among fungal species has been demonstrated frequently (Shearer, 1995), including among *Ophiostoma* species (Klepzig, 1998; K. Bleiker and D.L. Six, unpublished data), *Ophiostoma*, *Ceratocystiopsis*, and *Entomocorticium* (Klepzig and Wilkens, 1997; Klepzig, 1998), and *Ophiostoma* and yeasts (A.S. Adams and D.L. Six, unpublished data). The outcome of competitive interactions may determine which fungi the larvae will contact and ingest and consequently affect their development and survival. In the *D. frontalis* system, the two mycangial fungi, *Entomocorticium* sp. A and *C. ranaculosus*, and the nonmycangial antagonistic associate *O. minus* compete for uncolonized phloem during the initial colonization of the tree (Klepzig and Wilkens, 1997). In this case, *O. minus* is the superior competitor, colonizing uninfected phloem more rapidly than either *Entomocorticum* sp. A or *C. ranaculosus*. However, the two mycangial fungi differ greatly in their ability to maintain resources once they have been captured. *C. ranaculosus* is quickly overgrown by *O. minus,* while *Entomocorticium* sp. A is not overgrown and is able to maintain resources in the presence of *O. minus* (Klepzig and Wilkens, 1997). Thus, larvae growing in tissues colonized by *Entomocorticium* sp. A may receive protection from *O. minus*, while those growing in tissues colonized by *C. ranaculosus* are more susceptible to contact with the detrimental species.

Other Effects

Effects of association, both direct and indirect, may also benefit host beetles in ways other than through nutrition, transport, and protection. For example, yeasts isolated from *I. typographus* (L.) are capable of oxidizing *cis*-verbenol, an aggregation pheromone of the host, to verbenone, an antiaggregation pheromone (Leufven et al., 1984). The abundance of yeasts associated with individual beetles during host colonization increases during later stages of attack, coinciding with the period when release of antiaggregation pheromones is at its peak (Leufven and Nehls, 1986). Two yeasts, *P. pini* and *P. capsulata*, associated with *D. ponderosae* convert *trans*-verbenol, an aggregation pheromone, to verbenone, an antiaggregation pheromone, and this conversion may play a role in terminating mass attacks on trees by this beetle (Hunt and Borden, 1990).

FACTORS CONSTRAINING MUTUALISM IN BARK BEETLE–FUNGUS ASSOCIATIONS

The factors and conditions allowing the development and maintenance of mutualism are often complex but at a minimum require that there is (1) a low cost to resource exchange, (2) a gain (reward) for each partner through exchange of specialized resources, and (3) a strong probability of continual contact between the associates.

Costs

Costs of association are critical considerations when attempting to describe interactions between symbionts. However, costs are difficult to quantify in biological systems. Regardless, we can at least speculate as to what some may be. Structurally complex glandular mycangia are likely to incur costs to the beetle through energy expenditures in the production of the structure and secretions by glands; however, such costs are probably low given the relatively small size of mycangia and the small overall amount of secretions produced. Although less complex mycangia may incur lower costs, the costs of structurally complex glandular mycangia may be sufficiently offset through nutritional gains and other potential benefits to the beetle derived from maintaining specific fungi.

Costs to the fungi consist primarily of energy expended in production of mycelium ingested by larvae and of spores ingested by teneral adult beetles during maturation feeding. These losses are likely offset by increased resource availability, increased efficiency in transport, and lower overall spore production compared with fungi that do not rely on arthropod vectors. Noninsect-vectored fungi rely instead on wind or rain-splash dissemination that requires the production of extremely high numbers of spores, most of which die in transit or land on unsuitable substrates.

Other potential costs to the fungi include production of insect-adapted spores with a sticky adherent matrix and the production of nutrients useful to the host beetle. Because fewer spores are made, costs of producing more complex insect-adapted spores are likely to be insignificant. Likewise, the production or concentration of essential nutrients useful to the host insect may be minimal or come at no additional cost. In fact, production or concentration of nutrients by fungi may have been an important preadaptation that allowed some of these associations to form in the first place.

Rewards

The concept of rewards in mutualism is tightly linked to the magnitude of costs incurred. Costs must be minimal and rewards must exceed costs. Cost–benefit analyses are often complicated to conduct, especially in systems such as those involving bark beetles where rearing the insect and conducting manipulative experiments can be very difficult. Under these conditions, determining rewards and, consequently, assigning a mutualistic role to interacting organisms must be carefully considered. For example, in the *D. frontalis* system, at least one mycangial fungus, *Entomocorticium* sp. A, has strong positive effects on host-beetle fitness and is clearly mutualistic (Barras, 1973; Bridges, 1983; Goldhammer et al., 1990). For the other two common fungal associates, assignment of type of effect on the host is not so straightforward. The mycangial fungus, *C. ranaculosus*, does not greatly benefit host development and may actually function more as a commensal with the beetle. However, by displacing the more beneficial mycangial fungus it reduces potential benefits to the host and so may be considered to act antagonistically. A nonmycangial associate, *O. minus*, is highly antagonistic to developing beetle brood, which greatly lowers host fitness. Despite strong negative effects on beetle fitness, some workers have speculated that *O. minus*, which exhibits pathogenicity to host trees under certain circumstances, may still be mutualistic with the beetle by contributing to the decline and death of the tree during the early stages of colonization (Barras, 1970; Klepzig et al., 2001). However, *O. minus* is not required for successful colonization of the tree, and the beetle has a higher reproductive potential in its absence. Thus, this fungus should not be considered mutualistic with *D. frontalis*. Rather, when overall costs and benefits are considered, the fitness of *D. frontalis* would likely greatly increase in the complete absence of *O. minus*, and therefore *O. minus* should be considered an antagonistic associate.

CONTACT

Phytophagous bark beetles are in contact with associated mycelial fungi except on occasions when larval tunneling extends ahead of the growing front of the fungi. When this disassociation occurs, the fungi reestablish contact with the beetle in the pupal chamber (Whitney, 1971). Contact with yeasts is apparently always maintained (Whitney, 1971).

Progeny beetles probably develop most often with, and later disseminate, the same fungi as were inoculated into the tree by their parents. However, occasionally exceptions may occur. Galleries excavated by parental beetles and their brood may occasionally be colonized by a fungus originating in an adjacent gallery of a conspecific or other species of beetle, or from tissues infected by fungi carried by other arthropods. The parent-inoculated fungus may be joined or even replaced by another fungus when this occurs, resulting in an addition to or loss of the original symbiont and its potential replacement with another. If replacement occurs, it may be with another individual of the same fungus or by an individual of another species. The frequency of such replacements is unknown. However, fungi associated with one bark beetle species have been isolated from another when two or more beetle species cohabit the same trees. For example, *O. ips* can occasionally be isolated from the exoskeleton of *D. ponderosae* when it cohabits trees with *I. pini,* a common host of *O. ips* (D.L. Six, unpublished).

POTENTIAL ORIGINS AND CONSEQUENCES OF OLIGOPHILY

The potential for hosts to contact symbionts not of parental origin may be an important determinant of whether associations are monophilic or oligophilic. Replacement of, or addition to, established fungal symbionts with other fungi may at least in part explain why the majority of bark beetle species thus far investigated are oligophilic (Six and Paine, 1999). Because bark beetle–associated fungi may sometimes become disassociated from the host beetle for a period while growing within host-tree tissues, and because several species of fungi are often present in a bark beetle–colonized tree, the potential exists that contact with an established fungus will be lost and that contact with other fungi may occur. Simultaneous colonization of host trees by two or more scolytid species as well as by a large number of other potential arthropod vectors increases the probability of inoculation of a large community of fungi into the tree. While some partitioning of resources within the tree by insects occurs, galleries often intermingle, increasing the potential for contact of beetle progeny with nonparental fungi. Such exposures may have allowed host shifts, as well as exploitation of established mutualisms by nonbeneficial fungi, to occur (Six and Paine, 1999).

Despite the relatively high potential for contact between beetles and nonassociated fungi, it is apparent that a high degree of specificity exists in many associations. The mechanisms behind maintaining such specificity are unclear. However, such mechanisms are likely imperfect and may allow the invasion and parasitism of coevolved associations. For example, many mycangial *Dendroctonus* possess two mycangial fungi. For most, one associate appears to be highly coevolved with the host beetle, while the other appears to be the result of a more recent colonization event (Six and Paine, 1999). For coevolved associates, many appear to be mutualistic, while the more recent associates appear to be antagonistic or at least confer less benefit to the beetle than do the highly coevolved partners. The newer associates may be cheaters in the system, parasitizing existing associations to their benefit while conferring few or no benefits to the host.

Certain characteristics may increase the ability of particular groups of fungi to invade already established associations between beetles and fungi. For example, the sticky spores produced by ophiostomatoid fungi may account, at least in part, for their widespread association with scolytids. Furthermore, fungi closely related to established associates, and thus possessing similar attributes, may more easily invade established associations. For example, several beetles with coevolved *Ophiostoma* associates also possess *Ophiostoma* associates of more recent origin (Six and Paine, 1999).

Competitiveness may also play an important role in the invasion of established associations by new fungi. Highly competitive fungi may capture and retain host-plant resources (and consequently hosts) ahead of less competitive established associates. In aggressive beetle systems, many coevolved associates are nonpathogenic, while more recently associated fungi exhibit varying degrees of virulence to the host tree. Pathogenicity, by increasing competitiveness, may allow some fungi to invade established associations because they can grow in and capture still living or recently killed tree tissues ahead of nonpathogenic associates.

Oligophily is likely to have several important consequences for both the host beetles and the symbiotic fungi. For any given association, fungal associates vary greatly in their effects on host-beetle fitness and therefore are likely to differ in their influence on host-population dynamics. Additionally, interactions, including competition for hosts, among multiple fungal associates of a beetle species may determine the relative abundance of each in a population, which, in turn, may also influence host-population dynamics. Oligophily may be a less desirable state than monophily in associations that involve at least one mutualistic fungus. In such associations, for at least some beetles in a population the mutualist can be displaced by less beneficial associates or by detrimental fungi, resulting in lower overall fitness within a population. While cheaters are potentially detrimental, beetles may be unable to develop effective means of avoiding cheaters without also negatively impacting beneficial associates.

ACKNOWLEDGMENTS

I extend sincere gratitude to T.D. Paine, A. Adams, and K. Bleiker for their very helpful comments and suggestions for this chapter.

REFERENCES

Abrahamson, L.P. and Norris, D.M. (1970). Symbiontic interrelationships between microbes and ambrosia beetles. V. Amino acids as a source of nitrogen to the fungi in the beetle. *Ann. Entomol. Soc. Am.* **63:** 177–180.

Amman, G.D. (1972). Some factors affecting oviposition behavior of the mountain pine beetle. *Environ. Entomol.* **1:** 691–695.

Anderbrant, O. (1988). Survival of parent and brood adult bark beetles, *Ips typographus*, in relation to size, lipid content and re-emergence or emergence day. *Physiol. Entomol.* **13:** 121–129.

Atkins, M.D. (1969). Lipid loss with flight in the Douglas-fir beetle. *Can. Entomol.* **101:** 164–165.

Ayres, M.P., Wilkens, R.T., Ruel, J.J., Lombardero, M.J., and Vallery, E. (2000). Nitrogen budgets of phloem-feeding bark beetles with and without symbiotic fungi (Coleoptera: Scolytidae). *Ecology* **81:** 2198–2210.

Barras, S.J. (1970). Antagonism between *Dendroctonus frontalis* and the fungus *Ceratocystis minor*. *Ann. Entomol. Soc. Am.* **63:** 1187–1190.

Barras, S.J. (1973). Reduction of progeny and development in the southern pine beetle following removal of symbiotic fungi. *Can. Entomol.* **105:** 1295–1299.

Barras, S.J. and Hodges, J.D. (1974). Weight, moisture, and lipid changes during the life cycle of the southern pine beetle. USDA Forest Service Experimental Station Research Note SO-178.

Barras, S.J. and Perry, T. (1971). Gland cells and fungi associated with prothoracic mycangium of *Dendroctonus adjunctus* (Coleoptera: Scolytidae). *Ann. Entomol. Soc. Am.* **64:** 123–126.

Barras, S.J. and Perry, T. (1972). Fungal symbionts in the prothoracic mycangium of *Dendroctonus frontalis* (Coleoptera: Scolytidae). *Z. Angew. Entomol.* **71:** 95–104.

Batra, L.R. (1963). Ecology of ambrosia fungi and their dissemination by beetles. *Trans. Kansas Acad. Sci.* **66:** 213–236.

Batra, L.R. (1966). Ambrosia fungi: extent of specificity to ambrosia beetles. *Science* **153:** 193–195.

Beaver, R.A. (1986). The taxonomy, mycangia, and biology of *Hypothenemus curtipennis* (Schedl), the first known cryphaline ambrosia beetle (Coleoptera: Scolytidae). *Entomol. Scand.* **17:** 131–135.

Beaver, R.A. (1989). Insect-fungus relationships in the bark and ambrosia beetles. In *Insect–Fungus Interactions* (N. Wilding, N.M. Collins, P.M. Hammond, and J.F. Webber, Eds.), pp. 121–143. Academic Press, London.

Berryman, A.A. (1989). Adaptive pathways in scolytid-fungus associations. In *Insect–Fungus Interactions* (N. Wilding, N.M. Collins, P.M. Hammond, and J.F.Webber, Eds.), pp. 145–159. Academic Press, London.

Botterweg, P.F. (1982). Dispersal and flight behavior of the spruce beetle *Ips typographus* in relation to sex, size, and fat content. *Z. Angew. Entomol.* **94:** 466–489.

Botterweg, P.F. (1983). The effect of attack density on size fat content and emergence of the spruce beetle, *Ips typographus* L. *Z. Angew. Entomol.* **96:** 7–55.

Bridges, J.R. (1983). Mycangial fungi of *Dendroctonus frontalis* (Coleoptera: Scolytidae) and their relationship to beetle population trends. *Environ. Entomol.* **12:** 858–861.

Bridges, J.R., Marler, J.E., and McSparrin, B.H. (1984). A quantitative study of the yeasts associated with laboratory-reared *Dendroctonus frontalis* Zimm. (Coleopt., Scolytidae). *Z. Angew. Entomol.* **97:** 261–267.

Caird, R.W. (1935). Physiology of pines infested with bark beetles. *Bot. Gaz.* **96:** 709–733.

Callaham, R.Z. and Shifrine, M. (1960). The yeasts associated with bark beetles. *For. Sci.* **6:** 146–154.

Cassier, P., Levieux, J., Morelet, M., and Rougon, D. (1996). The mycangia of *Playpus cylindrus* Fab. and *P. oxyurus* Dufour (Coleoptera: Platypodidae) structure and associated fungi. *J. Insect Physiol.* **42:** 171–179.

Clarke, A.L., Webb, J.W., and Franklin, R.T. (1979). Fecundity of the southern pine beetle in laboratory pine bolt. *Ann. Entomol. Soc. Am.* **72:** 229–231.

Clayton, R.B. (1964). The utilization of sterols by insects. *J. Lipid Res.* **5:** 3–19.

Coppedge, B.R., Stephen, F.M., and Felton, G.W. (1995). Variation in female southern pine beetle size and lipid content in relation to fungal associates. *Can. Entomol.* **127:** 145–154.

Farris, S.H. and Funk, A. (1965). Repositories of symbiotic fungus in the ambrosia beetle, *Platypus wilsoni* Swaine (Coleoptera: Platypodidae). *Can. Entomol.* **97:** 527–536.

Fox, J.W., Wood, D.L., and Akers, R.P. (1993). Survival and development of *Ips paraconfusus* Lanier (Coleoptera: Scolytidae) reared axenically and with tree-pathogenic fungi vectored by cohabiting *Dendroctonus* species. *Can. Entomol.* **125:** 1157–1167.

French, J.R.J. and Roeper, R.A. (1975). Studies on the biology of the ambrosia beetle *Xyleborus dispar* (F.) (Coleoptera: Scolytidae). *Z. Angew. Entomol.* **78:** 241–247.

French, J.R.J., Robinson, P.J., Thornton, J.D., and Saunders, I.W. (1981). Termite-fungi interactions II. Response of *Coptotermes aciniformis* to fungus-decayed softwood blocks. *Mater. Organ.* **16:** 1–14.

Furniss, M.M., Woo, Y., Deyrup, M.A., and Atkinson, T.H. (1987). Prothoracic mycangium on pine-infesting *Pityoborus* spp. (Coleoptera: Scolytidae). *Ann. Entomol. Soc. Am.* **80:** 692–696.

Furniss, M.M., Solheim, H., and Christiansen, E. (1990). Transmission of blue-stain fungi by *Ips typographus* (Coleoptera: Scolytidae) in Norway spruce. *Ann. Entomol. Soc. Am.* **83:** 712–716.

Furniss, M.M., Harvey, A.E., and Solheim, H. (1995). Transmission of *Ophiostoma ips* (Ophiostomatales: Ophiostomataceae) by *Ips pini* (Coleoptera: Scolytidae) in ponderosa pine in Idaho. *Ann. Entomol. Soc. Am.* **88:** 653–660.

Furniss, R.L. and Carolin, V.M. (1977). Western Forest Insects. USDA Forest Service Miscellaneous Publication No. 1339.

Goldhammer, D.S., Stephen, F.M., and Paine, T.D. (1989). Average radial growth rate and chlamydospore production of *Ceratocystis minor*, *Ceratocystis minor* var. *barrasii*, and SJB 122 in culture. *Can. J. Bot.* **67:** 3498–3505.

Goldhammer, D.S., Stephen, F.M., and Paine, T.D. (1990). The effect of the fungi *Ceratocystis minor* (Hedgcock) Hunt, *Ceratocystis minor* (Hedgcock) Hunt var. *barrasii* Taylor, and SJB 122 on reproduction of the southern pine beetle, *Dendroctonus frontalis* Zimmermann (Coleoptera: Scolytidae). *Can. Entomol.* **122:** 407–418.

Grosmann, H. (1930). Beiträge zur Kenntnis der Lebensgemeinschaft zwischen Borkenkäfern und Pilzen. *J. Parasitenk.* **36:** 56–102.

Happ, G.M., Happ, C.M., and Barras, S.J. (1971). Fine structure of the prothoracic mycangium, a chamber for the culture of symbiotic fungi, in the southern pine beetle, *Dendroctonus frontalis*. *Tissue Cell* **3:** 295–308.

Happ, G.M., Happ, C.M., and Barras, S.J. (1976). Bark beetle fungal symbioses. II. Fine structure of a basidiomycetous ectosymbiont of the southern pine beetle. *Can. J. Bot.* **54:** 1049–1062.

Harrington, T.C. (1987). New combinations in *Ophiostoma* of *Ceratocystis* species with *Leptographium* anamorphs. *Mycotaxonomy* **28:** 39–43.

Harrington, T.C. (1993). Diseases of conifers caused by species of *Ophiostoma* and *Leptographium*. In *Ceratocytis and Ophiostoma: Taxonomy, Ecology, and Pathogenicity* (M.J. Wingfield, K.A. Seifert, and J.F. Webber, Eds.), pp. 161–172. APS Press, St. Paul, MN.

Hausner, G., Reid, J., and Klassen, G.R. (1993a). *Ceratocystiopsis*: a reappraisal based on molecular criteria. *Mycol. Res.* **97:** 625–633.

Hausner, G., Reid, J., and Klassen, G.R. (1993b). On the phylogeny of *Ophiostoma*, *Ceratocystis* s.s., and *Microascus*, and relationships within *Ophiostoma* based on partial ribosomal DNA sequences. *Can. J. Bot.* **71:** 1249–1265.

Holst, E.C. (1937). Aseptic rearing of bark beetles. *J. Econ. Entomol.* **30:** 676–677.

Hsiau, P.T.-W. (1996). The taxonomy and phylogeny of the mycangial fungi from *Dendroctonus brevicomis* and *Dendroctonus frontalis* (Coleoptera: Scolytidae). Ph.D. dissertation, Iowa State University, Ames.

Hunt, D.W.A. and Borden, J.H. (1990). Conversion of verbenols to verbenone by yeasts isolated from *Dendroctonus ponderosae* (Coleoptera: Scolytidae). *J. Chem. Ecol.* **16:** 1385–1397.

Klepzig, K.D. (1998). Competition between a biological control fungus, *Ophiostoma piliferum*, and symbionts of the southern pine beetle. *Mycologia* **90**: 69–75.

Klepzig, K.D. and Wilkens, R.T. (1997). Competitive interactions among symbiotic fungi of the southern pine beetle. *Appl. Environ. Microbiol.* **63:** 621–627.

Klepzig, K.D., Smalley, E.B., and Raffa, K.F. (1995). *Dendroctonus valens* and *Hylastes porculus* (Coleoptera: Scolytidae): vectors of pathogenic fungi (Ophiostomatales) associated with red pine decline disease. *Great Lakes Entomol.* **28:** 81–87.

Klepzig, K.D., Moser, J.C., Lombardero, F.J., Hofstetter, R.W., and Ayres, M.P. (2001). Symbiosis and competition: complex interactions among beetles, fungi and mites. *Symbiosis* **30:** 83–96.

Kok, L.T. (1979). Lipids of ambrosia fungi and the life of mutualistic beetles. In *Insect–Fungus Symbiosis: Nutrition, Mutualism, and Commensalism.* (L. Batra, Ed.), pp. 33–52. John Wiley & Sons, New York.

Kok, L.T. and Norris, D.M. (1973). Comparative sterol compositions of adult female *Xyleborus ferrugineus* and its mutualistic fungal ectosymbionts. *Comp. Biochem. Physiol.* **44:** 499–505.

Kok, L.T., Norris, D.M., and Chu, H.M. (1970). Sterol metabolism as a basis for mutualistic symbiosis. *Nature* **225:** 661–662.

Leach, J.G., Orr, L.W., and Christiansen, C. (1934). Interrelations of bark beetles and blue-staining fungi in felled Norway pine timber. *J. Agric. Res.* **49:** 315–341.

Leufven, A. and Nehls, L. (1986). Quantification of different yeasts associated with the bark beetle, *Ips typographus*, during its attack on a spruce tree. *Microb. Ecol.* **12:** 237–243.

Leufven, A., Bergstrom, G., and Falsen, E. (1984). Interconversion of verbenols and verbenone by identified yeasts isolated from the spruce bark beetle *Ips typographus. J. Chem. Ecol.* **10:** 1349–1361.

Levieux, J., Cassier, P., Guillaumin, L., and Roques, A. (1991). Structures implicated in the transportation of pathogenic fungi by the European bark beetle, *Ips sexdentatus* Boerner: ultrastructure of a mycangium. *Can. Entomol.* **123:** 245–254.

Lewinsohn, D., Lewinsohn, E., Bertagnolli, C.L., and Partridge, A.D. (1994). Blue-stain fungi and their transport structures on the Douglas-fir beetle. *Can. J. For. Res.* **24:** 2275–2283.

Lieutier, F., Vouland, G., Pettinetti, M., Garcia, J., Romary, P., and Yart, A. (1992). Defence reactions of Norway spruce (*Picea abies* Karst.) to artificial insertion of *Dendroctonus micans* Klug (Col., Scolytidae). *J. Appl. Entomol.* **114:** 174–186.

Livingston, R.L. and Berryman, A.A. (1972). Fungus transport structures in the fir engraver, *Scolytus ventralis* (Coleoptera: Scolytidae). *Can. Entomol.* **104:** 1793–1800.

Lu, K.C., Allen, D.G., and Bollen, W.B. (1957). Association of yeasts with Douglas-fir beetle. *For. Sci.* **3:** 336–343.

Malloch, D. and Blackwell, M. (1993). Dispersal biology of the ophiostomatid fungi. In *Ceratocystis and Ophiostoma: Taxonomy, Ecology, and Pathogenicity* (M.J. Wingfield, K.A. Seifert, and J.F. Webber, Eds.), pp. 195–206. APS Press, St. Paul, MN.

Maurer, P., Debieu, D., Malosse, C., Leroux, P., and Riba, G. (1992). Sterols and symbiosis in the leaf-cutting ant *Acromyrmex octospinosus* (Reich) (Hymenoptera, Formicidae: Attini). *Arch. Insect Biochem. Physiol.* **20:** 13–21.

Moein, S.I. and Rust, M.K. (1992). The effect of wood degradation by fungi on the feeding and survival of the West Indian drywood termite, *Cryptotermes brevis* (Isoptera: Kalotermitidae). *Sociobiology* **20:** 29–40.

Mondy, N. and Corio-Costet, M.-F. (2000). The response of the grape berry moth (*Lobesia botrana*) to a dietary phytopathogenic fungus (*Botrytis cinerea*): the significance of fungus sterols. *J. Insect Physiol.* **46:** 1557–1564.

Morales-Ramos, J.A., Rojas, M.G., Sittertz-Bhatkar, H., and Saldana, G. (2000). Symbiotic relationship between *Hypothenemus hampei* (Coleoptera: Scolytidae) and *Fusarium solani* (Moniliales: Tuberculariaceae). *Ann. Entomol. Soc. Am.* **93:** 541–547.

Moser, J.C. and Roton, L.R. (1971). Mites associated with southern pine beetles in Allan Parish, Louisiana. *Can. Entomol.* **103:** 1775–1798.

Moser, J.C., Perry, T.J., Bridges, J.R., and Yin, H.-F. (1995). Ascospore dispersal of *Ceratocystiopsis ranaculosis*, a mycangial fungus of the southern pine beetle. *Mycologia* **87:** 84–86.

Nakashima, T. (1975). Several types of mycetangia found on Platypodid ambrosia beetles (Col. Platypodidae). *Insect Matsum.* **7:** 1–69.

Noirot, C. and Quennedey, A. (1974). Fine structure of insect epidermal glands. *Ann. Rev. Entomol.* **19:** 61–80.

Norris, D.M. (1972). Dependence of fertility and progeny development of *Xyleborus ferrugineus* upon chemicals from its symbiotes. In *Insect and Mite Nutrition* (J.G. Rodriguez, Ed.), pp. 299–310. North Holland, Amsterdam.

Norris, D.M. (1979). The mutualistic fungi of *Xyleborus* beetles. In *Insect–Fungus Symbiosis* (L.R. Batra, Ed.), pp. 53–63. Allanheld, Osmun & Co., Montclair, NJ.

Norris, D.M. and Baker, J.K. (1967). Symbiosis: effects of a mutualistic fungus upon the growth and reproduction of *Xyleborus ferrugineus*. *Science* **156:** 1120–1156.

Norris, D.M., Baker, J.M., and Chu, H.M. (1969). Symbiontic interrelationships between microbes and ambrosia beetles. III. Ergosterol as the source of sterol to the insect. *Ann. Entomol. Soc. Am.* **62:** 413–414.

Paine, T.D. and Birch, M.C. (1983). Acquisition and maintenance of mycangial fungi by *Dendroctonus brevicomis* LeConte (Coleoptera: Scolytidae). *Environ. Entomol.* **12:** 1384–1386.

Paine, T.D., Raffa, K.F., and Harrington, T.C. (1997). Interactions among scolytid bark beetles, their associated fungi, and live host conifers. *Annu. Rev. Entomol.* **42:** 179–206.

Reid, R.W. (1962). Biology of the mountain pine beetle, *Dendroctonus monticolae* Hopkins, in the east Kootenay region of British Columbia. II. Behavior in the host, fecundity, and the internal changes in the female. *Can. Entomol.* **94:** 605–613.

Safranyik, L. (1976). Size- and sex-related emergence, and survival in cold storage, of mountain pine beetle adults. *Can. Entomol.* **108:** 209–212.

Schneider, I.A. and Rudinsky, J.A. (1969). Mycetangial glands and their seasonal changes in *Gnathotrichus retusus* and *G. sulcatus*. *Ann. Entomol. Soc. Am.* **62:** 39–43.

Shearer, C.A. (1995). Fungal competition. *Can. J. Bot.* **73(Suppl. 1):** S1259–S1264.

Shifrine, M. and Phaff, H.J. (1956). The association of yeasts with certain bark beetles. *Mycologia* **48:** 41–55.

Siemasko, W. (1939). Fungi associated with beetles in Poland. *Planta Pol.* **7:** 38–40.

Six, D.L. and Paine, T.D. (1996). *Leptographium pyrinum* is a mycangial fungus of *Dendroctonus adjunctus*. *Mycologia* **88:** 739–744.

Six, D.L. and Paine, T.D. (1997). *Ophiostoma clavigerum* is the mycangial fungus of the Jeffrey pine beetle, *Dendroctonus jeffreyi*. *Mycologia* **89:** 858–866.

Six, D.L. and Paine, T.D. (1998). Effects of mycangial fungi and host tree species on progeny survival and emergence of *Dendroctonus ponderosae* (Coleoptera: Scolytidae). *Environ. Entomol.* **27:** 1393–1401.

Six, D.L. and Paine, T.D. (1999). Phylogenetic comparison of ascomycete mycangial fungi and *Dendroctonus* bark beetles (Coleoptera: Scolytidae). *Ann. Entomol. Soc. Am.* **92:** 159–166.

Solheim, H. and Safranyik, L. (1997). Pathogenicity to Sitka Spruce of *Ceratocystis rufipenni* and *Leptographium abietinum*, blue-stain fungi associated with the spruce beetle. *Can. J. For. Res.* **27:** 1336–1341.

Spatafora, J.W. and Blackwell, M. (1994). The polyphyletic origins of ophiostomatoid fungi. *Mycol. Res.* **98:** 1–9.

Strongman, D.B. (1982). The relationship of some associated fungi with cold-hardiness of the mountain pine beetle, *Dendroctonus ponderosae* Hopkins. M.S. thesis, University of Victoria, B.C., Canada.

Svoboda, J.A., Thompson, M.J., Robbins, W.E., and Kaplanis, J.N. (1978). Insect sterol metabolism. *Lipids* **13:** 742–753.

Thompson, S.N. and Bennett, R.B. (1971). Oxidation of fat during flight of male Douglas-fir beetles, *Dendroctonus pseudotsugae*. *J. Insect Physiol.* **17:** 1555–1563.

Tsuneda, A. (1988). Pleomorphism in beetle-gallery conidial fungi and protoplast reversion. *Proc. Jpn. Acad. Ser. B* **64:** 135–138.

Tsuneda, A. and Hiratsuka, Y. (1984). Sympodial and annelidic conidiation in *Ceratocystis clavigera*. *Can. J. Bot.* **62:** 2618–2624.

Tsuneda, A., Murakami, S., Sigler, L., and Hiratsuka, Y. (1992). Schizolysis of dolipore-parenthosome septa in an arthrooconidial fungus associated with *Dendroctonus ponderosae* and in similar anamorphic fungi. *Can. J. Bot.* **71:** 1032–1038.

Viiri, H. (1997). Fungal associates of the spruce bark beetle *Ips typographus* L. (Col. Scolytidae) in relation to different trapping methods. *J. Appl. Entomol.* **121:** 529–533.

Webb, J.W. and Franklin, R.T. (1978). Influence of phloem moisture on brood development of the southern pine beetle (Coleoptera: Scolytidae). *Environ. Entomol.* **7:** 405–410.

Wetzel, J.M., Ohnishi, M., Fujita, T., Nakanishi, K., Naya, Y., Noda, H., and Sugiura, M. (1992). Diversity in steroidogenesis of symbiotic microorganisms from planthoppers. *J. Chem. Ecol.* **18:** 2083–2094.

Whitney, H.S. (1971). Association of *Dendroctonus ponderosae* (Coleoptera: Scolytidae) with blue-stain fungi and yeasts during brood development in lodgepole pine. *Can. Entomol.* **103:** 1495–1503.

Whitney, H.S. (1982). Relationships between bark beetles and symbiotic organisms. In *Bark Beetles in North American Conifers* (J.B. Mitton and K.B. Sturgeon, Eds.), pp. 183–211. University of Texas Press, Austin.

Whitney, H.S. and Farris, S.H. (1970). Maxillary mycangium in the mountain pine beetle. *Science* **167:** 54–55.

Whitney, H.S. and Blauel, R.A. (1972). Ascospore dispersion in *Ceratocystis* spp. and *Europhium clavigerum* in conifer resin. *Mycologia* **64:** 410–414.

Whitney, H.S. and Cobb, F.W., Jr. (1972). Non-staining fungi associated with the bark beetle *Dendroctonus brevicomis* (Coleoptera: Scolytidae) on *Pinus ponderosa*. *Can. J. Bot.* **50:** 1943–1945.

Whitney, H.S.R., Bandoni, J., and Oberwinkler, F. (1987). *Entomocorticium dendroctoni* gen. et sp. nov. (Basidiomycotina), a possible nutritional symbiote of the mountain pine beetle in lodgepole pine in British Columbia. *Can. J. Bot.* **65:** 95–102.

Wood, S.L. (1982). The bark and ambrosia beetles of North and Central America (Coleoptera: Scolytidae), a taxonomic monograph. Great Basin Natural Memoir 6. 1359 pp.

Wulff, J.L. (1985). Clonal organisms and the evolution of mutualism. In *Population Biology and Evolution of Clonal Organisms* (J.B.C. Jackson, L.W. Buss, and R.E. Cook, Eds.), pp. 437–466. Yale University Press, New Haven, CT.

Yearian, W.C., Gouger, R.J., and Wilkinson, R.C. (1972). Effects of the blue stain fungus, *Ceratocystis ips*, on development of *Ips* bark beetles in pine bolts. *Ann. Entomol. Soc. Am.* **65:** 481–487.

8 Symbiotic Relationships of Tephritids

Carol R. Lauzon

CONTENTS

INTRODUCTION

Members of the family Tephritidae include some of the world's most devastating agricultural pests such as *Ceratitis capitata* (Weidemann), the Mediterranean fruit fly. While much is known about the biology and behavior of many of these pests, little is known about the association these pests have with microorganisms, particularly with bacteria, and the possible roles bacteria play in their life history.

Tephritid pests destroy the marketability and palatability of a variety of host fruits and nuts once an adult female deposits her eggs into the host. The feeding of burrowing and voracious larvae is accompanied by microbial contamination and fruit rot and thus product loss. Introduction of pest tephritids into areas of favorable climate and host availability represents a major threat to agricultural communities worldwide. For example, establishment of the Mediterranean fruit fly in California would result in devastating losses in crops and personal, state, and national income and employment. It is estimated that if Japan alone refused importation of citrus from California, statewide loss of income would be approximately $618 million, and if U.S. national quarantine implementation occurred, each California family would lose approximately $3.6 billion of income (www.cdfa.ca.gov). Researchers are studying aspects of fruit-fly biology that may provide new avenues of prevention and control, yet the study of bacteria–tephritid relationships comes with numerous challenges associated with the complexities of understanding and manipulating multitrophic interactions and microbial ecology.

0-8493-1286-8/03/$0.00+$1.50

GENERAL TERMS FOR DESCRIBING MICROBE–TEPHRITID INTERACTIONS

Early studies on bacteria–tephritid interactions often described any bacteria isolated from life stages as symbionts. This description of the association between bacteria and fruit flies could be flawed because the bacteria may have been contaminants or not necessarily "living together" with the insects. Drew and Lloyd (1991) were both correct and responsible when they stated that the use of the terms symbionts and symbiosis in early studies implied that bacteria and fruit flies were engaged in "mutually obligatory relationships which often were not demonstrated." This assumption has been one of the biggest obstacles to understanding completely relationships between bacteria and tephritids. Therefore, before providing a clear description of bacteria–tephritid interactions we must avoid the tendency to make assumptions and acknowledge, at a minimum, four caveats. First, symbiosis can be defined broadly and simply as "living together" or be tightly and narrowly defined as "an obligatory relationship," with several variations of these two descriptions in between, such as commensalism. Therefore, the term symbiosis should be strictly defined in all cases. Second, bacterial relationships can be associated as (1) external extracellular, (2) internal extracellular, or (3) internal intracellular (Jones, 1984). Thus, bacteria can be and are associated with tephritids in a variety of ways and may have a variety of effects on the insect — negative, positive, or neutral. The third caveat is that members of the family Tephritidae display behaviors and use host plants in ways that are not consistent for all tephritid species. Therefore, an association for one species may not be true for all members of the Tephritidae or even true for species within a single genus. The fourth caveat is that the life history of tephritids includes aspects of plant, soil, vertebrate, and invertebrate microbiology. More than 10 years have passed since the last compilation of work was presented on tephritid–microbe interactions (the reader is referred to Barbosa et al., 1991). This chapter serves to expand on that work and thus present more demonstrative and definitive accounts of bacteria–tephritid associations. A few accounts of other microbe–tephritid interactions are briefly mentioned.

BACKGROUND: EXPOSURE, ACQUISITION, AND TRANSMISSION

The first published accounts of bacterial associations for tephritids began in the early 1900s when Petri (1909) described *Pseudomonas savastanoi*, a bacterial symbiont of the olive fly, *Dacus oleae* (Gmelin). Petri found that the extracellular bacterium inhabited the esophageal diverticulum and an invagination near the distal end of the female ovipositor. Petri speculated that during oviposition the eggs would be coated with bacteria that would eventually enter the micropyle of the eggs and remain there until the bacteria eventually became established within the larval midgut. Though Petri noted that the bacterium was transmitted from one generation to the next, no clear mechanism of this transmission was described, and no role of the bacterium was elucidated.

While a mechanism of transmission remains undefined to date, a possible role for a bacterial species emerged when, in laboratory studies using *D. oleae*, Hagen (1966) found that *P. savastanoi* hydrolyzed protein in olive flesh and inferred that the bacteria synthesized or provided amino acids necessary for larval development. Artificial larval diet containing antibiotics inhibited larval development (Tzanakakis et al., 1983), presumably because important bacteria were inhibited or eliminated; however, it was not determined that the effects were truly due to any lack of important bacteria. It is possible that the antibiotic was toxic to the larvae and exerted some negative effects on insect development, or perhaps the antibiotic interfered with the nutritive quality of larval diet. These questions can and should be addressed for bacteria associated with the olive fly, despite the findings of Luthy et al. (1983), who suggested that the reported symbiont of *D. oleae* was a case of bacterial misidentification. Regardless, *D. oleae* is a serious economic pest and affects olive production worldwide; in addition, enough evidence exists to suggest that bacteria are important in the life history of these flies.

The first purported symbiont of *Rhagoletis pomonella*, Walsh, the apple maggot fly, occurred when Allen and Riker (1932) isolated a bacterium reported as *Pseudomonas (Phytomonas) melopthora* in soft rot of apple tissue. Subsequently, Allen et al. (1934) reported that the bacterium was associated with apple maggot eggs and larvae within rotting host fruit. Continued work by others (i.e., Baerwald and Boush, 1968; Dean and Chapman, 1973; Rossiter et al., 1983) suggested that this microorganism had likely been misidentified and perhaps was not a symbiote of *R. pomonella*. It is evident that the problems associated with this research and that of the olive fly lay with the microbiological accuracy of the research.

Howard et al. (1985) suggested that while bacteria were associated with the apple maggot fly, no bacterium appeared to be in an obligate symbiotic relationship with the apple maggot fly. This contention was questioned when Lauzon (1991) isolated a bacterium, *Enterobacter agglomerans,* from oviposition sites in apple and eggs of *R. pomonella*. The bacterium was often isolated in pure culture prior to invasion by other microorganisms into the oviposition site and larval development. Although an obligatory relationship has yet to be determined between *E. agglomerans* and any tephritid, evidence is presented below that suggests that this bacterium and one other bacterial species, *Klebsiella pneumoniae*, are involved in the life history of several tephritid species.

Since the time of Howard's work, certain bacterial species have been routinely isolated from *R. pomonella,* and other pest tephritids, notably enteric and environmental species such as *Enterobacter* and *Klebsiella* spp. The commonality of isolation of these species from both temperate and tropical tephritids suggests that though bacterial associations may not be obligatory, these two bacterial genera likely play an important role in the life history and, importantly, in the ecology of these pests (Drew and Lloyd, 1991).

The common but not exclusive isolation of certain members of the family Enterobacteriaceae, notably *Enterobacter* and *Klebsiella* spp., internally from pest tephritids and oviposition sites (i.e., Petri, 1909; Allen et al., 1934; Rubio and McFadden, 1966; Hagen, 1966; Boush and Matsumura, 1967; Dean and Chapman, 1973; Rossiter et al., 1983; Howard et al., 1985; Girolami, 1986; Drew and Lloyd, 1987; Lauzon, 1991; Kuzina et al., 2001; Potter, 2001) has stimulated interest for some researchers to continue defining the possible relationships that might exist between tephritids and certain bacterial species. The relationships may range from bacteria serving as food and contributors of host-plant cues to being detoxifiers of synthetic and natural pesticides or plant defensive compounds. It is likely that bacteria provide diverse metabolic and physiological capabilities that impact tephritids in equally diverse ways. It is also possible that no one role may dominate under all environmental conditions. For example, a bacterium may give off odors indicative of quality food when it is actively growing within fecal material; then, the bacterium may subsequently serve as a digester of nitrogen when it resides in an insect gut. Also, one catabolic activity afforded by a bacterium could result in several benefits to a tephritid. For example, degrading host tissue could afford both food and easier movement for a larva throughout a host fruit. Hence, it may be more prudent to view microbial–tephritid associations in holistic, expanded ways than to rely on reductionist approaches from some previous research.

In the early 1970s, Dean and Chapman (1973) found large numbers of bacteria within the crop of *R. pomonella*. The possibility that bacteria might serve as a source of food and not play any obligatory symbiotic role was subsequently evaluated. Drew et al. (1983) found that bacteria served as an adequate breeding diet for *Bactrocera tryoni,* and as a continuation of this work Courtice and Drew (1984) initiated studies that attempted to explain the relationship between bacteria and adult fruit flies from a nutritional standpoint. They suggested that leaf-surface bacteria were important sources of adult food, and once flies ingested bacteria, the bacteria were lysed within the acidic, pH 3 gut and the cellular contents served as nutrients. In addition, Drew and Lloyd (1987) reported that bacteria colonizing *B. tryoni* were utilized by females to produce eggs. Furthermore, in the same paper they linked members of the genera *Enterobacter* and *Klebsiella* spp. with *B. tryoni* and the pest's host plant. They found that prior to oviposition into host fruit, *B. tryoni* would regurgitate these bacteria on the fruit surface, reverse their position, and oviposit at the site of the bacterial droplet. The bacteria would then be transported into the fruit and become localized around the eggs. These bacteria were found to facilitate softening of the host tissue and provide a nutritious environment for developing eggs and larvae. Lloyd et al. (1986) also found that the bacteria would consistently migrate some distance ahead of a burrowing larva, presumably digesting fruit tissue and thus making available an easier course of movement in the fruit for the larva with a complementary availability of nutrients.

Similarly, Lauzon (1991) found that an *Enterobacter* sp. was deposited with eggs into apple flesh by *R. pomonella*; however, the bacteria were deposited not via regurgitation but directly during oviposition. While Ratner and Stoffolano (1984) suggested that fecal material may be the source of enteric bacteria within the oviposition site, Potter (2001) visualized bacteria attached to the internal aspect of ovipositors from *R. completa* and also isolated *Enterobacter* spp. from this region. Perhaps the direct deposition of bacteria through the ovipositor is typical for *Rhagoletis* spp. and atypical for all tephritid pests such as *B. tryoni*. Also contrary to the biology of *B. tryoni*, Hendrichs et al. (1990) found that a mix of phylloplane bacteria — those bacteria that typically reside on leaf surfaces — did not support substantial egg production in *R. pomonella*.

BACTERIA IN THE LIFE HISTORY OF TEPHRITIDS: A LINK TO NITROGEN PROVISIONING AND OTHER METABOLIC ACTIVITIES

It makes sense that bacteria deposited into an oviposition site would provide or facilitate a favorable environment for developing eggs and larvae in many different host fruits and nuts. Most host fruits, for example, have dense, compact tissue that is difficult to penetrate and also lacks nitrogen, a necessary component for egg and larval development. Howard et al. (1985) hypothesized that bacteria gaining entrance into fruit tissue were likely pectinolytic. Thus, one can easily envision that larval migration is made easier and is less energy expensive if the tissue is degraded and softened by bacteria. The *E. agglomerans* strain isolated by Lauzon in apple tissue was confirmed to possess pectinolytic capabilities (Lauzon et al., 1988).

Lack of bioavailable nitrogen in nature is a recurring theme for many living organisms on our planet. The deposition of bacteria during oviposition could feasibly provide adequate nitrogen to developing eggs and larvae on the basis of bacterial biomass alone. Most host fruits contain little nitrogen, and after water, protein is the most abundant constituent of bacteria.

Drew (1988) also reported that essential amino acid content of host tissue increased in the presence of fruit-fly larvae. Miyazaki et al. (1968) found specifically that the amino acid methionine not present in sterile apple tissue was synthesized by a bacterium associated with *R. pomonella* in apple tissue and was bioavailable to larvae. Therefore, it is likely that bacteria within host fruit serve as direct and indirect sources of nitrogen (Fitt and O'Brien, 1985).

To create a more certain and consistent intake of nitrogen, some insects, e.g., termites, have overcome the nitrogen dilemma by forming symbiotic relationships with dinitrogen-fixing microorganisms (Potrikus and Breznak, 1977). Few studies have been conducted to determine if nitrogen fixation occurs within fruit fly guts. Howard et al. (1985) found no evidence to suggest that this activity occurred within the gut of *R. pomonella*; however, Murphy et al. (1988) found nitrogenase activity within *B. tryoni*. If some tephritids possessed such a relationship, it would certainly help those insects overcome the substantial demand for nitrogen for developing larvae and would be of benefit to females who must provision enough nitrogen for developing eggs. Honeydew, a main source of food for most adult tephritid flies in nature, is low in nitrogen (Hagen, 1958; Neilson and Wood, 1966; Nishida, 1980) and does not support maximum egg production. *E. agglomerans* and *Klebsiella* spp. do include strains that are known to fix atmospheric nitrogen, and *E. agglomerans* strains associated with *R. pomonella* have been shown to fix atmospheric nitrogen (C.R. Lauzon, unpublished data). Detailed experimentation is lacking, however, for determining the activity and extent of nitrogen fixation within tephritids.

ATTRACTION OF TEPHRITIDS TO ODORS OF BACTERIA OR ODORS PRODUCED BY BACTERIA

The idea that bacteria are important to tephritids and either serve as food or make the host fruit an environment for eggs and larvae more nutritious was emphasized when the *E. agglomerans* isolated from oviposition sites of *R. pomonella* was found to be attractive to *R. pomonella* flies in field studies (MacCollom et al., 1992). Gravid females appeared particularly attracted to the bacteria (MacCollom et al., 1994). The simultaneous deposition of specific bacteria during oviposition coupled to the attraction to these specific bacteria by *R. pomonella* makes for an intriguing possibility. The attraction to specific bacteria by fruit flies suggests that females that routinely feed on nitrogen-poor food are directed to additional nitrogen by bacteria. This navigation of females by bacteria to more nitrogen, including the nitrogen contained in bacteria themselves, may also benefit eggs and larvae for nutritional and developmental purposes while within the nutrient-insufficient host fruit.

Hodson (1943) and Gow (1954) were the first to describe the use of bacterial odors as attractants for fruit flies, with observations that certain tephritids were highly attracted to proteinaceous baits containing actively growing bacteria. The odors that emanated from baits contained ammonia, a known attractant to many insect species (e.g., Hribar et al., 1992; Taneja and Guerin, 1997). Though ammonia is considered a common attractant for fruit flies (Hodson, 1943; Drew and Fay, 1988; Hendrichs et al., 1990, 1993), in later studies, researchers examined the attraction of *R. pomonella* and other fruit flies to odors emitted from bacteria other than ammonia (i.e., Bateman and Morton, 1981; Drew and Fay, 1988; Robacker and Warfield, 1993; Martinez et al., 1994; Robacker et al., 1998; Epsky et al., 1998). Robacker and Moreno (1995) isolated attractive compounds from *Staphylococcus aureus*, and *K. pneumoniae* and *Citrobacter freundii* (Robacker and Bartelt, 1997) growing in nutrient broth. They methodically tested individual and combinations of the compounds for their attraction to *Anastrepha ludens* and examined volatiles from numerous other bacteria (Robacker et al., 1991, 1993, 1998). Although *S. aureus* is not known to be closely associated with *A. ludens*, odors produced by the bacterium were found to be particularly attractive to the flies. The attractive odors from *S. aureus* are also those commonly emitted from rotting carrion and from the degradation of other substrates containing proteins and lipids. Attractive odors from the degradation of organic material by bacteria is related to documented attraction of fruit flies to bird and other animal feces by Prokopy et al. (1993) and chemical profiles identified from bird feces by Epsky et al. (1997).

Undoubtedly, natural odors serve as important cues and indicators of food for foraging fruit flies in nature. Microbial action appears to contribute substantially to the ability of tephritid flies to locate organic compounds in nature. Prokopy et al. (1993) found that feces lost their initial ability to attract *R. pomonella* adults after treatment with antibiotics. *E. agglomerans* was isolated from the bird feces used in Prokopy's study, and, unlike the *S. aureus* used in Robacker's work, *E. agglomerans* has been found in close association with many fruit flies. As mentioned earlier, MacCollom et al. (1992) found that *R. pomonella* flies were attracted to traps containing washed cells of *E. agglomerans*. In particular, gravid females appeared to home in on the bacterial odors (MacCollom et al., 1994). This is particularly interesting because the attraction behavior of the flies was strong to cells of bacteria that were free of a medium. All previous and subsequent studies on fruit-fly attraction involved compounds produced by bacteria while the bacteria were catabolizing an organic substrate, for example, a protein bait (Gow, 1954), microbiological media (Robacker and Flath, 1995; Robacker and Bartlett, 1997; Robacker et al., 1998), or chicken (Epsky et al., 1997) or duck (Robacker et al., 2000) feces. The isolation, identification, and screening of chemicals emitted from the cells alone would presumably narrow down the numerous chemical candidates that may be serving as potent specific attractants. This was indeed the case when Epsky et al. (1998) isolated ammonia and 3-methyl-1-butanol as the primary attractive compounds for *A. suspensa* from medium-free cells of *E. agglomerans*. Robacker and Lauzon (2002) also isolated candidate attractants from *E. agglomerans* for Mexican fruit fly. Thus, these findings stimulated others to explore the attractive influences of *E. agglomerans* cells toward *R. pomonella* and other fruit fly species. *E. agglomerans* has been reported to be a mediate attraction of locusts (Dillon et al., 2000) and brown treesnakes (Jojola-Elverum et al., 2001).

Initially, however, others did not find the consistent behavioral responses to *E. agglomerans* described by MacCollom et al. (1992, 1994) for other fruit fly species. The contrary findings of Jang and Nishijima (1990) using Oriental fruit flies stimulated a closer examination by Lauzon of strains of *E. agglomerans* for potential differences in their ability to attract fruit flies. Lauzon et al. (1998) examined the attractive qualities of *E. agglomerans* isolated from a variety of sources that were associated with fruit flies, such as bird feces, foliar surfaces, and the gut of the fly, in an attempt to explain the contrary or inconsistent results of others. They found that *R. pomonella* were indeed preferentially attracted to certain strains of *E. agglomerans* and those strains that flies were attracted to contained catabolic capabilities, an enzyme, in their outer membranes not seen in nonattractive strains (Lauzon et al., 2000) identified as uricase. These findings provided an explanation for the inconsistent results reported by others (see Robacker and Lauzon, 2002, for an example of this concept) and suggested another example of attraction between fruit flies and nutrition.

TEPHRITID–MICROBE–PLANT INTERACTIONS

While microbial odors assist or mediate foraging fruit flies to find valuable sources of nitrogen and bacteria provide nitrogen in nitrogen-poor host plants, Drew and Lloyd (1991) hypothesized that odors from "fruit-fly-type" bacteria, namely, *E. cloacae* and *K. oxytoca*, also played a role in host-plant location. They suggested that as fruit flies entered host plants, host plant structures accumulated *Enterobacter* and *Klebsiella* spp. They suggested that over time bacterial numbers would increase on the phylloplane and thus an increase in bacterial odors in the host-plant area would naturally follow. These odors would facilitate host location for nearby, migrating flies. While this phenomenon remains to be definitively proven, Raghu et al. (2002) and Lauzon et al. (unpublished data) have additional data that support this hypothesis. Also, Epsky et al. (1997) and Robacker and Lauzon (2002) have found that odors from *E. agglomerans* are considered typical fruit and plant odors. Mediation of host-plant location by bacterial odors is known to occur for other insect pests, such as the bark beetle (Brand et al., 1975), and thus it is feasible that bacteria influence fruit fly behavior in host detection and location.

Enterobacter and *Klebsiella* spp. populations on the phylloplane of tephritid host plants may be doing more than attracting fruit flies. They may be catabolizing plant allelochemicals that are either toxic to fruit flies or that may interfere with the nutritive quality of compounds ingested by foraging fruit flies. Jones (1984) and Dowd (1991) suggested that symbionts might be important in the detoxification of toxic and noxious compounds that insects encounter in nature. Howard et al. (1985) hypothesized that bacteria associated with *Rhagoletis* performed these functions. Detoxification of plant allelocompounds has been studied extensively for herbivorous insects (Appel, 1993a,b); however, information regarding detoxification strategies of nonfolivorous insects is scant. *E. agglomerans* isolated from *R. pomonella* has been shown to degrade and detoxify the organophosphate azinphosmethyl and phloridzin, a dihydroxychalcone found typically in *Malus* spp., to sublethal levels for *R. pomonella* adults (Lauzon et al., 1994). It is possible that this bacterium and other bacteria on or within host plants may be abating or remediating the toxicity of plant defensive compounds. The same may be true for bacteria that reside on the phylloplane within the host fruit or within the alimentary canal of tephritids. Bacterial utilization of highly toxic compounds may be the reason why certain fruit flies can survive within certain hosts. This type of involvement of microorganisms in plant allelochemistry and insect behavior is poignantly made by Berenbaum (1988) and expanded in her chapter in Barbosa and Letourneau (1988).

TEPHRITIDS AND INTERNAL EXTRACELLULAR BACTERIA

While most of the bacterial associations discussed thus far have been external in nature, it is important to address the repeated isolation of *Enterobacter* and *Klebsiella* spp. internally from the alimentary-canal organs of several tephritid species, for both adult and larval forms. Despite the numerous microbial species that *R. pomonella* consume through their feeding and foraging activities, only two internal extracellular bacterial species are routinely isolated from the alimentary canal organs, *Enterobacter* and *Klebsiella* spp. (Lauzon et al., 1998). Lauzon (1991) isolated these two species from all life stages of *R. pomonella*, and Potter (2001) isolated these two species from all life stages of *R. completa*. Similarly, these bacterial species have been isolated from the alimentary canal organs of other tephritids, such as *C. capitata* (Girolami, 1986, Marchini et al., 2002), *B. tryoni* (Drew and Lloyd, 1991), *A. ludens* (Rubio and McFadden, 1966; Kuzina et al., 2001), *A. suspensa* (C.R. Lauzon, unpublished data), and *R. completa* (Tsiropoulous, 1976). Lauzon and Potter (1999) have found that these two bacterial species form an extensive biofilm within the esophageal bulb, crop, and intestines as well as on the apical end of eggs within ovaries (Lauzon et al., 2002, unpublished data). This specialized, complex assemblage of two bacterial species further supports the hypothesis (Lauzon et al., 1998 and 2000) that *Enterobacter* and *Klebsiella* spp. jointly participate in the cycling of nitrogen. Keilin (1913) and Vijaysegaran et al. (1997) suggested that pores, acting as filters in the mouthparts of tephritids, regulated the size and thus the type of microorganisms that enter the tephritid alimentary canal. Although *Enterobacter* and *Klebsiella* spp. appear to dominate the alimentary canal biofilms, other microbial species may inhabit the biofilm and be viable but nonculturable species. Indeed, other bacterial species, such as *Escherichia coli*, have been isolated from fruit flies (i.e., Lauzon, 1991). It is likely that these bacteria are planktonic, transient species that are either digested or pass through the alimentary canal and exit the fly rather quickly. Currently, with collaborative efforts from Edouard Jurkevitch and Boaz Yuval of Hebrew University, we are using molecular techniques to define the bacterial communities that comprise the biofilms.

Few surveys of insect intestines include attempts to isolate anaerobic bacteria. Anaerobic bacteria are typical inhabitants of the intestinal environments of vertebrate species; however, Potter (2001) surveyed the intestines of *R. completa* Cresson, the walnut husk fly, and did not isolate any anaerobic bacteria from alimentary canal organs of adult and larval *R. completa*. If *Enterobacter* and *Klebsiella* spp. act as the sole important bacterial dual-species system within the fruit fly gut, then this simple system can be manipulated more easily and precisely than one that contains numerous species with numerous interactions (Banks and Bryers, 1991; Siebel and Characklis, 1991). Such research will facilitate a more comprehensive and complete understanding of the population dynamics of these bacteria within the fly gut and likely lead to a comprehensive understanding of the roles these bacteria play with the nutritional and, possibly, the reproductive physiology of the flies.

BIOFILMS WITHIN DIGESTIVE AND REPRODUCTIVE ORGANS AND TEPHRITIDS

In the past, the coordinated assembly of bacteria into a biofilm was generally thought to be a response of microorganisms to unfavorable or suboptimal environmental conditions (Atlas and Bartha, 1987). Today, most researchers contend that biofilms are the typical form of microbial communities and that the biofilm structure itself can afford much information about the functions that occur within the biofilm (Costerton et al., 1987, 1995). The microbial species that comprise the biofilm also provide information regarding the metabolic processes or population dynamics that occur within the community. In an attempt to learn more about the structure and function of the biofilms within tephritids, my laboratory initiated an examination of the alimentary organs of adult *R. pomonella, R. completa, A. ludens,* and *C. capitata* using electron and confocal laser scanning microscopy (C.R. Lauzon and

S.E. Potter, unpublished data). Dense biofilms were found within the midgut of these flies, with less complex but prevalent biofilms elsewhere in the alimentary tract. Biofilms were also found in all life stages of *R. completa* (Potter, 2001) and *C. capitata* (Lauzon et al., 2002).

Biofilms form within the adult tephritid within the first few hours post eclosion. We found that upon emergence, *C. capitata*, *A. ludens*, *R. pomonella*, and *R. completa* flies lack a peritrophic membrane and, thus exposed, lush carpets of microvilli extend into the lumen and with very few bacteria present. A peritrophic membrane, or matrix, forms within hours, and subsequently bacterial numbers increase. Initially, the bacteria can be seen studded primarily along the perimeter of the peritrophic membrane. Bacterial numbers continue to increase, and microcolonies and subsequently late-stage biofilms form. A complete biofilm is formed typically within a tephritid gut 12 to 18 h post eclosion. Planktonic bacteria, either ingested as natural food or from the outer layers of gut microcolonies, likely arrive at and leave these biofilms routinely.

To further characterize biofilm formation within the fruit fly gut, we used strains of *E. agglomerans* and *K. pneumoniae* that were transformed to express different colored fluorescent proteins (Peloquin et al., 2000, 2002) and monitored their establishment within the *R. completa* and *C. capitata* gut. Newly emerged flies fed on diet containing the bacteria, and we subsequently examined their alimentary canal and reproductive organs for the presence and arrangement of the bacterial species. We paid particularly close attention to the density and channels that defined the biofilm structures. To reiterate, the structure of a biofilm can lend important insight to the functions that occur within a biofilm. Though biochemical events are often the defining parameters of a biofilm, we cannot overlook the fact that physically these biofilms prevent nonindigenous or threatening microbial species from becoming established in the tephritid alimentary canal organs and provide a physical barrier to toxic or noxious compounds. Biofilms likely direct ingested materials that are not digested efficiently through the alimentary canal.

We found that *Enterobacter* and *Klebsiella* spp. colonized rapidly within the alimentary canals of both fruit fly species. Similarly, the different bacterial strains routinely colocalized, notably in areas bordering channels and pores that coursed throughout and meandered within the biofilms. There were areas within the biofilm where the two species did not colocalize, but typically the colocalized bacteria were present at the borders of the channels and pores. Indeed, this suggests that colocalization relates to the joint participation of *Enterobacter* and *Klebsiella* spp. in nitrogen cycling. Nitrogenous uric acid enters the midgut and *Enterobacter* spp. that produce enzymes, such as uricase, that catabolize the uric acid to urea. *Klebsiella* spp. typically produce urease that catabolizes the urea to ammonia or ammonium products that are assimilated across the fly epithelium and used in anabolic processes for the fly (Lauzon et al., 2000). Urease activity does occur within the midgut of *C. capitata* and *R. completa*, and this enzymatic action resulted in the production of ammonia or ammonium products produced by bacteria-degrading protein including uric acid (Lauzon and Potter, 1999).

Catabolic processes are unlikely to be the exclusive biochemical activities that occur with the gut biofilms. We have found that biofilms also exist in the crops of *C. capitata*, although their structure is not as extensive as those seen in the midgut of the flies. Though the crop is considered to be a holding chamber for ingested materials, Lu and Teal (2001) found that mating pheromone components were produced within the crop of *A. suspensa*, the Caribbean fruit fly. It is possible that bacteria within the crop may contribute to the synthesis of these compounds. *Enterobacter* and *Klebsiella* spp. have been isolated from the Caribbean fruit fly (C.R. Lauzon and N.D. Epsky, 1990, unpublished results). Epsky et al. (1998) found that *A. suspensa* were attracted to odors produced by *E. agglomerans*. It is likely that different biofilms in different areas of the alimentary tract of fruit flies are engaged in a variety of metabolic activities.

While examining the gut using our transformants we found that the bacteria also migrated to the ovaries and formed biofilms on the egg surface. Moreover, the transformants were detected in all life stages of *C. capitata* through two successive generations (C.R. Lauzon et al., 2002, unpublished results). The vertical transmission of these two bacterial species fortifies our hypothesis that these bacteria are important in the life history of pest tephritids and perhaps other tephritid species. We now see a continuum with the discoveries that female flies are particularly attracted to washed cells of *E. agglomerans* and, to some extent, *K. pneumoniae*. The bacteria are transmitted vertically to the eggs via ingestion of the bacteria by female flies, and the bacteria are present within oviposition sites of *R. pomonella* and carried throughout all life stages.

This scenario suggests that these bacteria are important in the life history of the flies. The work of Marri et al. (1996) and Marchini et al. (1997) supports this contention. They describe an antibacterial substance, Ceratotoxin A produced by female *C. capitata*, that is also found on the surface of *C. capitata* eggs. This compound exhibited killing action on *E. coli*. A substance that would inhibit or kill *E. coli* should theoretically do the same to the closely related enterics, *E. agglomerans* and *Klebsiella* spp.; however, Marri et al. (1996) found that *E. cloacae* was not as sensitive to the compound as *E. coli*. Therefore, it appears that some selection factors exist for these two bacterial species within the gut and on the egg surfaces.

Our current studies include further defining the architecture and other metabolic activities that occur within gut and egg biofilms, including using molecular techniques to ascertain if the bacteria that compose the biofilm contain viable but nonculturable bacterial species. Information on the species that comprise a biofilm can be just as important as the architecture of a biofilm in terms of structure and function. The microbial species that inhabit biofilms within female flies may be different from those found in male flies, and these differences could be due to nutritional or physiological requirements. Lauzon (1991) found that female apple maggot flies harbored a variety of strains of *E. coli* in their alimentary canal organs contrary to the dominance of *E. coli* strains typical of warm-blooded animal feces origin in male-fly digestive tracts. This difference could be due to feeding behavior, need, or a combination of the two.

MICROOGANISMS AS POTENTIAL BIOCONTROL AGENTS OF TEPHRITIDS

Fruit flies consume a variety of microorganisms as they feed on fecal material, rotting fruit, and honeydew. In these natural food sources, pathogens of vertebrates and invertebrates exist. Few accounts exist, however, describing pathogens that inflict disease for fruit flies, and the information that does exist (e.g., Jacques et al., 1969; Robinson and Hooper, 1989) does not report specific causative agents and associated pathogenicity. *Serratia marcescens* has been isolated from *C. capitata* Weidemann and *Dacus* (*Bactrocera*) *dorsalis* Hendel flies (Grimont and Grimont, 1978) and from diseased or dead *R. pomonella* (Lauzon et. al., 2002). *S. marcescens* has been considered a minor threat to insects in the past (Steinhaus, 1959); however, recent reports suggest that virulent strains of this bacterium exist (Farrar et al., 2001; Lauzon et al., 2002). *S. marcescens* cells cannot be considered a biocontrol agent for pest tephritids or other agricultural pests because it is a pathogen of humans and other animals (Holt and Krieg, 1992). Products such as toxins produced by *S. marcescens* may show some promise for fruit fly control. The use of toxins, toxic bacterial spores, or toxin-producing bacteria, such as *Bacillus thuringiensis*, has been tested for some fruit fly species (Robacker et al., 1996; Navrozidis et al., 2000). Although not routinely used for fruit fly control, a strain of *B. thuringiensis* isolated in Greece and applied to olives provided significant protection against *Bactrocera oleae* infestation (Navrozidis et al., 2000).

Viruses have been found to infect some fruit flies. Cricket Paralysis Virus was reported to be a potential biocontrol agent for *B. oleae*, the olive fruit (Manousis and Moore, 1987). A nonoccluded reovirus was also isolated from the olive fly and found to be pathogen *per os* and via injection (Anagnou-Veroniki et al., 1997). Its potential as a biocontrol agent, however, remains unclear. Most reports of virus–tephritid interactions describe and partially characterize viruses only. For example, several different reoviruses have been documented to infect *C. capitata* (Plus et al., 1981a,b), but field tests have not been implemented.

Wolbachia, the fascinating intracellular bacterium that exerts cytoplasmic incompatibility and other powerful effects, has also been found to inhabit cells of *Bactocera* spp. (Kittayapong et al., 2000). The promise of using *Wolbachia* spp. as biological control agents of agricultural pests has been discussed (O'Neill et al., 1997; Bourtzis and O'Neill, 1998) and may include control of tephritid pests. A comprehensive survey of pest tephritids for the presence of *Wolbachia* spp. and other intracellular species has not been performed. Similarly, pathogens, such as microsporidia, have not been vigorously sought after; however, they have been reported to cause disease in tephritids, such as the Oriental fruit fly, *Dacus dorsalis* Hendel (Fujii and Tamashiro, 1972).

USE OF PROBIOTICS IN MASS-REARING PROGRAMS AND STERILE INSECT TECHNIQUE

Contaminating microorganisms are a major problem at mass-rearing fruit fly facilities. These organisms have not been candidates for biological control agents because many of the microbial species that cause disease in insects would also cause disease in humans and other animals. Measures implemented to avert disease in insects also work to avoid contamination of diets and rearing substrates by microorganisms that cause disease or decrease the nutritive and environmental conditions of the substrates, i.e., pH, temperature, consistency, and viscosity. Antimicrobial agents are often added to larval and adult diets; however, expense, issues concerning antimicrobial resistance, and environmental-impact assessment of disposing diets containing these materials present real obstacles for corrective measures.

Complete and accurate monitoring and assessment of the health of insect life stages and the presence of microorganisms in rearing facilities is usually not done. This would accomplish many goals. It would maintain consistent production, detect presence of pathogens before they reach a level that would threaten quality assurance and control as well as output, decrease costs by using corrective chemicals or treatment only when necessary, and allow use of corrective measures specific for the pathogen. The lack of information about disease in tephritids limits our ability to diagnose and implement specific and effective treatments. This direction would also provide information about the production and use of acceptable pathogens for biological control.

A new applied aspect of tephritid–bacteria research is the use of endosymbionts as probiotics. Evidence is mounting that probiotic supplements play an important role in animal and human nutrition and health, and thus it stands to reason that similar effects should apply to insects, especially given the accounts of obligatory symbioses for certain insects. For example, *Sitophilus oryzae* have been shown to be unable to fly without their endosymbionts (Nardon and Grenier, 1991). Probiotics may hold much promise for improving the performance and fitness of mass-reared insects, particularly tephritids. Probiotics have been shown to inhibit pathogens and alter rates of apoptosis, a known effect of radiation exposure. Theoretically, probiotics would be added to adult and larval diets, and the normal microbiota would rapidly establish as biofilms within these stages. Normal metabolic and physical benefits would occur, radiation damage would be repaired, and, as an extra benefit, the endosymbionts would competitively exclude contaminants that routinely establish within the diets. Diet quality may be positively affected, and incidence of disease may decrease. This is the hope, of course, behind the use of probiotics. Current studies are under way to determine the true merits behind this idea, with preliminary data suggesting that probiotics may indeed be a positive addition to mass-rearing protocols and lead to improvements in SIT (sterile insect technique) effectiveness (Lauzon et al., 2000, unpublished).

CONCLUSIONS

Since the first accounts of tephritid–bacteria associations, we have moved from the basic observation of fruit fly attraction, to proteinaceous compounds containing bacteria, to those of complex dynamics. These observations include: (1) attraction behavior mediated by catabolic capabilities of bacteria, (2) selection of specific bacteria that establish within the fruit fly gut, (3) migration and establishment of ingested bacteria to egg surfaces, and (4) vertical transmission of certain bacteria throughout life stages and subsequent generations. We cannot forget that certain bacteria associated with tephritids may be involved in mediating host-plant location and detoxifying toxic plant compounds as well as facilitating larval feeding within host fruits. In summary, it is likely that specific bacteria are essential components of both the nutritional and reproductive physiology of tephritids, involve host plant–insect relationships, and overall are major contributors to the success or failure of tephritid insects.

ACKNOWLEDGMENTS

I thank Sarah Potter for her assistance with research and T.A. Miller and K. Bourtzis for their editorial comments. Work reported here was supported by grants from the California Department of Food and Agriculture, the California Citrus Research Board, and the United States Department of Agriculture.

REFERENCES

Allen, T.C. and Riker, A.J. (1932). A rot of apple fruit caused by *Phytomonas melopthora* N. sp., following invasion by the apple maggot. *Phytopathology* **22:** 557–571.

Allen, T.C., Pinckard, J.A., and Riker, A.J. (1934). Frequent association of *Phytomonas melopthora* with various stages in the life cycle of the apple maggot, *Rhagoletis pomonella*. *Phytopathology* **24:** 228–238.

Anagnou-Veroniki, M., Veyrunes, J-C., Kuhl, G., and Bergoin, M. (1997). A nonoccluded reovirus of the olive fly, *Dacus oleae*. *J. Gen. Virol.* **78:** 259–263.

Appel, H.M. (1993a). Phenolics in ecological interactions: the importance of oxidation. *J. Chem. Ecol.* **19:** 1521–1551.

Appel, H.M. (1993b). The chewing herbivore gut lumen: physiochemical conditions and their impact on plant nutrients, allelochemicals, and insect pathogens. In *Insect–Plant Interactions*, Vol. V (E.A. Bernays, Ed.), pp. 209–223. CRC Press, Boca Raton, FL.

Atlas, R.M. and Bartha, R. (1987). *Microbial Ecology: Fundamentals and Applications*. Benjamin Cummings, Menlo Park, CA.

Baerwald, R.J. and Boush, G.M. (1968). Demonstration of the bacterial symbiote *Pseudomonas melopthora* in the apple maggot, *Rhagoletis pomonella*, by fluorescent-antibody techniques. *J. Invertebr. Pathol.* **11:** 251–259.

Banks, M.K. and Bryers, S.D. (1991). Bacterial species dominance within a binary culture biofilm. *Appl. Environ. Microbiol.* **57:** 1974–1979.

Barbosa, P. and Letourneau, D.K. (1988). *Novel Aspects of Insect–Plant Interactions*. Wiley Interscience, New York.

Barbosa, P., Krischik, V.A., and Jones, C.G., Eds. (1991). *Microbial Mediation of Plant–Herbivore Interactions*. John Wiley & Sons, New York.

Bateman, M.A. and Morton, T.C. (1981). The importance of ammonia in proteinaceous attractants of fruit flies (Family: Tephritidae). *Aust. J. Agric. Res.* **32:** 883–903.

Berenbaum, M.R. (1988). Allelochemicals in insect-microbe-plant interactions; agent provocateurs in the coevolutionary arms race. In *Novel Aspects of Insect–Plant Interactions* (P. Barbosa and D. Letourneau, Eds.), pp. 97–123. John Wiley & Sons, New York.

Bourtzis, K. and O'Neill, S.L. (1998). *Wolbachia* infections and arthropod reproduction. *BioScience* **48:** 287–293.

Boush, G.M. and Matsumura, F. (1967). Insecticidal degradation by *Pseudomonas melopthora*, the bacterial symbiote of the apple maggot. *J. Econ. Entomol.* **60:** 918–920.

Brand, J.M., Bracke, J.W., Markovetz, A.J., Wood, D.L., and Browne, L.E. (1975). Production of verbenol pheromone by a bacterium isolated from bark beetles. *Nature* **254:** 136–137.

Costerton, J.W., Cheng, K.J., Geesey, G.G., Ladd, T.I., Nickel, J.C., Dasgupta, M., and Marrie, T.J. (1987). Bacterial biofilms in nature and disease. *Annu. Rev. Microbiol.* **41:** 435–464.

Costerton, J.W., Lewandowski, Z., Caldwell, D.E., Korber, D.R., and Lappin-Scott, H.M. (1995). Microbial biofilms. *Annu. Rev. Microbiol.* **49:** 711–745.

Courtice, A.C. and Drew, R.A.I. (1984). Bacterial regulation of abundance in tropical fruit flies (Diptera: Tephritidae). *Aust. Zool.* **21:** 251–268.

Dean, R.W. and Chapman, P.J. (1973). Bionomics of the apple maggot in eastern New York. *Search Agric. Entomol.* **3:** 1–62.

Dillon, R.J., Vennard, C.T., and Charnley, A.K. (2000). Pheromones: exploitation of gut bacteria in the locust. *Nature* **403:** 851.

Dowd, P.F. (1991). Symbiont-mediated detoxification in insect herbivores. In *Microbial Mediation of Plant–Herbivore Interactions* (P. Barbosa, V.A. Krischik, and C.G. Jones, Eds.), pp. 411–440. John Wiley & Sons, New York.

Drew, R.A.I. (1988). Amino acid increases in fruit infested by fruit flies of the family Tephritidae. *Zool. J. Linn. Soc.* **93:** 107–112.

Drew, R.A.I. and Faye, H.A. (1988). Elucidation of the roles of ammonia and bacteria in the attraction of *Dacus tryoni* (Froggatt) (Queensland fruit fly) to proteinaceous suspensions. *J. Plant Prot. Trop.* **5:** 127–130.

Drew, R.A.I. and Lloyd, A.C. (1987). Relationship of fruit flies (Diptera: Tephritidae) and their bacteria to host plants. *Ann. Entomol. Soc. Am.* **80:** 629–636.

Drew, R.A.I. and Lloyd, A.C. (1991). Bacteria in the life cycle of tephritid fruit flies. In *Microbial Mediation of Plant–Herbivore Interactions* (P. Barbosa, V. Krischik, and C. Jones, Eds.), pp. 441–465. John Wiley & Sons, New York.

Drew, R.A.I., Courtice, A.C., and Teakle, D.S. (1983). Bacteria as a natural source of food for adult fruit flies (Diptera: Tephritidae). *Oecologia (Berlin)* **60:** 279–284.

Epsky, N.D., Dueben, B.D., Heath, R.R., Lauzon, C.R., and Prokopy, R.J. (1997). Attraction of *Anastrepha suspensa* (Diptera: Tephritidae) to volatiles from avian fecal material. *Fla. Entomol.* **80:** 270–277.

Epsky, N.D., Heath, R.R., Dueben, B.D., Lauzon, C.R., Proveaux, A.T., and MacCollom, G.B. (1998). Attraction of 3-methyl-1-butanol and ammonia identified from *Enterobacter agglomerans* to *Anastrepha suspensa*. *J. Chem. Ecol.* **24:** 1867–1880.

Farrar, R.R., Martin, P.A., and Ridgeway, R.L. (2001). A strain of *Serratia marcescens* (Enterobacteriaceae) with high virulence *per os* to larvae of a laboratory colony of the corn ear worm (Lepidoptera: Noctuidae). *J. Entomol. Sci.* **36:** 380–390.

Fitt, G.P. and O'Brien, R.W. (1985). Bacteria associated with four species of Dacus (Diptera: Tephritidae) and their role in the nutrition of larvae. *Oecologia (Berlin)* **67:** 447–454.

Fujii, J.K. and Tamashiro, M. (1972). *Nosema tephritidae* sp. N., a microsporidium pathogen of the Oriental fruit fly, *Dacus dorsalis* Hendel. *Proc. Hawaiian Entomol. Soc.* **21:** 191–203.

Girolami, V. (1986). Mediterranean fruit fly associated bacteria: transmission and larval survival. In *Pest Control: Operations and Systems Analysis in Fruit Fly Management,* NATO ASI Series, Vol. G 11, (M. Mangel, Ed.), pp. 135–146. Springer-Verlag, Berlin.

Gow, P.L. (1954). Proteinaceous bait for the Oriental fruit fly, *Ceratitis capitata* Weid. II. Biology and control. *Bull. Soc. Found. Lett. Entomol.* **31:** 251–285.

Grimont, P. and Grimont, F. (1978). The genus *Serratia*. *Rev. Microbiol.* **32:** 221–248.

Hagen, K.S. (1958). Honeydew as an adult fruit fly diet affecting reproduction. *Proc. 10th Int. Congr. Entomol.* **3:** 25–30.

Hagen, K.S. (1966). Dependence of the olive fruit fly, *Dacus oleae*, larvae on symbiosis with *Pseudomonas savastanoi* for the utilization of olive. *Nature* **209:** 423–425.

Hendrichs, J., Lauzon, C.R., Cooley, S.S., and Prokopy, R.J. (1990). What kinds of foods do apple maggot flies need for survival and reproduction? *Mass. Fruit Notes* **55:** 3.

Hendrichs, J., Lauzon, C.R., Cooley, S.S., and Prokopy, R.J. (1993). Contribution of natural food sources to adult longevity and fecundity of *Rhagoletis pomonella* (Diptera: Tephritidae). *Ann. Entomol. Soc. Am.* **86:** 250–264.

Hodson, A.C. (1943). Lures attractive to the apple maggot. *J. Econ. Entomol.* **38:** 545–548.

Holt, J.G. and Krieg, N.R., Eds. (1992). *Bergey's Manual of Determinative Bacteriology.* Williams & Wilkins, Baltimore, MD.

Howard, D.J., Bush, G.L, and Breznak, J.A. (1985). The evolutionary significance of bacteria associated with *Rhagoletis*. *Evolution* **39:** 405–417.

Hribar, L.J., Leprince, D.J., and Foil, L.D. (1992). Ammonia as an attractant for adult *Hybomitra lasiophthalma* (Diptera: Tabanidae). *J. Med. Entomol.* **29:** 346–348.

Jacques, R.P., Neilson, W.T.A., and Huston, F. (1969). A bloating disease of adults of the apple maggot. *J. Econ. Entomol.* **62:** 850–851.

Jang, E.B. and Nishijima, K.A. (1990). Identification and attractancy of bacteria associated with *Dacus dorsalis* (Diptera: Tephritidae). *Environ. Entomol.* **19:** 1726–1751.

Jojola-Elverum, S.M., Shivik, J.A., and Clark, L. (2001). Importance of bacterial decomposition and carrion substrate to foraging brown treesnakes. *J. Chem. Ecol.* **27:** 1315–1331.

Jones, C.G. (1984). Microorganism as mediators of plant resource exploitation by insect herbivores. In *A New Ecology: Novel Approaches to Interactive Systems* (P.W. Price, C.N. Slobodchikoff, and W.S. Gaud, Eds.), pp. 53–99. John Wiley & Sons, New York.

Keilin, D. (1913). Sur les conditions de nutrition de certaines larvaes de Dipteres parasites de fruits. *C. R. Soc. Biol.* **74:** 24–26.

Kittayapong, P., Milne, J.R., Tigvattananont, S., and Baimai, V. (2000). Distribution of the reproduction-modifying bacteria, *Wolbachia*, in natural populations of Tephritid fruit flies in Thailand. *Sci. Asia* **26:** 93–103.

Kuzina, L.V., Peloquin, J.J., Vacek, D.C., and Miller, T.A. (2001). Isolation and identification of bacteria associated with adult laboratory Mexican fruit flies, *Anastrepha ludens* (Diptera: Tephritidae). *Curr. Microbiol.* **42:** 290–294.

Lauzon, C.R. (1991). Microbial ecology of a fruit fly pest, *Rhagoletis pomonella* (Walsh) (Diptera: Tephritidae). Ph.D. dissertation, University of Vermont, Burlington.

Lauzon, C.R. (2000). Co-localization and other interactions between enteric bacteria in the tephritid gut. Presentation, Joint Annual Meeting Entomological Society of Canadian Entomology Society of America, Montreal, Quebec.

Lauzon, C.R. and Potter, S.E. (1999). Biochemical evidence for the selection of certain enterobacteriaceae in two pest tephritids. Presentation, Entomolological Society of America National Meeting, Atlanta.

Lauzon, C.R., Rutkowski, A.A., Currier, W.W., and MacCollom, G.B. (1988). The role of a newly described microorganism in the life cycle of *Rhagoletis pomonella*. Proceedings 18th International Congress on Entomology, Vancouver, B.C.

Lauzon, C.R., Robert, B.J., Bussert, T.G., and Prokopy, R.J. (1994). Could bacteria in nature be detoxifying compounds for the apple maggot fly? *Mass. Fruit Notes* **59:** 1–3.

Lauzon, C.R., Sjogren, R.E., Wright, S.E., and Prokopy, R.J. (1998). Attraction of *Rhagoletis pomonella* (Diptera: Tephritidae) flies to odor of bacteria: apparent confinement to specialized members of Enterobacteriaceae. *Environ. Entomol.* **27:** 853–857.

Lauzon, C.R., Sjogren, R.E., and Prokopy, R.J. (2000). Enzymatic capabilities of bacteria associated with apple maggot flies: a postulated role in attraction. *J. Chem. Ecol.* **26:** 953–957.

Lauzon, C.R., Bussert, T.G., and Prokopy, R.J. A strain of *Serratia marcescens* pathogenic to *Rhagoletis pomonella* (Diptera: Tephritidae). *Eur. J. Entomol.* In press.

Lloyd, A.C., Drew, R.A.I., Teakle, D.S., and Howard, A.C. (1986). Bacteria associated with some *Dacus* species (Diptera: Tephritidae) and their host fruit in Queensland. *Aust. J. Biol. Sci.* **39:** 361–368.

Lu, F. and Teal, P.E.A. (2001). Sex pheromone components in oral secretions and crop of male Caribbean fruit flies, *Anastrepha suspensa* (Loew). *Arch. Insect Biochem. Physiol.* **48:** 144–154.

Luthy, P., Struder, D., Jaquet, F., and Yamvrias, C. (1983). Morphology and *in vitro* cultivation of the bacterial symbiote of *Dacus oleae*. *Mitt. Schweiz. Entomol. Ges.* **56:** 67–72.

MacCollom, G.B., Lauzon, C.R., Weires, R.W., and Rutkowski, A.A., Jr. (1992). Attraction of adult apple maggot (Diptera: Tephritidae) to microbial isolates. *J. Econ. Entomol.* **85:** 83–87.

MacCollom, G.B., Lauzon, C.R., Payne, E.B., and Currier, W.W. (1994). Apple maggot (Diptera: Tephritidae) trap enhancement with washed bacterial cells. *Environ. Entomol.* **23:** 354–359.

Manousis, T. and Moore, N.F. (1987). Cricket paralysis virus, a potential control agent for the olive fruit fly, *Dacus oleae* Gmel. *Appl. Environ. Microbiol.* **53:** 142–148.

Marchini, D., Marri, L., Rosetto, M., Manetti, A., and Dallai, R. (1997). Presence of antibacterial peptides on the laid egg chorion of the medfly *Ceratitis capitata*. *Biochem. Biophys. Res. Commun.* **240:** 657–663.

Marchini, D., Rosetto, M., Dallai, R., and Marri, L. (2002). Bacteria associated with the oesophageal bulb of the medfly *Ceratitis capitata* (Diptera: Tephritidae). *Curr. Microbiol.* **44:** 120–124.

Marri, L., Dallai, R., and Marchini, D. (1996). The novel antibacterial peptide ceratotoxin A alters permeability of the inner and outer membrane of *Escherichia coli* K-12. *Curr. Microbiol.* **33:** 40–43.

Martinez, A.J., Robacker, D.C, Garcia, J.A., and Esau, K.L. (1994). Laboratory and field olfactory attraction of the Mexican fruit fly (Diptera: Tephritidae) to metabolites of bacterial species. *Fla. Entomol.* **77:** 117–126.

Miyazaki, S., Boush, G.M., and Baerwald, R.J. (1968). Amino acid synthesis by *Pseudomonas melopthora*, bacterial symbiote of *Rhagoletis pomonella* (Diptera). *J. Insect Physiol.* **14:** 513–518.

Murphy, K.M., Teakle, D.S., and McRae, I.C. (1988). Kinetics of colonization of adult Queensland fruit flies (*Bactrocera tryoni*) by dinitrogen-fixing alimentary tract bacteria. *Appl. Environ. Microbiol.* **60:** 2508–2517.

Nardon, P. and Grenier, A. (1991). *Symbiosis as a Source of Evolutionary Innovation,* pp. 153–169. MIT Press, Cambridge, MA.

Navrozidis, E.I., Vasara, E., Karamanlidou, G., Salpiggidis, G.K., and Koliasis, S.I. (2000). Biological control of *Bactrocera oleae* (Diptera: Tephritidae) using a Greek *Bacillus thuringiensis* isolate. *J. Econ. Entomol.* **93:** 1657–1661.

Neilson, W.T.A. and Wood, F.A. (1966). Natural source of food of the apple maggot. *J. Econ. Entomol.* **59:** 997–998.

Nishida, T. (1980). Food system of tephrid flies in Hawaii. *Proc. Hawaiian Entomol. Soc.* **23:** 245–254.

O'Neill, S.L., Hoffman, A.A., and Werren, J.H. (1997). *Influential Passengers: Inherited Microorganisms and Arthropod Reproduction.* Oxford University Press, New York.

Peloquin, J.J., Kuzina, L.V., Lauzon, C.R., and Miller, T.A. (2000). Transformation of internal extracellular bacteria isolated from *Rhagoletis completa* Cresson gut with enhanced green fluorescent protein. *Curr. Microbiol.* **40:** 367–371.

Peloquin, J.J., Lauzon, C.R., Potter, S.E., and Miller, T.A. (2002). Transformed bacterial symbionts reintroduced to and detected in host gut. *Curr. Microbiol.* **45:** 41–45.

Petri, L. (1909). Ricerche sopra i batteria intestinali della mosca olearia. *Mem. Staz. Pat. Veg.* Roma, 1–29.

Plus, N., Gissman, L., Veyrunes, J.C., Pfister, H., and Gateff, E. (1981a). Reoviruses of *Drosophila* and *Ceratitis* populations and of *Drosophila* cell lines: a possible new genus of the Reoviridae family. *Ann. Virol. Inst. Pasteur* **132:** 261–270.

Plus, N., Veyrunes, J.C., and Cavalloro, R. (1981b). Endogenous viruses of *Ceratitis capitata* Wied. J.R.C. Ispra strain, and of *C. capitata* permanent cell lines. *Ann. Virol. Inst. Pasteur* **132:** 91–100.

Potrikus, C.J. and Breznak, J.A. (1977). Nitrogen-fixing *Enterobacter agglomerans* isolated from guts of wood-eating termites. *Appl. Environ. Microbiol.* **33:** 392–399.

Potter, S.E. (2001). Bacterial associations in the walnut husk fly, *Rhagoletis completa* Cresson. Master's thesis, California State University, Hayward.

Prokopy, R.J., Cooley, S.S., Galarza, L., Bergweiler, C., and Lauzon, C.R. (1993). Bird droppings compete with bait sprays for *Rhagoletis pomonella* (Walsh) flies (Diptera: Tephritidae). *Can. Entomol.* **125:** 413–422.

Raghu, S., Clarke, A.R., and Bradley, J. (2002). Microbial mediation of fruit fly–host plant interactions: is the host plant the "centre of activity"? *Oikos* **97:** 319–328.

Ratner, S.S. and Stoffolano, J.G., Jr. (1984). Ultrastructural changes of the esophageal bulb of the adult female apple maggot, *Rhagoletis pomonella* (Walsh) (Diptera: Tephritidae). *Can. Entomol.* **125:** 413–422.

Robacker, D.C. and Bartelt, R.J. (1997). Chemicals attractive to Mexican fruit fly from *Klebsiella pneumoniae* and *Citrobacter freundii* cultures sampled by solid-phase microextraction. *J. Chem. Ecol.* **23:** 2897–2915.

Robacker, D.C. and Flath, R.A. (1995). Attractants from *Staphylococcus aureus* cultures for the Mexican fruit fly, *Anastrepha ludens. J. Chem. Ecol.* **21:** 1861–1874.

Robacker, D.C. and Lauzon, C.R. (2002). Purine metabolizing capability of *Enterobacter agglomerans* affects volatiles production and attractiveness to Mexican fruit fly. *J. Chem. Ecol.* **28:** 1549–1563.

Robacker, D.C. and Moreno, D.S. (1995). Protein feeding attenuates attraction of Mexican fruit flies (Diptera: Tephritidae) to volatile bacterial metabolites. *Fla. Entomol.* **78:** 62–69.

Robacker, D.C. and Warfield, A.J. (1993). Attraction of both sexes of Mexican fruit fly, *Anastrepha ludens*, to a mixture of ammonia, methylamine, and putresine. *J. Chem. Ecol.* **19:** 2999–3016.

Robacker, D.C., Garcia, J.A., Martinez, A.J., and Kaufman, M.G. (1991). Strain of *Staphylococcus* attractive to laboratory strain *Anastrepha ludens* (Diptera: Tephritidae). *Ann. Entomol. Soc. Am.* **84:** 555–559.

Robacker, D.C., Warfield, A.J., and Albach, R.F. (1993). Partial characterization and HPLC isolation of bacteria-produced attractants for Mexican fruit fly, *Anastrepha ludens. J. Chem. Ecol.* **19:** 543–557.

Robacker, D.C., Martinez, A.J., Garcia, J.A., Diaz, M., and Romero, C. (1996). Toxicity of *Bacillus thuringiensis* to Mexican fruit fly (Diptera: Tephritidae). *J. Econ. Entomol.* **89:** 104–110.

Robacker, D.C., Martinez, A.J., Garcia, J.A., and Bartelt, R.J. (1998). Volatiles attractive to the Mexican fruit fly (Diptera: Tephritidae) from eleven bacteria taxa. *Fla. Entomol.* **81:** 497–508.

Robacker, D.C., Garcia, J.A., and Bartelt, R.J. (2000). Volatiles from duck feces attractive to Mexican fruit fly. *J. Chem. Ecol.* **26:** 1849–1867.

Robinson, A.S. and Hooper, G., Eds. (1989). In *Fruit Flies: Their Biology, Natural Enemies, and Control*, Vol. 3B, pp. 121–129. Elsevier, Amsterdam.

Rossiter, M.A., Howard, D.J., and Bush, G.L. (1983). Symbiotic bacteria of *Rhagoletis pomonella*. In *Fruit Flies of Economic Importance* (R. Cavalloro, Ed.), pp. 77–82. Balkema, Rotterdam, the Netherlands.

Rubio, R.E.P. and McFadden, M. (1966). Isolation and identification of bacteria in the digestive tract of the Mexican fruit fly, *Anastrepha ludens. Ann. Entomol. Soc. Am.* **59:** 1015–1016.

Siebel, M.A. and Characklis, W.G. (1991). Observations of binary populations. *Biotechnol. Bioeng.* **37:** 778–789.

Steinhaus, E.A. (1959). *Serratia marcescens* Bizio as an insect pathogen. *Hilgardia* **28:** 351.

Taneja, J. and Guerin, P.M. (1997). Ammonia attracts the haematophagous but *Triatoma infestans:* behavioural and neurophysiological data on nymphs. *J. Comp. Physiol. A* **181:** 21–34.

Tsiropoulous, G.J. (1976). Bacteria associated with the walnut husk fly, *Rhagoletis completa. Environ. Entomol.* **5:** 83–86.

Tzanakakis, M.E., Prophetou, D.A., Vassilou, G.N., and Papadopoulos, J.J. (1983). Inhibition of larval growth of *Dacus oleae* by topical application of streptomycin to olives. *Entomol. Hellenica* **1:** 65–70.

Vijaysegaran, S., Walter, G.H., and Drew, R.A.I. (1997). Mouthpart structure, feeding mechanisms, and natural food sources of adult *Bacterocera* (Diptera: Tephritidae). *Ann. Entomol. Soc. Am.* **90:** 2.

9 Symbionts Affecting Termite Behavior

Kenji Matsuura

CONTENTS

INTRODUCTION

Symbioses between higher organisms and microorganisms often create opportunities for the exploitation of new food resources that are abundant but have not been used effectively; this opens new niches. Although cellulose in the form of dead plants is a superabundant potential resource, digesting it is not an easy task for most animals. With the aid of intestinal symbionts, termites successfully use dead plants as food resources. This wood-decomposition niche had been occupied only by various microorganisms, including bacteria, fungi, and protozoa, before the ancestors of termites began exploiting it. In this sense, we can say that termites live in the world of microorganisms. Therefore, how they get along with various microorganisms is one of the most important selection pressures on termites favoring the evolution of behavioral and physiological adaptations.

The interactions between termites and microorganisms vary from total dependence of the termite on the microorganism for food at one end of the spectrum to total dependence of the microorganism on the termite at the other end, with many degrees of mutualism in between. In his excellent treatise on coevolution, Thompson (1982) pointed out that high physical stress, such as nutrient-poor environments, increases the probability of novel mutualisms between species because small inputs of nutrients from a symbiont can significantly increase the growth rates of hosts. The digestive processes of termites are driven largely by symbiotic relationships with intestinal microorganisms, which produce various digestive enzymes (Breznak, 1984; Breznak and Brune, 1994). Besides the difficulties in decomposing hard cell-wall components, dead plants have a low nitrogen:carbon ratio. To use dead plants as food resources termites must solve the "carbon-nitrogen balance

0-8493-1286-8/03/$0.00+$1.50

problem" (Higashi et al., 1992); they have done so through association with microorganisms (Potrikus and Breznak, 1981; Higashi et al., 1992).

The mutualistic association between the Macrotermitinae and the basidiomycete *Termitomyces* has been the fascinating subject of most investigations of termite–fungus relationships. It is well known that members of the Macrotermitinae culture *Termitomyces* on feces substrates. Cultivation of fungus combs results in high assimilation of food because the fungus is high in nitrogen and other nutrients and possesses digestive enzymes (Sands, 1969; Wood and Thomas, 1989; Rouland et al., 1991). Nutritional relationships with microorganisms are not limited to these typical symbiotic relationships. Even various decay fungi surrounding termites may render poor foods palatable for termites by detoxifying the allelochemicals in wood, predigesting wood products, enhancing nitrogen and other nutrients, and enhancing moisture content (Waller and LaFage, 1987).

Most published works on the interactions of termites and microorganisms have concentrated on nutritional relationships and pathogens. Recent research has shown that microorganisms are much more important to the ecology and evolution of termites than previously realized. Here I discuss the symbioses with microorganisms that affect termite behavior, based primarily on the discovery of novel termite–bacteria and termite–fungus interactions.

NESTMATE RECOGNITION MEDIATED BY INTESTINAL BACTERIA

NESTMATE RECOGNITION IN EUSOCIAL INSECTS

The ability to distinguish nestmates from strangers is vital in maintaining the integrity of insect societies (Hölldobler and Wilson, 1990). Throughout the evolutionary history of social insects, intense selection pressure has led to heightened recognition abilities because the absence of nestmate recognition makes a colony vulnerable to social parasitism by various types of nest invaders (Wallis, 1964). Many eusocial insects, including social Hymenoptera (ants, bees, and wasps) and Isoptera (termites), are able to distinguish nestmates from nonnestmates (reviewed in Vander Meer et al., 1998), even though these groups have independently evolved their eusociality (Wilson, 1971). Over the past few decades, a considerable number of studies have been conducted to investigate the mechanisms of nestmate recognition in social insects. The origin and nature of the cues used in nestmate recognition have been especially well studied in Hymenopterans, and it is known that cuticular hydrocarbons play an important role in nestmate recognition among ants and wasps (Bonavita-Cougourdan et al., 1987; Lorenzi et al., 1996; Vander Meer and Morel, 1998; Lenoir et al., 2001).

It has been hypothesized that cuticular hydrocarbons may also serve as chemical cues for nestmate recognition in termites. To play a role in the signature of a colony these molecules would have to be shared by all castes in the same pattern in any given colony. In termites, however, the cuticular hydrocarbon profiles vary among castes within a colony (Haverty et al., 1996). Su and Haverty (1991) found that agonistic behavior of *Coptotermes formosanus* colonies was not correlated with any patterns of intercolonial variation in cuticular hydrocarbons. Therefore, cuticular-hydrocarbon composition itself cannot be fully responsible for nestmate recognition, although cuticular hydrocarbons can play a role in interspecific recognition (Bagnères et al., 1991; Takahashi and Gassa, 1995). Currently, no ethological and chemical evidence supports the use of hydrocarbons in nestmate recognition by termites in a way similar to that of hymenopterans (Clément and Bagnères, 1998). Unfortunately, among the flurry of studies that have focused on the analysis of chemical profiles, the mechanism of nestmate recognition in termites has rarely been discussed in terms of their biological characteristics. Isoptera and Hymenoptera are quite different in their life histories, feeding habits, and the evolutionary process of sociality, so the analogy of Hymenoptera cannot always be applied to Isoptera.

Much attention has recently focused on exogenous factors, such as digestive compounds (Su and Haverty, 1991; Shelton and Grace, 1997), instead of endogenous factors, including cuticular hydrocarbons. Shelton and Grace (1997) suggested that environmental cues were part of a multicomponent system for nestmate recognition in *C. formosanus*. The authors hypothesized that termites of different colonies fed on different food sources in varying proportions and that variation in diet would be reflected in a variation in digestive components, which could be used in nestmate recognition.

Nest Odor and Nestmate Recognition

In the evolutionary history of nestmate recognition, the ability to recognize own nests via chemoreception may be an ancestral characteristic that served as a preadaptation for the evolution of nestmate recognition (Gamboa et al., 1986; Breed et al., 1995; Singer et al., 1998). In wasps and honeybees, the nest plays a critical and intermediate role for the acquisition of recognition cues by colony members (Singer and Espelie, 1992; Breed et al., 1998). Singer and Espelie (1992, 1996) found that newly emerged wasps, *Polistes metricus*, learned and acquired the colony-specific hydrocarbon profile from the nest surface, which enabled them to not only recognize the natal nest but also discriminate between nestmates and nonnestmates.

Considering the evolutionary process of nestmate recognition, it is also reasonable that the chemical cues used in nestmate recognition in termites would originate from nest odor. Worker termites recycle their feces to use as construction material for the gallery, the inner wall of the nest, and other areas. The use of feces as nesting material also has a hygienic significance. It is known that damp-wood termites cover their nest chambers and gallery systems with fecal pellets, which contain antibiotic substances, to reduce the risk of infections by naturally occurring pathogens (Rosengaus et al., 1998). Therefore, compounds in their feces may adhere to the termites and may provide cues for nestmate recognition. As mentioned earlier, the digestive components in feces are also a likely recognition cue. However, colony-specific nestmate recognition is not fully explained by the variation in digestive components derived from the variation in food sources, for example, woody species, because neighboring colonies in similar environments should feed on similar food resources. Nestmate recognition should be more important between neighboring colonies than between distant colonies in different environments.

Colony-Specific Composition of Intestinal Bacteria and Formation of Nest Odor

Matsuura (2001) focused on the colony-specific composition of intestinal bacteria as a possible origin of colony-specific nest odors. Previous studies revealed diverse bacterial communities present in the gut of *Reticulitermes* spp. (Schultz and Breznak, 1978; Ohkuma and Kudo, 1996; Kudo et al., 1998). Although our understanding of the role of bacteria in the metabolic cycle of lower termites is poor, several possible beneficial roles have been postulated: termites obtain nutrition from cellulose digestion (Breznak and Brune, 1994), methanogenesis and acetogenesis from H_2 and CO_2 (Brauman et al., 1992), recycling of uric-acid nitrogen (Potrikus and Breznak, 1981), and nitrogen fixation (Breznak et al., 1973). Not all intestinal bacteria are transmitted from the mother colony; some are acquired from foods (Waller and LaFage, 1987). Moreover, the reproductive rate of each termite symbiont depends specifically on the food quality (Mannesmann, 1972). The fungal stage of decay may be especially important because of changing nutritional composition and detoxification of wood. Therefore, the composition of intestinal bacteria is considered to reflect subtle environmental differences among colonies. Because of frequent food exchange, the gut microbiota are transferred among nestmates (Waller and LaFage, 1987), which probably results in the colony-specific composition of intestinal bacteria.

To examine the composition of intestinal bacteria, *Reticulitermes speratus* workers were collected from nine colonies in the field. The workers were sterilized and dissected to remove the gut without rupturing it. They were then homogenized, dropped onto LB agar plates, and incubated aerobically for 72 h at 30°C. Bacterial colonies were counted after being categorized according to their color, shape, and size. Each type of bacterial colony was purified by plating on LB agar plates, and its morphology

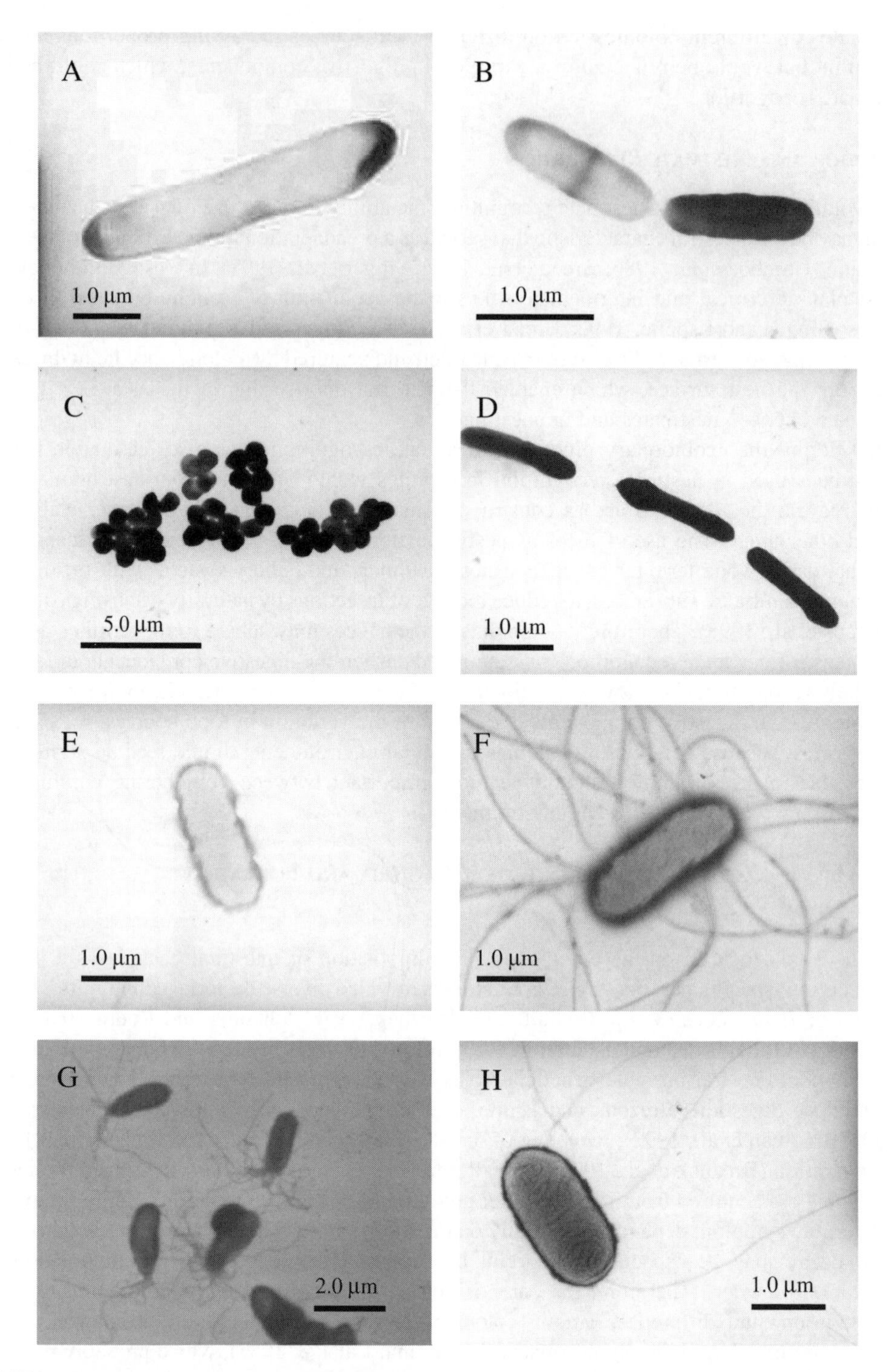

FIGURE 9.1 The intestinal bacteria of *Reticulitermes speratus*. Each type of bacterial colony growing on the LB agar plates (A to H) was purified and morphology examined under an electron microscope (Matsuura, 2001). Bacterial types were consistent with those indicated in Figure 9.2.

was examined in detail under an electron microscope (Figure 9.1). The composition of bacterial colonies grown on the LB agar plates differed significantly among termite colonies (Figure 9.2).

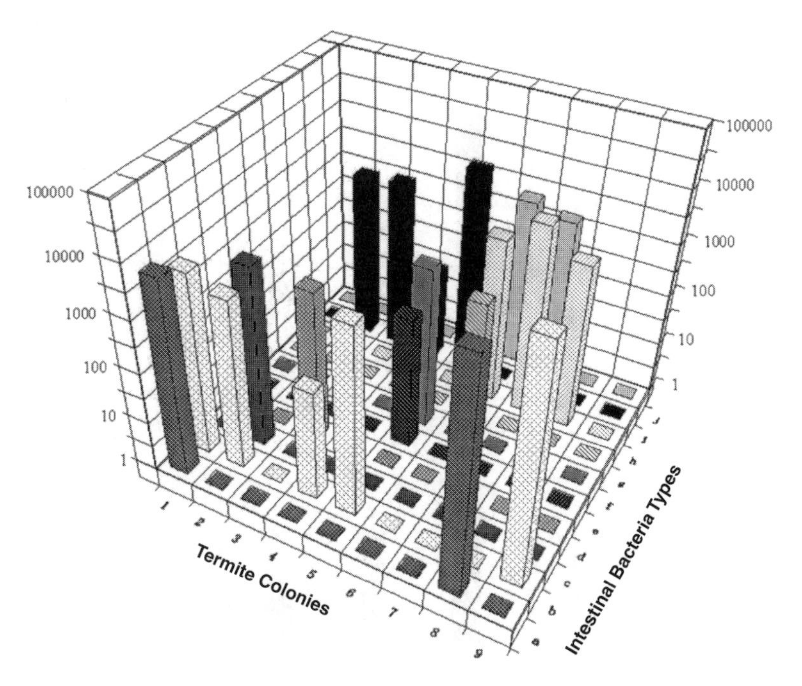

FIGURE 9.2 The composition of bacterial colonies grown on the LB agar plates was highly significantly different among termite colonies (Wilks' $\lambda = 0.0030$, Rao's r $(80, 465) = 8.71$, $p < 0.0001$).

Does Composition of Intestinal Bacteria Mediate Nestmate Recognition?

If recognition cues are based on bacterial composition, we can predict that termites that have adsorbed an unfamiliar bacterial odor on their bodies will be attacked by nestmates. To test this prediction, three colonies were collected in the field. About 500 workers of the three colonies were separated from each and maintained in the laboratory on mixed sawdust bait. A portion of the culture colony was extracted and isolated on the following two baits for 3 days. Fifteen workers were maintained on either mixed sawdust bait containing aqueous extracts of unfamiliar bacteria isolated from another colony (hereafter abbreviated as BEB) or on bait containing only distilled water (DWB). Three days later, they were marked and returned to the original culture colony. The survival rates of introduced termites maintained on BEB were significantly lower than the survival rates of those maintained on DWB (Figure 9.3A). This bacterial-odor-absorbing test clearly showed that the adsorption of an unfamiliar bacterial odor on the body surface altered nestmate recognition.

Moreover, recognition behavior toward nestmates changed when the composition of bacteria was changed artificially using antibiotics (Kanamycin, Carbenicillin, and Ampicillin). The level of agonism was significantly greater among groups given different antibiotics than among those given the same antibiotics. Apart from variations in the total level of agonism, the colonies showed a common tendency to be more aggressive toward termites derived from a different antibiotic group than toward termites derived from the same antibiotic group (Figure 9.3B).

Colony Fusion

It is necessary to remember that what we can observe by experiment is not nestmate recognition but discrimination. Waldman et al. (1988) emphasized the distinction between these two terms. *Discrimination* denotes differential treatment of conspecifics correlated with kinship. *Recognition*, which is a series of internal and essentially unobservable physiological events, may occur without any behavioral response. A social insect society is defined as "open" when members accept alien conspecific individuals and "closed" when they act aggressively toward members

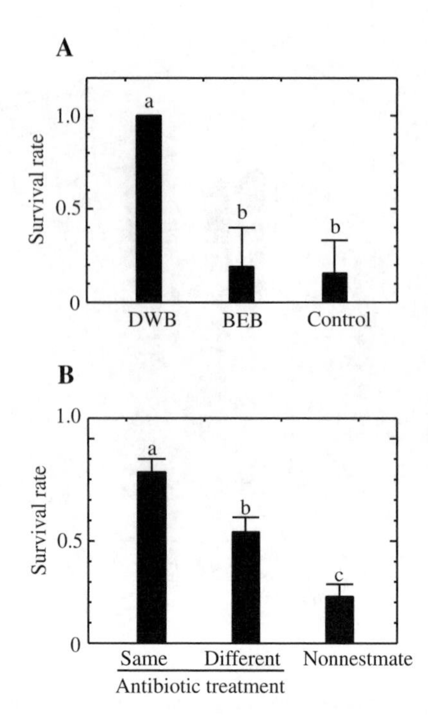

FIGURE 9.3 (A) The agonistic behavior against nestmates that had adsorbed an unfamiliar odor of bacteria obtained from nonnestmates. Bars indicate SE for three colonies. DWB; workers fed for 3 days on bait containing distilled water. BEB: workers fed for 3 days on bait containing bacterial extracts obtained from nonnestmates and returned to the colony. Control: workers of another colony. Groups with different letters above columns are significantly different based on Scheffe's test. (B) The agonistic behavior toward nestmates if the composition of the intestinal bacteria was changed by addition of antibiotics to the bait. The survival rate of the introduced termites was compared by Scheffe's test. All groups are significantly different from each other (Matsuura, 2001).

of a neighboring nest (Wallis, 1964). It has been assumed that insect societies are always closed, i.e., members always act aggressively toward individuals of other nests if they recognize the intruders as nonnestmates.

In some termite species, however, the level of intercolonial aggression within a species varies from one nest to another. Bulmer et al. (2001) investigated the genetic organization of *R. flavipes* colonies using polymorphic allozymes and double-stranded conformation polymorphism (DSCP) analysis and demonstrated that large colonies contained two or more unrelated queens. This suggests that the number of queens and nestmate relatedness change with colony age and size. Matsuura and Nishida (2001) suggested that *R. speratus* colonies facultatively changed their agonistic response toward nonnestmates based on the costs and benefits of colony fusion. A colony-pairing bioassay showed that members of the host colony tended to attack and kill intruders that had a higher nymph ratio (number of nymphs/number of workers) than the host colony had. In contrast, the intruders were accepted if they had a lower nymph ratio than the host colony. After colony fusion, workers were unable to distinguish nonnestmate nymphs from nestmate nymphs and reared both equally. Acceptance of a higher-nymph-ratio colony might be disadvantageous because the workers must rear nonrelative nymphs additionally. However, acceptance of a lower-nymph-ratio colony could be beneficial for the host colony because it would gain additional labor force to feed its own nymphs. Colony fusion reduces wasteful cost of fighting between colonies. Moreover, colony fusion should be an adaptive way of outbreeding. In termites, replacement or supplementary reproductives may be reared within a nest following the death of the primary reproductive or during the formation of nest buds (Roisin, 1993). These secondary reproductives reproduce by inbreeding and continue the colony. Such an inbreeding cycle will lead to inbreeding depression. Termites outbreed through the swarming of alates, but the swarming involves considerable risk of predation. Colony fusion, or the selective acceptance of nonnestmates, may be a better way of outbreeding without predation risk. These recent findings suggest that the process leading from nestmate recognition to agonistic responses in termites is much more complex than previously believed.

Microorganisms live not only in the guts of termites but also in nesting material (wood). The decay fungi and bacteria in the wood may also contribute to the formation of nest odor. Many questions still remain unanswered. What chemicals are actually used as cues in nestmate recognition? How does the colony state, such as colony genetic structure, caste composition, nutritional status, and colony size, affect the agonistic response?

"TERMITE BALLS": THE EGG-MIMICKING SCLEROTIA OF A FUNGUS

Matsuura et al. (2000) reported "egg mimicry" by a fungus as a novel insect–fungus relationship. Egg protection is one of the most essential behaviors of social insects. When termite workers recognize the eggs laid by queens, they bring the eggs together and heap them so as to care for them. In nests of *R. speratus*, workers make several piles of eggs in this way (Figure 9.4A). Brown balls (called "termite balls") similar to eggs in size and smooth texture are frequently found within these egg piles (Figure 9.4B). DNA-sequence analysis of the internal transcribed spacers (ITS) and the 5.8S ribosomal RNA gene of the nuclear ribosomal repeat unit identified these balls as the sclerotia of the fungus *Fibularhizoctonia* sp. nov., which is phylogenetically closest to decay fungi,

FIGURE 9.4 (Color figure follows p. 206.) The "termite balls." (A) Termites unwittingly caring for termite balls, that is, egg-mimicking sclerotia of the fungus. (B) Close-up of the egg piles in the nursery nest. There are oval-shaped and transparent eggs and spherical-shaped brown termite balls.

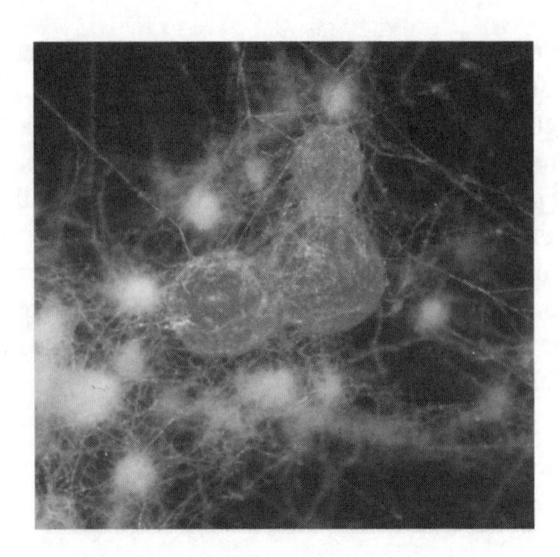

FIGURE 9.5 The formation of termite balls on potato dextrose agar.

Athelia spp. The sclerotia are tough balls of densely packed filaments that germinate into fungal colonies under favorable conditions (Figure 9.5).

A wide-ranging survey of Japan showed that most *R. speratus* colonies, regardless of subspecies (*speratus*, *leptolabralis*, or *kyushuensis*), had termite balls in their egg piles; however, no termite balls were found in the nests of *R. okinawanus*, which is distributed on Okinawa Island. The percentage of colonies with sclerotia differed significantly among the tree species harvested by the termites. Almost all *R. speratus* colonies nesting in rotten pine wood (*Pinus densiflora*) had sclerotia in their egg piles; in several colonies, the number of sclerotia exceeded the number of true eggs. However, only 40% of the colonies nesting in rotten Japanese cedar (*Cryptomeria japonica*) had sclerotia. This was partly explained by the fungistatic substances found in fresh cedar wood, which inhibit fungal growth.

Egg-Mimicking Sclerotia

In birds, the color and pattern of the eggshell are the most important sign stimuli by which parents can recognize their eggs. Therefore, eggs of brood-parasitic cuckoos resemble the eggs of their hosts in the color and pattern of the eggshell. Unlike birds, however, *Reticulitermes* workers cannot recognize their eggs visually because they live in the dark and do not have eyes. Instead, egg-carrying tests using dummy eggs made of glass beads or sea sand, coated or not coated with egg-derived chemicals, demonstrated that termites recognize their eggs by a combination of chemical cues and the spherical shape and size of the eggs (Table 9.1).

Therefore, both morphological and chemical camouflage would be required to mimic termite eggs. Compared to the sclerotia of closely related fungi, *Athelia arachnoidea* and *A. epiphylla*, termite balls are smaller than those of other fungi, and their diameter precisely matches the small diameter of termite eggs. In termite balls, the variation in diameter is much smaller than in other fungi (Figure 9.6). Moreover, termite balls are smooth textured, similar to termite eggs, while those of other fungi are rougher. This suggests that termite balls are morphologically similar to termite eggs.

In an experiment, more than 90% of the sclerotia that were collected from the egg piles of a termite nest were piled up with the eggs, but sclerotia that had been washed with ether were rarely piled. The washed sclerotia recovered their piling rates if they were coated with egg wax again. These results suggest that sclerotia have chemicals on their surface that are used in termite-egg recognition.

Additional evidence also suggests that termite balls mimic termite eggs. As mentioned previously, *R. okinawanus*, a closely related species of *R. speratus* found on Okinawa Island, does not have termite balls in its egg piles. It is clear that *R. okinawanus* has no contact with termite balls. Nevertheless, when termite balls were experimentally given to *R. okinawanus* workers, they could

TABLE 9.1
Comparison of Piled-Up Rates for Termite Eggs and Dummy Eggs

Materials	Treatments	Piled-Up Rates ± SE
Termite egg (0.36 mm)	None	0.998 ± 0.043[a]
	Washed	0.542 ± 0.104[b]
Glass beads (0.4 mm)	0 eq.	0.080 ± 0.049[c]
	0.6 eq.	0.251 ± 0.049[bc]
	12 eq.	0.471 ± 0.117[b]
Glass beads (0.6 mm)	0 eq.	0[c]
	12 eq.	0[c]
Sea sand (40 mesh)	0 eq.	0[c]
	12 eq.	0[c]

Termite eggs are oval, and workers always grasp the short side of the eggs when carrying them. The 0.4-mm glass beads are almost the same size as the short diameter of termite eggs and have a spherical shape. The 0.6-mm glass beads have a spherical shape but are larger than termite eggs and termite balls. The 40-mesh sea sand consists of the same material as the glass beads but does not have a spherical shape. They were coated with or without egg wax extracted with ethyl ether. Significant differences exist between different lowercase letters of the alphabet at 0.05 level by Scheffe's test.

eq. = Equivalent of the egg chemical applied on each dummy material.

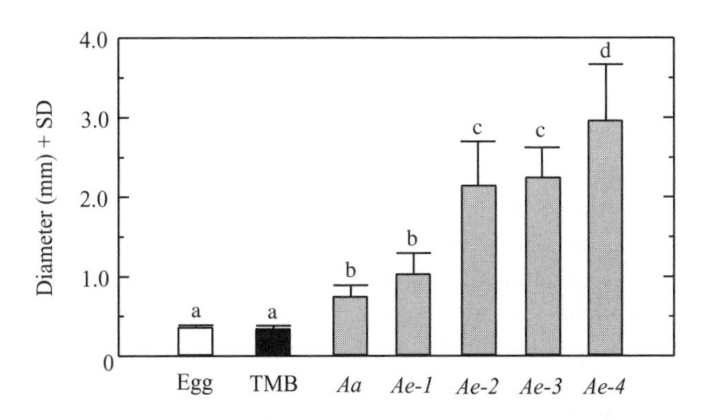

FIGURE 9.6 Size comparison of termite eggs and termite balls (sclerotia of the TMB) and other closely related sclerotium-forming fungi. *Aa*, *Athelia archnoidea*; *Ae*, *Athelia epiphylla* (four isolates). There were significant differences between different superscript letters at the 0.05 level by Scheffe's test. The sclerotia of the TMB were significantly smaller than those of the other fungi, and their diameter was the same as the short diameter of termite eggs.

not distinguish the balls from their eggs and carried them in the same way as did *R. speratus*. This provides convincing evidence that termite balls mimic termite eggs.

Interaction between the Termites and the Fungus

Eggs and sclerotia are manipulated by worker termites and are therefore smeared on the surface with saliva. The saliva of workers has been reported to inhibit fungal germination (Thomas, 1987); thus, workers actually inhibit the germination of sclerotia in egg piles. Additionally, damp-wood termites line nest chambers and galleries with fecal pellets, which have fungistatic properties (Rosengaus et al., 1998) that protect the nest from naturally occurring fungi. However, over the long term, termites do not completely inhibit sclerotial germination and fungal growth, and they remove from the egg piles sclerotia that have shrunk or deformed. These sclerotia then germinate and grow on termite excretions in the corner of the nest. Transportation of sclerotia by termites is thus probably indispensable for *Fibularhizoctonia* sp. nov., because this fungus does not fruit spores. By mimicking eggs, the fungus can be transported by termites and gain a competitor-free habitat more quickly than other fungi.

Without the inhibition by workers, sclerotia in the egg piles germinate and absorb the eggs around them; however, if their germination is inhibited, the sclerotia enhance egg survival. One interpretation of this result is that bacteriostatic and fungistatic substances in the sclerotia might protect termite eggs from pathogens; this fungus forms an antagonistic zone against some other fungi. Termites can protect their eggs from pathogens with their own saliva, and the fungus can also produce an antifungal agent. This interaction may be beneficial for both sides because each antifungal and antibacterial compound cannot be completely effective against all fungi and bacteria. It is possible that the antifungal and antibacterial compounds found in sclerotia and in termites complement each other. Hence, the *Reticulitermes-Fibularhizoctonia* relationship is mutualistic rather than parasitic.

The Evolutionary Process of Egg Mimicry by the Sclerotia

Consideration of the evolutionary processes of the egg-mimicking sclerotia reveals that mimetic weeds provide an informative example of the evolution of mimicry. The rye weed *Secale montanum*, with its large seeds and hard spindle, resembles the wheat *Triticum*. One race of the wild rye began to grow as a weed in wheat fields. The rye seeds that resembled wheat seeds were harvested with the wheat, sown in the following year, and given human protection. This unintentional domestic selection promoted the mimicry of rye weed in wheat fields (Wickler, 1968).

Similar processes may be true of the early stages in the development of fungal egg mimicry. Dummy-egg tests showed that termites carried a few dummy eggs without chemicals if they were spherical and similar in size to the eggs. Therefore, it is possible that this termite–fungus interaction began when termites mistook the sclerotia of the ancestral fungus for their eggs and brought them to their egg piles. This fungus exploited a novel niche in the termite nest and moved to distant new homes when the termite colonies propagated by budding and migrated to neighboring woods. This size selection of the sclerotia by the termites would promote morphological egg mimicry, which might then be accompanied by the development of chemical mimicry.

Workers suppress the germination of sclerotia, and termite-egg survival increases in the presence of sclerotia only if the workers tend to the eggs. When the workers are experimentally removed, the sclerotia in the piles of eggs germinate and absorb the eggs around them. Hence, the sclerotia are "double-edged swords" for termites. It is often accepted that mutualisms are derived evolutionarily from initially antagonistic interactions between species (Thompson, 1982). Roughgarden (1975) assumed that symbiotic mutualisms passed through a parasitic phase. It is possible that the egg-mimicking fungus was initially an egg parasite that was harmful to the eggs.

Unfortunately, egg-mimicking fungi have not yet been found in the nests of other species. Zoberi and Grace (1990) isolated 40 species of fungi in a survey of the mycoflora associated with *R. flavipes* in Canada, but they did not find any sclerotium-forming fungi. It seems that interactions with this egg-mimicking fungus have arisen recently in evolutionary history because it is not common, even in closely related termites. However, the lack of observations of egg-mimicking fungi in other termites may be partly attributed to the difficulty in finding egg piles because they are often protected deep inside the nest and the season of egg production is limited. Future discoveries of similar egg-mimicking sclerotia in other termite species will enable us to perform comparative studies and estimate the evolutionary process of termite–fungus interactions based on egg recognition by termites.

CONCLUSION

Adapting to the selection pressures of microorganisms appears to have been of primary importance in the social evolution of insects (Hamilton, 1987; Schmid-Hempel, 1998; Rosengaus et al., 1999a). Many excellent works by Rosengaus and her colleagues have clarified the mechanisms of resistance to infection by bacterial and fungal pathogens in termites, from various angles. Rosengaus et al. (1999b) reported that the damp-wood termite *Zootermopsis angusticollis* performed "pathogen alarm behavior" when it detected the spores of the pathogenic fungus *Metarhizium anisopliae*. This vibrational alarm signal conveyed information about the presence of pathogens to nearby nestmates and reduced the risk of disease within the nest. This finding indicates that termites are very sensitive to the presence of pathogens. Termites are therefore also likely able to detect the presence of various kinds of microorganisms in the nest through the perception of chemical products. This capacity of termites should be somewhat associated with their ability to distinguish bacterial products as cues in nestmate recognition. Throughout evolutionary history, the selection pressure of pathogens must have favored a high sensitivity to potential pathogens, which served as a preadaptation for the evolution of nestmate recognition. Distinguishing nestmates from nonnestmates is similar in kind to the problem of distinguishing self from nonself in immune defense (Coombe at al., 1984). It can be speculated that the pathways involved in immune defense may overlap those involved in producing recognition markers, like the role of certain key substances in vertebrates (Schmid-Hempel, 1998).

Considering the characteristic biology of termites, I have emphasized the importance of environmental factors in nestmate recognition by termites. The influence of environmental factors on intercolony agonism has also been realized in social Hymenopterans. Liang and Silverman (2000) demonstrated that the Argentine ant *Linepithema humile* acquired chemical cues for nestmate recognition from insect prey. Diet altered both the recognition cues present on the cuticular surface and the response of nestmates to the new colony odor and resulted in aggression among former nestmates reared on different insect prey.

In studies of nestmate recognition and egg-mimicking fungi, it has been suggested that termites use the chemicals produced by microorganisms as signals or defensive chemicals. The sequestration of metabolites has been observed in various insects. A number of aposematic butterfly species sequester unpalatable or toxic substances from their host plants rather than manufacturing their own defensive substances (reviewed by Nishida, 2002). Currie et al. (1999) showed that fungus-cultivating ants used antibiotic bacteria to control garden parasites. They demonstrated that a filamentous bacterium of the genus *Streptomyces*, which produces antibiotics targeted to suppress the growth of the specialized garden parasite fungus *Escovopsis*, is carried on the ant cuticle. This system is reminiscent of the idiom "fight fire with fire."

Although significant advances have been made in our understanding of the biology of termites, our knowledge of termite–microorganism relationships remains limited. As mentioned at the beginning of this chapter, the habits of termites are enmeshed with various microorganisms. All of these symbionts, not just the mutualists, interact to some degree with termites, involving both benefits and costs for each agent. Studies of the interactions between termites and potential symbiotic microorganisms will reveal many more types of relationships.

REFERENCES

Bagnères, A.G., Killian, A., Clément, J.L., and Lange, C. (1991). Interspecific recognition among termites of the genus *Reticulitermes*: evidence for a role for the cuticular hydrocarbons. *J. Chem. Ecol.* **17:** 2397–2420.

Bonavita-Cougourdan, A., Clément, J.L., and Lange, C. (1987). Nestmate recognition: the role of cuticular hydrocarbons in the ant *Camponotus vagus* Scop. *J. Entomol. Sci.* **22:** 1–10.

Brauman, A., Kane, M.D., Labat, M., and Breznak, J.A. (1992). Genesis of acetate and methane by gut bacteria of nutritionally diverse termites. *Science* **257:** 1384–1387.

Breed, M.D., Garry, M.F., Pearce, A.N., Hibbard, B.E., Bjostad, L.B., and Page, R.E., Jr. (1995). The role of wax comb in honey bee nestmate recognition. *Anim. Behav.* **50:** 489–496.

Breed, M.D., Leger, E.A., Pearce, A.N., and Wang, Y.J. (1998). Comb wax effects on the ontogeny of honey bees nestmate recognition. *Anim. Behav.* **55:** 13–20.

Breznak, J.A. (1984). Biochemical aspects of symbiosis between termites and their intestinal microbiota. In *Invertebrate–Microbial Interactions* (J.M. Anderson, A.D. Rayner, and D.W.H. Walton, Eds.), pp. 173–203. Cambridge University Press, Cambridge, U.K.

Breznak, J.A. and Brune, A. (1994). Role of microorganisms in the digestion of lignocellulose by termites. *Annu. Rev. Entomol.* **39:** 453–487.

Breznak, J.A., Brill, W.J., Mertins, J.W., and Coppel, H.C. (1973). Nitrogen fixation in termites. *Nature* **244:** 577–580.

Bulmer, M.S., Adams, E.S., and Traniello, J.F.A. (2001). Variation in colony structure in the subterranean termite *Reticulitermes flavipes*. *Behav. Ecol. Sociobiol.* **49:** 236–243.

Clément, J.L. and Bagnères, A.G. (1998). Nestmate recognition in termites. In *Pheromone Communication in Social Insects* (R.K. Vander Meer, M. Breed, M. Winston, and K.E. Espelie, Eds.), pp. 79–103. Westview Press, Boulder, CO.

Coombe, D.R., Ey, P.L., and Jenkin, C.R. (1984). Self/non-self recognition in invertebrates. *Q. Rev. Biol.* **59:** 231–255.

Currie, C.R., Scott, J.A., Summerbell, R.C., and Malloch, D. (1999). Fungus-growing ants use antibiotic-producing bacteria to control garden parasites. *Nature* **398:** 701–704.

Gamboa, G.J., Reeve, H.K., Ferguson, I.D., and Wacker, T.L. (1986). Nestmate recognition in social wasps: the origin and acquisition of recognition odors. *Anim. Behav.* **34:** 685–695.

Hamilton, W.D. (1987). Kinship, recognition, disease, and intelligence: constraints of social evolution. In *Animal Societies: Theories and Facts* (Y. Ito, J.L. Brown, and J. Kikkawa, Eds.), pp. 81–102. Japanese Science Society Press, Tokyo.

Haverty, M.I., Grace, J.K., Nelson, L.J., and Yamamoto, R.T. (1996). Intercaste, intercolony, and temporal variation in cuticular hydrocarbons of *Coptotermes formosanus* Shiraki (Isoptera: Rhinotermitidae). *J. Chem. Ecol.* **22:** 1813–1834.

Higashi, M., Abe, T., and Burns, T.P. (1992). Carbon-nitrogen balance and termite ecology. *Proc. R. Soc. London (B)* **249:** 303–308.

Hölldobler, B. and Wilson, E.O. (1990). *The Ants*. Harvard University Press, Cambridge, MA.

Kudo, T., Ohkuma, M., Moriya, S., Noda, S., and Ohtoko, K. (1998). Molecular phylogenetic identification of the intestinal anaerobic microbial community in the hindgut of the termite, *Reticulitermes speratus*, without cultivation. *Extremophiles* **2:** 155–161.

Lenoir, A., D'Ettorre, P., Errard, C., and Hefetz, A. (2001). Chemical ecology and social parasitism in ants. *Annu. Rev. Entomol.* **46:** 573–599.

Liang, D. and Silverman, J. (2000). "You are what you eat": diet modifies cuticular hydrocarbons and nestmate recognition in the Argentine ant, *Linepithema humile*. *Naturwissenschaften* **87:** 412–416.

Lorenzi, M.C., Bagnères, A.G., and Clement, J.L. (1996). The role of cuticular hydrocarbons in social insects: is it the same in paper-wasps? In *Natural History and Evolution of Paper-Wasps* (S. Turillazzi and M.J. West-Eberhard, Eds.), pp. 178–189. Oxford University Press, Oxford, U.K.

Mannesmann, R. (1972). Relationship between different wood species as a termite food source and the reproduction rate of termite symbionts. *Z. Angew. Entomol.* **72:** 116–128.

Matsuura, K. (2001). Nestmate recognition mediated by bacteria in a termite, *Reticulitermes speratus*. *Oikos* **92:** 20–26.

Matsuura, K. and Nishida T. (2001). Colony fusion in a termite: what makes the society "open"? *Insect Soc.* **48:** 378–383.

Matsuura, K., Tanaka, C., and Nishida, T. (2000). Symbiosis of a termite and a sclerotium-forming fungus: sclerotia mimic termite eggs. *Ecol. Res.* **15:** 405–414.

Nishida, R. (2002). Sequestration of defensive substances from plants by Lepidoptera. *Annu. Rev. Entomol.* **47:** 57–92.

Ohkuma, M. and Kudo, T. (1996). Phylogenetic diversity of the intestinal bacterial community in the termite *Reticulitermes speratus*. *Appl. Environ. Microbiol.* **62:** 461–468.

Potrikus, C.J. and Breznak, J.A. (1981). Gut bacteria recycle uric acid nitrogen in termites: a strategy for nutrient conservation. *Proc. Natl. Acad. Sci. U.S.A.* **78:** 4601–4605.

Roisin, Y. (1993). Selective pressures on pleometrosis and secondary polygyny: a comparison of termites and ants. In *Queen Number and Sociality in Insects* (L. Keller, Ed.), pp. 402–421. Oxford University Press, New York.

Rosengaus, R.B., Guldin, M.R., and Traniello, J.F.A. (1998). Inhibitory effect of termite fecal pellets on fungal spore germination. *J. Chem. Ecol.* **24:** 1707–1706.

Rosengaus, R.B., Traniello, J.F.A., Chen, T., Brown, J.J., and Karp, R.D. (1999a). Immunity in a social insect. *Naturwissenschaften* **86:** 588–591.

Rosengaus, R.B., Jordan, C., Lefebvre, M.L., and Traniello, J.F.A. (1999b). Pathogen alarm behavior in a termite: a new form of communication in social insects. *Naturwissenschaften* **86:** 544–548.

Roughgarden, J. (1975). Evolution of marine symbiosis: a simple cost-benefit model. *Ecology* **56:** 1201–1208.

Rouland, C., Lenoir, F., and Lepage, M. (1991). The role of the symbiotic fungus in the digestive metabolism of several species of fungus growing termites. *Comp. Biochem. Physiol A* **99:** 657–663.

Sands, W.A. (1969). The association of termites and fungi. In *Biology of Termites*, Vol.1 (K. Krishna and F.M. Weesner, Eds.), pp. 495–524. Academic Press, New York.

Schmid-Hempel, P. (1998). *Parasites in Social Insects*. Princeton University Press, Princeton, NJ.

Schultz, J.E. and Breznak, J.A. (1978). Heterotrophic bacteria present in hindguts of wood-eating termites *Reticulitermes flavipes* Kollar. *Appl. Environ. Microbiol.* **35:** 930–936.

Shelton, T.G. and Grace, J.K. (1997). Suggestion of an environmental influence on intercolony agonism of Formosan subterranean termite (Isoptera: Rhinotermitidae). *Environ. Entomol.* **26:** 632–637.

Singer, T.L. and Espelie, K.E. (1992). Social wasps use nest paper hydrocarbons for nestmate recognition. *Anim. Behav.* **44:** 63–68.

Singer, T.L. and Espelie, K.E. (1996). Nest surface hydrocarbons facilitate nestmate recognition for *Polistes metricus* Say. *J. Insect Behav.* **9:** 857–870.

Singer, T.L., Espelie, K.E., and Gamboa, G.J. (1998). Nest and nestmate discrimination in independent-founding paper wasps. In *Pheromone Communication in Social Insects* (R.K. Vander Meer, M. Breed, M. Winston, and K.E. Espelie, Eds.), pp. 79–103. Westview Press, Boulder, CO.

Su, N.Y. and Haverty, M.I. (1991). Agonistic behavior among colonies of the Formosan subterranean termite, *Coptotermes formosanus* SHIRAKI (Isoptera: Rhinotermitidae), from Florida and Hawaii: lack of correlation with cuticular hydrocarbon composition. *J. Insect Behav.* **4:** 115–128.

Takahashi, S. and Gassa, A. (1995) Roles of cuticular hydrocarbons in intra- and interspecific recognition behavior of two *Rhinotermitidae* species. *J. Chem. Ecol.* **21:** 1837–1845.

Thomas, R.J. (1987). Factors affecting the distribution and activity of fungi in the nest of Macrotermitinae (Isoptera). *Soil Biol. Biochem.* **19:** 343–349.

Thompson, J.N. (1982). *Interaction and Coevolution*. Wiley-Interscience, New York.

Vander Meer, R.K., Breed, M., Winston, M., and Espelie, K.E. (1998). *Pheromone Communication in Social Insects*. Westview Press, Boulder, CO.

Vander Meer, R.K. and Morel, L. (1998). Nestmate recognition in ants. In *Pheromone Communication in Social Insects* (R.K. Vander Meer, M. Breed, M. Winston, and K.E. Espelie, Eds.), pp. 79–103. Westview Press, Boulder, CO.

Waldman, B., Frumhoff, P.C., and Sherman, P.W. (1988). Problems of kin recognition. *Trends Ecol. Evol.* **3:** 8–13.

Waller, D.A. and LaFage, J.P. (1987). Nutritional ecology of termites. In *Nutritional Ecology of Insects, Mites, Spiders, and Related Invertebrates* (F. Slansky, Jr. and J.G. Rodriguez, Eds.), pp. 487–532. Wiley-Interscience, New York.

Wallis, D.I. (1964). Aggression in social insects. In *The Natural History of Aggression* (J.D. Carthy and F.J. Ebeling, Eds.), pp. 15–22. Academic Press, New York.

Wickler, W. (1968). *Mimicry in Plants and Animals*. Weidenfeld & Nicolson, London.

Wilson, E.O. (1971). *The Insect Societies*. Belknap Press, Cambridge, MA.

Wood, T.G. and Thomas, R.J. (1989). The mutualistic association between Macrotermitinae and *Termitomyces*. In *Insect–Fungus Interactions* (N. Wilding, N.M. Collins, P.M. Hammond, and J.F. Webber, Eds.), pp. 69–92. Academic Press, New York.

Zoberi, M.H. and Grace, J.K. (1990). Fungi associated with the subterranean termite *Reticulitermes flavipes* in Ontario. *Mycologia* **82:** 289–294.

10 Symbiosis of Microsporidia and Insects

Philip Agnew, James J. Becnel, Dieter Ebert, and Yannis Michalakis

CONTENTS

INTRODUCTION

A symbiotic relationship is one in which two dissimilar organisms live in close association with one another. This does not imply that they necessarily live together in harmony; symbiosis also applies to host–parasite relationships. The relationships formed by the microsporidia are extreme forms of symbiosis: these unicellular eukaryotes have become obligate intracellular parasites that cannot live independently outside of their host-cell environment.

In essence, microsporidia infect host cells, exploit what is there to replicate within them, produce spores, and then transmit themselves to other cells either within the same host or to a new host. However, the diversity of ways that they achieve this and the diversity of host species

0-8493-1286-8/03/$0.00+$1.50

they exploit indicate the number of the evolutionary pathways these organisms have managed to successfully negotiate in adapting to the needs of their own life histories and those of their hosts. The course of their evolution toward an obligate intracellular lifestyle is marked by several departures from what is considered to be a "standard" eukaryotic cell. Furthermore, their transition to a parasitic lifestyle is associated with a unique manner of infecting host cells that separates them from other eukaryotes.

The aim of this chapter is to illustrate aspects of the symbiotic relationship between the microsporidia and their hosts. Most examples are drawn from relationships involving invertebrate hosts, with additional material taken from those that infect vertebrates. The bias toward invertebrate host material is appropriate as the microsporidia have been described mainly as pathogens of insects. Indeed, the first species to be described, *Nosema bombycis* (Naegeli, 1857), was from an insect. This species was later shown by Pasteur (1870) to be the causative agent of pébrine, a disease responsible for serious economic harm to the commercial silk industry at the time; a related species, *N. apis*, still causes similar problems for commercial honey producers today. In fact, a substantial proportion of our knowledge about the basic biology and ecology of the microsporidia is based on studies related to their potential for controlling insects of either a medical or economic importance (Sweeney and Becnel, 1991). However, much of our knowledge about what goes on at the molecular level has been derived from studies on the microsporidia, which have been found to be important opportunist pathogens of immunocompromised humans.

PHYLOGENETIC POSITION AND GENOMIC PROPERTIES

The microsporidia have occupied the attention of taxonomists and systematists for some time and have been subject to several reclassifications at all levels of organization, from the species level up to that of their phylum's affinity with other eukaryotes. In this section we discuss their phylogenetic position and properties of their genome that reveal valuable information about the changes that have occurred during their transition to an intracellular and parasitic lifestyle.

ANCIENT EUKARYOTES

Early studies on the morphological characteristics of microsporidian cells found that they lacked several components of a "standard" eukaryotic cell, e.g., the organelles of mitochondria and peroxisomes, suggesting they were primitive and had branched off the eukaryote lineage before their acquisition (see Table 10.1 for further details). The interpretation of a distant origin was

TABLE 10.1
Cytological Structures Considered Standard for Eukaryote Cells That Are Absent in Microsporidian Cells

Mitochondria
Centrioles
Flagella
Peroxisomes
Hydrogenosomes
Glycosomes
Nutrient storage granules
A conventional Golgi apparatus
80S ribosomes

supported by molecular data showing that microsporidia possessed not the typical 80S ribosomes of eukaryotes but rather the 70S ribosomes more representative of the prokaryotes (Ishihara and Hayashi, 1968). The structure of their 5.8S rRNA was also viewed as prokaryote-like (Vossbrinck and Woese, 1986). Phylogenetic studies based on small and large subunit rRNA, elongation factors EF-1α and EF-2, and isoleucyl-tRNA synthetase also tended to place the microsporidia near the base of the eukaryotic tree, reinforcing the notion of their ancient origin (Vossbrinck et al., 1987; Keeling and McFadden, 1998).

Not-So-Ancient Eukaryotes

This picture began to change as evidence started to accumulate in favor of a much more recent origin of the microsporidia; there was also evidence suggesting that the traits considered primitive were more likely to be derived characteristics. For example, their amitochondrial origin was strongly challenged when genes coding for heat-shock protein 70 (hsp70) were detected in two species of microsporidia (Germot et al., 1997; Hirt et al., 1997). These genes were once part of a mitochondrial genome and are likely to have been transferred and incorporated into the genome of their host during the transition from symbiont to fully-fledged organelle. Their presence reveals that the ancestors of the microsporidia had mitochondria. The prokaryote-like 5.8S rRNA could also be interpreted as a derived state due to deletions in rRNA (Cavalier-Smith, 1993).

The basal phylogenetic placement of the microsporidia in the studies above is likely to be an artifact due to the high degree of divergence in their sequences and a rapid rate of evolution, which creates long branch lengths on phylogenetic trees (Keeling and McFadden, 1998). Long branches tend to cluster together, whether related or not, thereby increasing the probability that microsporidia and other long-branched eukaryotes will be found at the base of a phylogenetic tree and artificially clustered with long-branched prokaryotes (Keeling and McFadden, 1998).

As molecular data continue to accumulate, it appears that the microsporidia are located in the crown of the eukaryotes and closely related to the fungi, if not directly descended from them. Evidence for this is found in the similarities between the hsp70 sequence data of the microsporidia and the fungi (Hirt et al., 1999). Furthermore, data from a number of structural proteins, such as the α- and β-tubulins, place the microsporidia as a clearly defined lineage within the main fungal radiation (Keeling et al., 2000). A detailed treatment of the molecular data available on the microsporidia can be found in Weiss and Vossbrinck (1999).

Comparisons with the Fungi

Fungi constitute an extremely large and diverse group of heterotrophic organisms devoid of chlorophyll that have a cell wall, are nonmotile (some species have motile reproductive cells), and reproduce by means of a tremendous variety of spore types (Alexopopoulos et al., 1996). Fungi are usually filamentous and multicellular, with glycogen as the primary carbohydrate-storage product (trehalose in yeast and lichens). Obligate parasitic fungi infect plants, animals, and, in some cases, even other fungi.

The microsporidia are a large group of strictly obligate, intracellular parasites that infect most animal groups (from protists to humans) but are not known to infect plants or fungi (Becnel and Andreadis, 1999; Vávra and Larsson, 1999). Only the spores of microsporidia are walled, and all spores contain large amounts of trehalose. Vegetative growth is by nonmotile amoeba-like stages (often multinucleate) with simple plasma membranes. Although variable in some respects, all microsporidian spores are definitively and uniquely characterized by a coiled polar filament. At germination, the polar filament is inverted to become a tube for transport of the sporoplasm into the host cell.

Recent interest in comparisons of microsporidia and fungi is the result of molecular evidence that places microsporidia as a sister group of the fungi (Keeling et al., 2000). However, these comparisons are confounded by widespread convergence and rapid divergence of these groups. Therefore, the following table is an attempt, in a very general way, to examine some of the similarities and differences between microsporidia and fungi related to their basic biology and characteristics that are more closely aligned with parasitism (Table 10.2). It is hoped that this will stimulate investigations on both groups to help clarify the many points where information is unclear or lacking.

THE MICROSPORIDIA: DEGENERATE EUKARYOTES

Microsporidia were previously considered to be primitive eukaryotes partly because they lacked several "standard" components of a eukaryotic cell (see Table 10.1). Phylogenetic data provide a means to refute this possibility. Furthermore, this pattern of degeneration is widely observed in other organisms, eukaryotic and prokaryotic, that have made the transition to either a mutualistic or parasitic intracellular lifestyle (Andersson and Kurland, 1998; Moran and Wernegreen, 2000). In essence, selection acts to eliminate any redundancy caused by overlapping functions between the two organisms, leaving only genes that serve some essential function. In the case of parasitic relationships, these genes will include those responsible for ensuring the parasite's transmission among its hosts.

GENOMIC DATA

The impressive reductionism in the physical components of microsporidian cells is matched by that of their genomes. Microsporidia possess some of the smallest eukaryotic genomes known, currently ranging from 2.3 to 19.5 megabases (Mb), and show great economy in their genetic material relative to free-living organisms. Much of the data on this topic has been derived from studies on *Encephalitozoon cuniculi*, an important opportunist pathogen of immunocompromised humans whose entire genome has recently been sequenced (Katinka et al., 2001).

The haploid genome of *E. cuniculi* consists of 2.9 Mb and is considerably smaller than some prokaryotic genomes; for example, estimates for strains of the bacterium *Escherichia coli* vary from 4.6 to 5.6 Mb. Furthermore, comparably compact genomes are found in related species that also infect humans and other mammals — *Encephalitozoon hellem* 2.5 Mb, *E. intestinalis* 2.3 Mb (data taken from Méténier and Vivarès, 2001).

The genome of *E. cuniculi* itself is made from 11 linear chromosomes that range in length from 217 to 315 kilobases and harbor a total of approximately 2000 protein-coding genes (Katinka et al., 2001). Important features that reveal its compactness include the paucity of duplicated genes and lack of simple sequence repeats, minisatellite arrays, and known transposable elements in its chromosome cores. Its intergenic spacing regions are short, having a mean and minimum length of 129 and 29 base pairs, respectively. As a consequence, protein-coding DNA sequences occupy approximately 90% of the chromosome cores. Its rRNA operons are unusually short, due to deletions in transcribed spacers and coding regions. Interestingly, there is evidence that the 16S rRNA is shortened from early to late diverging species of *Amblyospora* and relatives (Baker et al., 1997).

There is also evidence of extensive reduction in the size of individual genes, as reflected in the number of amino acids forming their proteins. The genome-wide extent of these reductions was shown by a comparison of the lengths of 350 homologous proteins from *E. cuniculi* and those of another eukaryote that had been entirely sequenced — that of the free-living yeast *Saccharomyces cerevisiae*. Roughly 85% of the *E. cuniculi* proteins were shorter, and on average 15% shorter (Katinka et al., 2001). For example, genes for the receptor proteins of kinesin and SRP are up to one third shorter than those known for other eukaryotes (Biderre et al., 1999).

Future developments in the field of comparative genomics promise to be revealing, both with respect to other eukaryotes and within the microsporidia: the few data that currently exist suggest that the genomes of microsporidia infecting invertebrates are larger and vary considerably in size

TABLE 10.2
A Comparison of Major Features of the Microsporidia and Fungi

Characteristics	Microsporidia	Fungi
	Cytological Structures	
Mitochondria	Not demonstrated	Present
Perixosomes	Absent	Present
Lysosomes	Absent	Present
Paramural bodies	Present	Present
Golgi	Lack stacked cisternae	Variable, some stacked, some unstacked
Endoplasmic reticulum	Smooth and rough	Smooth and rough
Ribosomes	Prokaryote-size (70S) consisting of large (23S) and small (16S) subunits; lack 5.8S but homolog found at beginning of 23S subunit	Typical eukaroytic size of 80S
Cytoskeleton	Microtubles, keratin filaments, probably actin and intermediate filaments	Microtubules, actin, and intermediate filaments
Centrioles	Absent, spindle attachment to "spindle plaque"; small "polar bodies" located near the spindle plaque	Present only in Chytridiomycota (composed of nine triplets), replaced by spindle pole bodies (SPBs) in most true fungi; multivesicular bodies associated with the SPB
	Physiology	
Trehalose	Present in large amounts in spores of most species examined and is perhaps involved in germination	A reserve disaccharide of fungi (esp. yeast and lichens)
Chitin	Present in endospore	Present in cell and spore walls
Glucans	?	Present
	Motion	
Flagella	Absent	Present only in Chytridiomycota
	Invasion	
Resting stage (spore)	Polymorphic, uninucleate, and binucleate spores. Some species produce only one type of spore, some produce as many as four types	Sexually (oospores, zygospores, ascospores, basidiospores) and asexually (conidia or sporangiospores) are produced, all uninucleate; species with one up to four types of spores
Spore wall	Exospore (which can be layered) and endospore	Variable, two to five layers
Mechanism	Typically, spore germination occurs in the gut by eversion (under tremendous pressure) of the polar filament, which becomes a hollow tube for transfer of the sporoplasm to the cytoplasm of a host cell	Spores attach to surface and produce a penetration germ tube that gains access through enzymatic and mechanical activity
	Replication	
Nuclear arrangement	Uninucleate and diplokaryotic (stable arrangement of paired haploid nuclei)	Mostly uninucleate, some with a dikaryon (two unpaired haploid nuclei)
Mitosis	Intranuclear, nuclear envelopes do not break down during division	Intranuclear
Meiosis	Occurs in both uninucleate (diploid) and diplokaryotic nuclei (each diploid), synaptonemal complexes of typical eukaryotic type	Occurs only in uninucleate nuclei (diploid); not present in imperfect fungi

(*continued*)

TABLE 10.2 (CONTINUED)
A Comparison of Major Features of the Microsporidia and Fungi

Characteristics	Microsporidia	Fungi
Vegetative growth	Binary fission and multiple fission of wall-less plasmodia	Walled hyphae develop by apical growth, in yeast by budding or binary fission
Sporulation	Varies from bisporous sporogony that produces two spores from each sporont to polysporous sporogony producing many spores; can involve meiosis or nuclear dissociation of diplokarya	Quite variable
	Reproduction	
Asexual species	Two basic forms: one uninucleate (haploid) throughout development and produces uninucleate spores, the other diplokaryotic (diploid) throughout development and produces diplokaryotic spores	Uninucleate (haploid) throughout development producing uninucleate spores
Sexual species	Typically an alternation of uninucleate development producing haploid spores (some involving meiosis) and diplokaryotic development producing diplokaryotic spores; some involve an obligate intermediate host	Alternation of haploid and diploid cell states, usually includes meiosis and produces only uninucleate spores; some involve an intermediate host
	Other Features	
Habitat	Obligatory parasites	Facultative and obligatory parasites
Hosts	Only animals, protists to humans	Plants, animals, and other fungi
Tissues	Intracellular, mainly in cytoplasm; many species tissue specific with some causing systemic infections	Exocellular in pathogenic associations
Pathogens	None known	Infected by viruses and other fungi
Drug reaction	Some fungicides and antibiotics effective	Fungicides

relative to *E. cuniculi* and its close relatives, e.g., *N. locustae*, 5.4 Mb; *N. pyrausta*, 10.5 Mb; and *N. bombycis*, 15.3 Mb (data taken from Méténier and Vivarès, 2001). Whether or not these differences are due mainly to protein-coding sequences of DNA remains to be established. Identifying matching or mismatching genes from different genomes should shed light on some of the functional constraints faced by different species.

To summarize, several organelles that characterize a "standard" eukaryotic cell are missing in the microsporidia. Contrary to being a primitive state due to an early separation from the eukaryotic lineage, it now seems likely that these organelles were lost at a much later date and during their transition toward a parasitic intracellular lifestyle. This loss of organelles is paralleled by the loss of nonessential genetic material, leading to some microsporidia possessing some of the most compact eukaryotic genomes known. Current phylogenetic estimates place the microsporidia as being closely related to the fungi, if not derived from within them. These two groups share some characteristics but not others.

ASPECTS OF INTRACELLULAR LIFE

Almost all species of microsporidia undergo their growth and production of spores within the host cell's cytoplasm. The only exceptions to this pattern are the few species that develop within the nuclear cavity of their host cell, e.g., *Nucleospora* spp. (Elston et al., 1987). No microsporidia develop outside of their host cells. This pattern strongly suggests that these parasites require their host environment for the uptake of energy and nutrients. This suspicion is reinforced by data from *E. cuniculi* that show it possesses few genes for amino-acid biosynthesis and those required for the tricarboxylic-acid cycle, the respiratory electron-transport chain, and fatty-acid synthesis have not been detected (Méténier and Vivarès, 2001).

Cytological evidence shows that the initial developmental stages (meronts) can take resources directly from the host's cytoplasm by endocytosis (Vávra and Maddox, 1976). However, this cannot be the case in the later developmental stages of many species, as a structure forms around the developing parasite and physically separates it from the host's cytoplasm (forming a parasitophorous vacuole). Depending on the relationship involved, this structure may be secreted by the host, the parasite, or both (Cali and Takvorian, 1999). The functional significance of these differences is unknown.

A conspicuous feature observed in ultrastructural studies that implicates the host cell as a provider of resources is the physical attraction of host organelles, notably mitochondria and endoplasmic reticulum, to the interface of the structure enveloping the developing parasite (Figure 10.1). This physical proximity suggests the parasite is exploiting its host cell's resources. Furthermore, there are documented cases of gap junctions being formed with host mitochondria (e.g., Terry et al., 1999). These intercellular channels provide the means for the exchange of ions and macromolecules from host to parasite, e.g., iron, fatty acids, cholesterol, and ATP. The parasite's potential for energy uptake is further supported by the identification of four ADP/ATP carrier protein genes in *E. cuniculi* (Vivarès and Méténier, 2001). These carriers are homologs to those found in other obligate intracellular organisms known to import host ATP, e.g., *Rickettsia* and *Plasmodium*.

The microsporidia, however, are probably not entirely dependent on host ATP for their source of energy. In particular, the use of trehalose for ATP production, via glycolysis, is thought to be of importance (Weidner et al., 1999). Insect fat body cells are potentially rich sources of this resource; they have a central role in an insect's metabolism because they convert sugars into carbohydrates (glucose and trehalose) and lipids. The parasite's use of trehalose would correlate with the frequent observation of reduced amounts of granular carbohydrates in the cytoplasm of infected host cells (Weiser, 1976). Infected cells also often show reduced amounts of lipids. The microsporidia cannot be directly metabolizing this resource themselves, as that would require oxidative respiration by mitochondria. However, as insects cannot convert lipids back into carbohydrates, they must instead pass via the tricarboxylic-acid cycle within mitochondria. It is at this point that the microsporidia can exploit this source of the host's energy reserves.

Much still remains to be discovered about the metabolism of the microsporidia. For example, what are the parasite's protein requirements and where do they come from? Do they, as some

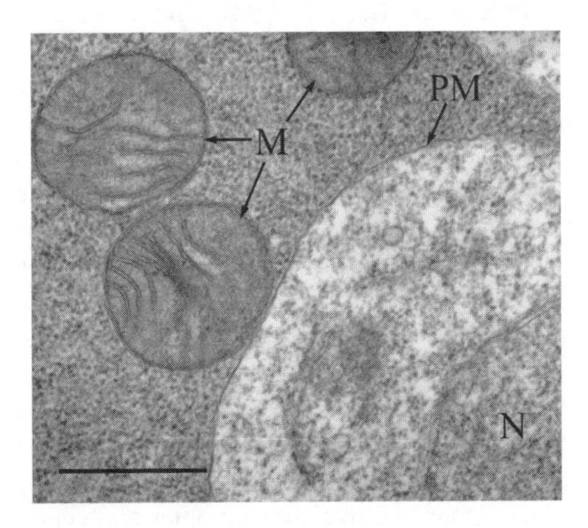

FIGURE 10.1 Several host mitochondria (M) near the plasma membrane (PM) of a uninucleate (N) meront. Bar = 1 μm.

authors suggest (Scanlon et al., 2000), manipulate activity of the host cell via polyamines secreted into its cytoplasm? Perhaps most intriguingly, it seems possible that the microsporidia have a unique form of energy metabolism catabolized by different enzymes than those of other amitochondrial eukaryotes (Vivarès and Méténier, 2001).

To summarize, the evidence suggests that microsporidia depend entirely on their host cells for their metabolic requirements. Resource uptake may be partially achieved by endocytosis, but a more important mechanism is probably by direct importation across the cell membrane from host organelles recruited into close physical proximity to the parasite or its surrounding parasitophorous vacuole. The exploitation of host-derived ATP and trehalose seems likely; less is currently known about how other metabolic requirements are satisfied. Progress in this area can be expected as further elements arise from the decoding of the *E. cuniculi* genome: to date, only approximately half of its 2000 protein-coding genes have been assigned a putative functional role.

MACROEVOLUTIONARY VARIATION AND PATTERNS OF TRANSMISSION

Despite the simplicity of the generic microsporidian life cycle outlined in the introduction, the diversity of ways that this has been adapted is impressive. Currently there are some 143 described genera of microsporidia, encompassing well over 1000 species. Almost half of these genera, 69, have an insect as the type host (Becnel and Andreadis, 1999). In this section, we focus on a limited range of examples to illustrate some of this diversity and adaptations these parasites have made to enhance their transmission success, both within and among individual hosts. However, first we will briefly describe the unique apparatus that enables them to infect host cells and that provides the basis for classifying the microsporidia apart from other eukaryotes.

SPORES AND THE POLAR FILAMENT

Like many other organisms, microsporidia produce spores for protecting themselves against harsh environmental conditions. These spores are protein-walled and typically small in size, varying from 1 to 10 μm in length depending on the species involved. Some species show adaptations to increase their contact with hosts, e.g., appendages to aid spore floatation for those infecting aquatic larvae (Vávra and Maddox, 1976), but none possesses its own means of propulsion. Species that infect terrestrial hosts typically produce spores that can withstand prolonged periods of desiccation and cold temperatures, while those from aquatic hosts may become inviable after only a few hours of being dried or held at 4°C (Undeen and Becnel, 1992). Up to four different types of spores may be produced during the course of a life cycle, with each type having a specific role in the transmission of the parasite (Johnson et al., 1997). However, the defining feature of the microsporidia is not their spores but the polar filament apparatus they enclose.

The polar filament is effectively a hollow tube anchored at the anterior of the spore and coiled up inside it (Figure 10.2). The number of coils varies among species and spore types, but when fully unfurled it can easily reach beyond a dozen times the length of the spore itself (Figure 10.3). When the appropriate environmental conditions trigger germination, this filament begins to evert itself from its point of attachment, turning inside out as it extends into the environment outside the spore. This unravelling is explosive, having been recorded with a mean acceleration and velocity of 500 μm/sec^2 and 105 μm/sec, respectively (Frixione et al., 1992). In many species, the force for this propulsion is thought to derive from an osmotic process increasing the turgor within the cell. Once externalized, the polar filament becomes known as the polar tube. It is thought that this tube is capable of piercing adjacent host cells and entering directly into their cytoplasm. As pressure continues to develop within the spore, the nucleus and other contents get forced down the polar tube, distending it as they go, until they pass out of the end of the tube. If the polar tube has pierced a host cell, these contents (the sporoplasm) are injected directly into the host cell's cytoplasm.

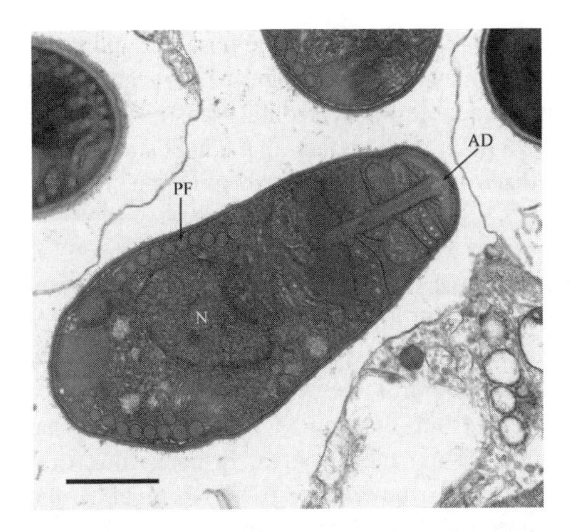

FIGURE 10.2 Immature spore of *Hazardia miller* showing the anchoring disc (AD) of the coiled polar filament (PF) and the single nucleus (N). Bar = 1 μm.

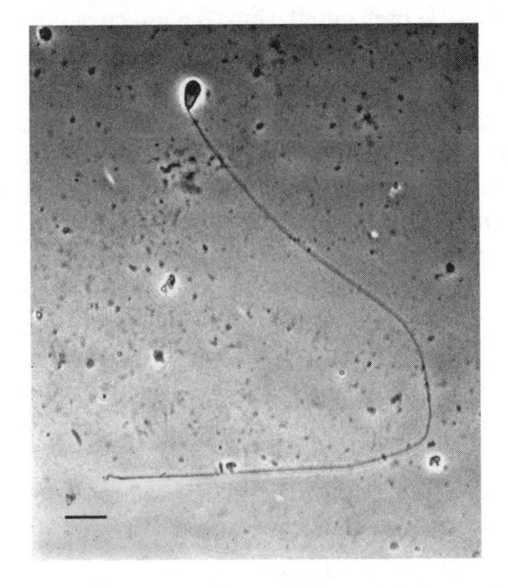

FIGURE 10.3 Germinated spore of *E. aedis* with the polar tube fully everted. Bar = 10 μm.

From this point, the parasite's intracellular development can begin. No other eukaryotic parasite possesses a similar means of entering its host cell that combines elements of a harpoon and a hypodermic needle.

Many details about the processes of spore formation, germination, and host-cell infection await clarification (Vávra and Larsson, 1999). Rapid progress in this area is predicted as the spores and their polar tubes are the likely targets for chemotherapeutic intervention and are under intensive study.

INTRAHOST TRANSMISSION

The predominant route of infection for most microsporidia is probably via spores ingested from the host's environment along with its food. This also means that epithelial cells lining the host's gut are the tissues most directly exposed to the risk of infection.

For microsporidia with a relatively simple life cycle, infection of a host gut cell is followed by a succession of developmental stages, during which proliferation occurs, before spores are produced.

The accumulation of spores leads to rupturing of the host cell and the dissemination of spores into the gut lumen. From here they may infect other gut cells or get passed out into the environment with feces and thus become available for horizontal transmission to other hosts. Such infections are usually chronic and have a debilitating effect on the host's fecundity and longevity. An example of such a microsporidian, discussed elsewhere in this chapter, is *Glugoides intestinalis* (Larsson et al., 1996).

Rather than completing their whole life cycle within a single cell of the host's gut epithelium, some microsporidia show intrahost transmission and spread themselves away from the original site of infection to other tissues within the host. Various mechanisms by which this happens have been proposed, but the most clearly documented is via the production of "early" spores (Iwano and Kurtti, 1995). These spores are often formed within 48 to 72 h post infection and are morphologically different from the spores responsible for the initial infection via the environment — "environmental" or "late" spores. In particular, they have shorter polar tubes and thinner spore walls, suggesting they are more economical to manufacture both in terms of time and material. The functional role of these spores is to infect secondary cells within the same host, a process also known as "autoinfection." This may occur by the *in situ* germination of spores and the inoculation of their contents into neighboring cells. Alternatively, ruptured gut cells may disseminate these spores into the host's hemocoel and provide a means of infecting more distant tissues.

There are several reasons to consider why intrahost transmission might be advantageous. It increases the number of foci of infection and potentially the number of spores produced. A related advantage is that it may permit the parasite to maintain its presence by escaping from epithelial gut cells likely to be shed or digested during a moult or metamorphosis (Andreadis, 1985b). Transmission within the host can also give access to other tissues of a superior nutritional value for the parasite's growth, e.g., fat-body cells. However, migration away from epithelial cells can also physically limit the opportunities for subsequent transmission to other hosts via spores shed into the alimentary canal. The microsporidia seem to have found two main ways of solving this problem.

The first is to exploit the host's inner tissues for a significant production of spores, causing widespread pathology and provoking the host's death. These spores are subsequently released into the environment as the host cadaver decomposes. In examples of such cases, the host's fat body can literally be converted into a sack containing millions of spores (Becnel and Andreadis, 1999).

An alternative solution is not to seek horizontal transmission from infected individuals themselves but instead to do so from their offspring.

Vertical Transmission

Many microsporidia show a pattern of intrahost transmission that brings them into close physical proximity with the host's reproductive tissues and from where cells of the developing offspring can be infected. This may be achieved directly by spores circulating in the host hemocoel or indirectly via passage in mobile host cells that migrate toward reproductive tissues, e.g., oenocytes (Becnel et al., 1989).

There are at least two important consequences of adopting this transmission behavior. First, a parasite's transmission success relies on the reproductive success of its host. This constrains the negative effects that the parasite can have on its host's reproduction if it is not to adversely effect its own transmission success. The second important constraint associated with vertical transmission is that a host offspring's cytoplasm is uniparentally inherited from its mother. Thus, only females provide the physical means by which their offspring can be directly infected, with males effectively being a dead end for vertical transmission. Despite these constraints, the use of vertical transmission is phylogenetically widespread among the microsporidia (Dunn et al., 2001). We will focus on two contrasting examples of adaptation to these constraints.

Mixed Patterns of Vertical and Horizontal Transmission

Our first example involves microsporidia of the family Amblyosporidae that predominantly infect Dipteran hosts, most notably mosquitoes. Host larvae are horizontally infected when they ingest spores from their aquatic environment. Although not documented for all species, it is likely that they all produce "early" spores, enabling migration away from the gut epithelium (Johnson et al., 1997). Circulating oenocytes appear to be targeted by these spores (Becnel et al., 1989). Within these cells a third type of spore is produced. These spores are responsible for vertical transmission when they germinate in proximity to a female host's reproductive tissues. In some relationships, the production and germination of these spores is arrested until the female mosquito host provides hormonal cues that she is in the process of maturing a clutch of eggs (Hall and Washino, 1986). This mechanism helps to limit the energetic drain and physical damage that continued development and spore production would otherwise have on the host prior to its reproduction.

The exploitation of the vertically infected larvae by this family of microsporidia is remarkable (Kellen et al., 1965). In the most straightforward cases, all of a female's offspring are likely to be killed late in their juvenile development by a significant production of spores in their fat-body cells. This fourth type of spore (a meiospore) is infectious to an obligate intermediate host, usually a microcrustacean (Andreadis, 1985a). It is the spores released upon the death of the intermediate host that are responsible for horizontal transmission to mosquito larvae. In some relationships, however, only vertically infected male larvae are killed by the production of meiospores, whereas infected females experience "benign" infections, become adults, and transmit the parasite to their own offspring. Intermediate relationships also exist in which varying proportions of female larvae are killed (Kellen et al., 1965). This "late male killing" behavior is reportedly unique to these mosquito–microsporidian relationships and contrasts with "early male killing" behavior shown by several other vertically transmitted parasites in which infected males are killed at an early stage in their development (Hurst, 1991). The essential difference between these two types of parasites is that the microsporidia are more adept at exploiting male hosts for horizontal transmission, whereas the others are limited to increasing their success by vertical transmission alone, e.g., by freeing infected females from resource competition with their brothers (Hurst and Majerus, 1993).

The reasons for this diversity in the sex-specific development of these microsporidia (mechanism unknown) are thought to depend on the ecological conditions of particular relationships and the relative success to be derived from exploiting host females for vertical or horizontal transmission (Hurst, 1991; Agnew and Koella, 1999a). The "sparing" of females for vertical transmission is more likely to be favored in conditions where hosts occupy ephemeral habitats or where the presence of intermediate hosts is not assured. Vertical transmission in these cases has the advantage of enabling the parasite to disperse with its host into the temporally available sites where spores released from male larvae may have the opportunity to infect other host lineages.

An interesting member of this family of microsporidia is *Edhazardia aedis*, as it has no requirement for an intermediate host to complete its life cycle, and the developmental sequence leading to meiospore production is abortive and rarely completed (Becnel et al., 1989). This may reflect an adaptation to a habitat change by its host, the yellow fever mosquito *Aedes aegypti*, to ovipositing in urban habitats that lack the presence of intermediate hosts. Furthermore, it does not rely on a host cue for the production and germination of its vertically transmitting spores. In certain circumstances, this can lead to these spores germinating and for the developmental sequence leading to the production of horizontally transmitting spores occurring in the infected individual rather than in the next generation after vertical transmission. Consequently, although the parasite undergoes the same developmental cycle, its observed sequence of transmission may be horizontal-vertical or horizontal-horizontal (Becnel et al., 1989). The former sequence of transmission is more likely in environmental conditions that favor fast larval growth, while slow larval growth conditions favor the latter (Agnew and Koella, 1999b). These patterns of host exploitation may be adaptive. Fast larval growth is associated with the emergence of large and fecund female mosquitoes likely to offer the parasite substantial vertical-transmission success. In contrast, slow larval growth rates lead to the emergence of much smaller females that potentially offer much less vertical-transmission success. Prolonged larval growth also increases the potential for the parasite to complete its life cycle and kill its host before emerging, thereby granting the possibility of exploiting both males and females for horizontal transmission (Agnew and Koella, 1999b). While this pattern of development is not without risk to the parasite's own transmission success (Koella and Agnew, 1997), it appears to be adapted to maximizing its vertical- or horizontal-transmission success as a function of its host's life-history traits given the temporally unpredictable and spatially heterogeneous environments occupied by its host.

Further examples of the variation in life cycles of microsporidia that infect mosquitoes or insects in general can be found in Becnel (1994) and Becnel and Andreadis (1999), respectively.

Where Vertical Transmission Dominates

Our second example of adaptation by the microsporidia to the constraints imposed by vertical transmission is centered on those that are almost exclusively transmitted by this means, with horizontal transmission probably being a rare event.

Some microsporidia only known to have vertical transmission and that infect crustaceans of the genus *Gammarus* have been found to cause very little tissue pathology and no negative effects on their host's growth, reproduction, or longevity (Bulnheim and Vávra, 1968). This pattern fits expectations that vertical transmission will not have an adverse influence on the host's reproductive success. However, a lack of pathology does not indicate that these parasites have a limited effect on their hosts. In contrast, these microsporidia share a trait with several other vertically transmitted pathogens in that they actively augment their possibilities for cytoplasmic inheritance by manipulating the functional sex ratio of their hosts toward females (Dunn et al., 2001). In doing so, these infected matrilines are likely to increase in frequency relative to those producing offspring in a 50:50 sex ratio and will help maintain the parasite's presence in its host population.

To summarize, the microsporidia that infect insects are diverse. They produce a range of spore types that have different functional roles in either transmitting the parasite among different cells within an individual host or among different host individuals. The parasite's migration to inner tissues is associated with either development leading to an important production of spores, host death, and horizontal transmission or a developmental sequence leading to vertical transmission. The ways in which vertically infected offspring are exploited vary across different relationships. These differences appear to reflect adaptations by the microsporidia to gain transmission success as a function of the possibilities offered by the host and its ecological context.

MICROEVOLUTIONARY VARIATION

Thus far we have concentrated on variation among different microsporidia. In this section, we focus on cases of evolutionary change within a species. In particular, attention will center on traits linked with a change in the parasite's virulence.

An organism's fitness can be defined as a composite function of how its life history traits interact with its environment (Stearns, 1992). Various life history traits are often correlated with each other to some extent due to the pleiotropic action of individual genes influencing the expression of more than one trait. Of particular relevance are negative pleiotropic interactions such that an increase in fitness due to the increased expression of one trait may be annulled or even reversed by its effect on a correlated trait. In natural conditions, selection pressures may arrive from a variety of sources to keep in check the overexpression of negatively correlated traits and to help maintain genetic diversity for trait values.

Experimentally accentuating particular sources of selection while controlling for others can provide a useful means by which to identify which traits harbor genetic diversity and how they are correlated with other traits. Serial passage experiments have been particularly useful in this context (Ebert, 1998). In such cases, parasites are usually grown under well-defined conditions where their passage from one environment to the next is performed by the experimenter, e.g., by the injection or transfer of infected material. This latter condition has the effect of reducing constraints acting on the parasite's transmission success among hosts while also offering a selective advantage to genotypes that are proportionately well represented in the parasite population at the time of transfer. Thus, parasites are likely to find themselves in a changed environment with fewer counterbalancing sources of selection due to environmental heterogeneity and where selection pressure is concentrated more on their growth rates than on their ability to come into contact and infect other hosts via the environment.

Adaptation to Altered Conditions of Transmission

An example of a serial-transfer experiment involving a microsporidian is provided by a study of *N. furnacalis* maintained in cell-culture conditions (Kurtti et al., 1994). After 77 generations of serial passage via infected cells, the parasite showed faster rates of spread than it did after only four generations of culture. This change in growth rate was correlated with an increase in the production of "early" spores, used for autoinfection within host tissues, relative to that for the "late" or "environmental" type of spore used for transmission among hosts via the environment. Furthermore, these "environmental" spores showed reduced infection success and virulence when subsequently exposed to their natural host, the Asian corn borer *Ostrina furnacalis*. Thus, these results show the parasite's life-history traits adapting to the environmental conditions of serial culture transfer in such a way as to increase their rate of spread through host tissue cells; these changes were negatively correlated with their ability to exploit their original host via the use of their "environmental" spores. Such tradeoffs between the ability to exploit individual hosts and transmit among them have been reported for a number of disease agents and are thought to play a role in limiting the evolution of increased virulence in natural conditions (Ebert, 1998).

Another indication that microsporidia can change their relative investment in different spore types can be drawn from an artificial selection experiment on the vertical transmission success of *Amblyospora dyxenoides* in its mosquito host *Culex annulirostris* (Sweeney et al., 1989). This parasite would normally kill the majority of its host's vertically infected offspring with the production of meiospores destined for horizontal transmission to an intermediate host. However, the spores are not produced in some vertically infected larvae, which experience "benign" infections, and females can vertically transmit the infection to a second generation of larvae. At the beginning of the experiment, 10% of the clutches of vertically infected larvae showed "benign" infections in which there was no meiospore production. This figure increased to 70% after nine generations of selection for vertical transmission. The possibility that these results were influenced by changes in the host cannot be excluded, as they would also have been selected during the course of the experiment. Nonetheless, the results are interesting in view of current estimations placing the Amblyosporidae at a basal position within the microsporidian phylogeny and thus being representative of the life histories from which other genera were derived (Baker et al., 1997).

Intrahost Competition

The idea that virulence increases due to intrahost competition among parasites has experimental support from the relationship between *Glugoides intestinalis* and its water flea host *Daphnia magna* (Ebert and Mangin, 1997). Replicated lines of the host–parasite relationship were subjected to a regime of either high or low rates of host mortality for a period of 14 months. It was predicted that higher virulence would evolve in the high-mortality treatment due to the need for faster average rates of host exploitation. The opposite trend was observed, with virulence being higher in the low host-mortality environment. It appears that the low-host-mortality environment increased host exposure to infection and the probability of their cells receiving a double infection. The consequence of this was to place parasites in direct competition with each other for the resources of the host cell. Indeed, the size of the sporophorous vesicles containing the parasite's spores was smaller in doubly vs. singly infected gut cells, indicating host resource limitation. When tested in standardized conditions, parasites from the low-host-mortality treatment showed faster rates of growth and produced larger sporophorous vesicles. The hosts exposed to these parasites were also killed earlier and with a higher sporeload, indicating that intrahost competition had resulted in the selection for more virulent parasites. Interestingly, although the more virulent strains produced more spores, they had a lower degree of transmission success for a standard infective dose. This suggests a potential tradeoff between virulence and transmission success that may prevent the escalation of virulence in the absence of intrahost competition.

LOCAL ADAPTATION

A further aspect of serial-passage experiments is that pathogens are often exposed to limited genetic diversity of the host during the course of selection (Ebert, 1998). This increases the possibility of their evolving to exploit a particular host genotype, a process known as local adaptation. Evidence that this occurs in natural conditions was shown in an experiment involving strains of *G. intestinalis* isolated from geographically different populations of *D. magna* (Ebert, 1994b). This cyclically parthenogenetic host has the advantage of using mainly clonal reproduction during the course of a breeding season, hence increasing the temporal possibility of parasites encountering the same host genotype within a season. The experiment found for several traits (infection success, probability of killing host, number and transmission success of spores produced) that the different strains of parasite tested were best adapted to exploiting the sympatric host population they were regularly exposed to rather than hosts of unfamiliar genotypes. While not demonstrating local adaptation, other studies have found heterogeneous results when crossing different strains of host and microsporidian in standardized conditions (Ebert, 1994a; Schmid-Hempel and Loosli, 1998), giving rise to the possibility that genotype-by-genotype interactions exist in other relationships involving microsporidia, though this has not always been found to be the case (Woyciechowski and Krol, 2001).

GENETIC VARIATION

A key element in determining the rates of evolutionary change in microsporidian populations will be the amount of genetic variation present. The potential for coevolutionary interactions with some invertebrate hosts may be increased in cases where both have life cycles of a similar duration (Koella et al., 1998). However, very little is currently known with certainty as to how much variation exists or how it is maintained. Some of the studies cited above suggest that individual populations harbor enough variation to respond to experimental selection pressure and that interpopulation variation exists. Genetic diversity at the molecular level has been most intensively studied for species that infect humans. Variation in the number of ITS copies from strains originating from different host reservoirs has been found for some species (Mathis et al., 1999), and nucleotide heterogeneity exists in others (Rinder et al., 1997). How representative these data may be is very unclear. Meiosis has not been reported from these microsporidia, whereas it has from those of invertebrates (Becnel and Andreadis, 1999), prompting the suggestion that this function was lost during the course of their genome compaction (Vivarès and Méténier, 2001). Even where meiosis exists, the opportunity for outcrossing is possible only when individual host cells are infected by more than one strain, for which no data from natural conditions is available. Perhaps the only certainty is that many species can exploit individual hosts for the production of millions of spores, prompting expectations that mutation will result in a certain level of genetic variation.

One type of microevolutionary change that has not yet been reported is that of the evolution of drug resistance. Several benzimidazoles with antimicrosporidial activity are used in the clinical treatment of infections (Blanshard et al., 1992). Widespread use of such drugs will create the selection conditions for this type of response, though this may be countered if it is traded off with transmission success or if humans form only a limited fraction of the mammalian hosts these species exploit.

To summarize, only a few studies have demonstrated evidence of microevolutionary change either within or among populations of microsporidia. The results obtained have tested theoretical predictions and shed valuable empirical light on factors influencing evolution in host–parasite relationships. However, our ability to address their evolution in the context of population genetics is currently negligible for want of appropriate data.

HOST RESPONSES

The first general line of defense likely to be encountered by microsporidia that infect insects is that of the peritrophic membrane (or matrix) lining the host's alimentary canal. This saves epithelial cells from abrasion during food passage and provides a barrier to help physically separate them from parasites ingested with their diet. It seems plausible that the spatial chasm this creates between spores enmeshed in the membrane and living cells of the host may have had a role in the evolution of the spectacular polar-tube-invasion apparatus of the microsporidia.

There is little or no evidence to show that invertebrate hosts have an immune response that can suppress or clear a microsporidian infection once it has successfully reached a host cell's cytoplasm. Spores can potentially be encapsulated once within the host's hemocoel (Vávra and Undeen, 1970). However, such host responses are rarely reported. Even if this response is possible, it will be useful only if spores are encapsulated before discharging their contents into another cell. Furthermore, this response can do nothing to prevent infection if spores germinate *in situ* and directly inoculate their sporoplasm into a neighboring cell, thus without being exposed to the host's hemocoel.

Hosts may be able to reduce the costs of parasitism by means other than an immune response. One example is provided by an experiment involving crickets infected with *N. acridophagus* (Boorstein and Ewald, 1987). When given the choice, infected hosts preferentially moved to hotter rather than cooler environments and did so more than uninfected controls. When hosts were not given the choice, those that were maintained in the hotter environments suffered less from the costs of parasitism than those kept in cooler conditions. Thus, the infected hosts displayed an example of "behavioral fever" by choosing a temperature that had a more adverse effect on their infection than on themselves. We note parenthetically that there do not appear to be any examples of microsporidia specifically manipulating their host's behavior to promote their own transmission success.

The scope for infected hosts to change their environment or behavior to reduce the costs of parasitism may be fairly limited. A more general response predicted by evolutionary theory is that hosts can achieve this by altering their life-history traits. In particular, infected hosts are expected to lessen the costs they experience by bringing forward their reproductive schedule (Minchella, 1985). In essence, infected hosts should reproduce as much as they can while they can, even if this reallocation of resources would be otherwise detrimental to their fitness if uninfected. Several insect hosts have responded to parasitism in this manner, including the mosquito *Culex pipiens* when infected by *Vavraia culicis*. When infected as larvae, females brought forward their age at pupation relative to control females, even though this was at the expense of their adult size and potential fecundity (Agnew et al., 1999). In different experimental conditions, Reynolds (1970) found no difference in ages at pupation of infected and uninfected females but an increased egg production by infected females in their early adult life. Though the responses to *V. culicis* by the two populations of *C. pipiens* were different, they both result in bringing forward the host's reproductive schedule.

A possible example of a life history response was observed in a seminatural field experiment where bumblebee colonies became naturally infected with *N. bombi* (Imhoof and Schmid-Hempel, 1999). These colonies showed an increased production of sexual offspring, particularly males, relative to uninfected colonies. A particular advantage of producing males is that, even if they become infected while within the colony, they mate soon after leaving it and cannot paternally transfer infection to their offspring. In contrast, infected females (future queens) must overwinter before founding a new colony (prolonging her exposure to the parasite's costs) and a colony she is likely to contaminate via spores in her feces.

The production of male offspring can potentially be a useful device against maternally transmitted parasites that have a negative effect on a female's fitness: a female's sons, even if infected themselves, cannot transmit the infection to the next generation, and hence their descendants will be purged of the infection. Mangin et al. (1995) reported the fitness costs for *D. magna* of being infected by *Flabelliforma magnivora* (then described as *Tuzetia* sp.). This microsporidian has a vertical transmission success of approximately 100% but a negative effect on female fecundity of 30 to 60% and no direct horizontal transmission among individuals of its host species. Without competition with uninfected hosts, infected *Daphnia* can be maintained despite the high virulence of the parasite. However, if faced with competition from uninfected clones, both infected hosts and parasites go extinct rapidly (Ebert et al., 2000). The potential of sexual reproduction to purge infection from part of the descending lineage via a generation including male offspring provides a means for the host to avoid this fate.

To summarize, the first line of defense for insects against microsporidian infections arriving from the external environment is probably to reduce the proximity between ingested spores and epithelial cells lining the gut. Once the parasite has established its presence in host cells, this class of hosts seems to have limited means with which to directly combat the infection immunologically. As a consequence, hosts need to resort to alternative forms of response to limit the costs of an infection that cannot be suppressed or cleared. Bringing forward investment in reproductive success provides a solution to this problem for some hosts, while increased investment in the production of the sex less affected by the costs of parasitism may work for hosts able to manipulate the sex ratio of their offspring.

CONCLUDING REMARKS

Microsporidia are obligate symbionts. The evolutionary transition to endosymbiosis is marked by a general loss of cytological complexity and reduction in the size of their genomes that is matched by few other eukaryotes. The adaptation to a parasitic lifestyle is particularly identifiable by their polar filaments, which provide a unique means to infect host cells. Despite this highly specialized lifestyle, they have been highly successful, being among the most common parasites of arthropods as well as many other taxa including humans. The spectrum of their known transmission behavior ranges from purely vertical to purely horizontal, including examples with intermediate patterns of investment in both modes of transmission and where an intermediate host may be involved. The spores produced for different types of transmission are usually specific to their tasks and destined to transmit the parasite either within the same host or to different host individuals. Field and laboratory data show that they are capable of adapting to the particular conditions or constraints provided by their host and its environment and that they can do so rapidly. Progress at the molecular level has recently provided the means to surmount many of the technical problems posed by their small physical size and intracellular lifestyle. These developments are throwing valuable new light on many aspects of their biology and helping to put their diversity and evolutionary origin in perspective.

ACKNOWLEDGMENTS

The author's work is funded by an ATIPE grant awarded by the Centre National de la Recherche Scientifique to YM, the United States Department of Agriculture (JJB), and the Swiss Nationalfonds (DE).

REFERENCES

Agnew, P. and Koella, J.C. (1999a). Constraints on the reproductive value of vertical transmission for a microsporidian parasite and its female killing behaviour. *J. Anim. Ecol.* **68:** 1010–1019.

Agnew, P. and Koella, J.C. (1999b). Life history interactions with environmental conditions in a host-parasite relationship and the parasite's mode of transmission. *Evol. Ecol.* **13:** 67–91.

Agnew, P., Bedhomme, S., Haussy, C., and Michalakis, Y. (1999). Age and size at maturity of the mosquito *Culex pipiens* infected by the microsporidian parasite *Vavraia culicis. Proc. R. Soc. London (B)* **266:** 947–952.

Alexopopoulos, C.J., Mims, C.W., and Blackwell, M. (1996). *Introductory Mycology.* John Wiley & Sons, New York.

Andersson, S.G.E. and Kurland, C.G. (1998). Reductive evolution of resident genomes. *Trends Genet.* **6:** 263–268.

Andreadis, T.G. (1985a). Experimental transmission of a microsporidian pathogen from mosquitoes to an alternate copepod host. *Proc. Natl. Acad. Sci. U.S.A.* **82:** 5574–5577.

Andreadis, T.G. (1985b). Life cycle and epizootiology and horizontal transmission of *Amblyospora* (Microspora: Amblyosporidae) in a univoltine mosquito *Aedes stimulans. J. Invertebr. Pathol.* **46:** 31–46.

Baker, M.D., Vossbrinck, C.R., Becnel, J.J., and Maddox, J.V. (1997). Phylogenetic position of *Amblyospora* Hazard & Oldacre (Microspora: Amblyosporidae) based on small subunit rRNA data and its implication for the evolution of the Microsporidia. *J. Eukaryot. Microbiol.* **44:** 220–225.

Becnel, J. (1994). Life cycles and host-parasite relationships of Microsporidia in culicine mosquitoes. *Folia Parasitol.* **41:** 91–96.

Becnel, J.J. and Andreadis, T.G. (1999). Microsporidia in insects. In *The Microsporidia and Microsporidiosis* (M. Wittner, Ed.), pp. 447–501. ASM Press, Washington, D.C.

Becnel, J.J., Sprague, V., Fukuda, T., and Hazard, E.I. (1989). Development of *Edhazardia aedis* (Kudo, 1930) N.G., N. Comb. (Microsporida: Amblyosporidae) in the mosquito *Aedes aegypti* (L.) (Diptera: Culicidae). *J. Protozool.* **36:** 119–130.

Biderre, C., Méténier, G., and Vivarès, C.P. (1999). Sequencing of several protein-coding genes of the chromosome X from the microsporidian *Encephalitozoon cuniculi. J. Eukaryot. Microbiol.* **46:** 27S–28S.

Blanshard, C., Ellis, D.S., Tovey, D.G., Dowell, S., and Gazzard, B.G. (1992). Treatment of intestinal microsporidiosis with albendazole in patients with AIDS. *AIDS* **6:** 311–313.

Boorstein, S.M. and Ewald, P.W. (1987). Costs and benefits of behavioral fever in *Melanoplus sanguinipes* infected by *Nosema acridophagus. Physiol. Zool.* **60:** 586–595.

Bulnheim, H.-P. and Vávra, J. (1968). Infection by the microsporidian *Octosporea effeminans* sp.n., and its sex determining influence in the amphipod *Gammarus duebeni. J. Parasitol.* **54:** 241–248.

Cali, A. and Takvorian, P.M. (1999). Developmental morphology and life cycles of the microsporidia. In *The Microsporidia and Microsporidiosis* (M. Wittner, Ed.), pp. 85–128. ASM Press, Washington, D.C.

Cavalier-Smith, T. (1993). Kingdom protozoa and its 18 phyla. *Microbiol. Rev.* **57:** 953–994.

Dunn, A.M., Terry, R.S., and Smith, J.E. (2001). Transovarial transmission in the Microsporidia. *Adv. Parasitol.* **48:** 57–100.

Ebert, D. (1994a). Genetic differences in the interactions of a microsporidian parasite and four clones of its cyclically parthenogenetic host. *Parasitology* **108:** 11–16.

Ebert, D. (1994b). Virulence and local adaptation of a horizontally transmitted parasite. *Science* **265:** 1084–1086.

Ebert, D. (1998). Experimental evolution in parasites. *Science* **282:** 1432–1435.

Ebert, D. and Mangin, K.L. (1997). The influence of host demography on the evolution of virulence in a microsporidian gut parasite. *Evolution* **51:** 1828–1837.

Ebert, D., Lipsitch, M., and Mangin, K.L. (2000). The effect of parasites on host population density and extinction: experimental epidemiology with *Daphnia* and six microparasites. *Am. Nat.* **156:** 459–477.

Elston, R.A., Kent, M.L., and Harrell, L.H. (1987). An intranuclear microsporidium associated with acute anemia in the chinook salmon, *Oncorhynchus tshawytscha. J. Protozool.* **34:** 274–277.

Frixione, E., Ruiz, L., Santillán, M., de Vargas, L.V., Tejero, J.M., and Undeen, A.H. (1992). Dynamics of polar filament discharge and sporoplasm expulsion by microsporidian spores. *Cell Motility Cytoskeleton* **22:** 38–50.

Germot, A., Philippe, H., and Le Guyader, H. (1997). Evidence for loss of mitochondria in Microsporidia from a mitochondrial-type HSP70 in *Nosema locustae*. *Mol. Biochem. Parasitol.* **87:** 159–168.

Hall, D.W. and Washino, R.K. (1986). Sporulation of *Amblyospora californica* (Microspora: Amblyosporidae) in autogenous female *Culex tarsalis*. *J. Invertebr. Pathol.* **47:** 214–218.

Hirt, R.P., Healy, B., Vossbrinck, C.R., Canning, E.U., and Embley, T.M. (1997). A mitochondria Hsp70 orthologue in *Vairimorpha necatrix*: molecular evidence that microspora once contained mitochondria. *Curr. Biol.* **7:** 995–998.

Hirt, R.P., Logsdon, J.M.J., Healy, B., Dorey, M.W., Doolittle, W.F., and Embley, T.M. (1999). Microsporidia are related to fungi: evidence from the largest subunit of RNA polymerase II and other proteins. *Proc. Natl. Acad. Sci. U.S.A.* **96:** 580–585.

Hurst, G.D.D. and Majerus, M.E.N. (1993). Why do maternally inherited microorganisms kill males? *Heredity* **71:** 81–95.

Hurst, L.D. (1991). The incidences and evolution of cytoplasmic male killers. *Proc. R. Soc. London (B)* **244:** 91–99.

Imhoof, B. and Schmid-Hempel, P. (1999). Colony success of the bumble bee, *Bombus terrestris*, in relation to infections by two protozoan parasites, *Crithidia bombi* and *Nosema bombi*. *Insectes Sociaux* **46:** 233–238.

Ishihara, R. and Hayashi, Y. (1968). Some properties of ribosomes from the sporoplasm of *Nosema bombycis*. *J. Invertebr. Pathol.* **11:** 377–385.

Iwano, H. and Kurtti, T.J. (1995). Identification and isolation of dimorphic spores from *Nosema furnacalis* (Microspora: Nosematidae). *J. Invertebr. Pathol.* **65:** 230–236.

Johnson, M.A., Becnel, J.J., and Undeen, A.H. (1997). A new sporulation sequence in *Edhazardia aedis* (Microsporidia: Culicosporidiae), a parasite of the mosquito *Aedes aegypti* (Diptera: Culicidae). *J. Invertebr. Pathol.* **70:** 69–75.

Katinka, M.D., Duprat, S., Cornillot, E., Méténier, G., Thomarat, F., Prensier, G., Barbe, V., Peyretaillade, E., Brottier, P., Wincker, P., Delbac, F., El Alaoui, H., Peyret, P., Saurin, W., Gouy, M., Weissenbach, J., and Vivarès, C.P. (2001). Genome sequence and gene compaction of the eukaryote parasite *Encephalitozoon cuniculi*. *Nature* **414:** 450–453.

Keeling, P.J. and McFadden, G.I. (1998). Origins of microsporidia. *Trends Microbiol.* **6:** 19–23.

Keeling, P.J., Luker, M.A., and Palmer, J.D. (2000). Evidence from beta-tubulin phylogeny that microsporidia evolved from within the fungi. *Mol. Biol. Evol.* **17:** 23–31.

Kellen, W.R., Chapman, H.C., Clark, T.B., and Lindegren, J.E. (1965). Host-parasite relationships of some *Thelohania* from mosquitoes (Nosematidae: Microsporidia). *J. Invertebr. Pathol.* **7:** 161–166.

Koella, J.C. and Agnew, P. (1997). Blood-feeding success of the mosquito *Aedes aegypti* depends on the transmission route of its parasite *Edhazardia aedis*. *Oikos* **78:** 311–316.

Koella, J.C., Agnew, P., and Michalakis, Y. (1998). Coevolutionary interactions between host life histories and parasite life cycles. *Parasitology* **116:** S47–S55.

Kurtti, T.J., Ross, S.E., Liu, Y., and Munderloh, U.G. (1994). *In vitro* developmental biology and spore production in *Nosema furnacalis* (Microspora: Nosematidae). *J. Invertebr. Pathol.* **63:** 188–196.

Larsson, J.I.R., Ebert, D., Vávra, J., and Voronin, V.N. (1996). Redescription of *Pleistophora intestinalis* Chatton, 1907, a microsporidian parasite of *Daphnia magna* and *Daphnia pulex*, with establishment of the new genus *Glugoides* (Microspora, Glugeidae). *Eur. J. Protistol.* **32:** 251–261.

Mangin, K.L., Lipsitch, M., and Ebert, D. (1995). Virulence and transmission modes of two microsporidia in *Daphnia magna*. *Parasitology* **111:** 133–142.

Mathis, A., Tanner, I., Weber, R., and Deplazes, P. (1999). Genetic and phenotypic intraspecific variation in the microsporidian *Encephalitozoon hellem*. *Int. J. Parasitol.* **29:** 767–770.

Méténier, G. and Vivarès, C.P. (2001). Molecular characteristics and physiology of microsporidia. *Microbes Infect.* **3:** 407–415.

Minchella, D.J. (1985). Host life-history variation in response to parasitism. *Parasitology* **90:** 205–216.

Moran, N.A. and Wernegreen, J.J. (2000). Lifestyle evolution in symbiotic bacteria: insights from genomics. *Trends Ecol. Evol.* **15:** 321–326.

Pasteur, L. (1870). Étude sur la maladie des vers à soie. Gauthier–Villars, Paris.

Reynolds, D.G. (1970). Laboratory studies of the microsporidian *Plistophora culicis* (Weiser) infecting *Culex pipiens fatigans* Wied. *Bull. Entomol. Res.* **60:** 339–349.

Rinder, H., Katzwinkel-Wladarsch, S., and Loscher, T. (1997). Evidence for the existance of genetically distinct strains of *Enterocytozoon bieneusi*. *Parasitol. Res.* **83:** 670–672.
Scanlon, M., Shaw, A.P., Zhou, C.J., Visvesvara, G.S., and Leitch, G.J. (2000). Infection by microsporidia distrupts the host cell cycle. *J. Eukaryot. Microbiol.* **47:** 525–531.
Schmid-Hempel, P. and Loosli, R. (1998). A contribution to the knowledge of *Nosema* infections in bumble bees, *Bombus* spp. *Apidologie* **29:** 525–535.
Stearns, S.C. (1992). *The Evolution of Life Histories*. Oxford University Press, Oxford, U.K.
Sweeney, A.W. and Becnel, J.J. (1991). Potential of microsporidia for the biological control of mosquitoes. *Parasitol. Today* **7:** 217–220.
Sweeney, A.W., Doggett, S.L., and Gullick, G. (1989). Laboratory experiments on infection rates of *Amblyospora dyxenoides* (Microsporidia: Amblyosporidae) in the mosquito *Culex annulirostris*. *J. Invertebr. Pathol.* **53:** 85–92.
Terry, R.S., Dunn, A.M., and Smith, J.E. (1999). Segregation of a microsporidian parasite during host cell mitosis. *Parasitology* **118:** 43–48.
Undeen, A.H. and Becnel, J.J. (1992). Longevity and germination of *Edhazardia aedis* (Microspora, Amblyosporidae) spores. *Biocontrol Sci. Technol.* **2:** 247–256.
Vávra, J. and Larsson, J.I.R. (1999). Structure of the Microsporidia. In *The Microsporidia and Microsporidiosis* (M. Wittner, Ed.), pp. 7–84. ASM Press, Washington, D.C.
Vávra, J. and Maddox, J.V. (1976). Methods in microsporidiology. In *Comparative Pathobiology*, Vol. 1 (L.E.J. Bulla and T.C. Cheng, Eds.), pp. 281–319. Plenum Press, New York.
Vávra, J. and Undeen, A.H. (1970). *Nosema algerae* n. sp. (Cnidospora, Microsporida) a pathogen in a laboratory colony of *Anopheles stephensi* Liston (Diptera, Culicidae). *J. Protozool.* **17:** 240–249.
Vivarès, C.P. and Méténier, G. (2001). The microsporidian *Encephalitozoon*. *BioEssays* **23:** 194–202.
Vossbrinck, C.R., Maddox, J.M., Friedman, S., Debrunner-Vossbrinck, B.A., and Woese, C.R. (1987). Ribosomal RNA sequence suggests microsporidia are extremely ancient eukaryotes. *Nature* **326:** 411–414.
Vossbrinck, C.R. and Woese, C.R. (1986). Eukaryotic ribosomes that lack a 5.8S RNA. *Nature* **320:** 287–288.
Weidner, E., Findley, A.M., Dolgikh, V., and Sokolova, J. (1999). Microsporidian biochemistry and physiology. In *The Microsporidia and Microsporidiosis* (M. Wittner, Ed.), pp. 172–195. ASM Press, Washington, D.C.
Weiser, J. (1976). Microsporidia in invertebrates: host–parasite relations at the organismal level. In *Comparative Pathobiology*, Vol. 1 (L.E.J. Bulla and T.C. Cheng, Eds.), pp. 163–201. Plenum Press, New York.
Weiss, L.M. and Vossbrinck, C.R. (1999). Molecular biology, molecular phylogeny, and molecular diagnostic approaches to the microsporidia. In *The Microsporidia and Microsporidiosis* (M. Wittner, Ed.), pp. 129–171. ASM Press, Washington, D.C.
Woyciechowski, M. and Krol, E. (2001). Worker genetic diversity and infection by *Nosema apis* in honey bee colonies. *Folia Biol. (Krakow)* **49:** 107–112.

11 A New Bacterium from the Cytophaga-Flavobacterium-Bacteroides Phylum That Causes Sex-Ratio Distortion

Andrew R. Weeks and Johannes A.J. Breeuwer

CONTENTS

INTRODUCTION

Many endosymbionts can cause reproductive abnormalities in their hosts, but none are as well known as bacteria from the genus *Wolbachia*. Over the last decade, *Wolbachia* have gained increasing notoriety due largely to their extremely high prevalence in arthropods and the numerous phenomena in which they have been implicated as causing in their hosts. *Wolbachia* have been found to infect mites, crustaceans, nematodes, and insects, where infection has been estimated to be as high as 76% of all insect species (Stouthamer et al., 1999; Jeyaprakash and Hoy, 2000). Of the extraordinary phenotypes that *Wolbachia* are assumed to cause in their hosts, four reproductive manipulations seem to be the most common: parthenogenesis induction, feminization of genetic males, male killing, and cytoplasmic incompatibility (see Stouthamer et al., 1999, for a review of these phenotypes). Each of these reproductive phenotypes is used to enhance transmission and spread of this maternally inherited bacterium through host populations. Other bacteria and microsporidia can cause male killing and feminization within hosts (Hurst and Jiggins, 2000; Bandi et al., 2001), but no single genus of endosymbionts has been found to cause more than one of these four reproductive phenotypes, and together with their high prevalence in arthropods this has made *Wolbachia* unique.

Recently, however, a new undescribed bacterium from the Cytophaga-Flavobacterium-Bacteroides (CFB) phylum has been implicated in causing feminization and parthenogenesis induction in its hosts (Weeks et al., 2001; Zchori-Fein et al., 2001). Evidence we present here also suggests that this bacterium infects other arthropod and nematode hosts and can co-occur with *Wolbachia* in a single host. These findings raise several questions about current research into endosymbionts, including *Wolbachia*, that have been presumed to cause various phenomena within their hosts.

0-8493-1286-8/03/$0.00+$1.50

This chapter considers the limited information known about this bacterium and the direction for future research. As this bacterium has not been formally described, throughout this chapter we will refer to this bacterium (and its similar strains) as the CFB-BP bacterium (after *Brevipalpus phoenicis*, the first host where the bacterium was shown to manipulate reproduction). We first review the cases where this CFB-BP bacterium has been implicated as causing host sex-ratio distortion. We then consider the association in the literature between the unique cellular ultrastructure of CFB-BP and infection in several arthropods and nematodes. We discuss other known endosymbionts from the CFB phylum and the phylogenetic placement of CFB-BP relative to other CFB bacteria. Finally, we discuss the significance of these findings and point to directions future research should take.

HAPLOID FEMALE PARTHENOGENESIS IN THE PHYTOPHAGOUS MITE *BREVIPALPUS PHOENICIS*

It had previously been thought that no species within the animal kingdom lived exclusively in a haploid state. While haplodiploidy (haploid males and diploid females) has evolved many times within animals, haploid females had never been found (Mable and Otto, 1998). However, in the early 1980s, a group of Dutch cytogeneticists, while studying chromosome evolution within the economically important mite superfamily, the Tetranychoidea, proposed that they had found the first animal species that existed exclusively within the haploid state (Pijnacker et al., 1980).

The researchers conducted a series of experiments trying to determine the chromosome number and ploidy level of the privet mite *Brevipalpus obovatus* (Acari: Tenuipalpidae). At this time, determining chromosome numbers and ploidy levels in animals was generally thought to be a relatively simple task, by comparing chromosome-banding patterns using DNA-specific stains. Many species within the superfamily Tetranychoidea had previously been karyotyped, and it was found that haplodiploidy was ancestral in this superfamily (Helle et al., 1970), which meant determining the karyotype of *B. obovatus* should have been relatively simple (by comparing chromosome numbers in males and females). But for several reasons determining the ploidy level of *B. obovatus* was no simple task.

This species, along with its two close relatives *B. phoenicis* and *B. californicus*, reproduces by thelytokous parthenogenesis (Helle et al., 1972), which meant male and female chromosome numbers could not be compared. In addition, *B. obovatus*, *B. phoenicis*, and *B. californicus* have two chromosomes (Figure 11.1A), with no morphologically distinguishing characters between them. It was first thought that this represented their diploid state simply because no animal has ever been found to live exclusively in a haploid state (Helle et al., 1972). Later, it was found that a close sexual relative within the same genus, *B. russulus*, had a haploid chromosome number of two (males), while females had four chromosomes (diploid) (Pijnacker et al., 1980). Reexamination of G-banding patterns of chromosomes in eggs of *B. obovatus* together with the cytological results from *B. russulus* led Pijnacker et al. (1980) to propose that *B. obovatus* females were haploid.

They then followed these astonishing claims with results suggesting infrequent *B. obovatus* males that could be induced by irradiation (Helle and Bolland, 1972) had the same DNA content in cells as females (two chromosomes), and, further, that patterns of oogenesis showed that a premeiotic doubling occurred prior to meiosis (Pijnacker et al., 1981). These classical cytological techniques provided substantial evidence that *B. obovatus* (and likely *B. phoenicis* and *B. californicus*) were haploid female parthenogens. However, such an extraordinary claim required unequivocal evidence, and therefore their conclusions did not receive widespread publicity or support (Norton et al., 1993; Wrensch et al., 1994). This potential phenomenon was not investigated further for another 20 years, despite one of these species (*B. phoenicis*) being a major agricultural pest (Kennedy et al., 1996).

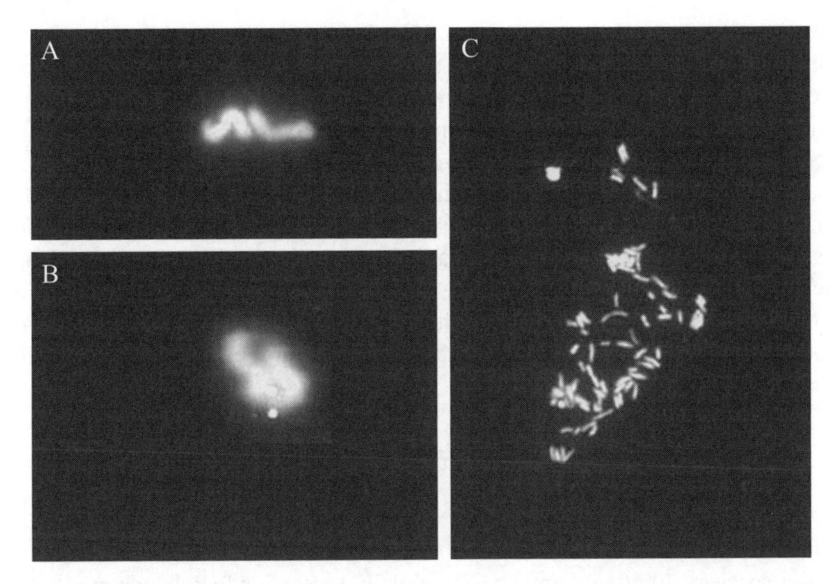

FIGURE 11.1 (A) Metaphase chromosomes from *B. phoenicis* eggs stained with YOYO-1, showing the presence of two chromosomes in a mitotic division. (B) DAPI-counterstained metaphase chromosomes are shown after hybridization to an 18S rDNA probe (bright spots). (C) The CFB-BP bacterium stained with YOYO-1 from a single 2-day-old egg of *B. phoenicis*. For fixing and staining procedures see Weeks, A.R., Marec, F., and Breeuwer, J.A.J. (2001). *Science*, **292:** 2479–2482.

Recently, however, Weeks et al. (2001) provided the unequivocal evidence needed to show that *B. phoenicis* did indeed live entirely in a haploid state. Using a fluorescent dye that stains both DNA and RNA, they showed that only one nucleolar organizing region (NOR) was present during early prophase mitotic divisions in 2-day-old eggs of *B. phoenicis*. If eggs were diploid, two NORs would have been found (as chromosomes are homologous). They then showed, using fluorescent *in situ* hybridization (FISH) with rDNA probes, that there was only one corresponding rDNA region within metaphase chromosomes (Figure 11.1B). To further show that *B. phoenicis* was haploid, they genotyped 450 individual females for nine polymorphic microsatellite loci and found no heterozygous individuals at any locus. Together, the data provide a compelling case for female haploid parthenogenesis.

But how could such a system evolve? While studying chromosomes in eggs of *B. phoenicis*, Weeks et al. (2001) found large numbers of a rod-shaped microorganism (Figure 11.1C), and subsequent sequence analysis of the 16S rDNA showed this microorganism to be an undescribed bacterium from the CFB phylum. After tetracycline treatment, *B. phoenicis* females produced a significantly greater proportion of male progeny than untreated females. Using polymerase chain reaction (PCR), it was shown that male progeny were not infected with the bacterium, while female progeny from both treated and untreated females were infected with the bacterium. The evidence indicated that this bacterium (the CFB-BP bacterium) caused unfertilized haploid eggs, which would normally develop as males, to develop as females, thereby providing an answer to how female haploidy could occur in this mite. Feminizing microorganisms had been found previously (most notably microsporidia); however, *Wolbachia* is the only other known bacterium that can induce this reproductive phenotype (Bandi et al., 2001). In addition, this is the first time host feminization has been found outside a heterogametic reproductive system (Weeks et al., 2001).

While the cause of female haploidy in *B. phoenicis* has been determined, numerous questions remain. For instance, how did this system evolve? Are the infrequent males produced by *B. phoenicis* in the field functional, and can they fertilize the haploid females? Field populations of *B. phoenicis* are highly variable, consisting of many different clones (Weeks et al., 2000); however, nothing is known about whether this variation has arisen from occasional sexual reproduction, polyphyletic origins of parthenogenesis, or simply mutation. If males are functional, then this mite would truly be extraordinary as it can avoid the well-known twofold costs of sexual reproduction, but occasional sexual reproduction would garner most of its benefits.

Are *B. obovatus* and *B. californicus* also infected with the CFB-BP bacterium? By designing CFB-BP-specific primers for the 16S rDNA, we have found that a population of *B. obovatus* from Riverside, CA is also infected with a similar strain of the CFB-BP bacterium (A.R. Weeks and R. Stouthamer, unpublished data). We have not tested *B. californicus* yet, and confirmation that other populations of *B. obovatus* are also infected has yet to be determined. The early work on these two species (Helle and Bolland, 1972; Helle et al., 1972; Pijnacker et al., 1980; 1981) together with the data from *B. phoenicis* and the presence of the CFB-BP bacterium in a population of *B. obovatus* suggests that they are also haploid female parthenogens that are feminized by the CFB-BP bacterium. However, this remains to be confirmed.

PARTHENOGENESIS IN THE WHITEFLY PARASITOID GENUS *ENCARSIA*

The economically important parasitoid genus *Encarsia* (Hymenoptera: Aphelinidae) presently contains 267 valid species, although many species remain undescribed (Babcock et al., 2001). There are numerous reports of species within this genus that are asexual (thelytokous parthenogens) or species that contain thelytokous populations (Woolley and Heraty, 1999; Hunter and Woolley, 2001). *Wolbachia* have previously been found to cause parthenogenesis induction (PI) in *E. formosa* (Zchori-Fein et al., 1992) and were also thought to cause PI in *E. hispida* (Hunter, 1999), as treatment of both species with antibiotics resulted in male production in the subsequent generation. While *Wolbachia* infection in *E. formosa* has been confirmed through PCR with *Wolbachia*-specific primers (Van Meer et al., 1995), *Wolbachia* infection in *E. hispida* was never established, but it was assumed because *Wolbachia* was the only known endosymbiont at that time that caused PI and could be cured through treatment with antibiotics.

Recently, however, it has been shown that similar strains of the CFB-BP bacterium infect many thelytokous species of *Encarsia* (Zchori-Fein et al., 2001). In fact, Zchori-Fein et al. (2001) have shown an association between thelytokous parthenogenesis within *Encarsia* and infection with similar strains of the CFB-BP bacterium. Samples from five *Encarsia* species that reproduce by thelytokous parthenogenesis were found to be infected with different strains of the CFB-BP bacterium (Table 11.1), while samples from five different sexual species were not infected. They also showed that the same thelytokous population of *E. hispida* that had in a previous experiment produced males upon antibiotic treatment (Hunter, 1999) was actually infected with a strain of the CFB-BP bacterium and not *Wolbachia*.

In another species that has both sexual and thelytokous populations (*E. pergandiella*), individuals from one sexual and one asexual population were infected with different strains of the CFB-BP bacterium, while individuals from another sexual population were not infected. To determine if the strain of the CFB-BP bacterium was causing PI in the thelytokous population of *E. pergandiella*, the authors treated individuals with tetracycline, expecting to find males in the subsequent generation. But what they observed instead was a change in host selection behavior with no subsequent progeny developing. The authors attributed the behavioral difference to infection by the CFB-BP bacterium, which they assumed caused PI within this population.

While CFB-BP seems to have an association with thelytokous parthenogenesis in *Encarsia*, the authors have not directly shown that it causes parthenogenesis. Within *E. hispida*, where the association is strongest, they link their results showing infection with the CFB-BP bacterium with that of another study showing that male progeny are produced when females are fed tetracycline (Hunter, 1999) and conclude that it causes parthenogenesis within this species. However, it has not been shown that tetracycline treatment cures the CFB-BP bacterium in females of *E. hispida* or that the subsequent male progeny are also uninfected. In addition, it must be shown that the CFB-BP bacterium is the only

TABLE 11.1
Known/Suspected Hosts of the CFB-BP Bacterium Based on Transmission Electron Microscopy (TEM) or through PCR with CFB-BP-Specific Primers for the 16S rDNA

Species	TEM	PCR16S rDNA	Phenotype	Ref.
		Phylum Arthropoda		
Class Arachnida				
Acari				
Tenuipalpidae				
Brevipalpus phoenicis	?	+	Feminization	Weeks et al., 2001
B. obovatus	?	+	Feminization*	Weeks and Stouthamer, unpublished data
B. californicus	?	?	Feminization*	Weeks et al., 2001
Phytoseiidae				
Metaseiulus occidentalis	+	+	Unknown	Hess and Hoy, 1982; Weeks and Stouthamer, unpublished data
Ixodidae				
Ixodes scapularis	+	+	Unknown	Kurtti et al., 1996
Class Insecta				
Hymenoptera				
Aphelinidae				
Encarsia pergandiella	+	+	Parthenogenesis*	Zchori-Fein et al., 2001
E. berlesei	?	+	Parthenogenesis*	Zchori-Fein et al., 2001
E. citrina	?	+	Parthenogenesis*	Zchori-Fein et al., 2001
E. protransvena	?	+	Parthenogenesis*	Zchori-Fein et al., 2001
E. hispida	?	+	Parthenogenesis*	Zchori-Fein et al., 2001
E. perniciosi	?	+	Parthenogenesis*	Zchori-Fein et al., 2001
Homoptera				
Aleyrodidae				
Bemisia tabaci (biotype A)	+	+	Unknown	Costa et al., 1995; Weeks and Stouthamer, unpublished data
Cicadellidae				
Helochara communis	+	?	Unknown	Chang and Musgrave, 1972
		Phylum Nematoda		
Class Secernentea				
Tylenchida				
Heteroderidae				
Globodera rostochiensis	+	?	Unknown	Shepherd et al., 1973
Heterodera goettingiana	+	?	Unknown	Shepherd et al., 1973
H. glycines	+	?	Unknown	Endo, 1979

* = Suspected phenotypes; + = evidence exists; ? = no information is available.

endosymbiont that is strictly associated with thelytokous females of *E. hispida*. Similarly, as the authors mention, they cannot exclude the possibility that the host selection behavior they observed after treating *E. pergandiella* females with tetracycline was due to the tetracycline itself. One control to determine this may be to compare the host-selection behavior of virgin sexual female *E. pergandiella*.

While the direct link between CFB-BP infection and thelytokous parthenogenesis in *Encarsia* still awaits confirmation, other circumstantial evidence is accumulating. Giorgini (2001) found that tetracycline treatment of females of the thelytokous species *E. meritoria* resulted in mostly male progeny. The same species has also been found to produce males after females are exposed to high temperatures (Giorgini, 2001). This species is a close sister species to *E. hispida*, with some researchers considering both species synonymous (Schauff et al., 1996). Giorgini (2001) also reported that thelytokous females of *E. protransvena* produced significantly more male progeny after females were exposed to high temperatures (31°C) but not tetracycline treatment. Both *E. hispida* and *E. protransvena* thelytokes are infected with the CFB-BP bacterium (Table 11.1).

CELLULAR ULTRASTRUCTURE OF THE CFB-BP BACTERIUM AND OTHER INFECTED HOSTS

Kurtti et al. (1996) were the first to identify the CFB-BP bacterium in the blacklegged tick, *Ixodes scapularis*. They isolated the bacterium in tick cell culture and described its ultrastructure using transmission electron microscopy (TEM). In addition, they sequenced part of the 16S rDNA of this bacterium, and the sequence confirmed its similarity to the CFB-BP bacterium. Zchori-Fein et al. (2001) also described the cellular ultrastructure of the CFB-BP bacterium they had found in *E. pergandiella* using TEM. The cells of the CFB-BP bacterium are Gram-negative and rod-shaped (but can appear coccoid) and measure between 1 and 5 μm in length and 0.4 and 0.7 μm in width. Cells are in direct contact with host cytoplasm, which has an outer envelope that measures 30 to 35 nm in width and is comprised of two lipid bilayers. They seem to resemble rickettsia bacteria in some respects but are longer and contain a unique structure inside their cells. Their cytoplasm contains a parallel array of filamentous rods between 12 and 17 nm in diameter (Figure 11.2). Cross sections showed these filaments to be electron dense and separated from each other by 6 to 7 nm. The filaments are attached to the plasma membrane and always run across the short diameter of the cell, never running along its length. They also rarely extend completely across the diameter of the cell.

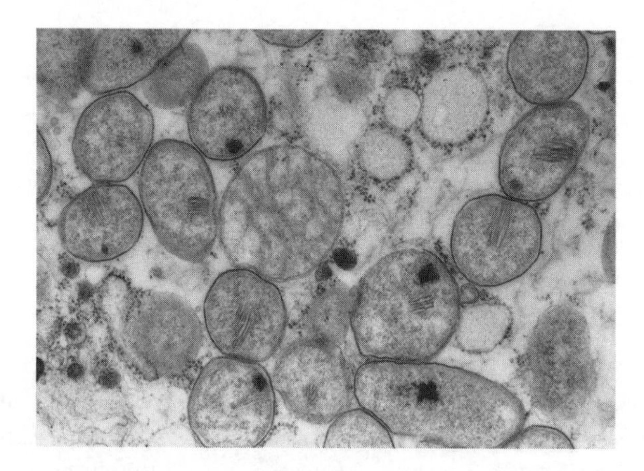

FIGURE 11.2 Endosymbiotic bacterium (CFB-BP) observed through a transmission electron microscope in the mycetocytes of *B. tabaci* biotype A. The four white arrows point to the unique filamentous rods present inside cells of the endosymbiont. (Image courtesy of Heather Costa, Department of Entomology, University of California, Riverside.)

These filamentous rods have been described in other endosymbiotic bacteria. They were first described from bacteria found in host cells of the leafhopper *Helochara communis* (Chang and Musgrave, 1972) and several plant nematode species (Shepherd et al., 1973; Endo, 1979). They have also been found in the economically important predatory mite *Metaseiulus occidentalis* (Hess and Hoy, 1982), and in two biotypes of the whitefly *Bemisia tabaci* (Costa et al., 1995). Figure 11.2 shows a TEM picture of the ultrastructure of these bacteria in the A biotype of *B. tabaci*.

Are these endosymbiotic bacteria also the CFB-BP bacterium, and therefore are these filamentous rods characteristic of this bacterium? Using CFB-BP-specific 16S rDNA primers in PCR, we have recently confirmed that individuals of the A biotype of *B. tabaci* are infected with the CFB-BP bacterium (Weeks and Stouthamer, unpublished). In addition, we have found that all individuals from a population of *M. occidentalis* are also infected with the CFB-BP bacterium (A.R. Weeks and R. Stouthamer, unpublished). We therefore suggest that these filamentous rod-like structures are in fact characteristic of the CFB-BP bacterium, although this awaits confirmation in the nematodes and the leafhopper. In Table 11.1 we have compiled a list of suspected or confirmed infections of the CFB-BP bacteria based on either TEM observations of the unique cellular ultrastructure or confirmation of infection with PCR.

Hess and Hoy (1982) found two bacterial endosymbionts in their study on *M. occidentalis*. The first endosymbiont (which they called type A) was found in all mites examined. This endosymbiont (which had the CFB-BP unique cellular ultrastructure) is likely to be the CFB-BP bacterium. The second endosymbiont that they found (type B) was present in two thirds of all mites examined and resembled a rickettsia bacterium. Johanowicz and Hoy (1996) have since identified this endosymbiont as *Wolbachia*. Based on the work of Hess and Hoy (1982), it is likely that individuals of *M. occidentalis* can be doubly infected with both *Wolbachia* and the CFB-BP bacterium. We have confirmed this prediction using both *Wolbachia*- and CFB-BP-specific primers in individuals from a population of *M. occidentalis* reared on prey not infected with either bacterium (*Tetranychus cinnabarinus*). Further, Johanowicz and Hoy (1998) have shown that heat treatment (33°C) of *Wolbachia*-infected female *M. occidentalis* cures the *Wolbachia* infection after five generations. Importantly, they also showed that crosses between cured females and *Wolbachia*-infected males resulted in cytoplasmic incompatibility (CI), a phenotype commonly associated with *Wolbachia* (Stouthamer et al., 1999). This raises several important questions: (1) Were the mites also infected with the CFB-BP bacterium? (2) Was the CFB-BP bacterium also cured by heat treatment? (3) What is causing the observed CI, *Wolbachia* or the CFB-BP bacterium? We will return to this case of double infection in the conclusion.

ENDOSYMBIONTS OF THE CFB PHYLUM

As the name suggests, the CFB phylum is a poorly characterized phylum that contains bacterial species showing an extremely diverse range of physiological and morphological characters (Reichenbach, 1991). Species from the CFB phylum have been found in most environments including the human gut, soil, fresh and seawater, and activated sludge (Horn et al., 2001). There are also several important pathogenic bacteria within this phylum that are associated with diseases in humans (Flaherty et al., 1984) and in ducks (Van Damme et al., 1999). While most endosymbiotic bacteria are found within the proteobacteria (Moran and Telang, 1998), several are known from the CFB. It has been known for over a century that cockroaches harbor intracellular bacteria in their mycetocytes, and recently Bandi et al. (1994) showed that these bacteria were Flavobacteria from the CFB. A similar strain has also been found in mycetocytes of the Australian termite, *Mastotermes darwiniensis* (Bandi et al., 1995). Several strains of another Flavobacterium have also been identified within two ladybird species, *Coleomegilla maculata* and *Adonia variegata* (Hurst et al., 1997, 1999). These bacteria have been shown to cause male killing in their respective ladybird hosts. Phylogenetic analysis places these endosymbionts in a monophyletic group in the CFB (Figure 11.3).

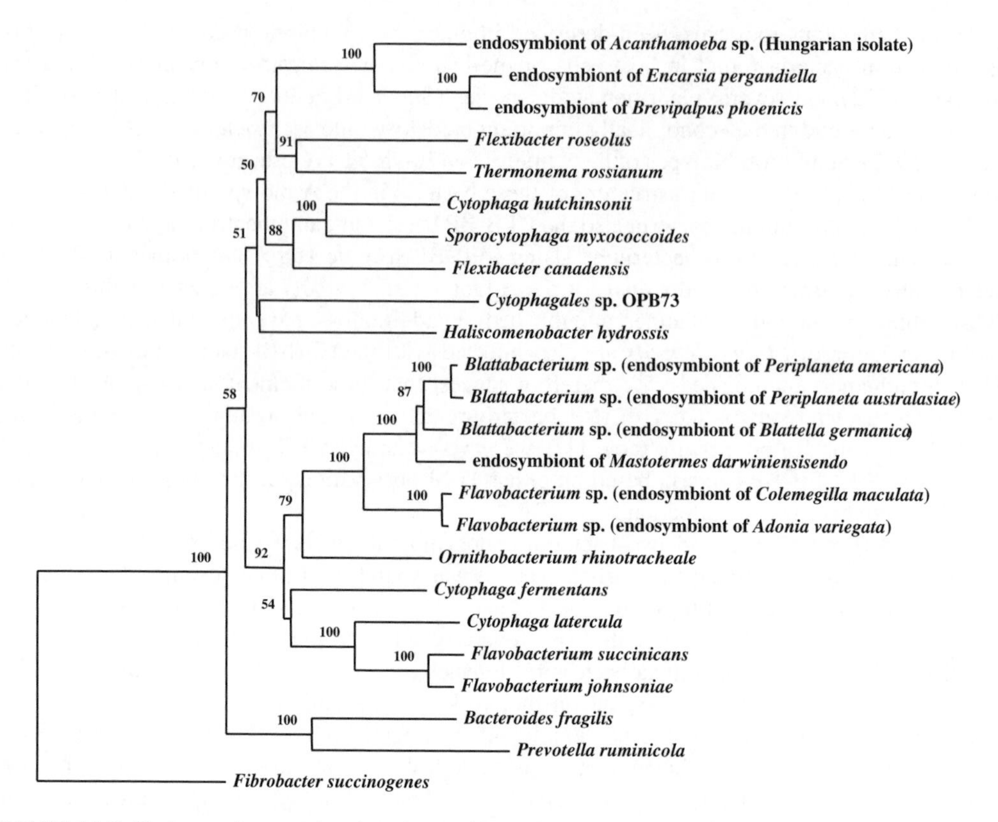

FIGURE 11.3 Phylogenetic tree showing the position of known endosymbiotic bacteria relative to other representative members of the Cytophaga-Flavobacterium-Bacteriodes phylum, as determined from 16S rDNA sequence (1279 bp). Tree was constructed using DAMBE (Xia and Xie, 2001) by maximum parsimony using default parameters, with *Fibrobacter succinogenes* used as an outgroup. Bootstrap values (percent) indicate the results of 1000 bootstrap replicates, with only those greater than 50% shown.

While many *Acanthamoeba* species contain endosymbionts from the proteobacteria and *Chlamydia*, recently two different *Acanthamoeba* species were found to harbor intracellular bacteria from the CFB that cannot be cultured outside their hosts (Horn et al., 2001). More interestingly, based on 16S rDNA, these endosymbionts form a monophyletic group with the CFB-BP bacteria (Figure 11.3). This monophyletic group has no known close bacterial relatives, with the next nearest bacterium having less than 82% 16S rDNA sequence similarity. There is even less similarity between this group and the endosymbiotic Flavobacteria mentioned above, suggesting that obligate endosymbiotic relationships have arisen at least twice within the CFB.

A phylogenetic analysis based on the known 16S rDNA sequences from the *Acanthamoeba* endosymbionts and the CFB-BP bacteria, using *Flexibacter roseolus* as an outgroup, has shown that the seven strains of the CFB-BP bacteria are also monophyletic (Figure 11.4). 16S rDNA sequence similarity between these two groups is less than 90%. Divergence between the seven strains of the CFB-BP bacterium is less than 3%. Horn et al. (2001) propose that the endosymbionts from *Acanthamoeba* represent a new genus because bacteria that have less than 95% sequence similarity in their 16S rDNA are thought to represent species from different genera (Ludwig et al., 1998). Under this criterion, the strains of CFB-BP also fall into a new separate genus from the *Acanthamoeba* endosymbionts. It would be interesting to determine if the *Acanthamoeba* endosymbionts also had the same unique cellular ultrastructure of the CFB-BP bacterium.

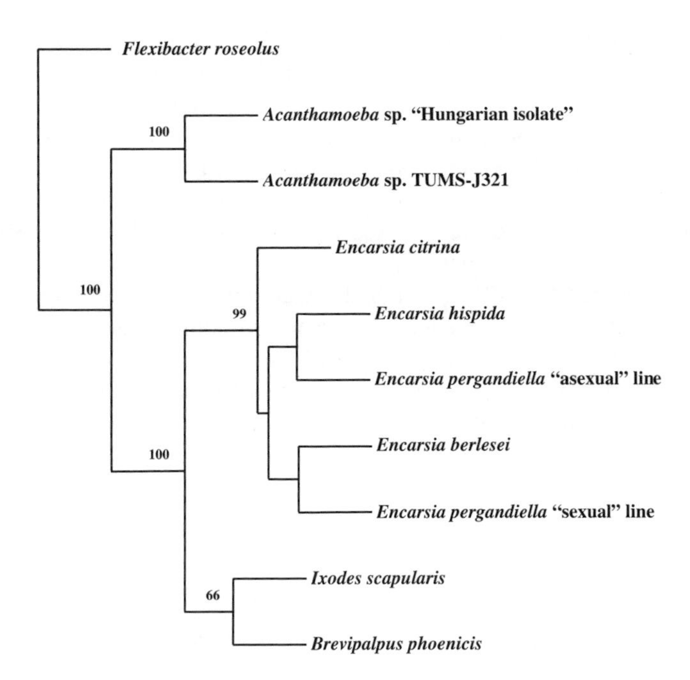

FIGURE 11.4 Phylogenetic tree of CFB-BP strains and the *Acanthamoeba* endosymbionts based on 768 bp of the 16S rDNA (host names are used to indicate strains). The tree was generated using DAMBE (Xia and Xie, 2001) by maximum parsimony using default parameters with the closest known bacterial relative as an outgroup (*F. roseolus*). Bootstrap values (percent) indicate the results of 1000 bootstrap replicates, with only those greater than 50% shown.

CONCLUSIONS

Over the last 15 years a vast shift has occurred in our understanding of how endosymbionts manipulate their hosts' reproduction to enhance their own transmission. This change in perception can be attributed largely to one microorganism, *Wolbachia*, which has been repeatedly implicated in altering its hosts' reproduction in many different ways. But has the increasing popularity of *Wolbachia* come at a cost?

The finding that the CFB-BP bacterium can induce feminization in a mite and its association with parthenogenesis in a parasitoid genus questions the current perception that *Wolbachia* is unique in its ability to induce multiple reproductive phenotypes in its hosts. We presented evidence here that the CFB-BP bacterium had been discovered previously, infecting other arthropod and nematode hosts. How many arthropod hosts it infects is unknown, but the fact that it has been found in such diverse taxa as ticks, mites, whiteflies, parasitoids, leafhoppers, and nematodes, mostly by accident, suggests that it will be found to infect many more invertebrates. Interestingly, it has been found in both asexual organisms (*B. phoenicis, B. obovatus, Encarsia* sp.) and several sexual species (*M. occidentalis, Bemisia tabaci*) where reproductive incompatibilities have been documented previously (Hoy and Cave, 1988; Perring et al., 1993; Costa et al., 1995). Can CFB-BP also induce other reproductive phenotypes, such as CI?

One of the more interesting results is the double infection between *Wolbachia* and the CFB-BP bacterium in the predatory mite *M. occidentalis.* This finding, however, should not really be that surprising. Different *Wolbachia* strains, some from very diverse lineages, have been known for some time to occur in a single host (Breeuwer et al., 1992; Perrot-Minnot et al., 1996). What this finding highlights is the "cost" of *Wolbachia*'s popularity that we referred to previously. Advances in molecular biology since the mid-1990s have resulted in the routine use of several *Wolbachia*-specific primers to screen for *Wolbachia* infection. If *Wolbachia* is found, researchers typically cure the host of the infection and then document the effects. The potential for other endosymbionts to cause effects is often disregarded. This is what occurred for *M. occidentalis*, where Johanowicz and Hoy (1998) implicated *Wolbachia* as causing CI, even though Hess and Hoy (1982) had previously shown the presence of another endosymbiont (CFB-BP) in *M. occidentalis* as well as *Wolbachia*. How common are double infections between *Wolbachia* and other endosymbionts, and are any host phenotypes previously attributed to *Wolbachia* caused by another endosymbiotic organism? We advocate a more thorough approach to documenting the effects of endosymbionts, as Weeks et al. (2002) have recently outlined.

Finally, research on the CFB-BP endosymbiotic bacterium is still in its early stages. The future is an open book considering the scant information we currently possess. Many lessons can be learned from research conducted on *Wolbachia* over the last decade. In addition, the *Wolbachia* genome projects currently being undertaken will provide a comparative genomics perspective that will greatly enhance our understanding of the mechanisms endosymbionts use to manipulate host reproduction.

REFERENCES

Babcock, C.S., Heraty, J.M., De Barro, P.J., Driver, F., and Schmidt, S. (2001). Preliminary phylogeny of *Encarsia* Forster (Hymenoptera: Aphelinidae) based on morphology and 28S rDNA. *Mol. Phylogenet. Evol.* **18:** 306–323.

Bandi, C., Damiani, G., Magrassi, L., Grigolo, A., Fani, R., and Sacchi, L. (1994). Flavobacteria as intracellular symbionts in cockroaches. *Proc. R. Soc. London (B)* **257:** 43–48.

Bandi, C., Dunn, A.M., Hurst, G.D.D., and Rigaud, T. (2001). Inherited microorganisms, sex-specific virulence and reproductive parasitism. *Trends Parasitol.* **17:** 88–94.

Bandi, C., Sironi, M., Damiani, G., Magrassi, L., Nalepa, C.A., Laudani, U., and Sacchi, L. (1995). The establishment of intracellular symbiosis in an ancestor of cockroaches and termites. *Proc. R. Soc. London (B)* **259:** 293–299.

Breeuwer, J.A.J., Stouthamer, R., Barns, S.M., Pelletier, D.A., Weisburg, W.G., and Werren, J.H. (1992). Phylogeny of cytoplasmic incompatibility microorganisms in the parasitoid wasp genus *Nasonia* (Hymenoptera: Pteromalidae) based on 16S ribosomal DNA sequences. *Insect Mol. Biol.* **1:** 25–36.

Chang, K.P. and Musgrave, A.J. (1972). Multiple symbiosis in a leafhopper, *Helochara communis* Fitch (Cicadellidae: Homoptera): envelopes, nucleoids and inclusions of the symbiotes. *J. Cell Sci.* **11:** 275–293.

Costa, H.S., Wescot, D.M., Ullman, D.E., Rosell, R., Brown, J.K., and Johnson, M.W. (1995). Morphological variation in *Bemisia* endosymbionts. *Protoplasma* **189:** 194–202.

Endo, B.Y. (1979). The ultrastructure and distribution of an intracellular bacterium-like microorganism in tissues of larvae of the soybean cyst nematode, *Heterodera glycines*. *J. Ultrastruct. Res.* **67:** 1–14.

Flaherty, D.K., Deck, F.H., Hood, M.A., Liebert, C.A., Singleton, F.L., Winzenburger, P.A., Bishop, K., Smith, I.R., Bynum, L.M., and Witmer, W.B. (1984). A *Cytophaga* species endotoxin as a putative agent of occupation-related lung disease. *Infect. Immun.* **43:** 213–216.

Giorgini, M. (2001). Induction of males in thelytokous populations of *Encarsia meritoria* and *Encarsia protransvena*: a systematic tool. *BioControl* **46:** 427–438.

Helle, W., and Bolland, H.R. (1972). Artificial induction of males in a thelytokous mite species by means of x-rays. *Entomol. Exp. Appl.* **15:** 395–396.

Helle, W., Bolland, H.R., and Gutierrez, J. (1972). Minimal chromosome number in false spider mites (Tenuipalpidae). *Experientia* **28:** 707.

Helle, W., Gutierrez, J., and Bolland, H.R. (1970). A study on sex-determination and karyotypic evolution in *Tetranychidae*. *Genetica* **41:**21–32.

Hess, R.T., and Hoy, M.A. (1982). Microorganisms associated with the spider mite predator *Metaseiulus* (= *Typhlodromus*) *occidentalis*: electron microscope observations. *J. Invertebrate Pathol.* **40:** 98–106.

Horn, M., Harzenetter, M.D., Linner, T., Schmid, E.N., Muller, K.D., Michel, R., and Wagner, M. (2001). Members of the *Cytophaga-Flavobacterium-Bacteroides* phylum as intracellular bacteria of acanthamoebae: proposal of "*Candidatus* Amoebophilus asiaticus." *Environ. Microbiol.* **3:** 440–449.

Hoy, M.A. and Cave, F.E. (1988). Premating and postmating isolation among populations of *Metaseiulus occidentalis* (Nesbitt) (Acarina: Phytoseiidae). *Hilgardia* **56:** 1–20.

Hunter, M.S. (1999). The influence of parthenogenesis-inducing *Wolbachia* on the oviposition behaviour and sex-specific developmental requirements of autoparasitoid wasps. *J. Evol. Biol.* **12:** 735–741.

Hunter, M.S. and Woolley, J.B. (2001). Evolution and behavioral ecology of heteronomous Aphelinid parasitoids. *Annu. Rev. Entomol.* **46:** 251–290.

Hurst, G.D.D. and Jiggins, F.M. (2000). Male-killing bacteria in insects: mechanisms, incidence, and implications. *Emerging Infect. Dis.* **6:** 329–336.

Hurst, G.D.D., Hammarton, T.C., Bandi, C., Majerus, T.M.O., Bertrand, D., and Majerus, M.E.N. (1997). The diversity of inherited parasities of insects: the male-killing agent of the ladybird beetle *Coleomegilla maculata* is a member of the Flavobacteria. *Genet. Res.* **70:** 1–6.

Hurst, G.D.D., Bandi, C., Sacchi, L., Cochrane, A.G., Bertrand, D., Karaca, I., and Majerus, M.E.N. (1999). *Adonia variegata* (Coleoptera: Coccinellidae) bears maternally inherited Flavobacteria that kill males only. *Parasitology* **118:** 125–134.

Jeyaprakash, A. and Hoy, M.A. (2000). Long PCR improves *Wolbachia* DNA amplification: wsp sequences found in 76% of sixty-three arthropod species. *Insect Mol. Biol.* **9:** 393–405.

Johanowicz, D.L. and Hoy, M.A. (1996). *Wolbachia* in a predator–prey system: 16S ribosomal DNA analysis of two Phytoseiids (Acari: Phytoseiidae) and their prey (Acari: Tetranychidae). *Ann. Entomol. Soc. Am.* **89:** 435–441.

Johanowicz, D.L. and Hoy, M.A. (1998). Experimental induction and termination of non-reciprocal reproductive incompatibilities in a parahaploid mite. *Entomol. Exp. Appl.* **87:** 51–58.

Kennedy, J.S., van Impe, G., Hance, T., and Lebrun, P. (1996). Demecology of the false spider mite, *Brevipalpus phoenicis* (Geijskes) (Acari, Tenuipalpidae). *J. Appl. Entomol.* **120:** 493–499.

Kurtti, T.J., Munderloh, U.G., Andreadis, T.G., and Magnarelli, L.A. (1996). Tick cell culture isolation of an intracellular prokaryote from the tick *Ixodes scapularis*. *J. Invertebr. Pathol.* **67:** 318–321.

Ludwig, W., Strunk, O., Klugbauer, S., Klugbauer, N., Weizenegger, M., Neumaier, J., Bachleitner, M., and Schleifer, K.H. (1998). Bacterial phylogeny based on comparative sequence analysis. *Electrophoresis* **19:** 554–568.

Mable, B.K. and Otto, S.P. (1998). The evolution of life cycles with haploid and diploid phases. *BioEssays* **20:** 453–462.

Moran, N.A. and Telang, A. (1998). Bacteriocyte-associated symbiosis of insects. *BioScience* **48:** 295–304.

Norton, R.A., Kethley, J.B., Johnston, D.E., and O'Connor, B.M. (1993). Phylogenetic perspectives on genetic systems and reproductive modes of mites. In *Evolution and Diversity of Sex Ratio in Insects and Mites* (D.L. Wrensch and M.A. Ebbert, Eds.), pp. 8–99. Chapman & Hall, New York.

Perring, T.M., Rodriguez, R.J., Farrar, C.A., and Bellows, T.S. (1993). Identification of a whitefly species by genomic and behavioral studies. *Science* **259:** 74–77.

Perrot-Minnot, M.J., Guo, L.R., and Werren, J.H. (1996). Single and double infections with *Wolbachia* in the parasitic wasp *Nasonia vitripennis*: effects on compatibility. *Genetics* **143:** 961–972.

Pijnacker, L.P., Ferwerda, M.A., Bolland, H.R., and Helle, W. (1980). Haploid female parthenogenesis in the false spider mite *Brevipalpus obovatus* (Acari: Tenuipalpidae). *Genetica* **51:** 211–214.

Pijnacker, L.P., Ferwerda, M.A., and Helle, W. (1981). Cytological investigations on the female and male reproductive system of the parthenogenetic privet mite, *Brevipalpus obovatus* (Donnadieu) (Phytoptipalpidae, Acari). *Acarologia* **22:** 157–163.

Reichenbach, H. (1991). The order *Cytophagales*. In *The Prokaryotes* (A. Balows, A.G. Truper, M. Dworkin, W. Harder, and K.H. Schleifer, Eds.), pp. 3631–3675. Springer-Verlag, New York.

Schauff, M.E., Evans, G.A., and Heraty, J.M. (1996). A pictorial guide to the species of *Encarsia* (Hymenoptera: Aphelinidae) parasitic on whiteflies (Homoptera: Aleyrodidae) in North America. *Proc. Entomol. Soc. Wash.* **98:** 1–35.

Shepherd, A.M., Clark, S.A., and Kempton, A. (1973). An intracellular micro-organism associated with tissues of *Heterodera* spp. *Nematologica* **19:** 31–34.

Stouthamer, R., Breeuwer, J.A.J., and Hurst, G.D.D. (1999). *Wolbachia pipientis*: microbial manipulator of arthropod reproduction. *Annu. Rev. Microbiol.* **53:** 71–102.

Van Damme, P., van Canneyt, M., Segers, P., Ryll, M., Kohler, B., Ludwig, W., and Hinz, K.H. (1999). *Coenonia anatina* gen. nov., sp. nov., a novel bacterium associated with respiratory disease in ducks and geese. *Int. J. Syst. Bacteriol.* **49:** 867–874.

Van Meer, M.M.M., van Kan, F.J.M.P., Breeuwer, J.A.J., and Stouthamer, R. (1995). Identification of symbionts associated with parthenogenesis in *Encarsia formosa* and *Diplolepis rosae*. *Proc. Sect. Exp. Appl. Entomol. Neth. Entomol. Soc.* **6:** 81–86.

Weeks, A.R., van Opijnen, T., and Breeuwer, J.A.J. (2000). AFLP fingerprinting for assessing intraspecific variation and genome mapping in mites. *Exp. Appl. Acarol.* **24:** 775–793.

Weeks, A.R., Marec, F., and Breeuwer, J.A.J. (2001). A mites species that consists entirely of haploid females. *Science* **292:** 2479–2482.

Weeks, A.R., Reynolds, K.T., and Hoffmann, A.A. (2002). *Wolbachia* dynamics and host effects: what has (and has not been) been demonstrated? *Trends Ecol. Evol.* **17:** 257–262.

Woolley, J.B. and Heraty, J.M. (1999). *Encarsia* of the world: an electronic catalog and database. http://hymenoptera.tamu.edu/

Wrensch, D.L., Kethley, J.B., and Norton, R.A. (1994). Cytogenetics of holokinetic chromosomes and inverted meiosis: keys to the evolutionary success of mites, with generalizations on eukaryotes. In *Ecological and Evolutionary Analyses of Life-History Patterns* (M.A. Houck, Ed.), pp. 282–343. Chapman & Hall, New York.

Xia, X. and Xie, Z. (2001). DAMBE: data analysis in molecular biology and evolution. *J. Heredity* **92:** 371–373.

Zchori-Fein, E., Roush, R.T., and Hunter, M.S. (1992). Male production by antibiotic treatment in *Encarsia formosa* (Hymenoptera: Aphelinidae), an asexual species. *Experentia* **48:** 102–105.

Zchori-Fein, E., Gottlieb, Y., Kelly, S.E., Brown, J.K., Wilson, J.M., Karr, T.L., and Hunter, M.S. (2001). A newly discovered bacterium associated with parthenogenesis and a change in host selection behavior in parasitoid wasps. *Proc. Natl. Acad. Sci. U.S.A.* **98:** 12555–12560.

12 Inherited Microorganisms That Selectively Kill Male Hosts: The Hidden Players of Insect Evolution?

Gregory D.D. Hurst, Francis M. Jiggins, and Michael E.N. Majerus

CONTENTS

INTRODUCTION

Contrary to the practice commonly found in genetics laboratories, geneticists in the early and mid-20th century worked frequently with insects taken directly from natural populations. They would cross these, using results to infer the genetic basis of the natural variation they observed. In the course of such experiments, many workers found isofemale lines that produced strongly female-biased sex ratios. Hubert Simmonds, a tropical entomologist, investigating the wing pattern polymorphism of the eggplant butterfly, *Hypolimnas bolina*, incidentally observed lines giving all-female broods (Simmonds, 1923). Ya Ya Lus, investigating elytral pattern polymor-

0-8493-1286-8/03/$0.00+$1.50

phism in *Adalia bipunctata*, the two-spot ladybird, incidentally observed isofemale lines giving all-, or near all-, female broods, in this case associated with lowered egg hatch rate (Lus, 1947). Many workers investigating the genetics of traits in *Drosophila* observed heritable production of female-biased sex ratios, with the affected female again producing eggs with lower hatch rates than found in "normal" females (Magni, 1953; Cavalcanti and Falcao, 1954; Carson, 1956; Malogolowkin, 1958).

Later work established that the cause of each of these female biases was not a nuclear gene but an inherited microorganism living and replicating within the cytoplasm of its host's cells. In each case, the distortion was shown to be associated with the death of male embryos. The causative agents thus became known as son killers or, now more commonly, male killers.

By 1991, a clear pattern had emerged from the cases of male killing on record. There appeared to be two classes of sex-ratio distortion (Hurst, 1991). First, there was distortion associated with the death of male embryos or first-instar larvae. In these cases, the cause was bacterial, which was curable with antibiotics. Second, there was distortion associated with the death of later-instar male larvae. In this case, the agent observed was eukaryotic, members of the phylum Microspora.

In addition to the differences in the agent responsible, work around this time also emphasized differences in the transmission biology of these pathogens. While the embryonic male-killing bacteria were very short-lived in the natural environment and their infectious transmission was very rare, the microsporidia that killed late-instar male larvae were hardy and capable of further transmission following death of their male host. In fact, the death of the male host was associated with the liberation of dispersal spores into the environment. This has led workers to regard the two "strategies" as differing in evolutionary logic, albeit with a common theme (Hurst, 1991) — that these pathogens, which are found in egg cytoplasm but not in sperm, are maternally inherited. Vertical transmission is impossible through a male host, making the death of male hosts at worst neutral for the symbiont. However, the advantage of male-host death differs between the two classes. The principal advantage gained by embryonic male killers is kin-selective, infected female siblings of killed males gain from the death of their brothers. Male-killer-infected neonate female ladybird larvae gain resources by feasting on the soma of their dead male siblings. In contrast, late larval male killing is not adaptive in terms of increasing female host survival. Rather, it allows the pathogen to disperse out of a male host, from which vertical transmission is impossible.

This taxonomic division between differing strategists should not be considered absolute, and it may also be premature to conclude that the list is final. While the hardiness of Microspora in the wild may predispose them to infectious transmission from males (see Chapter 10), there is no obvious obstacle that would prevent microsporidia from being embryonic male killers, and indeed some are known to feminize their hosts (see Chapter 11). Further, there is no reason that cytoplasmic viruses could not cause sex-ratio distortion.

In this review, we first examine the natural history of these two types of sex-ratio distortion, keeping the broad distinction between late and early male killing, which we believe to be intellectually sound, even if likely to be only an approximation of the bacteria/microsporidia lines found to date. We note that the trait of male-specific lethality has evolved independently on many occasions and argue that they are likely to be more common than previously thought, at least in terms of incidence across taxa. With this commonness comes the potential for them to be important in the evolution of a wide range of host taxa. We therefore assess their potential importance in this sphere. How have the biased sex ratios they produce altered the pattern of sexual selection in their hosts? Have hosts evolved to prevent their action, and if so, how? Could they have caused the extinction of host populations or even species? Are they the force that has driven the diversification of insect sex-determination systems? In summary, we propose that these and other sex-ratio-distorting microorganisms may be important hidden players in arthropod evolution.

LATE MALE KILLING

Although the two types of male killing — defined by the timing of male death — appear to be neatly divisible on the basis of the taxonomy of the male-killing agent, bacteria killing early and microsporidia late, the division is best made based on the relative importance of different transmission modes. In early male killing, transmission is almost exclusively vertical. However, late male killing involves a combination of vertical and horizontal transmission, with both being important in the population dynamics of the pathogen on an ecological time scale.

Examples of late male killers are confined to one group of single-celled eukaryotes, the microsporidia. These have been extensively studied in mosquito hosts, which were until recently the only known hosts of late male killers. To date, more than two dozen cases of late male killing by microsporidia have been recorded in mosquitoes. In only a few of these has vertical transmission been demonstrated, with the efficiency of transmission varying from about 0.5 (50% of progeny infected) in *Aedes stimulans* (Andreadis, 1985) to more than 0.9 in *Culex salinarius* (Andreadis and Hall, 1979). Reported prevalence levels typically range from 0.02 to 0.4. Early workers noted that they could not be maintained purely by transovarial transmission at the observed rates and therefore suggested that horizontal transmission must be occurring (Andreadis and Hall, 1979; Lord, et al., 1981). This was subsequently demonstrated in a number of cases (Andreadis, 1985; Avery, 1989; Sweeney et al., 1985, 1988; Becnel, 1986).

In the simplest scenario, the microsporidia in female mosquitoes are transovarially transmitted to offspring. Those in males, for which vertical transmission is not an option, may be horizontally transmitted to other hosts either via larval cannibalism or when spores are released into the water following the death of their host, which typically occurs when the larva is in its final instar. Released spores may be ingested by other mosquito larvae either directly or within a copepod intermediary, which may become infected by the spores and subsequently preyed upon by the mosquito larvae (Becnel and Sweeney, 1990). Both male and female larvae can be infected horizontally. Neither shows ill effects following novel infection within their own lives. Indeed, no phenotypic effects of the microsporidia have been reported for newly infected males. These males complete their life cycle normally, and the microsporidia that they now contain die when their host dies. Newly infected female hosts also complete their life cycle normally, but these pass the microsporidian into their eggs. The males in this following generation may be killed while females continue to transmit the symbionts vertically.

In nature, the situation is often more complex than this simplest scenario. Some microsporidians cause the death of both host sexes, while others kill neither (see Chapter 10). Kellen et al. (1965) detailed four types of infection of mosquitoes by microsporidia, the types being differentiated largely on the level of pathogenic effects of the microsporidia in the host sexes (Table 12.1).

Kellen et al. (1965) argue that Type III infections are a primitive state. In laboratory experiments, Sweeney et al. (1989) showed that the survival rate of female *C. incidens* and *C. inornata* to adulthood was less than 2% and that this figure varied between species. Moreover, the survival rate responded to artificial selection, implying a genetic component to the *modus operandi* of the microsporidian: either killing female hosts or being transovarially transmitted. One may then speculate that Type III infections will evolve into Type II, then Type I, and, perhaps, finally Type IV infections, as vertical transmission becomes predominant over horizontal transmission. The relative efficiencies of horizontal and vertical transmission will depend not only on the pathological characteristics of the microsporidian but also on host density and the availability of intermediate hosts.

The variations between the types detailed by Kellen et al. (1965) beg several questions. How do the microsporidians kill their hosts? How do some selectively kill male rather than female hosts? Why do host deaths usually occur precisely in the fourth larval instar? Why do different microsporidians employ different strategies, some being pathogenic to both sexes, some just to males, and some to neither sex? Not all of these questions have clear answers. In particular, the methods by

TABLE 12.1
Classification of Sex-Related Pathogenicity of Microsporidia in Mosquitoes

	Effect on Male Hosts	Effect on Female Hosts
Type I	Death in fourth instar due to progressive infection following sporogeny	Transovarial transmission; no reduction in fecundity of egg hatch rate
Type II	Death in fourth instar due to progressive infection following sporogeny	Some but not all infected females die in the fourth instar due to progressive infection following sporogeny; not reported whether surviving females transmit transovarially
Type III	Death in fourth instar due to progressive infection following sporogeny	Most die in fourth instar due to progressive infection following sporogeny; however, a few survive sporogeny to adulthood and transmit symbiont transovarially
Type IV	Infection limited to small regions of thoracic and abdominal adipose tissue; not lethal	Infection limited to small regions of thoracic and abdominal adipose tissue; not lethal; transovarially transmitted

Source: Data from Kellen, W.R. et al. (1965). *J. Invertebr. Pathol.* **7:** 161.

which microsporidians determine the sex of their host and by which they cause their host's death await discovery.

Hurst (1991) argued that the timing of host death was crucial if the death of the host was coupled with horizontal transmission. By causing death in the fourth larval instar the microsporidian maximizes its transmission rate. He suggests, justifiably, that horizontal transmission needs to occur in water. Further, he argues that the pupal case, the last aquatic stage of the life cycle, might act as a barrier to symbiont release. In consequence, he states that causing host death in the final larval instar, but not before, maximizes the number of microsporidian spores released into the water to be taken up by copepods that would then vector the microsporidian if preyed upon by other mosquito larvae.

Before last year, all known cases of late male killing involved microsporidia parasites in mosquito hosts. The recent observation of late male killing in the oriental tea tortrix moth, *Homona magnanima*, suggests it is premature to confine the phenotype of late male killing to aquatic insects. In the oriental tea tortrix, larval mortality was observed in all instars, with greatest mortality during the third instar (Morimoto et al., 2001). Total larval mortality in lines producing mainly females was around 50%, compared to 10% in normal lines. Feeding homogenate of dead larvae from female-biased lines to uninfected larvae successfully effected horizontal transmission. The trait was resistant to antibiotic treatment. The combination of the timing of male death with the ability to transfer horizontally makes a compelling case for the idea that late male killing occurs in the terrestrial environment, with the male killing similar in logic to that in mosquitoes.

A general consideration of insect pathology further suggests it is premature to confine late male killing to microsporidia as etiologic agents (Majerus, 2002; Stouthamer et al., 2002). Vertical transmission of pathogenic microorganisms in insects was first recorded by Louis Pasteur (1870), who found that the microsporidian responsible for pebrine disease in the silkworm, *Bombyx mori*, was transmitted both horizontally from dead or dying larvae and vertically in the cytoplasm of the eggs. Since then, vertical and horizontal transmission has been recorded in other parasites, such as nuclear polyhedrosis viruses, which are known to persist for over 25 years within polyhedra crystals exposed on plant material. If then ingested by appropriate lepidopteran larvae, the crystals are broken down in the gut and the virus migrates to cells where it replicates. As we will show, cannibalism and intraspecific consumption play an important part in the dynamics of early male killing. It is possible that intraspecific consumption or cannibalism also has a role to play in late male killing if horizontal transmission is facilitated. In some genera of moths (*Spodoptera, Mamestra, Melanchra*), larvae are attracted to fresh corpses of viral-killed conspecifics when the integument ruptures and then feed on the liquified remains, thereby imbibing high doses of the viral pathogen.

EARLY MALE KILLING

Early male killers present several contrasts to late male killers. First, all early male killers described to date are bacterial. They have been recorded from a phylogenetically wide range of hosts. Most significantly, early male killers are maternally transmitted, and horizontal or paternal transmission is usually rare or absent. This is known both directly from experiments that have looked for horizontal transmission (see Hurst and Majerus, 1993 for review) and indirectly as male-killing bacteria have been observed to be in linkage disequilibrium with maternally inherited host genes in the mitochondria (von der Schulenburg et al., 2002). The lack of horizontal transmission is probably at least in part linked to the fact that the causal agents are bacterial. As has been noted elsewhere in this book, inherited bacteria are generally very refractory to culture and show poor survival outside of host cells. While horizontal transmission on blood–blood contact is feasible (though likely to be limited in rate by the frequency of contact), transmission through the environment is unlikely.

The lack of horizontal transmission of early male killers out of the dead male embryo suggests both that there is some other function to male lethality and that some other force is maintaining the agents in the population. It is thought that these male killers are maintained because the death of males increases the survival and reproductive success of their sisters. These females bear the same bacterium by descent and will transmit it vertically. The increase in female fitness resulting from the death of males has been termed "fitness compensation" and can occur for three reasons: a reduction in the rate of inbreeding and consequent fitness losses through inbreeding depression, resource reallocation, and a reduction in the cannibalism of females themselves (Skinner, 1985; Werren, 1987; Hurst, 1991; Hurst et al., 1992). These are not mutually exclusive.

If a host species is prone to inbreeding, then male killing can decrease the rate at which infected females inbreed (they simply have no brothers to mate with) (Werren, 1987). If inbreeding is deleterious, then this will increase the reproductive success of infected females, too. The avoidance of inbreeding therefore represents a benefit to male killing that may occur in many different types of host, although it is perhaps unlikely to be a common benefit, as high rates of inbreeding are generally uncommon outside the Hymenoptera and other haplodiploid taxa. Within the Hymenoptera, inbreeding is more common, although the haplodiploid genetic system within this group means inbreeding depression is rarely severe.

The spread of a male killer as a result of resource reallocation relies on the resources made available to females from the death of males being preferentially available to infected females. The greatest benefit will accrue to surviving progeny of infected females when offspring from one mother occur close together. Thus, species that produce large clutches of eggs together will be more prone to the spread of early male killers than those that disperse their offspring. The benefits associated with resource reallocation can be fairly large and vary among arthropod taxa depending on their ecology. From this feature alone, the expectation is that the distribution of male-killer hosts will not be random taxonomically.

HOST DIVERSITY

Early-male-killing bacteria have been recorded from a taxonomically diverse array of hosts from five orders of insects and from two species of mite. Therefore, mechanistic constraints do not appear to confine them to a narrow range of arthropod taxa. Instead, their distribution is determined primarily by the ecological factors discussed above (Hurst and Majerus, 1993). Hot spots for male killing are known in the milkweed bugs (Hemiptera: Lygaeidae), nymphalid butterflies, particularly of the genus *Acraea* (Jiggins et al., 2001a), and the ladybird beetles (Coleoptera: Coccinellidae) (Majerus and Hurst, 1997). These hot spots are associated with aspects of host ecology that make male killing beneficial. The best-studied group in this regard is the ladybirds.

Three aspects of the biology of aphidophagous ladybirds underlie the high incidence of male-killer infection in this group. First, aphidophagous ladybirds lay eggs in tight clutches. Second, aphids are prone to rapid population increases and crashes so that they are a highly ephemeral prey for ladybirds. Third, ladybirds are highly cannibalistic; in particular, they indulge in sibling-egg cannibalism/consumption. These facts are not independent; prey ephemerality promotes sibling-egg cannibalism, which in turn causes rapid embryonic development (Majerus and Majerus, 1997). The result is that neonate ladybird larvae are very small and have minimal energy reserves when they hatch. Starvation rates of neonate aphidophagous coccinellid larvae are often very high (Banks, 1955, 1956; Wratten, 1973). Comparison of neonate larvae from clutches laid by male-killer-infected and uninfected *A. bipunctata* females showed that the former survived half as long again as the latter after dispersal from their natal egg clutch when denied food or water. Furthermore, larvae from male-killed clutches were larger at dispersal and could subdue a greater size range of aphids and travel further in search of food before dying from starvation than larvae from normal clutches (Hurst, 1993). The greater resources consumed by such larvae before they dispersed from their egg clutches led to more rapid development and higher likelihood of survival to first ecdysis.

Similar or greater advantages resulting from resource reallocation via sibling-egg consumption have been shown in other coccinellids. Here, then, by their sacrificial suicide, the bacteria in male eggs increase the fitness of clonally identical copies of themselves in female siblings of their hosts: an exquisite and extreme case of kin selection.

A second advantage to male killing dependent on the cannibalistic behavior of neonate ladybird beetle larvae entails the reduction in the probability that slow-developing female larvae in male-killed clutches will be cannibalized by faster-developing siblings. This reduced probability is a consequence of both the smaller number of larvae that hatch and the greater number of unhatched eggs available to early-hatching larvae in a male-killed clutch than in a normal one. Evidence that this reduction in cannibalism leads to an increase in the number of female larvae that hatch in male-killed compared to normal clutches has been obtained for two species of coccinellid, *A. bipunctata* and *Coccinula sinensis* (Hurst, 1993; Majerus, 2001).

A third potential advantage to male killing arising from larval cannibalism has yet to be verified by empirical study. Not only do larvae consume unhatched eggs in their clutch before dispersal, but they will also eat conspecific larvae once they have dispersed, particularly if other manageable prey is scarce. In interactions between two larvae, assuming that neither is restricted by ecdysis, the larger larva usually wins and eats the smaller (Majerus, 1994). As larvae from male-killed clutches are on average larger than normal larvae when they disperse, the larvae from male-killed clutches are likely to gain another cannibalistic advantage when clutches of eggs are laid close together and at the same time by infected and uninfected females.

The cannibalistic behavior of aphidophagous ladybirds, coupled with their habit of laying eggs in batches and the ephemerality of their prey, underlies their susceptibility to male-killing symbionts. In this case, the advantage from male killing results primarily from the redistribution of resources from the killed males to their sisters that have a high likelihood of carrying the same male-killer lineage of bacteria.

The evolutionary rationale underlying the widespread occurrence of male killers in aphidophagous coccinellids is well understood. In other species in which early male killers have been recorded, reasons are less clear, although circumstantial evidence based on the ecologies of the host insects suggests a resource advantage to male killing in some cases (Hurst and Majerus, 1993). However, some cases still represent something of a conundrum. The male-killing bacteria found in *Drosophila* are one case; a resource advantage associated with male killing is not obvious. The butterfly *Danaus chrysippus* represents another ambiguous case (Jiggins et al., 2000a). Here, eggs are laid singly, and antagonistic interactions between sibling larvae appear unlikely; thus, a significant resource advantage from the death of infected males to infected female siblings appears impossible. Clearly, further studies of potential sources of fitness compensation are needed in these butterflies. In general, it is proposed that these sporadic cases are maintained from advantages associated with reduced rates of inbreeding, but little or no data are available to corroborate this.

Potential Other Hosts of Early Male Killers

Early-male-killing bacteria have been reported from a diverse range of insect hosts, with some families of insect appearing to be particularly prone to male-killer invasion as a consequence of their ecology and behavior. It is likely that the list of hosts is, as yet, extremely incomplete. Certainly many other instances of male-killer infection will be detected in the known hot spots. For example, it is possible to make a rough estimate of the number of Coccinellidae likely to harbor male killers. Taking into account the approximate number of species in the family (6000), the proportion that are aphidophagous, lay eggs in clutches, and indulge in sibling-egg consumption/cannibalism (0.2) and the proportion of such species assayed for male killers that have been found to bear them (0.59) produce a figure of more than 500 host species. Furthermore, this figure does not take into account incomplete ascertainment, which in species where the number of matrilines assayed were low may be significant due to the low prevalence of many male killers.

The characteristics thought to promote male-killer invasion in known "hot-spot" groups can be used to predict other groups likely to bear male killers. The necessity in models of male killing for benefits accruing from the death of males to be preferentially conferred upon infected compared to uninfected females means that fitness compensation sufficient to allow the persistence of a male killer is most likely in species in which antagonistic interactions among siblings are common. These antagonistic interactions may be direct through cannibalism or indirect through competition for nutrients or other resources. A number of groups of insects and other arthropods, from which male killers have not previously been reported, would appear to be likely candidates for male-killer infection. These include aphidophagous lacewings (Neuroptera), web spinners of the order Embioptera, some gall-forming Diptera and Hymenoptera, some predatory Hemiptera, some mantids, and many spiders. On average, sibling interactions are more common in aposematic species than in other species simply because of the high proportion of aposematic species that remain in family groups during their immature stages. Consequently, male killers are more likely to occur in groups with a high proportion of aposematic species.

The Systematic Diversity of Male-Killing Bacteria

In all cases where the causative agents of early male killing have been identified, they have proved to be bacteria. Initially these studies consisted mostly of curing the trait using antibiotics or heat, and further identification was hampered because the bacteria could not be cultured. However, the advent of polymerase-chain-reaction (PCR)-based technology, particularly the sequencing of 16S ribosomal DNA, has revealed male-killing bacteria to be remarkably diverse.

The male-killing bacteria belong to various distantly related bacterial taxa (Table 12.2). From these data we can conclude that male killing must have evolved at least six times — twice in the α-proteobacteria, once in the γ-proteobacteria, twice in the *Spiroplasma*, and once in the Flavobacteria. These independent origins of male killing have occurred in bacterial groups with a variety of different ecologies. The members of all these bacterial groups are associated with arthropods, but they differ both in whether sister associations are mutualistic or parasitic and in whether they are vertically or horizontally transmitted (Hurst and Jiggins, 2000). The majority of species of *Rickettsia* and *Spiroplasma* are arthropod-vectored diseases of either plants (*Spiroplasma*) or vertebrates (*Rickettsia*) (Whitcomb, 1980; Winkler, 1990). In contrast, other arthropod symbionts in the genus *Wolbachia* are mostly vertically transmitted parasites that manipulate the reproduction of their hosts, often by distorting the primary sex ratio (Stouthamer et al., 1999).

TABLE 12.2
The Systematic Diversity of Early Male-Killing Bacteria and Their Hosts

	Bacterium	Host	Ref.
γ-Proteobacteria	*Arsenophonus nasoniae* (Enterobacteriaceae)	*Nasonia vitripennis* (Hymenoptera)	Werren et al., 1986
α-Proteobacteria	*Rickettsia typhi* relative (Rickettsiaceae)	*Adalia bipunctata* (Coleoptera)	Werren et al., 1994
		Adalia decempunctata (Coleoptera)	von der Schulenburg et al., 2001
		Brachys tesselatus (Coleoptera)	Lawson et al., 2001
	Wolbachia	*Adalia bipunctata* (Coleoptera)	Hurst et al., 1999c
		Tribolium madens (Coleoptera)	Fialho and Stevens, 2000
		Acraea encedon (Lepidoptera)	Hurst et al., 1999c
		Acraea encedana (Lepidoptera)	Jiggins et al., 2000b
		Drosophila bifasciata (Diptera)	Hurst et al., 2000
		Hypolimnas bolina (Lepidoptera)	Dyson et al., 2002
Flavobacteria	*Blattabacterium* relative	*Coleomegilla maculata* (Coleoptera)	Hurst et al., 1997b
		Adonia variegata (Coleoptera)	Hurst et al., 1999b
Mollicutes	*Spiroplasma ixodetis* relative	*Adalia bipunctata* (Coleoptera)	Hurst et al., 1999a
		Harmonia axyridis (Coleoptera)	Majerus et al., 1999
		Danaus chrysippus (Lepidoptera)	Jiggins et al., 2000a
	Spiroplasma poulsonii	*Drosophila willistoni* (Diptera)	Williamson et al., 1999

Finally, *Blattabacterium* is a vertically transmitted mutualist found in cockroaches and termites (Bandi et al., 1994), and *Arsenophonus nasoniae* is related to inherited bacteria isolated from the assassin bug *Triatoma* (Hypsa and Dale, 1997) and bacteria of unknown transmission biology from the psyllid *Diaphorina citri* (Subandiyah et al., 2000) and whitefly *Aleurodicus dugesi* (Spaulding and von Dohlen, 2001).

The male-killing taxa also differ in their location within the host. For example, *Spiroplasma poulsonii* is largely extracellular, while *Wolbachia* and *Rickettsia* are both found predominantly within the host-cell cytoplasm. Unlike mutualistic bacteria, which are often found in specialized tissues or organs, there is no evidence that male killers are restricted to specific tissues, and they are typically present in the hemocytes (Hurst et al., 1996a).

To what extent does the sample in Table 12.2 reflect the total diversity of male-killing bacteria? An indication of their total diversity can be gained by reviewing studies that first established the presence of male killers from their phenotypic effects and then identified the bacterium (as opposed to vice versa). This is shown graphically in Figure 12.1, which plots the rate at which new bacterial taxa of male killers have been discovered against the rate at which new male killers have been

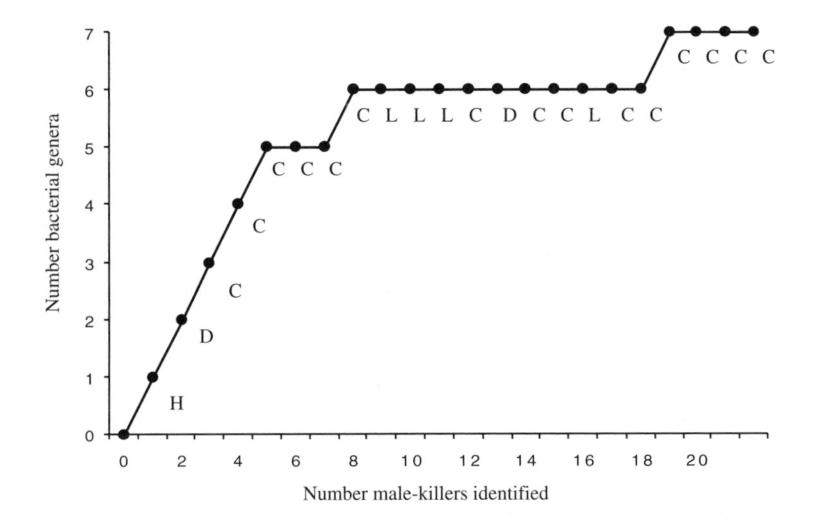

FIGURE 12.1 The rate at which new genera of male-killing bacteria have been described plotted against the number of male killer–host interactions where the bacterium has been identified. Two *Spiroplasma* clades are included as separate genera as this genus is paraphyletic. The date of the description of a new male killer was taken as the date when DNA sequences were published, which confirmed its taxonomic identity. The identity of the host is included as C = Coleoptera, D = Diptera, H = Hymenoptera, and L = Lepidoptera.

identified. Two conclusions can be drawn from this graph and Table 12.2. First, the majority of recent identifications belong to bacterial taxa already known to contain male killers, and therefore the total diversity of male killers in insects is unlikely to be orders of magnitude greater than that already described. Second, when new orders of insects are investigated, they are often found to be infected with bacteria closely related to those found in different orders. Therefore, there is surprisingly little evidence of any systematic variation in the causative agents of male killing across host taxa. If our unpublished records are included, then representatives of all the major bacterial groups in Table 12.2 have been recorded killing males in the best-studied insect family, the Coccinellidae.

Another pattern evident in Table 12.2 is that closely related bacteria are commonly found in distantly related hosts. Therefore, these bacteria must have switched hosts during their evolution. For example, the spiroplasmas that cause male killing in beetles (*A. bipunctata*, *Harmonia axyridis*) and butterflies (*D. chrysippus*) are phylogenetically more similar than their hosts (von der Schulenburg et al., 2000), suggesting that horizontal transmission between species has occurred. Artificial horizontal transmission of the *H. axyridis* male killer into the related beetle *A. bipunctata* produces a male-killing phenotype. Therefore, while horizontal transmission of early male killers is probably of little importance over ecological time, it may be crucial over evolutionary time. Does this mean that all male killers can survive and kill males across many different insect taxa? The only experimental study investigating the ability to transmit over wide phylogenetic distances of host range suggests that this is not the case. When *Spiroplasma poulsonii* and related bacteria that are male killers in the *D. willistoni* species group were injected into other species in the genus *Drosophila,* the majority (although not all) of the recipient species stably transmitted the infection and expressed the trait (Williamson and Poulson, 1979). However, in the more distantly related housefly, *Musca domestica,* the bacterium persisted in the hemolymph but was not transmitted transovarially (Williamson and Poulson, 1979). When the bacterium was injected into species from different insect orders, it was rapidly eliminated from the hemolymph (Williamson and Poulson, 1979). Thus, although transmission between disparate hosts can occur, as witnessed by the *Spiroplasma* in beetles and butterflies (Hurst et al., 1999a; Jiggins et al., 2000a), increasing phylogenetic distance between hosts does represent an increasing barrier to horizontal transmission.

Another intriguing question posed by the data in Table 12.2 is whether the mechanisms of male killing are the same in different host and bacterial taxa. This reflects perhaps the largest gap in our understanding of male killers, but two points are worthy of mention. First, related male-killing bacteria can be found in hosts with very different sex-determination mechanisms. Both beetles and butterflies contain related strains of *Spiroplasma* and *Wolbachia* that kill male hosts. These hosts have very different chromosomal sex-determination systems, beetles being male heterogametic and butterflies female heterogametic. This suggests that either these bacteria can have adapted to new sex-determination systems or that they use some common element of sex determination. This would be intriguing, given the wide diversification of insect sex-determination systems observed: to date there are no "common" mechanisms known. Second, it is unknown whether male death occurs because bacteria detect host sex and then express a second "toxin" gene or whether the bacteria systemically produce a toxin that acts only in males, interfering, for instance, with male sex determination. In this latter context, it is notable that many mutations in the sex-determination pathway are not embryonic lethal. In *Drosophila*, for instance, mutations altering *Sxl* expression are embryonic lethal, but those involved with somatic sex and dosage compensation are not. Thus, direct interference by bacteria will not produce male death. If the bacterium is interfering with sex determination itself, it must be doing so in a fairly fundamental fashion, at the level of the key switch genes. Otherwise, male killing is a response to differences in the expression of genes within the sex-determination pathway without itself being interference of it.

Variation in the Prevalence of Early Male Killers in Natural Populations

Male-killing bacteria are present in a wide variety of species, but their prevalence varies among species. Prevalence is commonly low in drosophilids (1 to 20% of females infected), varies from low (7%) to high (50%+) in ladybirds (Hurst et al., 1993; Majerus et al., 1998), and varies from low (4%) to extremely high (>90% females infected) in Lepidoptera (Geier et al., 1978; Jiggins et al., 1998). Intraspecific variation can also be profound. In the chrysomelid *Gastrolina depressa*, the ladybird *H. axyridis*, and the butterfly *D. chrysippus*, some populations are uninfected, but 40 to 60% of females are infected in others (Chang et al., 1991; Majerus et al., 1998; Jiggins et al., 2000a). In *A. encedon*, prevalence varies from 65 to 95% over relatively short ranges.

What is the cause of this variation in prevalence? The prevalence of male-killing bacteria in natural populations depends on the vertical-transmission efficiency of the male killer, the direct fitness effects of male killers on female hosts, and the extent of fitness compensation. In a minimal model in the absence of inbreeding effects (Hurst et al., 1997a), invasion is possible if $b > 1/(a(1 + s)) > 1$, where b is the benefit accruing to infected females as a result of the death of males (a benefit exists where $b > 1$), a is the vertical transmission of the male killer ($0 < a \leq 1$), and s is the direct effect of bearing a male killer on female fitness ($s > 0$ is a beneficial effect and $s < 0$ is a deleterious effect of infection). Following invasion, equilibrium prevalence is given by $p^* = (ab(1 + s) - 1)/(b(1 + as) - 1)$.

Therefore, these three factors are the most likely causes of the variation in prevalence we observe within and among species. We can investigate each of these in turn.

Transmission Fidelity

As is the case for late male killers, early male killers are not transmitted to all the offspring of infected females. Although some male killers have high vertical-transmission efficiencies (>0.99), laboratory assessments of the vertical-transmission efficiencies of most male killers give lower values, typically in the range 0.8 to 0.95 (Hurst and Majerus, 1993). Their rate of transmission contrasts with the higher figures found for *Wolbachia* that induce cytoplasmic incompatibility and the lower figures found for late male killers.

Variation in the rate of vertical transmission is observed in different male-killing bacteria found within a single host species. For instance, in *A. bipunctata* different male killers have different

TABLE 12.3
Vertical Transmission Efficiencies of Male-Killing Bacteria in Six Matrilines from Moscow

Line	Number of Families	Total Progeny	Sex Ratio (proportion males)	Mean Vertical Transmission Efficiency
Mos 3 (*Rickettsia*)	5	123	0.220	0.719
Mos 6 (*Wolbachia* Z)	6	222	0.126	0.856
Mos 9 (*Spiroplasma*)	6	130	0.015	0.984
Mos 18 (*Wolbachia* Y)	7	168	0.012	0.988
Mos 33 (*Spiroplasma*)	4	170	0.006	0.994
Mos 35 (*Spiroplasma*)	5	199	0	1

Source: Data from Majerus, M.E.N., von der Schulenburg, J.H.G., and Zakharov, I.A. (2000). *Heredity* **84:** 605.

vertical-transmission efficiencies within the same host (Table 12.3). This suggests that the vertical-transmission efficiency of these bacteria is a function of the male killer itself and not of the host. In contrast, the *Wolbachia* strain that infects both *A. encedana* and *A. encedon* in Uganda, which appear to be identical (no differences observed in more than 5000 nucleotides of bacterial and mitochondrial DNA sequenced), have a vertical-transmission efficiency of 0.96 in the former and 1.0 in the latter, suggesting that vertical-transmission efficiency may in part be a function of the host. In reality, the variation in transmission efficiencies is likely to be a product of interaction between the symbiont, its host, and the environment.

Inefficient transmission is thought to be the main factor that results in stable polymorphisms of infected and uninfected females in a population (Hurst, 1991). If a male killer can invade a population and is transmitted with perfect fidelity to the next generation, it will spread to fixation, potentially driving the host extinct (see below). However, if transmission is imperfect, then uninfected females and males are produced by infected mothers, which can result in a stable polymorphism of infected and uninfected females. The causes of this imperfect transmission are unknown, but possible explanations include environmental curing (Hurst et al., 2000), some form of host immune response, or a possible tradeoff between transmission efficiency and some other aspect of bacterial fitness (see below).

Direct Effects on Female Host Survival and Fecundity

Male killers rely on the survival and reproduction of female hosts to be transmitted and are therefore selected to have low virulence in female hosts. While assessments of the direct fitness effects of harboring male killers are sparse, in the majority of those that have been undertaken, costs rather than benefits have been found. In three species of ladybird beetle carrying different male killers (*A. bipunctata* — *Rickettsia*; *H. axyridis* — *Spiroplasma*; *Adonia variegata* — Flavobacterium), costs in terms of decreased rate of egg laying, higher levels of infertility, and shorter adult life span have been observed (Matsuka et al., 1975; Hurst et al., 1994, 1999). In *D. willistoni*, fertility and female embryo survival are reduced (Ebbert, 1991), and in *S. littoralis* and *E. postvittana*, fecundity is reduced (Geier et al., 1978; Brimacombe, 1980).

Recent work on *Wolbachia* in nematodes and insects has suggested the presence of positive physiological advantages to possessing inherited bacteria, even ones that are reproductive parasites. This is certainly a possibility for male-killing bacteria, although definitive data are yet to be obtained. More rapid larval development has been reported in *D. willistoni* and *D. nebulosa* infected with *S. poulsoni* (Malogolowkin-Cohen and Rodriguez-Pereira, 1975; Ebbert, 1991), though it should be noted that infection is associated with an increased frequency of host sterility in this interaction.

Why is it that experiments have generally found a cost to infection, when a "Darwinian Demon" male killer would have either no cost or a benefit to infection? One possible evolutionary scenario is that the virulence of the male killer is correlated to some other trait affecting parasite fitness, such as transmission efficiency (Hurst et al., 1994). If transmission efficiency and bacterial titer are correlated, it is easy to see how natural selection could produce a costly bacterium because this was more efficiently transmitted. It may also explain why transmission efficiency is rarely perfect.

Level of Fitness Compensation

As discussed earlier, fitness compensation is the increase in lifetime fecundity of females that results from the death of males. This is the principal factor responsible for driving up male-killer frequency. While it is easy to get a broad view of differences in the magnitude of this factor (it is fairly noncontentious that the incidence and prevalence of male killing are much higher in aphidophagous coccinellids than in *Drosophila* because the former show egg cannibalism), it is hard to gain any absolute measure (Hurst and Majerus, 1993). This is because the magnitude of this factor is ecologically dependent. How much eating an egg increases ladybird survival depends on the local aphid density, which varies greatly within and between years and over space. Measurement must be performed in the field and varies from being very hard to measure to nearly impossible.

An idea of the magnitude of these effects can be gained from scoring prevalence of the agent and then measuring transmission efficiency and costs of infection. What is left is the likely magnitude of any direct benefit. For instance, 7% of *A. bipunctata* in Cambridge, U.K. are infected with male-killing *Rickettsia*, which has a transmission fidelity of around 87%. To achieve this prevalence with this transmission efficiency requires that an infected female produce around 1.16 times as many surviving daughters as an uninfected female (Hurst et al., 1993). Given that there is around a 10% cost of infection in the absence of any benefit from the death of males (Hurst et al., 1994), this leaves us estimating an increase in female fitness resulting from male death of around 28%.

This process was also conducted for the male-killing *Wolbachia* in *A. encedana*, where prevalence is very high (96% of females infected). Jiggins et al. (2000b) obtained an estimate of the vertical-transmission efficiency of the *Wolbachia* male killer in *A. encedana* by collecting larval nests from the wild and testing individual larvae for *Wolbachia* presence or absence. The vertical-transmission efficiency was estimated to be 0.96. On the basis of this estimate, and assuming that the male killer was at equilibrium, the lifetime reproductive output of infected relative to uninfected females was estimated to be at least half as great again (minimum estimate within 95% confidence intervals = 1.55). This is an extraordinarily high estimate, and it left the authors to conclude that there might be a direct positive effect of infection on female fitness.

Perhaps more tractable is the measurement of the effect of male killing on losses from inbreeding depression. In only two cases have data been sought specifically on this point. In *A. bipunctata*, the level of inbreeding in the wild was measured from egg clutches and found to be rare (Hurst et al., 1996b). In the gypsy moth, *Lymantria dispar*, which is host to a male killer with high vertical-transmission efficiency (Higashiru et al., 1999), a similar approach again showed inbreeding levels to be low. With the establishment of neutral genetic markers in a wide variety of species, measurement of inbreeding rates in the field should become more routine. It should be possible to test whether the spatial variation in male-killer prevalence in *D. chrysippus* or *D. bifasciata* is associated with variation in inbreeding rates.

In conclusion, it is clear there is variation in the magnitude of the effect that male death has on sibling-female survival, and this will be responsible for some of the variation in prevalence observed within species. However, variation in prevalence of different male killers within a locality, as is observed in *A. bipunctata* from Moscow (see below), indicates that other factors above (transmission efficiency, cost) are also important in producing the variation in prevalence we observed among and within species.

Coexistence of Male Killers within a Population

Minimal models of male-killer dynamics, based on vertical-transmission efficiency, direct-fitness effects on female hosts, and fitness compensation, predict that long-term coexistence of multiple male killers in a host population is not possible. Yet this prediction is undermined by the finding of four different male killers (a *Rickettsia*, a *Spiroplasma*, and two *Wolbachia*) in a Muscovite population of *A. bipunctata* (Majerus et al., 2000) and the observation of distinct *Wolbachia* strains in *A. encedon* (Jiggins et al., 2001b).

It is possible that these are stable polymorphisms maintained by natural selection. For example, Randerson et al. (2000a) have shown that two male killers can coexist if the host evolves partial resistance to male killers with the higher product of transmission efficiency and cost. Although no evidence of male-killer suppression has been obtained in *A. bipunctata*, and the parameter space representing coexistence is limited in the Randerson et al. model, their result does suggest that at least two male killers may coexist at equilibrium in a host population. Multiple male killers could also be maintained due to spatial heterogeneity, with different male killers favored in different patches coupled with migration between these patches. Alternatively, we may be observing a dynamic system that rarely, if ever, reaches equilibrium. Simulations suggest that the replacement of one male killer by another more competitive male killer can sometimes take thousands of generations. The four male killers in Muscovite *A. bipunctata* may be the result of migration, temporal variance in the selective pressures on each male-killer strain, and slow reduction in prevalence of the less-fit strains.

Consequences of Male-Killing Bacteria for the Dynamics and Evolution of Their Hosts

Male-killing bacteria can be common within a species. As they become more common, the sex ratio in the population becomes more female biased, and this may have effects on female fertility and alter patterns of sexual selection. Additionally, they can produce very strong selection for genes in the host population that prevent their action or transmission or otherwise ameliorate the effects of their presence. Here they may become very important evolutionary agents because of the sheer magnitude of selective pressures they can generate. We argue that prevalence can reach such extreme levels that they can select for genes that are usually very costly. As such, they have the potential to be a real source of evolutionary novelty, for instance, in the evolution of sex-determination systems.

Could Male-Killing Bacteria Lower Population Size or Cause Population Extinction?

In 1967, Hamilton (1967) noted that selfish genetic elements that distort the sex ratio could result in the extinction of entire populations or species through lack of the sex against which the selfish genetic element drove. He considered the case of a meiotically driving X chromosome and noted that as the driving element reached higher and higher frequency, the sex ratio became increasingly female-biased, and there would come a point at which there were insufficient males to fertilize all the females within the population. Ultimately, the population would progressively shrink in size, making possible the extinction of the population bearing the element.

Hatcher et al. (1999) have examined in greater detail the effects of inherited parasites that feminize their hosts. They noted that the point at which population decline occurred depended on two factors. The first factor is the extent to which a male can mate multiply. If a single male is able to fertilize many females, then female unmatedness will occur only at extreme population-sex-ratio biases. The second factor is the strength of density-dependent factors that regulate population size. The effects of unmatedness of females due to lack of males would be buffered at the population level by decreased density-dependent mortality. In simple terms, larvae from different females usually compete, and this interference causes a certain degree of larval mortality or reduced fecundity of adults. The unmatedness of some females in extremely sex-biased populations is therefore directly compensated at the population level by increased survival and fecundity of the progeny of others.

The above papers explicitly modeled cases of meiotic drive and feminization, but the principle (that sex-ratio-distorting elements transmitted with high fidelity and with strong sex-ratio distortion can damage populations) holds true across classes of selfish genetic element. Indeed, Hamilton used the case of the female-biased sex ratios in *A. encedon* as evidence for his notion that driving sex chromosomes could cause population-level effects. This system is now known to be one where strongly female-biased population sex ratios are associated with high prevalence of male-killing *Wolbachia*, not sex-chromosome drive. But how commonly might male killers place a population at risk of extinction?

It is hard to evaluate how commonly male killers cause population extinction, as this is an outcome that we would be unlikely to observe. Indeed, Stouthamer et al. (2002) argue that this unobservable process may be a likely outcome. They take as evidence for this the observation that parthenogenesis-inducing microorganisms can become fixed in natural populations. They note that the parallel dynamic acting in the case of a sex-ratio distorter, rather than a distorter of sexuality, would be population extinction, although they caution that what is true of parthenogenesis induction does not necessarily hold true of all classes of sex-ratio distorter. However, there are two reasons to believe this process is unlikely to occur for a male-killing microorganism. First, the drive associated with male death (up to a 55% increase in female sibling survival) is less than the doubling of daughter numbers present in parthenogenesis induction. Male killers, therefore, reach lower prevalence for a given level of transmission efficiency, and this will lower their probability of causing population damage. It is notable in this context that there are cases of male killers that have invaded a population and settled at relatively low equilibrium prevalence without the evolution of resistance genes and that hosts with male-killer prevalence in the range of 60 to 90% are rather rare. Second, selection at the point of invasion for resistance to the action or transmission of a male killer is higher than for an organism that induces parthenogenesis. A female host infected with a parthenogenesis inducer is fully viable and produces around the same number of progeny as uninfected females, while a male-killer-infected individual produces around 50% the number of progeny of an uninfected individual, even in the absence of direct costs on fecundity. Thus, selection for resistance against parthenogenesis inducers will evolve more slowly, relying solely on Fisherian benefits for producing males in populations made female-biased following the spread of sex-ratio distorters. This would increase the likelihood that a male killer with strong drive would remain polymorphic in the population.

It is easier to examine whether the shortage of males is limiting the population size in extant male killers. The majority of male killers tend to be found in fewer than 50% of females, and because males can usually fertilize far in excess of two females, there will be little risk of population extinction. However, there are a few exceptional cases where the prevalence of the male killer is so high that there are significant numbers of unmated females present. However, even where female reproduction is limited by the shortage of males, the population size may not be greatly reduced. This is because the male killer will also both increase the number of females and reduce levels of competition between males and females, and both of these effects will tend to oppose population extinction due to unfertilized eggs. In a survey of insects, the rate of population growth in the absence of density dependence, R_0, was generally observed to lie between 1.3 and 13 (with some species with higher values) (Hassell et al., 1976). For the common case, where R_0 is greater than 2, a male killer will significantly reduce population viability only when female fertility is reduced by one half or more. Given the capacity of males to mate multiply, this will require an extremely high male-killer prevalence.

Sex-Biased Populations and Sexual Selection

In most animals, males have the potential to reproduce faster than females, which has led to the evolution of males that compete for access to females and females that may choose between alternative males (Bateman, 1948). The direction and intensity of this sexual selection depends on the operational sex ratio, which is the ratio of males to females in the population that are available to mate (Emlen and Oring, 1977). If male-killing bacteria bias the population sex ratio toward females, this will also make the operational sex ratio more female-biased. In turn, this is expected to cause a reduction in both the choosiness of females and intensity of competition among males. In extreme cases, it could potentially result in a reversal of sexual selection, with choosy males and competing females. The bacterial prevalence required to reverse the sex roles would depend on the potential rates at which males and females could reproduce given free access to the opposite sex.

The impact of male-killing bacteria on sexual selection has been most comprehensively studied in *A. encedon* (Jiggins et al., 2000c). These effects are likely to be particularly marked in this species — in addition to the populations being very female-biased, the life span and mating rate of males are limited. In southern Uganda, the bacterial prevalence varies over a few kilometers from less than 80% to more than 97% of females being infected. This variation was exploited to investigate how the mating behavior of the butterflies changes with the population sex ratio. In the most female-biased populations, virgin females aggregate at hilltops or other landmarks in large swarms where they vigorously chase other butterflies and exhibit typical butterfly "mate-acceptance" behavior. Once females have mated, they leave the swarm. Similarly, males released in the swarm rapidly mate and then leave the site. This behavior strongly suggests that females are competing for males. Moreover, the swarming of females resembles a common male behavior in insects, where males form swarms in which they compete for females.

Therefore, there is strong evidence that the male-killing infection results in large numbers of unmated females in the population competing for access to males. It is, however, unclear if there has been any evolutionary change in response to the population sex ratio or whether we are observing a behavioral response by the butterflies to the shortage of males. It is also unknown whether male killers have directly influenced the choice of mates made by males in natural populations. Theoretical studies suggest that selection will favor males that choose to mate with females that do not carry the infection (Randerson et al., 2000b), but field studies of *A. encedon* have failed to provide any support for this hypothesis.

This example is clearly atypical of male killers, which typically cause far smaller shifts in the population sex ratio and in which such dramatic reversals of behavior are unlikely. However, we do expect slight shifts in many sexually selected traits such as female choosiness, male–male competition, and sperm allocation in many other species.

Resistance to the Action or Transmission of Male-Killing Bacteria

The fact that male-killing bacteria kill male hosts means that they are always likely to be detrimental to their host, notwithstanding any positive effects of compensation on female host survival or any direct physiological benefit that accrues from bacterial anabolic activity. Selection will therefore favor the spread of host nuclear genes that prevent either the transmission or action of these bacteria.

Resistance genes will spread provided they do not have a cost that exceeds the benefit arising from the production of surviving sons from infected females. Given that the benefit of preventing male-killer action or transmission is positively related to parasite prevalence, while the cost is not, modifiers that prevent the action of male killers that naturally exist at low prevalence will spread only if they have low cost, while modifiers that prevent the action of male killers that naturally exist at high prevalence may have higher costs.

Resistance genes that invade a population initially will go to fixation if uninfected individuals with the resistance genes have the same fitness as those without (i.e., the modifier is cost free). If the modifier has a cost, however, it may arrive at a polymorphic equilibrium in the population. In the former case, the male-killing bacterium is likely to become lost, unless either the host resistance gene in question is of low efficiency or if counteradaptation occurs on the part of the bacterium. It has been noted that the presence of resistance genes can, in theory, stabilize multiple male killers within a population (Randerson et al., 2000a). However, the parameter space in which multiple infections are maintained is small, and it is questionable whether resistance genes really do represent the factor stabilizing multiple male-killer systems. The coexistence of multiple symbionts may in fact indicate some other incorrect assumption within current models. For example, as yet we do not have the data to determine whether any male killer is maintained at a stable equilibrium for a significant period of time.

The population genetic theory relating to the spread of resistance to male-killing bacteria is currently much better developed than empirical studies of resistance in natural populations. What is established is that polymorphism for resistance to transmission of male killers is known in some systems but does not exist in others. Resistance to the transmission of *S. poulsonii* has been observed in *D. willistoni*. Malogolowkin (1958) observed that certain strains of *D. willistoni* were refractory to *Spiroplasma* transmission. Further, resistance to the transmission of the unknown male killer of *D. prosaltans* has been recorded, with host resistance being explained by a single gene and the resistant allele being recessive (Cavalcanti et al., 1957). However, searches of *D. bifasciata* lines for resistance to male-killer transmission or action proved fruitless; the male-killing *Wolbachia* was transmitted and killed males perfectly in all 38 lines tested (Hurst et al., 2001).

We can therefore state that resistance to the transmission of male-killing bacteria can but does not necessarily evolve. This conclusion parallels the more extensive research that has been undertaken for other selfish genetic elements. The genetic mechanism underlying resistance is completely unknown, and it would seem that knowledge of it is some way off. Parallels with other host–parasite interactions (Carius et al., 2001) suggest that we may need to look beyond a host genotype that resists the transmission of all parasite strains to a host genotype × parasite genotype interaction. There is some suggestion for such an interaction in *D. willistoni*, where the cost of infection shows strong host × parasite interaction effects (Ebbert, 1991). Parallels with other systems also suggest that resistant genotypes may act at the point where the parasite crosses epithelia, for example, into the developing egg. It is notable that *S. poulsonii* fails in other species of fly not through inability to live in the host but from an inability to be vertically transmitted into the eggs. However, one area in which these interactions are likely to differ from other parasite–host associations is in the role of the innate immune system. While these systems are enabled and function effectively against bacteria that are injected into an insect, studies in *Drosophila* indicate they are not induced against inherited bacteria (Bourtzis et al., 2000). Resistance is likely to involve different mechanisms and produce new insights into wholly novel pathways of immunity in insects.

The study of resistance to the action of these agents is even less developed but potentially as interesting. Resistance to the action of male-killing bacteria, genes that prevent infected males from dying, could take two forms. First, the genes could repress the killing mechanism of the bacterium. Second, the genes could represent alterations in the sex-determination system of the host, preventing the bacterium from cueing in on host sex. Male killers must interact with the sex-determination system at some level, and the ability of some male killers to act early in embryogenesis suggests that the elements of host sex used are early in the sex-determination cascade and are not peripheral late-acting parts of sex determination (Hurst and Jiggins, 2000).

The interaction of male-killing bacteria with the host's sex-determination system produces the tempting hypothesis that these elements may actually drive the evolution of these systems in insects. Insect sex determination shows wide diversification, with the genes involved changing despite their function remaining constant. Mutations in sex-determination pathways are usually highly deleterious, so how could a novel system ever have spread? Male-killing bacteria represent a potent selective force in this context. If a population has a male killer at high prevalence, then mutations within the sex-determination pathway that lower the fitness of uninfected males or females (producing "bad" sex determination) may spread if they "save" infected males from death. Just as feminizing bacteria can induce the spread of novel sex-determining genes, so might male killers. It is notable here that male killers have a higher cross-species incidence than feminizers, potentially making them a more widely important agent in driving the evolution of sex-determining systems. This line of thought is exciting but clearly begs empirical work. How do male killers detect host sex? Can we find natural variation in host resistance to the action of male-killing agents? There is suggestive evidence, with reports of loss and recovery of male killing within lineages, but more data are needed.

In addition to alterations in sex-determination systems, male killers may also facilitate the evolution of parthenogenetic reproduction. The initial mutation producing parthenogenetic reproduction is usually very deleterious, and although spontaneous occasional parthenogenetic reproduction is fairly common across insect taxa, it is usually very inefficient and is more of a curiosity than an adaptive feature. However, in a population where male-killing bacteria are at high prevalence, and females suffer a cost from forced virginity or lack of sperm, the production, even inefficient, of parthenogenetic progeny may become adaptive. It is most likely to spread if it is conditional on lack of sperm, that is to say, sexual reproduction is used if sperm are available.

SUMMARY

It is now established that male-killing bacteria occur in a wide range of arthropods. Late male killing was until recently thought of as being associated with aquatic insects, but the recent discovery in a moth suggests many more cases may be revealed in terrestrial species. Early male killing has long been noted as a phenomenon observed in a wide array of hosts and is now recognized as being phylogenetically widespread within bacteria: if a group of bacteria is vertically transmitted in insects, the presence of a male-killing strain in that group is likely.

We now understand the population biology of male-killing bacteria in its most rudimentary form. The pace and precision of theoretical work has, as ever, outstripped empirical work, and empirical case studies are now beginning to challenge our understanding of male-killer population biology gained from modeling and are refining our understanding. The high prevalence achieved by male killers in some species despite inefficient vertical transmission and the coexistence of different male-killing strains within a single population can both be explained on current models but stretch the assumptions to such a point that the alternative hypothesis — that current understanding is missing some aspect of their population biology — is becoming increasingly tenable to a pragmatic observer. It would certainly seem premature to say we understand the population biology of these elements, and deficiencies in our ability to measure certain parameters are preventing a total understanding of why prevalence varies among associations.

Finally, the high prevalence achieved by male killers in certain species makes them tempting prospects for driving important evolutionary changes. Host extinction may happen but is perhaps unlikely to be common. Alterations in the behavioral ecology of reproduction are likely but may be considered of narrow import, given the general lability of this kind of trait over evolutionary time. Perhaps the most important changes are those that would usually be too costly to spread — modifications in sex-determination system. Parthenogenesis may equally spread and then become refined more easily within a species where male killers are common. Here is a big idea, but the need for definitive case studies remains.

REFERENCES

Andreadis, T.G. (1985). Life cycle and epizootiology and horizontal transmission of Amblyospora (Microspora: Amblyosporidae) in a univoltine mosquito. *J. Invertebr. Pathol.* **46:** 31–46.

Andreadis, T.G. and Hall, D.W. (1979). Significance of transovarial infections of *Amblyospora* sp. (Microspora: Thelohaniidae) in relation to parasite maintenance in the mosquito *Culex salinarius. J. Invertebr. Pathol.* **53:** 424–426.

Avery, S.W. (1989). Horizontal transmission of Parathelohania (Protozoa: Microsporidia) to *Anopheles quadrimaculatus* (Diptera: Culicidae). *J. Invertebr. Pathol.* **53:** 424–426.

Bandi, C., Damiani, G., Magrassi, L., Grigolo, A., Fani, R., and Sacchi, L. (1994). Flavobacteria as intracellular symbionts in cockroaches. *Proc. R. Soc. London (B)* **257:** 43–48.

Banks, C.J. (1955). An ecological study of Coccinellidae associated with *Aphis fabae* on *V. faba*. *Bull. Entomol. Res.* **46:** 561–587.

Banks, C.J. (1956). Observations on the behaviour and mortality in Coccinellidae before dispersal from the egg shells. *Proc. R. Entomol. Soc. London A* **31:** 56–61.

Bateman, A.J. (1948). Intra-sexual selection in *Drosophila*. *Heredity* **2:** 349–368.

Becnel, J.J. (1986). Microsporidian sexuality in culicine mosquitoes. In *Fundamental and Applied Aspects of Invertebrate Pathology* (R.A. Samson, J.M. Vlak, and D. Peters, Eds.), pp. 331–334. Foundation of the 4th International Colloquium for Invertebrate Pathology, Wageningen, the Netherlands.

Becnel, J.J. and Sweeney, A.W. (1990). *Amblyospora trinus* n. sp. (Microsporida: Amblyosporidae) in the Australia mosquito *Culex halifaxi* (Diptera: Culicidae). *J. Protozool.* **37:** 584–592.

Bourtzis, K., Pettigrew, M.M., and O'Neill, S.L. (2000). *Wolbachia* neither induces nor suppresses transcripts encoding antimicrobial peptides. *Insect Mol. Biol.* **9:** 635–639.

Brimacombe, L.C. (1980). All-female broods in field and laboratory populations of the Egyptian cotton leafworm, *Spodoptera littoralis* (Boisduval) (Lepidoptera: Noctuidae). *Bull. Entomol. Res.* **70:** 475–481.

Carius, H.J., Little, T.J., and Ebert, D. (2001). Genetic variation in a host–parasite association: potential for coevolution and frequency-dependent selection. *Evolution* **55:** 1136–1145.

Carson, H.L. (1956). A female producing strain of *D. borealis* Patterson. *Drosophila Information Service* **30:** 109–110.

Cavalcanti, A.G.L. and Falcao, D.N. (1954). A new type of sex-ratio in *Drosophila prosaltans* Duda. *Proc. Ninth Int. Congr. Genet.* **2:** 1233–1235.

Cavalcanti, A.G.L., Falcao, D.N., and Castro, L.E. (1957). "Sex-ratio" in *Drosophila prosaltans* — a character due to interaction between nuclear genes and cytoplasmic factors. *Am. Nat.* **91:** 327–329.

Chang, K.S., Shiraishi, R., Nakasuji, F., and Morimoto, N. (1991). Abnormal sex ratio condition in the Walnut leaf beetle, *Gastrolina depressa* (Coleoptera: Chrysomelidae). *Appl. Entomol. Zool.* **26:** 299–306.

Dyson, E.M., Kamath, M.K., and Hurst, G.D.D. (2002). *Wolbachia* infection associated with all-female broods in *Hypolimnas bolina* (Lepidoptera: Nymphalidae): evidence for horizontal tranfer of a butterfly male killer. *Heredity* **88:** 166–171.

Ebbert, M. (1991). The interaction phenotype in the *Drosophila willistoni* — spiroplasma symbiosis. *Evolution* **45:** 971–988.

Emlen, S.T. and Oring, L.W. (1977). Ecology, sexual selection and the evolution of mating systems. *Science* **197:** 215–223.

Fialho, R.F. and Stevens, L. (2000). Male-killing *Wolbachia* in a flour beetle. *Proc. R. Entomol. Soc. London B* **267:** 1469–1474.

Geier, P.W., Briese, D.T., and Lewis, T. (1978). The light brown apple moth *Epiphyas postvittana* (Walker). 2. Uneven sex ratios and a condition contributing to them in the field. *Aust. J. Ecol.* **3:** 467–488.

Hamilton, W.D. (1967). Extraordinary sex ratios. *Science* **156:** 477–488.

Hassell, M.P., Lawton, J.H., and May, R.M. (1976). Patterns of dynamical behaviour in single-species populations. *J. Anim. Ecol.* **45:** 471–486.

Hatcher, M.J., Taneyhill, D.E., Dunn, A.M., and Tofts, C. (1999). Population dynamics under parasitic sex ratio distortion. *Theor. Pop. Biol.* **56:** 11–28.

Higashiru, Y., Ishihara, M., and Schaefer, P.W. (1999). Sex ratio distortion and severe inbreeding depression in the gypsy moth *Lymantria dispar* L. in Hokkaido, Japan. *Heredity* **83:** 290–297.

Hurst, L.D. (1991). The incidences and evolution of cytoplasmic male killers. *Proc. R. Soc. London (B)* **244:** 91–99.

Hurst, G.D.D. (1993). Studies of biased sex-ratios in *Adalia bipunctata* L. Ph.D. thesis, University of Cambridge, Cambridge, U.K.

Hurst, G.D.D. and Jiggins, F.M. (2000). Male-killing bacteria in insects: mechanisms, incidence and implications. *Emerging Infect. Dis.* **6:** 329–336.

Hurst, G.D.D. and Majerus, M.E.N. (1993). Why do maternally inherited microorganisms kill males? *Heredity* **71:** 81–95.

Hurst, G.D.D., Majerus, M.E.N., and Walker, L.E. (1992). Cytoplasmic male killing elements in *Adalia bipunctata* (Linnaeus) (Coleoptera: Coccinellidae). *Heredity* **69:** 8–91.

Hurst, G.D.D., Majerus, M.E.N., and Walker, L.E. (1993). The importance of cytoplasmic male killing elements in natural populations of the two spot ladybird, *Adalia bipunctata* (Linnaeus) (Coleoptera: Coccinellidae). *Biol. J. Linn. Soc.* **49:** 195–202.

Hurst, G.D.D., Purvis, E.L., Sloggett, J.J., and Majerus, M.E.N. (1994). The effect of infection with male-killing *Rickettsia* on the demography of female *Adalia bipunctata* L. (two spot ladybird). *Heredity* **73:** 309–316.

Hurst, G.D.D., Walker, L.E., and Majerus, M.E.N. (1996a). Bacterial infections of hemocytes associated with the maternally inherited male-killing trait in British populations of the two spot ladybird, *Adalia bipunctata. J. Invertebr. Pathol.* **68:** 286–292.

Hurst, G.D.D., Sloggett, J.J., and Majerus, M.E.N. (1996b). Estimation of the rate of inbreeding in a natural population of *Adalia bipuncata* L. (Coleoptera: Coccinellidae) using a phenotypic indicator. *Eur. J. Entomol.* **93:** 145–150.

Hurst, G.D.D., Hurst, L.D., and Majerus, M.E.N. (1997a). Cytoplasmic sex ratio distorters. In *Influential Passengers: Microbes and Invertebrate Reproduction.* (S.L. O'Neill, A.A. Hoffmann, and J.H. Werren, Eds.), pp. 125–154. Oxford University Press, Oxford, U.K.

Hurst, G.D.D., Hammarton, T.C., Majerus, T.M.O., Bertrand, D., Bandi, C., and Majerus, M.E.N. (1997b). Close relationship of the inherited parasite of the ladybird, *Coleomegilla maculata*, to *Blattabacterium*, the beneficial symbiont of the cockroach. *Genet. Res.* **70:** 1–9.

Hurst, G.D.D., von der Schulenberg, J.H.G., Majerus, T.M.O., Bertrand, D., Zakharov, I.A., Baungaard, J., Volkl, W., Stouthamer, R., and Majerus, M.E.N. (1999a). Invasion of one insect species, *Adalia bipunctata*, by two different male-killing bacteria. *Insect Mol. Biol.* **8:** 133–139.

Hurst, G.D.D., Bandi, C., Sacchi, L., Cochrane, A., Bertrand, D., Karaca, I., and Majerus, M.E.N. (1999b). *Adonia variegata* (Coleoptera: Coccinellidae) bears maternally inherited Flavobacteria that kill males only. *Parasitology* **118:** 125–134.

Hurst, G.D.D., Jiggins, F.M., von der Schulenburg, J.H.G., Bertrand, D., West, S.A., Goriacheva, I.I., Zakharov, I.A., Werren, J.H., Stouthamer, R., and Majerus, M.E.N. (1999c). Male-killing *Wolbachia* in two species of insect. *Proc. R. Soc. London (B)* **266:** 735–740.

Hurst, G.D.D., Johnson, A.P., von der Schulenburg, J.H.G., and Fiyama, Y. (2000). Male-killing *Wolbachia* in *Drosophila*: a temperature sensitive trait with a threshold bacterial density, *Genetics* **156:** 699–709.

Hurst, G.D.D., Jiggins, F.M., and Robinson, S.J.W. (2001). What causes inefficient transmission of male-killing *Wolbachia* in *Drosophila. Heredity* **87:** 220–226.

Hypsa, V. and Dale, C. (1997). *In vitro* culture and phylogenetic analysis of "Candidatus *Arsenophonus triatominarum*," an intracellular bacterium from the triatomine bug, *Triatoma infestans. Int. J. Syst. Bacteriol.* **47:** 1140–1144.

Jiggins, F.M., Hurst, G.D.D., and Majerus, M.E.N. (1998). Sex ratio distortion in *Acraea encedon* (Lepidoptera: Nymphalidae) is caused by a male-killing bacterium. *Heredity* **81:** 87–91.

Jiggins, F.M., Hurst, G.D.D., Jiggins, C.D., von der Schulenburg, J.H.G., and Majerus, M.E.N. (2000a). The butterfly *Danaus chrysippus* is infected by a male-killing *Spiroplasma* bacterium. *Parasitology* **120:** 439–446.

Jiggins, F.M., Hurst, G.D.D., Dolman, C.E., and Majerus, M.E.M. (2000b). High prevalence of male-killing *Wolbachia* in the butterfly *Acraea encedana. J. Evol. Biol.* **13:** 495–501.

Jiggins, F.M., Hurst, G.D.D., and Majerus, M.E.N. (2000c). Sex ratio distorting Wolbachia causes sex role reversal in its butterfly host. *Proc. R. Soc. London (B)* **267:** 69–73.

Jiggins, F.M., Bentley, J.K., Majerus, M.E.N., and Hurst, G.D.D. (2001a). How many species are infected with *Wolbachia*? Cryptic sex ratio distorters revealed by intensive sampling. *Proc. R. Soc. London (B)* **268:** 1123–1126.

Jiggins, F.M., Hurst, G.D.D., von der Schulenburg, J.H.G., and Majerus, M.E.N. (2001b). Two male-killing *Wolbachia* strains coexist within a population of the butterfly *Acraea encedon. Heredity* **86:** 161–166.

Kellen, W.R., Chapman, H.C., Clark, T.B., and Lindegren, J.E. (1965). Host-parasite relationships of some Thelohania from mosquitoes (Nosematidae: Microsporidia). *J. Invertebr. Pathol.* **7:** 161–166.

Lawson, E.T., Mousseau, T.A., Klaper, R., Hunder, M.D., and Werren, J.H. (2001). Rickettsia associated with male-killing in a buprestid beetle. *Heredity* **86:** 497–505.

Lord, J.C., Hall, D.W., and Ellis, E.A. (1981). Life cycle of a new species of Amblyospora (Microspora: Amblyosporidae) in the mosquito *Aedes taeniorhynchus I. J. Invertebr. Pathol.* **37:** 66–72.

Lus, Y.Y. (1947). Some aspects of the population increase in *Adalia bipunctata* 2. The strains without males, *Dokl. Akad. Nauk S.S.S.R.* **57:** 951–954.

Magni, G.E. (1953). 'Sex-ratio': a non-Mendelian character in *Drosophila bifasciata. Nature* **172:** 81.

Majerus, M.E.N. (1994). *Ladybirds*, Harper-Collins.

Majerus, T.M.O. (2001). The Evolutionary Genetics of Male-killing in the Coccinellidae. Ph.D. thesis, University of Cambridge, Cambridge, U.K.

Majerus, M.E.N. (2002). *Moths.* HarperCollins, London.

Majerus, M.E.N. and Hurst, G.D.D. (1997). Ladybirds as a model system for the study of male-killing endosymbionts. *Entomophaga* **42:** 13–20.

Majerus, M.E.N. and Majerus, T.M.O. (1997). Cannibalism among ladybirds. *Bull. Am. Entomol. Soc.* **56:** 235–248.

Majerus, T.M.O., Majerus, M.E.N., Knowles, B., Wheeler, J., Bertrand, D., Kuznetsov, V.N., Ueno, H., and Hurst, G.D.D. (1998). Extreme variation in the prevalence of inherited male-killing microorganisms between three populations of *Harmonia axyridis* (Coleoptera: Coccinellidae). *Heredity* **81:** 683–691.

Majerus, T.M.O., von der Schulenburg, J.H.G., Majerus, M.E.N., and Hurst, G.D.D. (1999). Molecular identification of a male-killing agent in the ladybird *Harmonoia axyridis* (Pallas) (Coleoptera: Coccinellidae). *Insect Mol. Biol.* **8:** 551–555.

Majerus, M.E.N., von der Schulenburg, J.H.G., and Zakharov, I.A. (2000). Multiple causes of male-killing in a single sample of the two spot ladybird, *Adalia bipunctata* (Coleoptera: Coccinellidae) from Moscow. *Heredity* **84:** 605.

Malogolowkin, C. (1958). Maternally inherited "sex-ratio" condition in *Drosophila willistoni* and *Drosophila paulistorum. Genetics* **43:** 274–286.

Malogolowkin-Cohen, C. and Rodriguez-Pereira, M.A.Q. (1975). Sexual drive of normal and SR flies of *Drosophila nebulosa. Evolution* **29:** 579–580.

Matsuka, M., Hashi, H., and Okada, I. (1975). Abnormal sex-ratio found in the lady beetle, *Harmonia axyridis* Pallas (Coleoptera: Coccinellidae), *Appl. Entomol. Zool.* **10:** 84–89.

Morimoto, S., Nakai, M., Ono, A., and Kunimi, Y. (2001). Late male-killing phenomenon found in a Japanese population of the Oriental tea tortrix, *Homona magnanima* (Lepidoptera: Tortricidae). *Heredity* **87:** 435–440.

Pasteur, L. (1870). Études sur la maladie des vers à soie. Gauthier–Villars, Paris.

Randerson, J.P., Smith, N.G.C., and Hurst, L.D. (2000a). The evolutionary dynamics of male-killers and their hosts. *Heredity* **84:** 152–160.

Randerson, J.P., Jiggins, F.M., and Hurst, L.D. (2000b). Male killing can select for male mate choice: a novel solution to the paradox of the lek. *Proc. R. Soc. London (B)* **267:** 867–874.

Simmonds, H.W. (1923). All female families of *Hypolimnas bolina* L., bred in Fiji by H.W. Simmonds, *Proc. R. Entomol. Soc. London* 1923, ix.

Skinner, S.W. (1985). Son-killer: a third extrachromosomal factor affecting sex ratios in the parasitoid wasp *Nasonia vitripennis*, *Genetics* **109:** 745–754.

Spaulding, A.W. and von Dohlen, C.D. (2001). Psyllid endosymbionts exhibit patterns of co-speciation with hosts and destabilizing mutation in ribosomal RNA. *Insect Mol. Biol.* **10:** 57–67.

Stouthamer, R., Breeuwer, J.A.J., and Hurst, G.D.D. (1999). *Wolbachia pipientis*: microbial manipulator of arthropod reproduction, *Annu. Rev. Microbiol.* **53:** 71–102.

Stouthamer, R., Hurst, G.D.D., and Breeuwer, J.A.J. (2002). Sex ratio distorters and their detection. In *Sex Ratios: Concepts and Research Methods* (I.C.W. Hardy, Ed.), pp. 195–215. Cambridge University Press, Cambridge, U.K.

Subandiyah, S., Nikoh, N., Tsuyumu, S., Somowiyarjo, S., and Fukatsu, T. (2000). Complex endosymbiotic microbiota of the citrus psyllid *Diaphorina citri* (Homoptera: Psylloidea). *Zool. Sci.* **17:** 983–989.

Sweeney, A.W., Hazard, E.I., and Graham, M.F. (1985). Intermediate host for an *Amblyospora* sp. (Microspora) infecting the mosquito *Culex annulirostris. J. Invertebr. Pathol.* **46:** 98–102.

Sweeney, A.W., Graham, M.F., and Hazard, E.I. (1988). Life cycle of *Amblyospora dyxenoides* sp. nov. in the mosquito *Culex annulirostris* and the copepod *Mesocyclops albicans. J. Invertebr. Pathol.* **51:** 46–57.

Sweeney, A.W., Doggett, S.L., and Gullick, G. (1989). Laboratory experiments on infection rates of *Amblyospora dyxenoides* (Microspora: Ambylosporidae) in the mosquito *Culex annulirostris*, *J. Invertebr. Pathol.* **53:** 85–92.

von der Schulenburg, J.H.G., Majerus, T.M.O., Dorzhu, C., Zakharov, A., Hurst, G.D.D., and Majerus, M.E.N. (2000). Evolution of male-killing *Spiroplasma* (Procaryotes: Molicutes) inferred from ribosomal spacer sequences. *J. Gen. Appl. Microbiol.* **46:** 95–98.

von der Schulenburg, J.H.G., Habig, M., Sloggett, J.J., Webberley, K.M., Bertrand, D., Hurst, G.D.D., and Majerus, M.E.N. (2001). Incidence of male-killing *Rickettsia* spp. (alpha-proteobacteria) in the ten spot ladybird beetle *Adalia decempunctata* L. (Coleoptera: Coccinellidae). *Appl. Environ. Microbiol.* **67:** 270–277.

von der Schulenburg, J.H.G., Hurst, G.D.D., Tetzlaff, D., Booth, G.E., Zakharov, I.A., and Majerus, M.E.N. (2002). History of infection with different male-killing bacteria in the two spot ladybid beetle *Adalia bipunctata* revealed through mitochondrial DNA sequence analysis. *Genetics* **160:** 1075–1086.

Werren, J.H. (1987). The coevolution of autosomal and cytoplasmic sex ratio factors, *J. Theor. Biol.* **124:** 317–334.

Werren, J.H., Skinner, S.W., and Huger, A.M. (1986). Male-killing bacteria in a parasitic wasp. *Science* **231:** 990–992.

Werren, J.H., Hurst, G.D.D., Zhang, W., Breeuwer, J.A.J., Stouthamer, R., and Majerus, M.E.N. (1994). Rickettsial relative associated with male killing in the ladybird beetle (*Adalia bipunctata*), *J. Bacteriol.* **176:** 388–394.

Whitcomb, R.F. (1980). The genus *Spiroplasma*. *Annu. Rev. Microbiol.* **34:** 677–709.

Williamson, D.L. and Poulson, D.F. (1979). Sex ratio organisms (Spiroplasmas) of *Drosophila*. In *The Mycoplasmas* (R.F. Whitcomb and J.G. Tully, Eds.), pp. 175–208. Academic Press, New York.

Williamson, D.L., Sakaguchi, B., Hakett, K.J., Whitcomb, R.F., Tully, J.G., Carle, P., Bove, J.M., Adams, J.R., Konai, M. and Henegar, R.B. (1999). *Spiroplasma poulsonii* sp. nov., a new species associated with male-lethality in *Drosophila willistoni*, a neotropical species of fruit fly. *Int. J. Syst. Bacteriol.* **49:** 611–618.

Winkler, H.H. (1990). *Rickettsia* species (as organisms). *Annu. Rev. Microbiol.* **44:** 131–153.

Wratten, S.D. (1973). The effectiveness of the coccinellid beetle, *Adalia bipunctata* (L.) as a predator of the lime aphid. *Eucallipterua tiliae* L. *J. Anim. Ecol.* **42:** 785–802.

13 *Wolbachia pipientis*: Impotent by Association

Stephen L. Dobson

CONTENTS

EVOLUTION OF *WOLBACHIA* REPRODUCTIVE PARASITES

Two evolutionary trajectories are commonly recognized for obligate bacterial symbionts that are vertically inherited. As the success of vertically inherited infections depends on host fitness, bacterial variants with an increasingly benign or beneficial relationship with their host have a selective advantage, resulting in a trend toward commensalism or mutualism, respectively (Ewald, 1987; Fine, 1975). Examples of mutualistic *Wolbachia* infections occur in nematodes (Hoerauf et al., 1999; Langworthy et al., 2000) and a wasp (Dedeine et al., 2001). However, additional *Wolbachia* infections have followed an alternative evolutionary trajectory of reproductive parasitism.

Reproductive parasitism can evolve when the bacterial infection is inherited exclusively through a single sex of the host population. Thus, direct selective pressure on the symbiont occurs only through the sex that is responsible for transmission. Multiple examples of reproductive parasitism can be observed in infections of obligate intracellular *Wolbachia* bacteria. As *Wolbachia* is maternally transmitted to embryos via female-host cytoplasm, males are an evolutionary dead end for *Wolbachia* infections, and there is no direct selection for beneficial symbiosis in infected males. On the contrary, examples occur in which *Wolbachia* infections decrease male fitness. Cytoplasmic incompatibility (CI) (Hoffmann and Turelli, 1997), male killing (Hurst et al., 1999), parthenogenesis (Stouthamer et al., 1993), and feminization (Rousset et al., 1992) all provide examples of *Wolbachia* symbioses in which infected female reproductive success is increased at the expense of infected male hosts.

The *Wolbachia* reproductive manipulations of parthenogenesis and feminization act to increase infection frequency by escalating the number of daughters produced by infected females. Parthenogenesis-inducing *Wolbachia* infections are predicted to spread in a bisexual population if parthenogenic females produce more infected daughters (Stouthamer et al., 2001). With nearly perfect maternal transmission of infections, parthenogenesis-inducing *Wolbachia* infections can spread to

0-8493-1286-8/03/$0.00+$1.50

fixation in the host population. The subsequent accumulation of host mutations can result in complete parthenogenesis that is irreversible despite removal of the *Wolbachia* infection (Zchori-Fein et al., 1992). *Wolbachia* infections that feminize genetic males in some isopod species act by suppressing an androgenic gland and result in reproductively competent females (Rigaud and Juchault, 1993). Male-killing *Wolbachia* infections can also be selected in host populations with strong antagonistic sibling interactions or with intraspecific competition for resources. For example, *Wolbachia*-infected daughters can receive an early nutritional boost by cannibalizing their unhatched male siblings (Hurst, 1991b; Majerus et al., 2000; Zakharov et al., 2000; von der Schulenburg et al., 2000, 2001). Male killing can also reduce the inbreeding rate in infected females (Werren, 1987).

Interestingly, the reproductive manipulations by *Wolbachia* are of interest as potential mechanisms of speciation. Three primary routes have been suggested. *Wolbachia*-induced host parthenogenesis can result in speciation via genetic drift in the asexual hosts. Genetic divergence can lead to daughter populations that are genetically incompatible due to deleterious gene combinations (Stouthamer et al., 1990; Zchori-Fein et al., 1992). Similarly, the post-zygotic reproductive barrier caused by bidirectional CI (discussed below) can facilitate host genetic divergence and speciation (Laven, 1967b; Breeuwer and Werren, 1990; Werren, 1997; Bordenstein et al., 2001; Wade, 2001). In a third proposed route, unidirectional CI can contribute to the reproductive isolation of two taxa (Werren et al., 1995b; Perrot-Minnot et al., 1996; Shoemaker et al., 1999).

CYTOPLASMIC INCOMPATIBILITY

Of the multiple types of *Wolbachia*-induced reproductive manipulations described to date, the form of reproductive parasitism known as CI has attracted a large portion of the scientific attention. This bias is historical in part, since CI was the first described *Wolbachia* phenotype (discussed below). The diversity of taxa in which CI occurs has also contributed to the level of research with this reproductive manipulation. A partial list of taxa in which CI has been described includes Coleoptera (Stanley, 1961; Hsiao and Hsiao, 1985b; Wade and Stevens, 1985), Diptera (Laven, 1967c; Trpis et al., 1981; Hoffmann et al., 1986; Solignac et al., 1994; Giordano et al., 1995), Homoptera (Noda, 1984), Hymenoptera (Saul, 1961; Richardson et al., 1987; Bordenstein et al., 2001), Lepidoptera (Brower, 1976), Acari (Breeuwer, 1997; Johanowicz and Hoy, 1998), and Isopoda (Michel-Salzat et al., 2001).

CI results in karyogamy failure and arrested development of early embryos in diploid insects (Callaini et al., 1997; Tram and Sullivan, 2002). In haplodiploid insect hosts with arrhenotokus parthenogenesis, CI-induced loss of the paternally contributed chromosomes results in all male broods (Dobson and Tanouye, 1998). Unidirectional CI can occur in matings between *Wolbachia*-infected males and uninfected females (Figure 13.1). The reciprocal cross and matings between individuals harboring similar infections are compatible. In host populations that include infected and uninfected individuals, CI provides a reproductive advantage to infected females because they can mate successfully with all male types. In contrast, uninfected females are incompatible with infected males, reducing their reproductive success. The advantage afforded to females by CI comes at the expense of infected males, which are incompatible with uninfected females.

As demonstrated both theoretically and empirically (Caspari and Watson, 1959; Fine, 1978; Stevens and Wade, 1990; Hurst, 1991a; Turelli and Hoffmann, 1991; Turelli, 1994; Hoffmann and Turelli, 1997), the reproductive advantage afforded by CI to infected females can result in population replacement, with the infected cytotype invading the host population and replacing the uninfected cytotype, known as "population replacement." The reproductive advantage afforded by CI to infected females can result in the spread of infections despite female fitness costs associated with *Wolbachia* infection (Turelli and Hoffmann, 1995; Fleury et al., 2000). Factors affecting the rate of population replacement are CI levels (egg hatch resulting from cytoplasmically incompatible crosses), *Wolbachia* effect on host-female fitness, and the fidelity of *Wolbachia* transmission (maternal transmission rate) (Prout, 1994; Turelli, 1994; Rousset and Solignac, 1995; Dobson et al.,

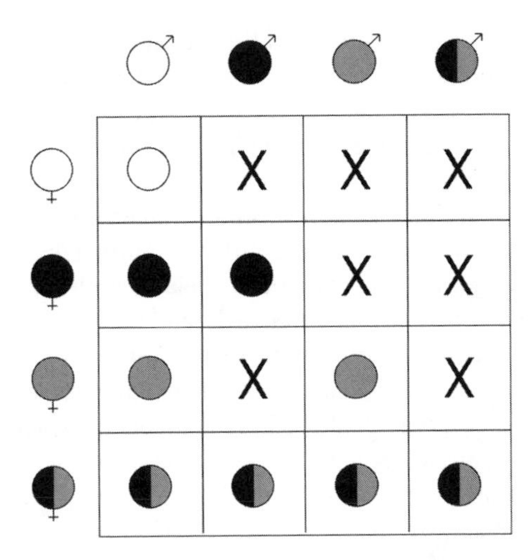

FIGURE 13.1 Pattern of egg hatch resulting from additive CI. Differing *Wolbachia* single infections are indicated by black or gray shading. Superinfection is indicated by black + gray shading. "X" indicates CI. CI provides a reproductive advantage to single-infected females relative to uninfected females (unshaded) because they can mate successfully with more male types. Similarly, superinfected females are at a reproductive advantage relative to single and uninfected females because they are compatible with all male types. The reproductive advantage afforded by CI to infected females can result in population replacement, with the infected cytotype driving into the host population and replacing the uninfected cytotype.

2002b). A general equation (Hoffmann et al., 1990; Turelli, 1994) describing *Wolbachia* infection frequency (p) at time t is given as:

$$p_{t+1} = \frac{p_t(1-\mu)(1-s_f)}{1 - s_f p_t - s_h p_t(1-p_t) - \mu s_h p_t^2(1-s_f)} \tag{13.1}$$

where $(1 - s_h)$ is the proportion egg hatch resulting from CI crosses, $(1 - \mu)$ is the fidelity of *Wolbachia* maternal transmission, and $(1 - s_f)$ is *Wolbachia* effects on host fecundity. The speed at which *Wolbachia* invades a host population (population-replacement rate) is directly related to the fidelity of maternal transmission and relative fecundity of infected host females and is inversely related to egg hatch resulting from incompatible crosses. Equation 13.1 may also be used to characterize the initial infection frequency required for *Wolbachia* to invade the host population (threshold infection frequency) and the equilibrium infection frequency following *Wolbachia* invasion. Infection levels below the required infection threshold do not invade and are lost from the host population. Additional models examining infection dynamics include *Wolbachia* effects on sperm competition (Wade and Chang, 1995), *Wolbachia* infections that do not cause CI (Hoffmann et al., 1996), CI-inducing *Wolbachia* infections that increase host fitness relative to uninfected hosts (Dobson et al., 2002b), and natural "curing" of infected individuals by heat, naturally occurring antibiotics, and diapause (Stevens, 1989; Karr, 1991; Stevens and Wicklow, 1992; Perrot-Minnot et al., 1996).

Although the mechanism responsible for CI has not yet been identified, a poison/antidote model has been used to explain the observed phenomenon (Werren, 1997). In this model, modification (*mod*) occurs on the male pronucleus, before *Wolbachia* are shed from maturing sperm (Presgraves, 2000). Embryos that are fertilized with modified sperm are arrested in early development unless the rescue (*resc*) function is expressed in eggs from *Wolbachia*-infected females. In recent studies examining cytoplasmically incompatible embryos of *Wolbachia*-infected *Nasonia vitripennis* wasps, researchers have observed a delayed breakdown of the paternally contributed pronuclear envelope (Callaini et al., 1997; Tram and Sullivan, 2002). The observations have led to the postulation that the resulting asynchrony between the maternal and paternal pronuclei is the primary defect in CI embryos. However, the patterns of additive unidirectional and bidirectional CI that have been described in *N. vitripennis* (Perrot-Minnot et al., 1996; Bordenstein et al., 2001) and other insects (Rousset and de Stordeur, 1994; Guillemaud et al., 1997; Rousset et al., 1999; Dobson et al., 2001) complicate the sole use of a timing model to explain observed CI phenomena.

As illustrated in Figure 13.1, differing *Wolbachia* infections can result in bidirectional CI, demonstrating that *mod* and *resc* interact in a specific manner, such that different infection types do not necessarily rescue the modification of a differing *Wolbachia* type (Bourtzis et al., 1998; Merçot and Poinsot, 1998). In addition, CI patterns associated with superinfection demonstrate that differing *mod/resc* mechanisms may act autonomously and additively, such that coinfections may result in novel patterns of additive unidirectional CI (see Figure 13.1; Merçot et al., 1995; Sinkins et al., 1995; Perrot-Minnot et al., 1996; Dobson et al., 2001; James et al., 2002). Thus, an additional "recognition factor" must be incorporated into the timing model to explain the failure of the maternal infection to influence paternal pronuclei that have been modified by different *Wolbachia* infection types. The proposed recognition factor could be tested by examining the timing of nuclear-envelope breakdown in *N. vitripennis* embryos that result from bidirectionally incompatible crosses (Perrot-Minnot et al., 1996). Comparison of nuclear-envelope-breakdown timing in crosses of superinfected males with the four female infection types would further elucidate an additive timing effect as an explanation for the pattern of CI observed with *Wolbachia* superinfection.

Following *Wolbachia* invasion, several evolutionary trajectories are possible. With infections that have an imperfect maternal transmission rate or a high host cost associated with *Wolbachia* modification, *Wolbachia* variants may be selected that can rescue but that have a reduced ability to induce CI or that do not cause CI (mod– $resc^+$) (Turelli, 1994; Bourtzis et al., 1998; Merçot and Poinsot, 1998). After the mod– $resc^+$ infection has become fixed in the host population, there is no selection to maintain the $resc^+$ phenotype. Thus, the host population can be invaded by a variant that does not rescue (mod– resc–) or by the uninfected cytotype leading to a cyclical pattern of invasion under some circumstances (Hurst and McVean, 1996; Hatcher, 2000).

An additional trajectory includes the evolution of *Wolbachia* variants with new compatibility types (Charlat et al., 2001). *Wolbachia* variants with a novel modification type that is incompatible with the infection of origin will have the same pattern of incompatibility (Table 13.1). Thus, modification-type variants are neutral and may spread by drift. If the infection frequency of the variant increases sufficiently, the conditions can permit the invasion of an additional variant with a novel rescue type (i.e., mod_B $resc_B$). Thus, novel CI types can emerge from ancestral infection types.

WOLBACHIA DESCRIPTION AND EARLY APPLIED SUPPRESSION STRATEGIES

Initially described by Hertig and Wolbach (1924), the species name *Wolbachia pipientis* was not defined for an additional 12 years (Hertig, 1936). The bacterium was reported as a rickettsia, appearing as "minute rods and coccoids" in the germ cell cytoplasm of *Culex pipiens* (common house mosquito). The infection was described as a "relatively harmless parasite…transmitted via

TABLE 13.1
Pattern of Cytoplasmic Incompatibility When the Type of Modification (mod) and Rescue (resc) Are Affected

		Male		
		$mod_A resc_A$	$mod_B resc_A$	Uninfected
Female	$mod_A resc_A$	+	–	+
	$mod_B resc_A$	+	–	+
	Uninfected	–	–	+

the egg." Attempts to culture the infection were unsuccessful (Hertig, 1936). Unfortunately, following the initial description, *W. pipientis* essentially disappeared from scientific literature for more than three decades.

Although the name *W. pipientis* failed to attract scientific attention in the years following its description, *Wolbachia*-induced embryo lethality became the research focus of multiple laboratories and field trials sponsored by the World Health Organization (WHO). *C. pipiens* was again the experimental subject when crosses intended to clarify *Culex* systematics identified intriguing patterns of incompatibility resulting from interstrain hybridizations (reviewed in Laven, 1967c). Crossing experiments revealed a complex pattern of both bidirectional and unidirectional incompatibility. Additional backcrossing experiments with the hybrid offspring demonstrated that compatibility types were determined by a maternally inherited factor, prompting Laven to coin the term "cytoplasmic incompatibility" to describe the crossing phenomena (Laven, 1967c). A total of 17 *Culex* crossing types were defined by Laven (Laven, 1967c). Contemporary with the *Culex*-crossing experiments, similar patterns of CI were also described in additional mosquitoes (Rozeboom and Kitzmiller, 1958), *Nasonia* (Mormoniella) *vitripennis* (Saul, 1961; Ryan and Saul, 1968), and *Tribolium confusum* (Stanley, 1961). However, *W. pipientis* failed to be incriminated as the etiological agent responsible for CI.

Not knowing the agent responsible for CI did not prevent researchers from proposing and testing applied strategies that employed CI and targeted the control of medically important *Culex* populations. *C. quinquefasciatus* Say (*C. pipiens fatigans* Wiedemann) is the primary vector for human filariasis in Southeast Asia. The proposed strategy was similar to conventional sterile insect technique (SIT) (Knipling, 1955) but employed releases of cytoplasmically incompatible males instead of irradiated or chemosterilized males (Barr, 1966; Laven, 1967a). The strategy attracted support from the WHO, resulting in field trials of the strategy in conjunction with its Filariasis Research Unit in Rangoon, Burma (Anonymous, 1964). Cytoplasmically incompatible males were repeatedly released into a relatively isolated population, with the goal of reducing the vector population and interrupting the transmission of filariasis. In one study, field releases of incompatible males were continued over a 12-week period in the village of Okpo, Burma. To assess strategy efficacy, *C. quinquefasciatus* egg rafts were field-collected and monitored for hatching. During the course of the experiment, raft egg hatching dropped from 95.7 to 0%. By the end of the study, all field-collected egg rafts failed to hatch (Laven, 1967a). The success of this strategy for the suppression of field populations led to the proposed adaptation of this strategy for additional targets including economically important pest species (Davidson, 1974; Pal and LaChance, 1974; Brower, 1979). However, the success of the CI and related strategy field tests were tainted by a public-relations scandal, in which the Indian press accused the WHO and involved governments of conducting biological warfare research (Anonymous, 1975, 1976; Hanlon, 1975; Tomiche, 1975; Wood, 1975; Curtis and Curtis, 1976; Curtis and Von Borstol, 1978; Walgate, 1978). Due in part to the allegations and negative attention, research with CI for the suppression of pest populations was not continued.

Of equal concern, scientific critics argued that the CI-suppression strategy was impractical due to the requirement that only males be released (Pal and Whitten, 1974). With traditional SIT strategies, releases are designed to consist primarily of males for maximum efficiency, but the absolute elimination of females from releases is not imperative to the success of the strategy (Knipling, 1998; Krafsur, 1998). In contrast, accidental female release in the CI-based SIT strategy could permit the establishment of the new cytotype in the field, resulting in a field population that was compatible with additional releases of males. Thus, the end result would be the replacement of the field population with the released vector population that harbored the new cytotype. Additional male releases would no longer be incompatible, permitting recovery of the host population. Although all-male releases were possible for small pilot experiments, the complete elimination of females was not considered practical on a scale required for large-area control or eradication programs (Pal and Whitten, 1974). Subsequently, interest in the applied use of CI shifted from population-suppression to population-replacement strategies, which employ *Wolbachia* to spread desired genes into a host population (e.g., harnessing *Wolbachia*-induced population replacement to spread genes conferring resistance to disease transmission; discussed below) (Curtis, 1992).

WOLBACHIA INCRIMINATION

Wolbachia research was reinvigorated when Janice Yen rediscovered Hertig's description and proposed that *W. pipientis* could be the cytoplasmically inherited agent responsible for CI (Yen and Barr, 1971). She subsequently demonstrated *Wolbachia* to be the etiological agent of CI in *C. pipiens* by removing the infection using tetracycline. Crosses of the aposymbiotic strain demonstrated the expected pattern of unidirectional incompatibility associated with CI (Yen and Barr, 1974).

Another significant advance in *Wolbachia* research occurred with the advent of the polymerase chain reaction (PCR) and molecular phylogenetics. Detection of *Wolbachia* infection had previously relied on electron microscopy of tissue sections. PCR provided a relatively quick, simple, and sensitive assay to detect infection. Sequencing and molecular phylogenetic comparison of PCR amplicons provided a valuable complement to prior morphological characterizations using electron microscopy (Hertig and Wolbach, 1924; Hertig, 1936) and allowed an improved taxonomic positioning of *Wolbachia*, placing *Wolbachia* in the α-subdivision of the proteobacteria (O'Neill et al., 1992; Rousset et al., 1992).

PCR and molecular phylogenetics also demonstrated that *C. pipiens* was not unique in its *Wolbachia* infection. As described above, similar patterns of CI had been described for additional species (Rozeboom and Kitzmiller, 1958; Saul, 1961; Stanley, 1961; Ryan and Saul, 1968; Kellen et al., 1981; Trpis et al., 1981; Hsiao and Noda, 1984; Hsiao, 1985a; Wade and Stevens, 1985; Hoffmann et al., 1986; O'Neill, 1989; Breeuwer and Werren, 1990; Hoffmann et al., 1990). Surveys using *Wolbachia*-specific PCR primers have revealed widespread *Wolbachia* infection in invertebrates, with estimates of infection rates ranging from 15 to 75% (Werren et al., 1995a; Jeyaprakash and Hoy, 2000; Werren and Windsor, 2000; Jiggins et al., 2001a). The many *Wolbachia* infection types have also been shown to be responsible for multiple types of host reproductive manipulations including CI, parthenogenesis (Stouthamer et al., 1993), feminization (Rousset et al., 1992), and male killing (Hurst et al., 1999). *Wolbachia* infections have been broadly described in nematodes as mutualistic symbionts, attracting medical attention as novel targets for the reduction of filarial pathology (Hoerauf et al., 2000; Bandi et al., 2001; Hoerauf et al., 2001; Taylor et al., 2001).

PHYLOGENETIC ANALYSES

Wolbachia infections are widespread in insects, with estimates ranging from 16 to 76% infection rates (Werren et al., 1995a; Jeyaprakash and Hoy, 2000; Werren and Windsor, 2000; Jiggins et al., 2001a). Early phylogenetic examination of *Wolbachia* infections from different invertebrate hosts demonstrated a monophyletic group within the α-subdivision of the proteobacteria (O'Neill et al., 1992; Rousset et al., 1992; Roux and Raoult, 1995). The initial comparisons of host and *Wolbachia* phylogeny indicated that infections had been acquired more than once by different insects and suggested horizontal transmission of infections between host species. Subsequent studies that employed additional *Wolbachia* gene sequences corroborated early results and allowed improved phylogenetic resolution of *Wolbachia* infections, including the subdivision of *Wolbachia* infections into multiple clades and subgroups (Bourtzis et al., 1994; Werren, 1997; Bandi et al., 1998; Zhou et al., 1998; Bazzocchi et al., 2000). To date, a cautious approach of designating infections as strains rather than species has been followed (Werren et al., 1995b).

Although it is agreed that the primary route of *Wolbachia* maintenance within the host population is vertical transmission via maternal cytoplasm, phylogenetic comparisons of *Wolbachia* and host demonstrate that naturally occurring, interspecific, horizontal transfer of infections can also occur (O'Neill et al., 1992; Rousset et al., 1992; Werren et al., 1995b). Examples of naturally occurring intraspecific transfer include the proposed movement of *Wolbachia* between hymenopteran parasitoids and their hosts (Werren et al., 1995b; Jochemsen et al., 1998; van Meer et al., 1999; Vavre et al., 1999). An interesting example of a natural route of intraspecific horizontal transmission has been infected and uninfected *Trichogramma* wasps that superparasitize the same host egg (Huigens et al., 2000).

The interpretation of *Wolbachia* phylogeny can be confounded by recombination (Werren et al., 1995b; Jiggins et al., 2001b; Werren and Bartos, 2001). Evidence for *Wolbachia* recombination is provided by comparisons of different regions of the *Wolbachia* genome, examining for congruency. Phylogenetic comparisons of the genomic regions are mutually incompatible. Therefore, the most parsimonious interpretation is genetic recombination between strains. *Wolbachia* genomic sequencing (Bandi et al., 1999; O'Neill, 1999; Slatko et al., 1999) will help to clarify the frequency and evolutionary significance of recombination among *Wolbachia* strains.

WOLBACHIA MORPHOLOGY AND INTERACTION WITH HOST CELLS

Earlier ultrastructural observations describe *Wolbachia* as small (0.5 to 1.5 μm) and pleomorphic, ranging from spherical to elongate (Wright and Barr, 1980; Kellen et al., 1981; Trpis et al., 1981; Larsson, 1983; Louis and Nigro, 1989; Ndiaye and Mattei, 1993; O'Neill et al., 1997; Popov et al., 1998; Taylor et al., 1999). *Wolbachia* cells are commonly observed to contain ribosomes and nucleic-acid fibrils (Yen and Barr, 1974; Beckett et al., 1978; Wright and Barr, 1980; Louis and Nigro, 1989; Ndiaye and Mattei, 1993). Although *Wolbachia* have been rarely observed without a host vacuole (i.e., exposed in the host-cell cytoplasm), *Wolbachia* are commonly observed within a three-layer membrane consisting of the bacterial plasma membrane, bacterial cell wall, and host vacuole membrane (Yen and Barr, 1974; Yen, 1975; Barr, 1982; Beckett et al., 1978; Kellen et al., 1981; Trpis et al., 1981; Larsson, 1983; Louis and Nigro, 1989; Ndiaye and Mattei, 1993; O'Neill et al., 1997; Popov et al., 1998; Taylor et al., 1999). During binary fission, *Wolbachia* appears elongate, presumably in prelude to cell division, which appears as a "dumbbell" shape (Wright et al., 1978; Wright and Barr, 1980; Larsson, 1983; Ndiaye and Mattei, 1993). During division, the vacuolar membrane follows the contours of the dividing cells, occasionally resulting in multiple *Wolbachia* contained within a single host vacuole (O'Neill et al., 1997). *Wolbachia* do not appear to be attacked by the host lysosome (Beckett et al., 1978). Both intracellular and extracellular morphs have been reported *in vitro* and *in vivo* (Ndiaye and Mattei, 1993; Ndiaye et al., 1995; Popov et al., 1998). With *in vitro* infections, intracellular *Wolbachia* were observed as round reticulate cells and extracellular bacteria as dense-cored cells, the latter being engulfed by a host cell in one observation (Popov et al., 1998). Intracellular bridges may also permit the flow of *Wolbachia* from one cell to another (Wright and Barr, 1980).

During mitosis, electron (Wright and Barr, 1980; Callaini et al., 1994; Kose and Karr, 1995; Ndiaye et al., 1995) and confocal (O'Neill and Karr, 1990; Braig et al., 1994; Callaini et al., 1994; Kose and Karr, 1995) microscopy reveal a close association between *Wolbachia* and host-centrosome-organized microtubules, suggesting that *Wolbachia* infections may employ the host spindle apparatus for segregation to daughter cells. At prophase and centrosome duplication, *Wolbachia* is loosely associated with centrosomes (Kose and Karr, 1995). During prometaphase/metaphase, however, *Wolbachia* become tightly clustered near the centrosome and spindle-pole asters. At the metaphase/anaphase transition, *Wolbachia* continue to associate with astral microtubules and redistribute away from the centrosomes, coincident with microtubule growth outward from spindle poles. By anaphase/telophase, *Wolbachia* is even more dispersed as the spindle elongates, forcing nuclei and associated spindle poles further apart. *Wolbachia* are not observed to associate with kinetochore-to-pole microtubules. By virtue of their association with the centrosome, *Wolbachia* infections segregate 1:1 during mitosis and therefore effectively behave like chromosomes. In embryos in which centrosomes are separated from nuclei (i.e., via inhibition of DNA replication using aphidicolin or mutations that block nuclear division), *Wolbachia* remains clustered around the centrosomes (Glover et al., 1990). The association of *Wolbachia* and astral microtubules is further suggested by experiments in which microtubule disruption with colchicine was observed to cause the partial disassociation of *Wolbachia* from the spindle poles, whereas cytochalasin B was observed to leave bacterial clusters intact (Callaini et al., 1994; Braeckman et al., 1997).

WOLBACHIA EFFECT ON HOST POPULATION SIZE

As described above, the reproductive advantage afforded to females by *Wolbachia*-induced CI allows for the spread of infections into the host population. Recently, an additional model has been developed to examine the effect of *Wolbachia* infection on host population size during natural invasions and artificial releases of *Wolbachia*-infected hosts (Dobson et al., 2002a). The model extends prior models by adding a density-dependent population-growth model and examining bidirectional CI with up to three different infection types and an uninfected cytotype. This model has been used to examine events occurring during both natural *Wolbachia* invasions and artificial releases conducted as part of applied population-suppression and replacement strategies. The model predicts a transient reduction in the host population size as a unidirectionally incompatible infection invades (Figure 13.2A). The temporary reduction results primarily from the reduced brood hatch in incompatible crosses. Therefore, the level of host-population reduction is greatest when the population harbors similar frequencies of both infected and uninfected individuals. As the frequency of one cytotype increases (i.e., due to invasion or elimination of the infection), the frequency of incompatible crosses decreases, and the host-population size recovers.

Once an infection has invaded and reached equilibrium, the carrying capacity of the infected host population may differ from the carrying capacity of the original uninfected population due to fitness effects associated with *Wolbachia* infection (Figure 13.2A). With increasing CI survivorship ($1 - s_h$), a reduced number of infected hosts is compensated by an increase in uninfected hosts such that the total host-population size at equilibrium remains constant (Dobson et al., 2002a). In contrast, increased fecundity costs associated with infection ($1 - s_f$) and number of uninfected eggs produced by infected females (μ) typically result in a reduced frequency of infected hosts at equilibrium and a lower carrying capacity. Variation in the parameters that affect density-dependent population growth and the initial host-population size have no effect on the initial *Wolbachia* infection frequency required for population replacement, cytoplasmic drive rates, or the equilibrium prevalence of the infection following invasion (Dobson et al., 2002a).

Interestingly, the type of intraspecific host competition has a significant effect on host-population dynamics during *Wolbachia* invasion (Dobson et al., 2002a). In host populations with high reproductive rates and scramble-type competition, the reduced brood hatch that results from incompatible matings yields fewer immatures. Due to reduced intraspecific competition, however, disproportionately more of the immatures survive. The increase in adult number that results from male releases is less pronounced in insect populations with lower reproductive rates and does not occur in insect populations with contest-type competition. Empirical tests are required to validate the predicted effects of *Wolbachia* invasion on host-population size.

APPLIED STRATEGIES EMPLOYING *WOLBACHIA* INFECTIONS

Applied strategies based on *Wolbachia*-induced CI and parthenogenesis have been proposed. As *Wolbachia* infections occur within a broad range of invertebrates, these strategies are potentially applicable to a variety of medically and economically important insects. Alternative strategies for controlling medically and economically important insects are of increasing interest due to environmental and

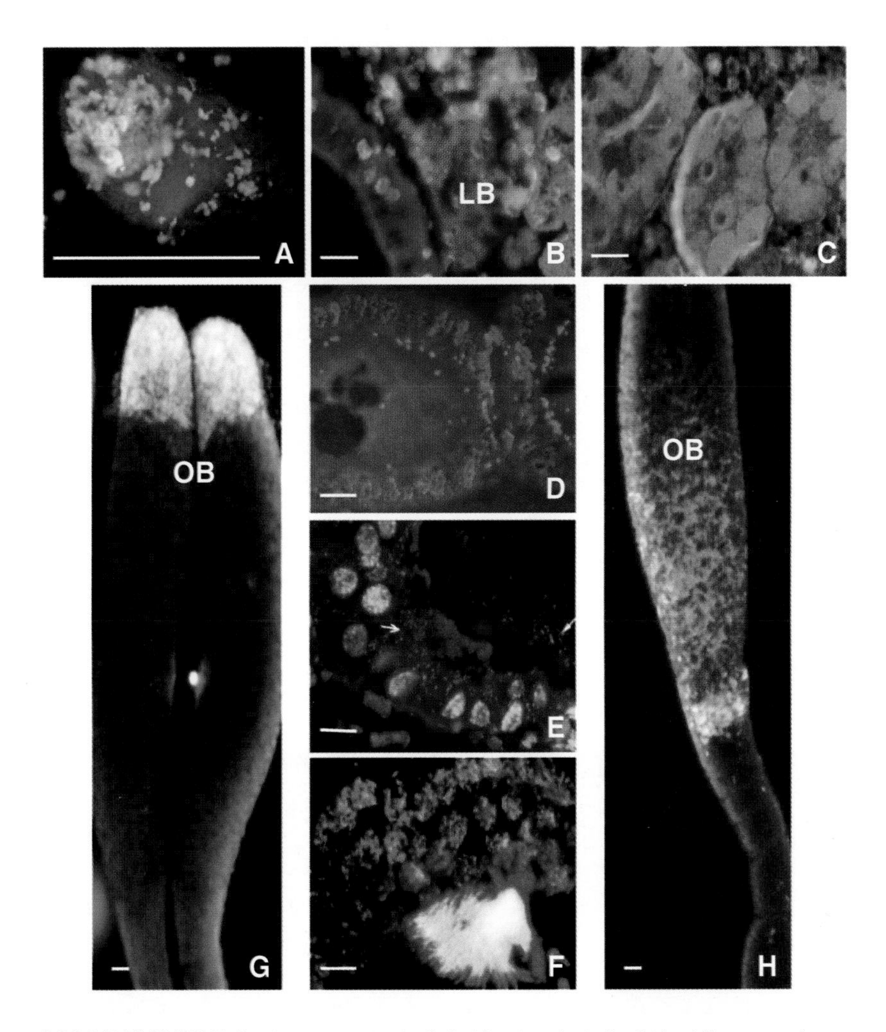

COLOR FIGURE 5.1 Fluorescent *in situ* hybridization (FISH) of *Sitophilus oryzae* intracellular bacteria. Specific oligonucleotide probes were designed by sequence alignment of *Wolbachia* and SOPE 16S rDNA. Two *Wolbachia* probes (W1, W2), 5′ end labeled with rhodamine, were used to increase the signals. The SOPE probe (S) was 5′ end labeled with rhodamine except in panel B (with fluorescein). Hybridization was performed as described by Heddi et al. (1999). Slides were mounted in Vectashield medium containing DAPI. (A) Bacteriocyte labeled with W1W2. (B) Larval bacteriome (LB) labeled with W1W2 and S (fluorescein). (C, D, E, F, and H) Adult mesenteric caeca bacteriomes, oocyte, follicular cells, testis, and ovary, respectively labeled with W1W2. (G) Ovary labeled with S. Scale bar = 10 μm.

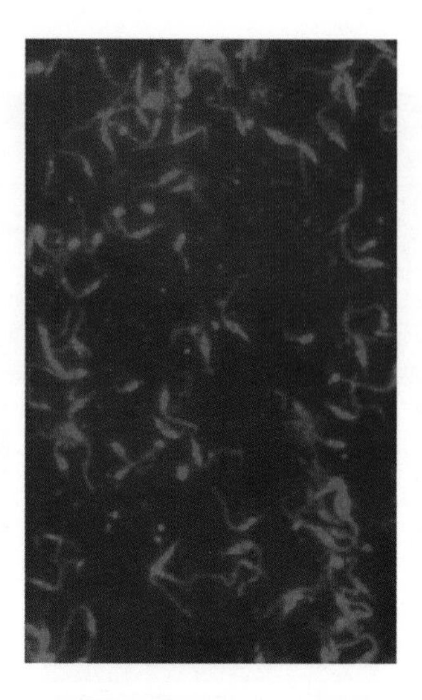

COLOR FIGURE 6.5 Immunofluorescence stain of "Y" strain *Trypanosoma cruzi* coated with the anti-Gp72 antibody, WIC 29.26 at 400× magnification.

COLOR FIGURE 7.1 Pupae of *Dendroctonus ponderosae* in fungal-lined pupal chambers.

COLOR FIGURE 7.3 Conidiophores (anamorph) of *Ophiostoma clavigerum* lining a pupal chamber of *Dendroctonus ponderosae.*

COLOR FIGURE 7.4 Ascomata (teleomorph) of *Ophiostoma montium* in parental galleries of *Dendroctonus ponderosae.*

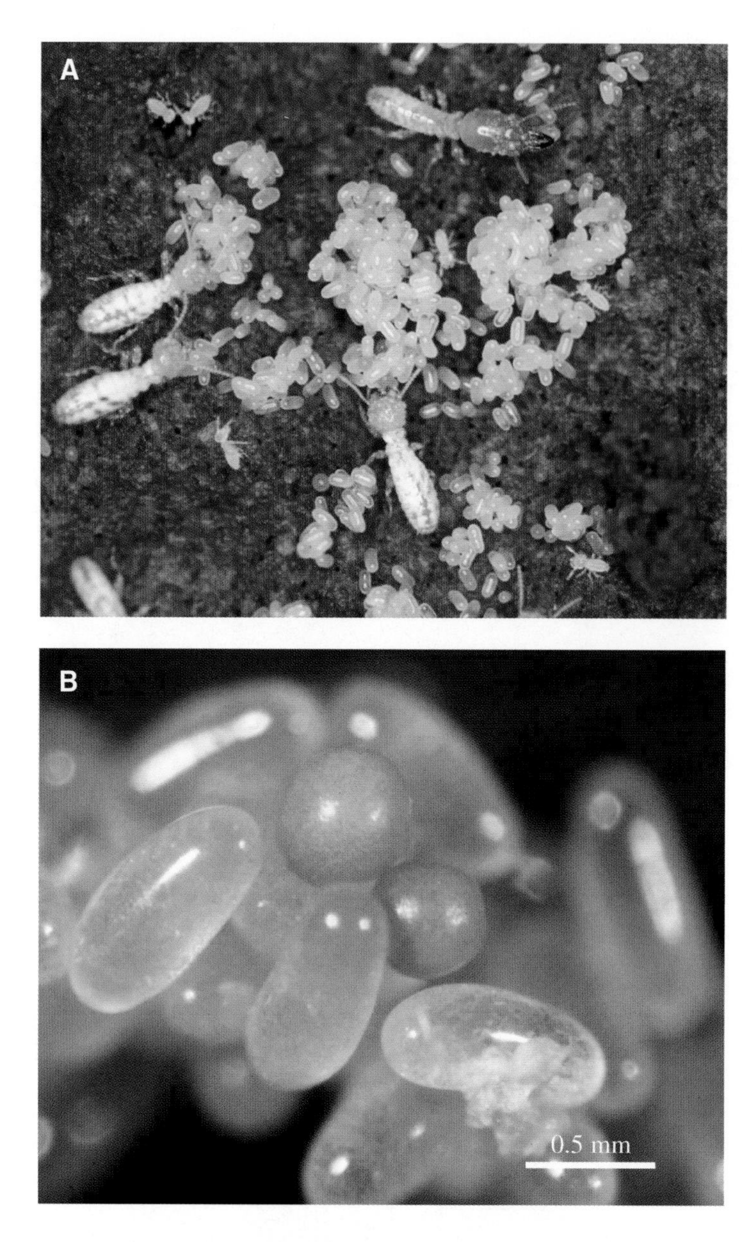

COLOR FIGURE 9.4 The "termite balls." (A) Termites unwittingly caring for termite balls, that is, egg-mimicking sclerotia of the fungus. (B) Close-up of the egg piles in the nursery nest. There are oval-shaped and transparent eggs and spherical-shaped brown termite balls.

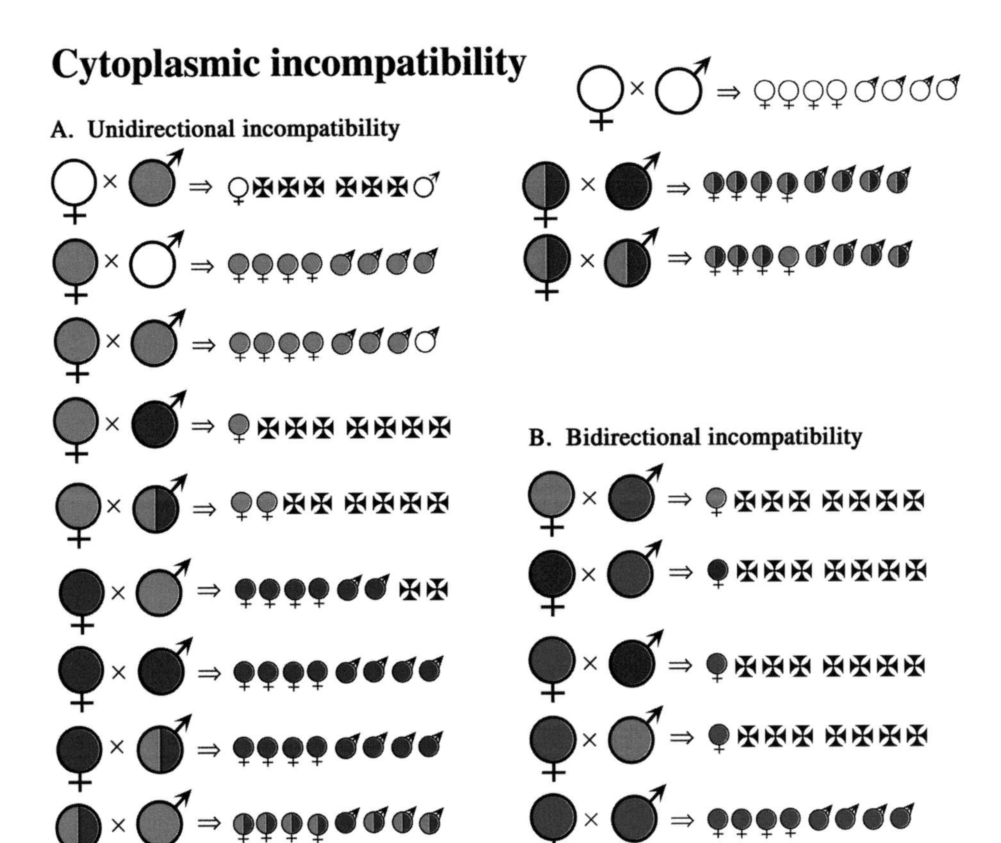

COLOR FIGURE 14.1 Crossing types of *Wolbachia* infections. Red, blue, and violet represent different *Wolbachia* strains. All infected strains are unidirectionally incompatible with uninfected strains; this is shown only for the red strain. The red and blue strains are unidirectionally incompatible with each other. Females of the red strain cannot rescue the modification of the blue strain, whereas females of the blue strain can rescue the modification of the red strain. The violet strain is bidirectionally incompatible with both the red and the blue strain; neither strain can rescue the other's modification. Females of the double-infected red–blue strain can rescue sperm from double-infected, both single-infected and the uninfected strains.

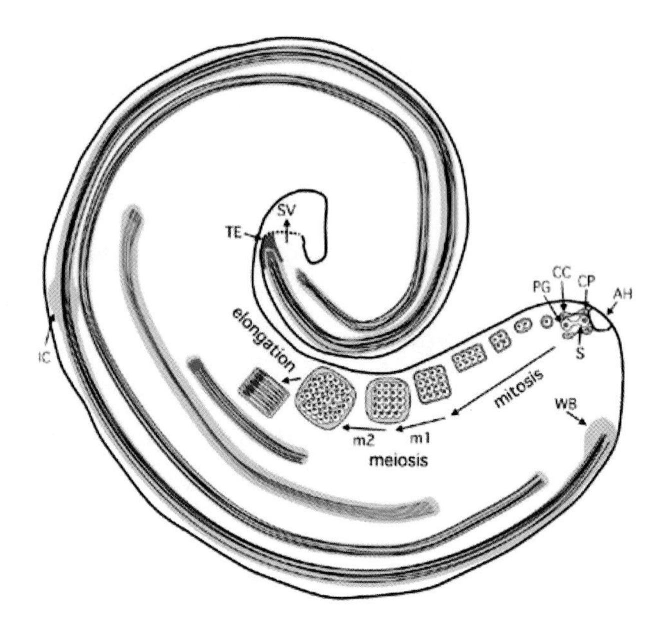

COLOR FIGURE 14.2 An idealized version of progressive stages of cyst development and their approximate location within an adult testis. Anchored to the apical hub (AH) are the gonial stem cells (S) as well as the cyst progenitor cells, which bud off to form a primary gonial cell (PG — white with red nuclei) surrounded by two somatically derived cyst cells (CC — gray). The primary gonial cell undergoes four rounds of mitotic division before entering meiosis. Cytokinesis is not complete in either mitosis or meiosis, the result being 64 interconnected haploid cells. The cyst cells (gray, nuclei not shown) do not undergo division; rather, they expand to form a continuous layer around the germ cells. Both cyst cells and germ cells comprise one cyst. Following the completion of meiosis, axoneme growth occurs, as the cyst elongates, eventually growing the entire length of the testis, with the sperm nuclei toward the seminal vesicle (SV). Individualization, the stripping away of the major mitochondrial derivative, as well as most cytoplasmic factors via the individualization complex (IC), a network of cytoskeletal factors, which is seen as a bulge proceeding along in the cyst pushing the stripped-away material into the waste bag (WB), a bulge in the tail end of the cyst. The nuclear end of the cyst becomes anchored to the terminal epithelium (TE), followed by the coiling of tightly packed sperm tails prior to liberation into the seminal vesicle.

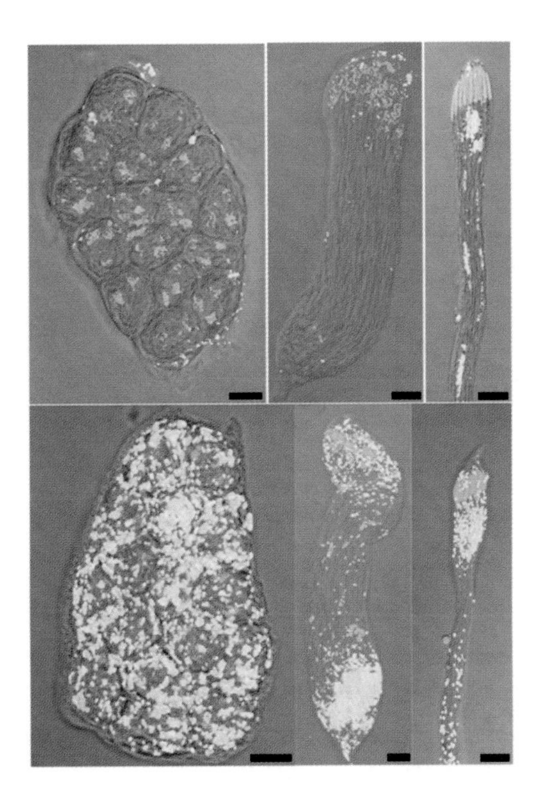

COLOR FIGURE 14.3 *Wolbachia* distribution and proliferation within cysts during meiosis and sperm elongation from *D. melanogaster* (DMB, top row) and *D. simulans* (DSR, bottom row). Each of three images shows progressively later stages of spermatogenesis: left, 16-cell spermatocyt; middle, beginning of spermatid elongation; right, apical region of fully elongated spermatid. DNA is shown as red (propidium iodide), and *Wolbachia* are seen as small punctate dots varying in color from yellow to green, depending on the relative staining intensity of the propidium and the FITC-labeled secondary antibody. In DMB at the 16-cell spermatocyst stage (top left), essentially no *Wolbachia* are present, while DSR (bottom left) display abundant *Wolbachia* (yellow–green), which are present in numbers great enough to obscure spermatocyte nuclei (red). During early stages of spermatid elongation (middle) DSR (bottom) contain a substantial number of *Wolbachia* at either end of the spermatid, while DMB (top) spermatids contain few, if any, bacteria. Fully elongated spermatids of both (right) contain *Wolbachia*, although DSR (bottom) have substantially greater numbers than DMB (top). All scale bars = 10 mm. [Adapted from Clark, M.E., Veneti, Z., Bourtzis, K., and Karr, T.L. (2002a). *Mech. Dev.* **111:** 3–15.]

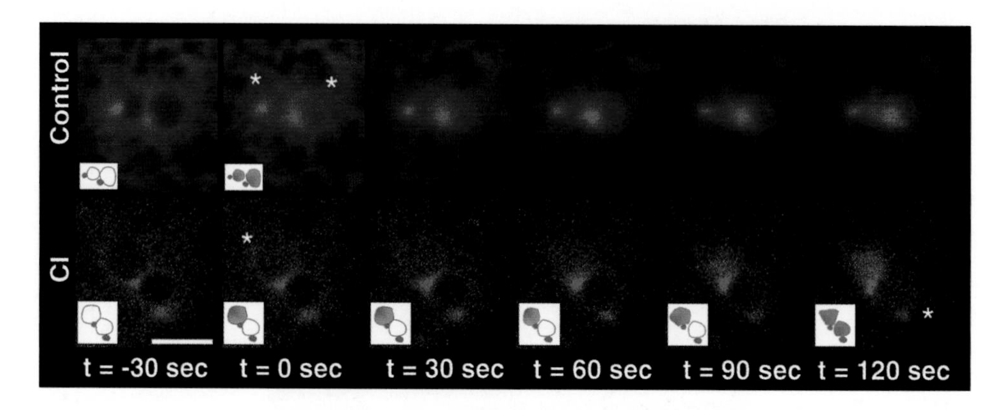

COLOR FIGURE 14.4 Nuclear-envelope breakdown (NEB) assessed by time-lapsed confocal microscopy of 0- to 1-h embryos injected with rhodamine-tubulin. When the nuclear envelope was intact, the nucleus appeared as a black circle surrounded by a ring of red (rhodamine-tubulin). During NEB rhodamine-tubulin invaded the nucleus (compare asterisks in upper and lower panels, $t = 0$ sec). The CI cross nucleus remains relatively devoid of tubulin (indicated by dark regions of the intact nucleus), indicating delayed NEB. Insets show schematized interpretation of NEB. [Adapted from Tram, U. and Sullivan, W. (2002). *Science* **296:** 1124–1126.]

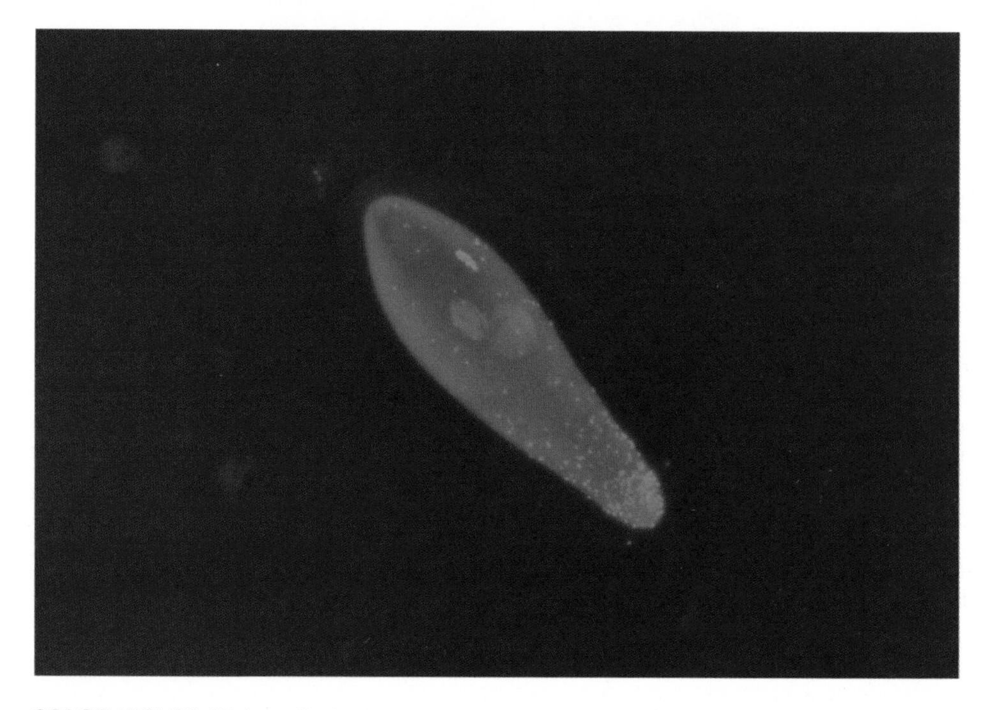

COLOR FIGURE 15.1 *Wolbachia* located at the posterior pole of a freshly laid *Trichogramma kaykai* egg. (Photo courtesy of Merijn Salverda.)

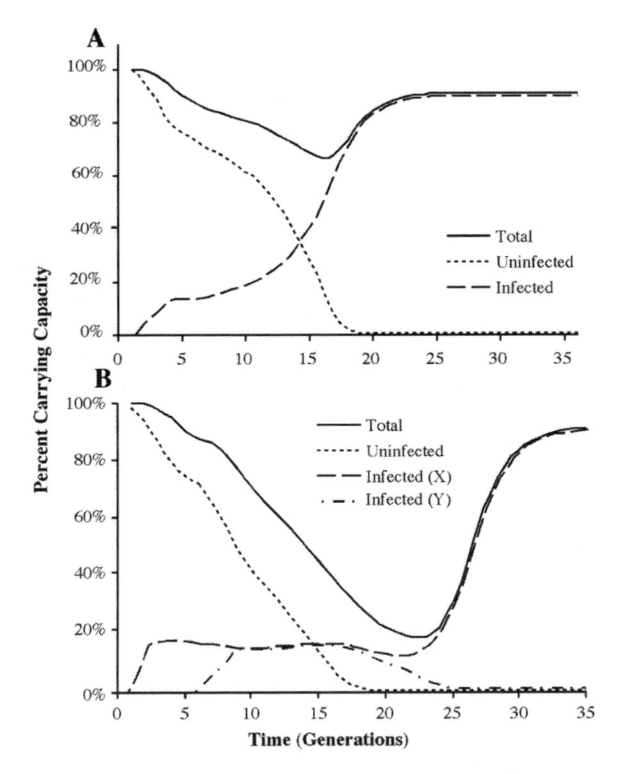

FIGURE 13.2 Model simulations examining (A) applied population replacement and (B) applied suppression strategies by cytoplasmically incompatible *Wolbachia* infections. (A) Following the introduction of *Wolbachia*-infected hosts into an uninfected population at generations one to three (one release/generation). The size of each release is equivalent to 5% of K_U (i.e., the carrying capacity of the uninfected host population) and consists of 50% females. The *Wolbachia* infection spreads into the host population in the subsequent generations, replacing the uninfected cytotype with the infected cytotype. The total population size decreases 33% to a minimum (N_{min}) during population replacement due primarily to the reduced hatching rates caused by CI. As the infection invades, the frequency of incompatible crosses decreases, and the total host population size recovers. However, the new carrying capacity of the infected population (K_I) is reduced relative to K_U due to CI and fecundity costs associated with *Wolbachia* infection. (B) This simulation is identical to that shown in (A), except that the host population size is further reduced by initiating releases of an additional, bidirectionally incompatible infection (Y). The additional releases of the Y infection are begun at generation six and repeated for a total of three generations (release size equivalent to 5% K_U per generation; 50% female). In this simulation, the host population is reduced by more than 83%. Host population size is shown as percent carrying capacity, which is calculated using the following formula: N_t/K_U, where N_t is the host population size at time t. For (A) and (B), the initial uninfected population is at carrying capacity; $\alpha = 0.00002$; $H_Z = 0.05$; $F_Z = 0.95$; and $\mu_Z = 0.03$. For (A), $R = 2.0$ and $\gamma = 1$.

public-health concerns associated with insecticide use and problems related to insecticide resistance. As discussed above, shortly after the description of CI in *C. pipiens* (common house mosquito), its potential applied use in insect-control strategies was recognized (Laven, 1967a). The early strategies were based on the release of cytoplasmically incompatible males, which would cause sterility when mated with field females. This research included field tests that successfully suppressed populations of *C. pipiens* by releasing cytoplasmically incompatible males (Laven, 1967a).

Despite the successful suppression of *C. pipiens* populations in field tests, work with this strategy was not continued due to political problems and scientific criticism. Scientific critics argued that the strategy was impractical due to the requirement that only incompatible males be released (Pal, 1974). With conventional SIT strategies, releases were designed to consist primarily of males for maximum efficiency (Knipling, 1998; Krafsur, 1998), but the absolute elimination of females from releases was not imperative to the success of the strategy. In contrast, females accidentally released as part of a CI strategy could permit the "released" *Wolbachia* infection type to become established in the host field population. Following this establishment, the efficacy of continued male releases for pest suppression would decline as the released infection spread through the field population, resulting in compatible crosses between field females and released males. Thus, the result would not be the eradication of the pest population but a transient suppression of the host population followed by population replacement, in which the original cytoplasm type (i.e., cytotype) was replaced with the cytotype of the released host strain. Although male-only releases were possible for small, pilot experiments (Laven, 1967a), the complete removal of females would not be practical on a scale required for large-area-control or eradication programs.

Subsequently, interest in the applied use of *Wolbachia*-induced CI has shifted from population-suppression to population-replacement strategies. Population-replacement strategies would employ the reproductive advantage afforded to infected females as a vehicle for spreading desired genes into a host population (Curtis, 1992; Turelli and Hoffmann, 1999; Sinkins and O'Neill, 2000). Instead of mass rearing and release of transgenic insects to replace field populations via simple dilution, population replacement via *Wolbachia* could drive desired transgenes (e.g., genes conferring refractoriness to pathogen transmission) into field populations from small release seedings. Furthermore, additive CI observed with *Wolbachia* superinfections (Sinkins et al., 1995; Dobson et al., 2001) suggests that *Wolbachia* population-replacement strategies can also be employed in host field populations that are naturally infected. The ability to generate superinfections via artificial transinfection (Braig et al., 1994; Rousset et al., 1999) has the potential to permit repeated population replacements within the same host population (Sinkins et al., 1995). However, population-replacement strategies have not been applied to date, due to the inability to genetically transform *Wolbachia* (Sinkins and O'Neill, 2000).

An additional strategy based on model predictions has been proposed for the suppression of insect populations using *Wolbachia* infections (Figure 13.2B). In brief, releases of *Wolbachia*-infected hosts can be used to artificially sustain an unstable coexistence that results when incompatible infections occur within different individuals in a single host population, allowing the host-population size to be reduced and maintained at low levels (Dobson et al., 2002a). Unlike previous CI-based suppression strategies, the new strategy would permit female releases. Simulations show that the vertical transmission of *Wolbachia* that occurs with infected female releases can permit multiple generations of control resulting from a single release, with the potential to increase cost efficacy. This novel suppression strategy employs the release of indigenous host insects and does not involve transgenic organisms, reducing technical and regulatory impediments to strategy implementation.

Additional applied strategies are based on the induction of host parthenogenesis by *Wolbachia* infections. Potential advantages of releasing parthenogenetic wasps as biological control agents are lower production costs associated with mass rearing of females, a higher growth rate in released populations, and improved ability to suppress the targeted pest population (Stouthamer, 1993).

CONCLUSION

Bacteria in the genus *Wolbachia* provide a useful model for examining the evolutionary interactions between host and maternally inherited symbionts, with examples ranging across the continuum between mutualism and parasitism. It is obvious that much work remains to be done in understanding currently known examples and with the description of additional infections and new types of reproductive manipulations. New avenues of research are possible with recently developed molecular techniques and genomic analyses, aiding the study of this fastidious bacterium. The attraction of additional researchers from other areas (e.g., *Drosophila* developmental biologists) will also facilitate progress.

In addition to understanding naturally occurring systems, research into the applied use of *Wolbachia* is also greatly needed. The resurgence of vector-borne diseases has led some to express pessimism at the prospect of control via existing means (Beaty, 2000; Aultman et al., 2001). Reasons for the intractability of arthropod-borne diseases to conventional control approaches are multifactorial and include a decline in the public-health infrastructure, the emergence of pesticide and drug resistance in arthropod vectors and parasites, and the legislated reduction of available pesticides due to environmental and public-health concerns. Thus, there is interest in the development of novel strategies to complement current control measures. Given recent advances that have moved the prospect of vector-replacement strategies into the realm of "near-term feasibility" (Handler and James, 2000; Aultman et al., 2001; Handler, 2001; Kokoza et al., 2001; Ito et al., 2002; Lycett and Kafatos, 2002), additional attention to appropriate and feasible *Wolbachia*-based population-replacement strategies is also needed.

REFERENCES

Anonymous. (1964). *Genetics of Vector and Insecticide Resistance*, p. 40. World Health Organization, Geneva.

Anonymous. (1975). Oh, New Delhi; Oh, Geneva. *Nature* **256:** 355–357.

Anonymous. (1976). WHO-supported collaborative research projects in India: the facts. *WHO Chronicle* **30:** 131–139.

Aultman, K.S., Beaty, B.J., and Walker, E.D. (2001). Genetically manipulated vectors of human disease: a practical overview. *Trends Parasitol* **17:** 507–509.

Bandi, C., Anderson, T.J.C., Genchi, C., and Blaxter, M.L. (1998). Phylogeny of *Wolbachia* in filarial nematodes. *Proc. R. Soc. London (B)* **265:** 2407–2413.

Bandi, C., Slatko, B., and O'Neill, S.L. (1999). *Wolbachia* genomes and the many faces of symbiosis. *Parasitol. Today* **15:** 428–429.

Bandi, C., Trees, A.J., and Brattig, N.W. (2001). *Wolbachia* in filarial nematodes: evolutionary aspects and implications for the pathogenesis and treatment of filarial diseases. *Vet. Parasitol.* **98:** 215–238.

Barr, A.R. (1966). Cytoplasmic incompatibility as a means of eradication of *Culex pipiens* L. *Proc. Calif. Mosquito Control Assoc.* **34:** 32–35.

Barr, A.R. (1982). The *Culex pipiens* complex. In *Recent Developments in the Genetics of Insect Disease Vectors* (K.S. Rai and S. Narcing), pp. 551–572. Stipes, Champaign, IL.

Bazzocchi, C., Jamnongluk, W., SL, O.N., Anderson, T.J.C., Genchi, C., and Bandi, C. (2000). *wsp* Gene sequences from the *Wolbachia* of filarial nematodes. *Curr. Microbiol.* **41:** 96–100.

Beaty, B.J. (2000). Genetic manipulation of vectors: a potential novel approach for control of vector-borne diseases. *Proc. Natl. Acad. Sci. U.S.A.* **97:** 10295–10297.

Beckett, E.B., Boothroyd, B., and Macdonald, W.W. (1978). A light and electron microscope study of rickettsia-like organisms in the ovaries of mosquitoes of the *A. scutellaris* group. *Ann. Trop. Med. Parasitol.* **72:** 277–283.

Bordenstein, S.R., O'Hara, F.P., and Werren, J.H. (2001). *Wolbachia*-induced incompatibility precedes other hybrid incompatibilities in *Nasonia*. *Nature* **409:** 707–710.

Bourtzis, K., Dobson, S.L., Braig, H.R., and O'Neill, S.L. (1998). Rescuing *Wolbachia* have been overlooked. *Nature* **391:** 852–853.

Bourtzis, K., Nirgianaki, A., Onyango, P., and Savakis, C. (1994). A prokaryotic *dnaA* sequence in *Drosophila melanogaster*: *Wolbachia* infection and cytoplasmic incompatibility among laboratory strains. *Insect Mol. Biol.* **3:** 131–142.

Braeckman, B., Simoens, C., Rzeznik, U., and Raes, H. (1997). Effect of sublethal doses of cadmium, inorganic mercury and methylmercury on the cell morphology of an insect cell line (*Aedes albopictus*, C6/36). *Cell Biol. Int.* **21:** 823–832.

Braig, H.R., Guzman, H., Tesh, R.B., and O'Neill, S.L. (1994). Replacement of the natural *Wolbachia* symbiont of *Drosophila simulans* with a mosquito counterpart. *Nature* **367:** 453–455.

Breeuwer, J.A.J. (1997). *Wolbachia* and cytoplasmic incompatibility in the spider mites *Tetranychus urticae* and *T. turkestani*. *Heredity* **79:** 41–47.

Breeuwer, J.A. and Werren, J.H. (1990). Microorganisms associated with chromosome destruction and reproductive isolation between two insect species. *Nature* **346:** 558–560.

Brower, J.H. (1976). Cytoplasmic incompatibility: occurrence in a stored product pest *Ephestia cautella. Ann. Entomol. Soc. Am.* **69:** 1011–1015.

Brower, J.H. (1979). Suppression of laboratory populations of *Ephestia cautella* (Walker) (Lepidoptera: Pyralidae) by release of males with cytoplasmic incompatibility. *J. Stored Prod. Res.* **15:** 1–4.

Callaini, G., Riparbelli, M.G., and Dallai, R. (1994). The distribution of cytoplasmic bacteria in the early *Drosophila* embryo is mediated by astral microtubules. *J. Cell Sci.* **107:** 673–682.

Callaini, G., Dallai, R., and Riparbelli, M.G. (1997). *Wolbachia*-induced delay of paternal chromatin condensation does not prevent maternal chromosomes from entering anaphase in incompatible crosses of *Drosophila simulans. J. Cell Sci.* **110:** 271–280.

Caspari, E. and Watson, G.S. (1959). On the evolutionary importance of cytoplasmic sterility in mosquitoes. *Evolution* **13:** 568–570.

Charlat, S., Calmet, C., and Mercot, H. (2001). On the mod resc model and the evolution of *Wolbachia* compatibility types. *Genetics* **159:** 1415–1422.

Curtis, C.F. (1992). Selfish genes in mosquitoes. *Nature* **357:** 450.

Curtis, C.F. and Von Borstol, R.C. (1978). Allegations against Indian research refuted. *Nature* **273:** 96.

Curtis, J. and Curtis, C.F. (1976). Mosquito programme. *New Sci.* **8 January:** 88–89.

Davidson, G. (1974). *Genetic Control of Insect Pests.* Academic Press, New York.

Dedeine, F., Vavre, F., Fleury, F., Loppin, B., Hochberg, M.E., and Boulétreau, M. (2001). Removing symbiotic Wolbachia bacteria specifically inhibits oogenesis in a parasitic wasp. *Proc. Natl. Acad. Sci. U.S.A.* **98:** 6247–6252.

Dobson, S.L. and Tanouye, M. (1998). Evidence for a genomic imprinting sex determination mechanism in *Nasonia vitripennis* (Hymenoptera: Chalcidoidea). *Genetics* **149:** 233–242.

Dobson, S.L., Marsland, E.J., and Rattanadechakul, W. (2001). *Wolbachia*-induced cytoplasmic incompatibility in single- and superinfected *Aedes albopictus* (Diptera: Culicidae). *J. Med. Entomol.* **38:** 382–387.

Dobson, S.L., Fox, C.W., and Jiggins, F.M. (2002a). The effect of *Wolbachia*-induced cytoplasmic incompatibility on host population size in natural and manipulated systems. *Proc. R. Soc. London (B)* **269:** 437–445.

Dobson, S.L., Marsland, E.J., and Rattanadechakul, W. (2002b). Mutualistic *Wolbachia* infection in *Aedes albopictus*: accelerating cytoplasmic drive. *Genetics* **160:** 1087–1094.

Ewald, P.W. (1987). Transmission modes and the evolution of the parasitism-mutualism continuum. *Ann. N.Y. Acad. Sci.* **503:** 295–306.

Fine, P.E.M. (1975). Vectors and vertical transmission: an epidemiologic perspective. *Ann. N.Y. Acad. Sci.* **266:** 173–194.

Fine, P.E.M. (1978). On the dynamics of symbiote-dependent cytoplasmic incompatibility in culicine mosquitoes. *J. Invertebr. Pathol.* **30:** 10–18.

Fleury, F., Vavre, F., Ris, N., Fouillet, P., and Boulétreau, M. (2000). Physiological cost induced by the maternally-transmitted endosymbiont *Wolbachia* in the *Drosophila* parasitoid *Leptopilina heterotoma. Parasitology* **121(Part 5):** 493–500.

Giordano, R., O'Neill, S.L., and Robertson, H.M. (1995). *Wolbachia* infections and the expression of cytoplasmic incompatibility in *Drosophila sechellia* and *D. mauritiana. Genetics* **140:** 1307–1317.

Glover, D.M., Raff, J., Karr, T.L., O'Neill, S.L., Lin, H., and Wolfner, M.F. (1990). Parasites in *Drosophila* embryos. *Nature* **348:** 117.

Guillemaud, T., Pasteur, N., and Rousset, F. (1997). Contrasting levels of variability between cytoplasmic genomes and incompatibility types in the mosquito *Culex pipiens. Proc. R. Soc. London (B)* **264:** 245–251.

Handler, A.M. (2001). A current perspective on insect gene transformation. *Insect Biochem. Mol. Biol.* **31:** 111–128.

Handler, A.M. and James, A.A. (2000). *Insect Transgenesis: Methods and Applications.* CRC Press, Boca Raton, FL.

Hanlon, J. (1975). Germ-war allegations force WHO out of Indian mosquito project. *New Sci.* **68:** 102–103.

Hatcher, M.J. (2000). Persistence of selfish genetic elements: population structure and conflict. *Trends Ecol. Evol.* **15:** 271–277.

Hertig, M. (1936). The Rickettsia. *Wolbachia pipientis* (Gen. et Sp.N.) and associated inclusions in the mosquito, *Culex pipiens*. *Parasitology* **28:** 453–486.

Hertig, M. and Wolbach, S.B. (1924). Studies on rickettsia-like micro-organisms in insects. *J. Med. Res.* **44:** 329–374.

Hoerauf, A., Nissen-Pahle, K., Schmetz, C., Henkle-Duhrsen, K., Blaxter, M.L., Buttner, D.W., Gallin, M.Y., Al-Qaoud, K.M., Lucius, R., and Fleischer, B. (1999). Tetracycline therapy targets intracellular bacteria in the filarial nematode *Litomosoides sigmodontis* and results in filarial infertility. *J. Clin. Invest.* **103:** 11–18.

Hoerauf, A., Volkmann, L., Hamelmann, C., Adjei, O., Autenrieth, I.B., Fleischer, B., and Buttner, D.W. (2000). Endosymbiotic bacteria in worms as targets for a novel chemotherapy in filariasis. *Lancet* **355:** 1242–1243.

Hoerauf, A., Mand, S., Adjei, O., Fleischer, B., and Buttner, D.W. (2001). Depletion of *Wolbachia* endobacteria in *Onchocerca volvulus* by doxycycline and microfilaridermia after ivermectin treatment. *Lancet* **357:** 1415–1416.

Hoffmann, A.A. and Turelli, M. (1997). Cytoplasmic incompatibility in insects. In *Influential Passengers: Inherited Microorganisms and Arthropod Reproduction* (S.L. O'Neill, A.A. Hoffmann, and J.H. Werren, Eds.), pp. 42–80. Oxford University Press, Oxford, U.K.

Hoffmann, A.A., Turelli, M., and Simmons, G.M. (1986). Unidirectional incompatibility between populations of *Drosophila simulans*. *Evolution* **40:** 692–701.

Hoffmann, A.A., Turelli, M., and Harshman, L.G. (1990). Factors affecting the distribution of cytoplasmic incompatibility in *Drosophila simulans*. *Genetics* **126:** 933–948.

Hoffmann, A.A., Clancy, D., and Duncan, J. (1996). Naturally-occurring *Wolbachia* infection in *Drosophila simulans* that does not cause cytoplasmic incompatibility. *Heredity* **76:** 1–8.

Hsiao, C. and Hsiao, T.H. (1985a). Rickettsia as the cause of cytoplasmic incompatibility in the alfalfa weevil, *Hypera postica*. *J. Invertebr. Pathol.* **45:** 244–246.

Hsiao, T.H. and Hsiao, C. (1985b). Hybridization and cytoplasmic incompatibility among alfalfa weevil strains. *Entomol. Exp. Appl.* **37:** 155–159.

Huigens, M.E., Luck, R.F., Klaassen, R.H.G., Maas, M.F.P.N., Timmermans, M.J.T.N., and Stouthamer, R. (2000). Infectious parthenogenesis. *Nature* **405:** 178–179.

Hurst, G.D.D., Jiggins, F.M., von der Schulenburg, J.H.G., Bertrand, D., West, S.A., Goriacheva, I.I., Zakharov, I.A., Werren, J.H., Stouthamer, R., and Majerus, M.E.N. (1999). Male-killing *Wolbachia* in two species of insect. *Proc. R. Soc. London (B)* **266:** 735–740.

Hurst, L.D. (1991a). The evolution of cytoplasmic incompatibility or when spite can be successful. *J. Theor. Biol.* **148:** 269–277.

Hurst, L.D. (1991b). The incidences and evolution of cytoplasmic male killers. *Proc. R. Soc. London (B)* **224:** 91–99.

Hurst, L.D. and McVean, G.T. (1996). Clade selection, reversible evolution and the persistence of selfish elements: the evolutionary dynamics of cytoplasmic incompatibility. *Proc. R. Soc. London (B)* **263:** 97–104.

Ito, J., Ghosh, A., Moreira, L.A., Wimmer, E.A., and Jacobs Lorena, M. (2002). Transgenic anopheline mosquitoes impaired in transmission of a malaria parasite. *Nature* **417:** 452–455.

James, A.C., Dean, M.D., McMahon, M.E., and Ballard, J.W.O. (2002). Dynamics of double and single *Wolbachia* infections in *Drosophila simulans* from New Caledonia. *Heredity* **88:** 182–189.

Jeyaprakash, A. and Hoy, M.A. (2000). Long PCR improves *Wolbachia* DNA amplification: wsp sequences found in 76% of sixty-three arthropod species. *Insect Mol. Biol.* **9:** 393–405.

Jiggins, F.M., Bentley, J.K., Majerus, M.E.N. and Hurst, G.D.D. (2001a). How many species are infected with *Wolbachia*? Cryptic sex ratio distorters revealed to be common by intensive sampling. *Proc. R. Soc. London (B)* **268:** 1123–1126.

Jiggins, F.M., von der Schulenburg, J.H.G., Hurst, G.D.D., and Majerus, M.E.N. (2001b). Recombination confounds interpretations of *Wolbachia* evolution. *Proc. R. Soc. London (B)* **268:** 1423–1427.

Jochemsen, P., Schilthuizen, M., and Stouthamer, R. (1998). Transmission of *Wolbachia* between Trichogramma species and their lepidopteran hosts? *Proc. Sect. Exp. Appl. Entomol. Neth. Entomol. Soc.* **9:** 131–135.

Johanowicz, D.L. and Hoy, M.A. (1998). Experimental induction and termination of non-reciprocal reproductive incompatibilities in a parahaploid mite. *Entomol. Exp. Appl.* **87:** 51–58.

Karr, T.L. (1991). Intracellular sperm/egg interactions in Drosophila: a three-dimensional structural analysis of a paternal product in the developing egg. *Mech. Dev.* **34:** 101–111.

Kellen, W.R., Hoffmann, D.F., and Kwock, R.A. (1981). *Wolbachia* sp. (Rickettsiales: Rickettsiaceae) a symbiont of the almond moth *Ephestia cautella* ultrastructure and influence on host fertility. *J. Invertebr. Pathol.* **37:** 273–283.

Knipling, E. (1955). Possibilities of insect control or eradication through the use of sexually sterile males. *J. Econ. Entomol.* **48:** 459–462.

Knipling, E.F. (1998). Sterile insect and parasite augmentation techniques: unexploited solutions for many insect pest problems. *Fla. Entomol.* **81:** 134–160.

Kokoza, V.A., Martin, D., Mienaltowski, M.J., Ahmed, A., Morton, C.M., and Raikhel, A.S. (2001). Transcriptional regulation of the mosquito vitellogenin gene via a blood meal-triggered cascade. *Gene* **274:** 47–65.

Kose, H. and Karr, T.L. (1995). Organization of *Wolbachia pipientis* in the *Drosophila* fertilized egg and embryo revealed by an anti-*Wolbachia* monoclonal antibody. *Mech. Dev.* **51:** 275–288.

Krafsur, E.S. (1998). Sterile insect technique for suppressing and eradicating insect population: 55 years and counting. *J. Agric. Entomol.* **15:** 303–317.

Langworthy, N.G., Renz, A., Mackenstedt, U., Henkle Duhrsen, K., Bronsvoort, M.B.D., Tanya, V.N., Donnelly, M.J., and Trees, A.J. (2000). Macrofilaricidal activity of tetracycline against the filarial nematode *Onchocerca ochengi*: elimination of *Wolbachia* precedes worm death and suggests a dependent relationship. *Proc. R. Soc. London (B)* **267:** 1063–1069.

Larsson, R. (1983). A rickettsia-like microorganism similar to *Wolbachia pipientis* and its occurrence in *Culex* mosquitoes. *J. Invertebr. Pathol.* **41:** 387–390.

Laven, H. (1967a). Eradication of *Culex pipiens fatigans* through cytoplasmic incompatibility. *Nature* **216:** 383–384.

Laven, H. (1967b). A possible model for speciation by cytoplasmic isolation in the *Culex pipiens* complex. *Bull. World Health Org.* **37:** 263–266.

Laven, H. (1967c). Speciation and evolution in *Culex pipiens*. In *Genetics of Insect Vectors of Disease* (J. Wright and R. Pal, Eds.), pp. 251–275. Elsevier, Amsterdam.

Louis, C. and Nigro, L. (1989). Ultrastructural evidence of *Wolbachia rickettsiales* in *Drosophila simulans* and their relationships with unidirectional cross incompatibility. *J. Invertebr. Pathol.* **54:** 39–44.

Lycett, G.J. and Kafatos, F.C. (2002). Medicine: anti-malarial mosquitoes? *Nature* **417:** 387–388.

Majerus, M.E.N., Hinrich, J., Schulenburg, G.V.D., and Zakharov, I.A. (2000). Multiple causes of male-killing in a single sample of the two-spot ladybird, *Adalia bipunctata* (Coleoptera: Coccinellidae) from Moscow. *Heredity* **84:** 605–609.

Merçot, H., Llorente, B., Jacques, M., Atlan, A., and Montchamp-Moreau, C. (1995). Variability within the Seychelles cytoplasmic incompatibility system in *Drosophila simulans*. *Genetics* **141:** 1015–1023.

Merçot, H. and Poinsot, D. (1998) [Rescuing *Wolbachia* have been overlooked] and discovered on Mount Kilimanjaro. *Nature* **391:** 853.

Michel-Salzat, A., Cordaux, R., and Bouchon, D. (2001). *Wolbachia* diversity in the *Porcellionides pruinosus* complex of species (Crustacea : Oniscidea): evidence for host-dependent patterns of infection. *Heredity* **87 (Part 4):** 428–434.

Ndiaye, M. and Mattei, X. (1993). Endosymbiotic relationship between a rickettsia-like microorganism and the male germ-cells of *Culex tigripes*. *J. Submicrosc. Cytol. Pathol.* **25:** 71–77.

Ndiaye, M., Mattei, X., and Thiaw, O.T. (1995). Extracellular and intracellular rickettsia-like microorganisms in gonads of mosquitoes. *J. Submicrosc. Cytol. Pathol.* **27:** 557–563.

Noda, H. (1984). Cytoplasmic incompatibility in allopatric field populations of the small brown planthopper, *Laodelphax striatellus*, in Japan. *Entom. Exp. Appl.* **35:** 263–267.

O'Neill, S.L. (1989). Cytoplasmic symbionts in *Tribolium confusum*. *J. Invertebr. Pathol.* **53:** 132–134.

O'Neill, S.L. (1999). *Wolbachia*: why these bacteria are important to genome research. *Microb. Comp. Genomics* **4:** 159.

O'Neill, S.L. and Karr, T.L. (1990). Bidirectional incompatibility between conspecific populations of *Drosophila simulans*. *Nature* **348:** 178–180.

O'Neill, S.L., Giordano, R., Colbert, A.M., Karr, T.L., and Robertson, H.M. (1992). 16S *rRNA* phylogenetic analysis of the bacterial endosymbionts associated with cytoplasmic incompatibility in insects. *Proc. Natl. Acad. Sci. U.S.A.* **89:** 2699–2702.

O'Neill, S.L., Pettigrew, M., Sinkins, S.P., Braig, H.R., Andreadis, T.G., and Tesh, R.B. (1997). *In vitro* cultivation of *Wolbachia pipientis* in an *Aedes albopictus* cell line. *Insect Mol. Biol.* **6:** 33–39.

Pal, R. (1974). WHO, ICMR programme of genetic control of mosquitoes in India. In *The Use of Genetics in Insect Control* (R. Pal and M.J. Whitten, Eds.), pp. 73–95. Elsevier North-Holland, Amsterdam.

Pal, R. and LaChance, L.E. (1974). The operational feasibility of genetic methods for control of insects of medical and veterinary importance. *Annu. Rev. Entomol.* **19:** 269–291.

Pal, R. and Whitten, M.J. (1974). *The Use of Genetics in Insect Control.* Elsevier North-Holland, Amsterdam.

Perrot-Minnot, M., Guo, L.R., and Werren, J.H. (1996). Single and double infections with *Wolbachia* in the parasitic wasp *Nasonia vitripennis*: effects on compatibility. *Genetics* **143:** 961–972.

Popov, V.L., Han, V.C., Chen, S.M., Dumler, J. S., Feng, H.M., Andreadis, T.G., Tesh, R.B., and Walker, D.H. (1998). Ultrastructural differentiation of the genogroups in the genus *Ehrlichia. J. Med. Microbiol.* **47:** 235–251.

Presgraves, D.C. (2000). A genetic test of the mechanism of *Wolbachia*-induced cytoplasmic incompatibility in *Drosophila. Genetics* **154:** 771.

Prout, T. (1994). Some evolutionary possibilities for a microbe that causes incompatibility in its host. *Evolution* **48:** 909–911.

Richardson, P.M., Holmes, W.P., and Saul, G.B., II. (1987). The effect of tetracycline on non-reciprocal cross incompatibility in *Mormoniella vitripennis (Nasonia vitripennis). J. Invertebr. Pathol.* **50:** 176–183.

Rigaud, T. and Juchault, P. (1993). Conflict between feminizing sex ratio distorters and an autosomal masculinizing gene in the terrestrial isopod *Armadillidium vulgare* Latr. *Genetics* **133:** 247–252.

Rousset, F. and de Stordeur, E. (1994). Properties of *Drosophila simulans* strains experimentally infected by different clones of the bacterium *Wolbachia. Heredity* **72:** 325–331.

Rousset, F. and Solignac, M. (1995). Evolution of single and double *Wolbachia* symbioses during speciation in the *Drosophila simulans* complex. *Proc. Natl. Acad. Sci. U.S.A.* **92:** 6389–6393.

Rousset, F., Braig, H.R., and O'Neill, S.L. (1999). A stable triple *Wolbachia* infection in *Drosophila* with nearly additive incompatibility effects. *Heredity* **82:** 620–627.

Rousset, F., Bouchon, D., Pintureau, B., Juchault, P., and Solignac, M. (1992). *Wolbachia* endosymbionts responsible for various alterations of sexuality in arthropods. *Proc. R. Soc. London (B)* **250:** 91–98.

Roux, V. and Raoult, D. (1995). Phylogenetic analysis of the genus *Rickettsia* by 16S rDNA sequencing. *Res. Microbiol.* **146:** 385–396.

Rozeboom, L.E. and Kitzmiller, J.B. (1958). Hybridization and speciation in mosquitoes. *Annu. Rev. Entomol.* **3:** 231–248.

Ryan, S.L. and Saul, G.B., II. (1968). Post-fertilization effect of incompatibility factors in *Mormoniella. Mol. Gen. Genet.* **103:** 29–36.

Saul, G.B., II. (1961). An analysis of non-reciprocal cross incompatibility in *Mormoniella vitripennis* (Walker). *Z. Vererbungsl.* **92:** 28–33.

van Meer, M.M.M., Witteveldt, J., and Stouthamer, R. (1999). Phylogeny of the arthropod endosymbiont *Wolbachia* based on the *wsp* gene. *Insect Mol. Biol.* **8:** 399–408.

Shoemaker, D.D., Katju, V., and Jaenike, J. (1999). *Wolbachia* and the evolution of reproductive isolation between *Drosophila recens* and *Drosophila subquinaria. Evolution* **53:** 1157–1164.

Sinkins, S.P. and O'Neill, S.L. (2000). *Wolbachia* as a vehicle to modify insect populations. In *Insect Transgenesis: Methods and Applications* (A.M. Handler and A.A. James, Eds.), pp. 271–287. CRC Press, Boca Raton, FL.

Sinkins, S.P., Braig, H.R., and O'Neill, S.L. (1995). *Wolbachia* superinfections and the expression of cytoplasmic incompatibility. *Proc. R. Soc. London (B)* **261:** 325–330.

Slatko, B.E., O'Neill, S.L., Scott, A.L., Werren, J.L., and Blaxter, M.L. (1999). The *Wolbachia* genome consortium. *Microb. Comp. Genomics* **4:** 161–165.

Solignac, M., Vautrin, D., and Rousset, F. (1994). Widespread occurrence of the proteobacteria *Wolbachia* and partial cytoplasmic incompatibility in *Drosophila melanogaster. C. R. Acad. Sci. Paris* **317:** 461–470.

Stanley, J. (1961). Sterile crosses between mutations of *Tribolium confusum* Duv. *Nature* **191:** 934.

Stevens, L. (1989). Environmental factors affecting reproductive incompatibility in flour beetles, genus *Tribolium. J. Invertebr. Pathol.* **53:** 78–84.

Stevens, L. and Wade, M.J. (1990). Cytoplasmically inherited reproductive incompatibility in *Tribolium* flour beetles: the rate of spread and effect on population size. *Genetics* **124:** 367–372.

Stevens, L. and Wicklow, D.T. (1992). Multispecies interactions affect cytoplasmic incompatibility in *Tribolium* flour beetles. *Am. Nat.* **140:** 642–653.

Stouthamer, R. (1993). The use of sexual versus asexual wasps in biological control. *Entomophaga* **38:** 3–6.

Stouthamer, R., Luck, R.F., and Hamilton, W.D. (1990). Antibiotics cause parthenogenic *Trichogramma* (Hymenoptera: Trichogrammatidae) to revert to sex. *Proc. Natl. Acad. Sci. U.S.A.* **87:** 2424–2427.

Stouthamer, R., Breeuwer, J.A., Luck, R.F., and Werren, J.H. (1993). Molecular identification of microorganisms associated with parthenogenesis. *Nature* **361:** 66–68.

Stouthamer, R., van Tilborg, M., de Jong, J.H., Nunney, L., and Luck, R.F. (2001). Selfish element maintains sex in natural populations of a parasitoid wasp. *Proc. R. Soc. London (B)* **268:** 617–622.

Taylor, M.J., Bilo, K., Cross, H.F., Archer, J.P., and Underwood, A.P. (1999). 16S rDNA phylogeny and ultrastructural characterization of *Wolbachia* intracellular bacteria of the filarial nematodes *Brugia malayi*, *B. pahangi*, and *Wuchereria bancrofti*. *Exp. Parasitol.* **91:** 356–361.

Taylor, M.J., Cross, H.F., Ford, L., Makunde, W.H., Prasad, G.B.K.S., and Bilo, K. (2001). *Wolbachia* bacteria in filarial immunity and disease. *Parasite Immunol.* **23:** 401–409.

Tomiche, F.J. (1975). The WHO and mosquitoes. *Nature* **257:** 175.

Tram, U. and Sullivan, W. (2002). Role of delayed nuclear envelope breakdown and mitosis in *Wolbachia*-induced cytoplasmic incompatibility. *Science* **296:** 1124–1126.

Trpis, M., Perrone, J.B., Reisseg, M., and Parker, K.L. (1981). Control of cytoplasmic incompatibility in the *Aedes scutellaris* complex. *J. Heredity* **72:** 313–317.

Turelli, M. (1994). Evolution of incompatibility-inducing microbes and their hosts. *Evolution* **48:** 1500–1513.

Turelli, M. and Hoffmann, A.A. (1991). Rapid spread of an inherited incompatibility factor in California *Drosophila*. *Nature* **353:** 440–442.

Turelli, M. and Hoffmann, A.A. (1995). Cytoplasmic incompatibility in *Drosophila simulans*: dynamics and parameter estimates from natural populations. *Genetics* **140:** 1319–1338.

Turelli, M. and Hoffmann, A.A. (1999). Microbe-induced cytoplasmic incompatibility as a mechanism for introducing transgenes into arthropod populations. *Insect Mol. Biol.* **8:** 243–255.

Vavre, F., Fleury, F., Lepetit, D., Fouillet, P., and Bouletreau, M. (1999). Phylogenetic evidence for horizontal transmission of *Wolbachia* in host-parasitoid associations. *Mol. Biol. Evol.* **16:** 1711–1723.

von der Schulenburg, J.H., Hurst, G.D., Huigens, T.M., van Meer, M.M., Jiggins, F.M., and Majerus, M.E. (2000). Molecular evolution and phylogenetic utility of *Wolbachia* ftsZ and wsp gene sequences with special reference to the origin of male-killing. *Mol. Biol. Evol.* **17:** 584–600.

von der Schulenburg, J.H.G., Habig, M., Sloggett, J.J., Webberley, K.M., Bertrand, D., Hurst, G.D.D., and Majerus, M.E.N. (2001). Incidence of male-killing *Rickettsia* spp. (α-proteobacteria) in the ten-spot ladybird beetle *Adalia decempunctata* L. (Coleoptera: Coccinellidae). *Appl. Environ. Microbiol.* **67:** 270–277.

Wade, M.J. (2001). Infectious speciation. *Nature* **409:** 675–677.

Wade, M.J. and Chang, N.W. (1995). Increased male fertility in *Tribolium confusum* beetles after infection with the intracellular parasite *Wolbachia*. *Nature* **373:** 72–74.

Wade, M.J. and Stevens, L. (1985). Microorganism mediated reproductive isolation in flour beetles (Genus *Tribolium*). *Science* **227:** 527–528.

Walgate, R. (1978). Research in third world countries: pugwash plans controls. *Nature* **272:** 8–9.

Werren, J.H. (1987). The coevolution of autosomal and cytoplasmic sex ratio factors. *J. Theor. Biol.* **124:** 317–334.

Werren, J.H. (1997). Biology of *Wolbachia*. *Annu. Rev. Entomol.* **42:** 587–609.

Werren, J.H. and Bartos, J.D. (2001). Recombination in *Wolbachia*. *Curr. Biol.* **11:** 431–435.

Werren, J.H., Windsor, D., and Guo, L.R. (1995a). Distribution of *Wolbachia* among neotropical arthropods. *Proc. R. Soc. London (B)* **262:** 197–204.

Werren, J.H. and Windsor, D.M. (2000). *Wolbachia* infection frequencies in insects: evidence of a global equilibrium? *Proc. R. Soc. London (B)* **267:** 1277–1285.

Werren, J.H., Zhang, W., and Guo, L.R. (1995b). Evolution and phylogeny of *Wolbachia*: reproductive parasites of arthropods. *Proc. R. Soc. London (B)* **261:** 55–63.

Wood, R.J. (1975). Mosquitoes. *Nature* **258:** 102.

Wright, J.D. and Barr, A.R. (1980). The ultrastructure and symbiotic relationships of *Wolbachia* of mosquitoes of the *Aedes scutellaris* group. *J. Ultrastruct. Res.* **72:** 52–64.

Wright, J.D., Sjöstrand, F.S., Portaro, J.K., and Barr, A.R. (1978). The ultrastructure of the Rickettsia-like microorganism *Wolbachia pipientis* and associated virus-like bodies in the mosquito *Culex pipiens*. *J. Ultrastruct. Res.* **63:** 79–85.

Yen, J.H. (1975). Transovarial transmission of rickettsia like microorganisms in mosquitoes. *Ann. N.Y. Acad. Sci.* **266:** 152–161.

Yen, J.H. and Barr, A.R. (1971). New hypothesis of the cause of cytoplasmic incompatibility in *Culex pipiens* L. *Nature* **232:** 657–658.

Yen, J.H. and Barr, A.R. (1974). Incompatibility in *Culex pipiens*. In *The Use of Genetics in Insect Control* (R. Pal and M.J. Whitten, Eds.), pp. 97–118. Elsevier North-Holland, Amsterdam.

Zakharov, I.A., Goryacheva, I.I., Shaikevich, E.V., Schulenburg, J.H., and Majerus, M.E.N. (2000). *Wolbachia*, a new bacterial agent causing sex-ratio bias in the two-spot ladybird *Adalia bipunctata* L. *Russ. J. Genet.* **36:** 385–388.

Zchori-Fein, E., Roush, R.T., and Hunter, M.S. (1992). Male production induced by antibiotic treatment in *Encarsia formosa* (Hymenoptera: Aphelinidae), an asexual species. *Experientia* **48:** 102–105.

Zhou, W., Rousset, F., and O'Neill, S.L. (1998). Phylogeny and PCR based classification of *Wolbachia* strains using *wsp* gene sequences. *Proc. R. Soc. London (B)* **265:** 509–515.

14 Cytoplasmic Incompatibility

Kostas Bourtzis, Henk R. Braig, and Timothy L. Karr

CONTENTS

INTRODUCTION

Cytoplasmic incompatibility (CI) is the most widespread and one of the most prominent features that *Wolbachia* endosymbionts impose on their hosts. In this chapter, we try to demonstrate that CI is not a peculiar phenotype of an obscure intracellular bacterium but a fundamental evolutionary trait. The distribution and mechanisms of *Wolbachia*-induced CI phenotypes are exemplified in the model *Drosophila*. The genetics and cell biology of this host–pathogen interaction will offer a better understanding of the early stages of fertilization and sterility.

CI SYSTEMS

CI is often seen as a mechanism that promotes the spread of the cytoplasmic factors causing it. This, however, is not accomplished, as one might expect, by increasing the likelihood of transmission of the cytoplasmic factors but rather by decreasing the proportion of progeny without those factors. CI functions as a wide variety of post-segregational killing (PSK) mechanisms or post-disturbance cell-killer systems that will kill whatever is not carrying the cytoplasmic factors. Since cells not having obtained or having lost these cytoplasmic factors are punished for withdrawal of a dispensable genetic element with no intrinsic adaptive value, these systems are also called addiction systems or modules.

0-8493-1286-8/03/$0.00+$1.50

CI systems are usually associated with means of horizontal transmission but have sometimes made their way (back) onto the chromosome and perhaps into the nucleus, so that there now exists a continuum of addiction systems ranging from plasmid systems, bacterial chromosomal systems, mitochondrial, and, perhaps, chloroplast systems in plants, chromosomal invertebrate systems, and *Wolbachia*-based arthropod systems. Table 14.1 provides an overview. The plasmid systems are the best-studied systems at the molecular level, whereas *Wolbachia*-induced CI is the most prominently investigated system at the phenotypic and population levels. CI systems are also known as toxin–antitoxin, poison–antidote, and modification–rescue systems. When the cytoplasmic factors are genes, they are also described as cytoplasmic genes, extrachromosomal genes, extranuclear genes, plasmagenes, or plasmons. These genes may be part of plasmids, viruses, endosymbionts, organelles, or chromosomes. Because cytoplasm is almost universally inherited through the female sex only, CI can be nuclear coded as long as the respective factors are expressed in the cytoplasm of the egg cell. This inevitable and exclusive association with the maternal cytoplasm during reproduction predestines these systems to manipulate the sex ratio of their hosts.

Intracellular Systems

Intracellular systems have evolved in at least three very disparate backgrounds. Intracellular addiction systems are very widespread in bacteria and presumably part of the great majority of plasmids, especially low-copy-number plasmids. They have also been described for plasmids of archaea, suggesting that the pairing of toxin and antitoxin in a functional cassette can be regarded as a very basic and successful evolutionary invention. A nuclear-coded cytoplasmic modification-and-rescue system called *Medea* has been detected in wild populations of a beetle species. So far it is unique in invertebrates, but very likely this only reflects our ignorance of invertebrate natural history. Incompatibility systems become obvious only in crosses between different populations. However, only in exceptional cases are different wild populations reared in the same lab and then mated with each other. For this simple reason most CI systems will escape detection. Host-based CI seems to be rare in invertebrates, but a unique system of cytoplasmic male sterility is common in plants. With the exception of a possible mutation in an inbred laboratory mouse strain called *scat*$^+$, CI systems have so far not been described in vertebrates.

Addiction Systems

The two essential components of intracellular systems in prokaryotes are a stable, long-lived toxin, poison, or modification factor expressed at low levels and a labile, short-lived antitoxin, antidote, or rescue factor expressed at high levels. The survival of the plasmid-bearing cells requires an autoregulation of the toxin expression (Engelberg-Kulka and Glaser, 1999). The modification factors are usually small proteins, but the rescue factors can be small, unstable antisense RNA molecules as well. The system relies on the fact that a progeny cell will always inherit cytoplasm during a cell division. Should a cell fail to inherit the cytoplasmic gene, a plasmid, for example, it will still have inherited the toxin–antitoxin components. Because the antitoxin is shorter lived than the toxin, this cell will eventually end up with the more stable component, the toxin, which will then kill the cell, as the doomed cell does not have the gene to synthesize the antitoxin. The short half-life of the antitoxins may be due to a low thermodynamic stability that keeps the antitoxins close to unfolding, after which they are rapidly degraded by cellular proteases such as Lon, ClpPX, ClpPA, or RNAses where the rescue factors are antisense RNAs. Originally, it was thought that these systems would directly stabilize the vertical transmission of the plasmids carrying them by adding an adaptive value, which led for some time to the term plasmid stabilization/stability system; however, addiction systems are pure selfish systems (Cooper and Heinemann, 2000).

The first *ccd* (now "controlled cell death," originally "couples cell division") system, encoded by the F plasmid of *Escherichia coli*, was identified in 1994 (Dao-Thi et al., 2002). The CcdB

TABLE 14.1
Cytoplasmic Incompatibility Systems

System	Plasmid or Organism	Modification Factor	Rescue Factor	Ref.
		Intracellular Mechanisms		
Plasmids in the Cytoplasm of Prokaryotes				
ccd	Plasmid F in *Escherichia coli*	CcdB (H, LetD) toxin	CcdA (G, LetA) antitoxin	Dao-Thi et al. 2002; van Melderen 2002
parDE	Plasmid RK2/RP4 in *E. coli* and many other bacteria	ParE	ParD	Jiang et al., 2002; Oberer et al., 2002
parD	Plasmid R1 in *E. coli*	Kid	Kis	Santos-Sierra et al., 2002; Hargreaves et al., 2002
pem	Plasmid R100 in *E. coli*	PemK	PemI	Jensen and Gerdes, 1995; Picardeau et al., 2001
hok/sok	Plasmid R1, NR1	Hok	sok antisense RNA	Nagel et al., 1999; Pedersen and Gerdes, 1999
par	Plasmid AD1 in *Enterococcus faecalis*	Fst	Antisense RNAII	Greenfield et al., 2001
stb	Plasmid R485 in *Morganella morganii*	StbE	StbD	Hayes, 1998
mvp	Plasmid pMYSH6000 in *Shigella flexerni*	MvpT (STBORF2)	MvpA (STBORF1)	Sayeed et al., 2000
hig	Plasmid Rts1 in *Escherichia coli*	HigB	HigA	Tian et al., 2001
pas	Plasmid pTF-FC2 in *Thiobacillus ferrooxidans*	PasB	PasA	Rawlings, 1999
ez	pSM19035 in *Streptococcus pyogenes*	Zeta	Epsilon	Meinhart et al., 2001
RM	Plasmids, viruses, transposons, and integrons	Restriction enzyme	DNA methylase	Kulakauskas et al., 1995; Naito et al., 1995; Kobayashi, 2001
Chromosomal Systems of Prokaryotes				
chpA/maz	*E. coli*	ChpAK (MazF)	ChpAI (MazE)	Santos-Sierra et al., 1997
ChpB	*E. coli*	ChpBK	ChpBI	Hazan et al., 2001; Sat et al., 2001
chpK	*Leptospira interrogans*	ChpK	ChpI	Picardeau et al., 2001
relBE	*E. coli, Haemophilus influenza, Vibrio cholerae,* plasmid p307, Gram-positive bacteria, archaea	RelE	RelB	Gronlund and Gerdes, 1999; Kristoffersen et al., 2000
din-yaf	*E. coli*	YafQ	DinJ	Gotfredsen and Gerdes, 1998
Prophages in the Cytoplasm of Prokaryotes				
phd/doc	P1	Doc	Phd	Gazit and Sauer, 1999; Hazan et al., 2001

(continued)

TABLE 14.1 (CONTINUED)
Cytoplasmic Incompatibility Systems

System	Plasmid or Organism	Modification Factor	Rescue Factor	Ref.
Mitochondria in the Cytoplasm of Higher Plants				
Ogura CMS	*Raphanus sativus* (Japanese radish)	Orf138	OrfB (subunit 8 of ATP synthase)	Terachi et al., 2001
Texas CMS	*Zea mays* (maize)	T-urf131	Orf221	Priolo et al., 1993; Rhoads et al., 1998
Polima CMS	*Brassica napus* (rapeseed)	Orf224	Subunit 6 of ATP synthase	L'Homme et al., 1997; Brown, 1999
Petunia CMS	*Petunia hybridia*	Pcf	NADH dehydrogenase subunit 3	Hanson et al., 1999
Petiolaris CMS	*Helianthus annuus* (sunflower)	Orf522	Alpha subunit of ATP synthase	Horn, 2002
Chromosomal Systems Working through the Maternal Cytoplasm in Invertebrates				
Medea	*Tribolium castaneum* (flour beetle)	Unknown	Unknown	Beeman and Friesen, 1999
Chromosomal Systems Working through the Maternal Cytoplasm in Vertebrates				
scat$^+$	*Mus musculus* (lab mouse)	Unknown	Unknown	Peters and Barker, 1993; Hurst, 1993
		Intercellular Mechanisms		
Chromosomal Systems in Prokaryotes				
Lantibiotics	*Lactococcus lactis*	Nisin	NisI lipoprotein	Kim et al., 1998
	Bacillus subtilis	Subtilin	SpaI lipoprotein	Stein et al., 2002
	Staphylococcus epidermidis	Epidermin	EpiF, -E, -G, GdmH	Hille et al., 2001
Plasmid systems in prokaryotes				
Bacteriocins	*E. coli*	Colicins	Colicins immunity proteins	Baba and Schneewind, 1998; Riley, 1998
	Pseudomonas aeruginosa	Pyocins	Pyocins immunity proteins	Parret and De Mot, 2002
	Klebsiella pneumoniae	Klebicin B	Klebicin B immunity protein	Riley et al., 2001
Bacteria in the Cytoplasm of Invertebrates				
CI	*Wolbachia pipientis*	Unknown	Unknown	See text

(LetD) toxin inactivates the host gyrase (topoisomerase II) by trapping it in an inactive DNA complex and preventing it from supercoiling the chromosomal DNA. A CcdB dimer binds to the central cavity of the *N*-terminal portion of the gyrase subunit A. If no antitoxin is present, cell growth is inhibited, causing filamentation, and the affected cell will eventually die. The antitoxin CcdA (LetA) displaces CcdB from the gyrase by forming a CcdA–CcdB complex, which detaches from the gyrase and restores its activity. The ParE toxin of the parDE system of the broad host-range plasmid RK2/RK2 is also a gyrase inhibitor. The F plasmid carries two additional addiction systems, *srn* (stable RNA degradation) and *flm* (F leading maintenance), which function as independent killing systems.

The Kid (killing determinant) and PemK (plasmid emergency maintenance killer) toxins of plasmids R1 and R100, which turned out to be identical, are part of another well-studied addiction module (Santos-Sierra et al., 2002). These toxins are inhibitors of DnaB, a crucial enzyme in DNA-chain elongation, and as such prevent the initiation of chromosomal-DNA replication. This will not kill the cell instantly but will preclude any further cell divisions, and the lineage will eventually be lost accidentally. The antitoxins Kis (killing suppression) and PemI (pem inhibitor) form complexes with their cognate toxins and thus neutralize their toxicity.

The *pas* (plasmid addiction system) system on the broad host-range plasmid pTF-FC2 originally found in *Thiobacillus ferrooxidans* is special in that it consists of three, rather than two, components (Rawlings, 1999). The third protein functions as an enhancer of toxin–antitoxin neutralization and in this way decreases the fitness costs of the system for its carrier. It is an interesting example of how a selfish element adapts to its host.

The mobility aspect of addiction systems becomes evident in the *E. coli* bacteriophage P1, which carries the phd/doc (prevents host death/death on curing) module. The precise cellular target of Doc is not yet known, but death occurs by cell-wide inhibition of protein synthesis. However, there is now evidence that this phage-borne addiction system acts through a chromosomal addiction system of *E. coli*, *mazEF*. The free Doc toxin inhibits the translation of the chromosomal MazE antitoxin, and this in turn leads to free MazF toxin, which either kills the cell directly or may initiate another "death cascade" (Hazan et al., 2001).

In yet another type of plasmid- and chromosome-encoded modification–rescue systems, the rescue factors are short, unstable, untranslated antisense RNAs of only some 60 nucleotides that inhibit the translation of very stable modification-factor-encoding mRNAs. In the best-studied system, *hoh/sok* on plasmid AD1 of *Enterococcus faecalis*, the *hok* (host killing) gene codes for a small hydrophobic toxin of 52 amino acids that presumably integrates into the cytoplasmic membrane with the C-terminus protruding into the periplasm (Gerdes et al., 1997). This and many related toxins of the Hok family functionally resemble a group of bacteriophage-encoded proteins known as holins that create pores in the cytoplasmic membrane. The *sok* gene (suppression of killing) constitutes the antitoxin. It codes for an antisense RNA that is complementary to the *hok* mRNA leader region. Like the *pas* system, a third gene, *mok* (modulation of killing), regulates *hok* translation. *hok* mRNA is stable and constitutively expressed from a weak promoter, whereas sok mRNA is unstable and expressed from a strong promoter. This system is quite different from the proteic addiction systems discussed earlier. The proteic systems are characterized by a bicistronic organization whereby one promoter controls the transcription of both toxin and antitoxin, in that precise order. The only exception to this rule so far is the *hig* (host inhibition of growth) addiction system on the Rts 1 plasmid of *Escherichia coli*. The toxin gene is upstream of the antitoxin gene, and the *hig* locus contains two promoters (Tian et al., 2001). The disequilibrium between toxin and antitoxin in the *hok/sok* system requires an additional dimension compared to the proteic systems in order to work. The toxin gene *hok* is transcribed as a full-length mRNA. Because of its secondary structure, the full-length form hides the binding site for the short, corresponding antisense RNA, the *sok* antitoxin gene. The secondary structure also prevents translation of the mRNA. The full-length mRNA is only slowly truncated at the 3′ end by exonucleases. This leads to an accumulation of the full-length mRNA in the cell. The truncation triggers a refolding of the mRNA into a configuration that allows both translation and antisense RNA binding. In plasmid-bearing cells, the truncated mRNA is rapidly bound by constitutively transcribed, unstable antisense RNA. The resulting double-stranded RNA is instantly degraded by RNase III. In plasmid-free cells, the unstable antisense RNA decays faster than truncation of the accumulated full-length toxin mRNA occurs. In the absence of antisense RNA, the truncated mRNA will be translated into the toxic peptide. The *E. coli* chromosome contains at least five such systems, all of which are no longer functional. However, regulation and control through antisense RNA might reflect an ancient and widespread system (Wagner and Simons, 1994; Eddy, 2001; Wagner and Flärdh, 2002).

The proteins of these toxin–antitoxin systems, most of which are only 33 to 130 amino acids long, show such a degree of sequence variation that it is difficult to identify characteristic motifs. This leaves only their small size as an indication of their presence. Experimentally, most addiction systems are sensitive to antibiotic and heat treatment. It is assumed that a strong positive selection is responsible for the rapid divergence of these systems. These systems presumably originated on the bacterial chromosome. This is substantiated by alternative systems, such as the *sop* system of plasmid F and the *par* system of prophage P1, which rely on a centromere-like region to achieve an equal and active distribution to all progeny cells more efficient than toxin–antitoxin systems. The moment a plasmid has acquired a toxin–antitoxin system, it is under selective pressure to become a unique toxin–antitoxin system. The post-segregational killing is aimed at the exclusion of competing cytoplasmic factors (Cooper and Heinemann, 2000). In a few cases, a chromosomal- and a plasmid-based system have been shown to interact, and the plasmid-coded toxin Kid can be neutralized by the chromosome-coded antitoxin ChpAI (Santos-Sierra et al., 1997, 1998). The ancestral function of the proteic toxin–antitoxin systems may lie in a stringent-relaxed response, which occurs when bacteria face starvation. The toxin may function as a protein-synthesis inhibitor during starvation. Continued starvation may lead to the killing of the cell, which would release nutrients, which in turn would permit neighboring cells of the same strain to survive until conditions improved. In the lysogenic state, the bacteriophage λ expresses the *rexB* gene, which prevents the degradation of the host chromosomal antitoxin MazE (ChpAI) and the prophage antitoxin Phd by specifically inhibiting ClpP proteases that otherwise would inactivate the two antitoxins; *rexB* acts here as an anticell death gene. This shows that a bacteriophage can override a modification–rescue system and prevent CI. This, again, can be overcome either by a point mutation or the insertion of an interposon in the *rexB* gene of the bacteriophage (Engelberg-Kulka et al., 1998; Engelberg-Kulka and Glaser, 1999).

One can envision an evolutionary scenario in which a chromosomal host system responsible for controlled or programmed cell death in cases of starvation evolves into a mobile, cytoplasmic, and parasitic system. Transposons, phages, or insertion-sequence activities that move modification–rescue systems onto new plasmids may also move these systems back onto the chromosome. Suddenly, a CI system has become a mutualistic system. This is in line with proposals that either an already existing addiction system of the rickettsial ancestor of mitochondria or the development of a new addiction system in the early mitochondria might have evolved into today's nuclear-encoded apoptosis systems (Kroemer, 1997; Ameisen, 2002).

In this context, it is also interesting that many of these bacterial toxin–antitoxin systems not only exhibit a very broad host range but are capable of functioning outside their kingdom as well. The chromosomal RelE toxin of *E. coli* does kill yeast cells when expressed from a yeast plasmid in *Saccharomyces cerevisiae.* When both toxin and antitoxin were placed on a different plasmid, the antitoxin was to some extent able to rescue RelB toxicity in the yeast. This proves that prokaryotic modification–rescue systems can work in eukaryotes. Expression of *relE* in a mammalian cell line leads to the inhibition of cell proliferation (Kristoffersen et al., 2000). In a human osteosarcoma cell line, RelE induces growth retardation and eventually cell death by apoptosis (Yamamoto et al., 2002). RelE, Shiga toxin (STX), and Microcin E492, a channel-forming bacteriocin from *Klebsiella pneumoniae*, are representatives of the few but increasing number of proteins that exert their effect on both prokaryotic and eukaryotic cells. Especially in invertebrates, where many taxa hold a wide variety of intracellular, symbiotic bacteria, these results make it increasingly possible that parts of these modification–rescue systems may have found their way into the nucleus of the symbiont's host.

Restriction–modification (RM) systems constitute by far the most universal post-segregational killing mechanism. The selective killing is a consequence of the dilution of the DNA methylase in plasmid-free cells, which acts as the antitoxin or rescue factor. The DNA is no longer sufficiently protected from the toxin in the form of the corresponding restriction enzyme. Over 4000 restriction and putative restriction enzymes and over 1000 methylases and putative methylases have been detected. RM systems have been identified in bacteria, archaea, and bacteriophages, as well as on plasmids and chromosomes. RM systems can move without a linked mobile element. Analyses of whole bacterial and archaeal genome sequences show that free-living bacteria such as *Helicobacter pylori* can carry as many as 19 different type II methylase genes alone, while bacteria with a more intracellular lifestyle have only a few methylase genes left, e.g., three in *Rickettsia conori*, two in *Buchneria* sp., and only one in *R. prowazekii* (Lin et al., 2001; Roberts and Macelis, 2001). In the context of CI, it is interesting that RM systems have a profound influence on the genome organization of their host. RM systems can comprise more than 4% of the total host genome. Host resistance against and host inactivation of RM systems are widespread. RM systems can pair modification and rescue factors of independent origin. Also, RM systems can characterize and maintain the integrity of a host strain (Handa et al., 2001; Kobayashi, 2001; Murray, 2002).

Medea Factors

For flour beetles, Medea is the acronym for Maternal Effect Dominant Embryonic Arrest (Beeman et al., 1992). Medea is also the mythological daughter of King Aeetes of Colchis. She married Jason, leader of the Argonauts, after she had helped him obtain the Golden Fleece from her father. When Jason deserted Medea for the daughter of King Creon of Corinth, Medea, in revenge, murdered Creon, his daughter, and her own two sons by Jason and took refuge with King Aegeus of Athens, whose wife she became. He later drove her away after her unsuccessful attempt to poison his son Theseus. In beetles, this means that the lethality is inherited from the mother and will kill all offspring that does not inherit a rescuing M allele from either parent. The toxin is always cytoplasmic and therefore maternal, whereas the antitoxin can be coded on a maternal or paternal allele. Medea breeds true through the female line and segregates in the male.

Four different Medea systems have been found in the red flour beetle, *Tribolium castaneum*. Individuals often carry multiple systems. Almost all populations carry the M-4 system. It is the only system in North American and European populations and can be found in about half of the populations. South American, Asian, and African populations often carry two or more systems, whereas Australian and Indian beetles are almost devoid of Medea. The Indian populations carry though the hybrid incompatibility factor H, which is absent from all other continents. This makes these populations unidirectionally incompatible to non-Indian populations. Again, the incompatibility is temperature sensitive (Thomson et al., 1995; Thomson and Beeman, 1999). The hybrid incompatibility factor causes the death of hybrids with a paternally derived H gene and a maternally derived Medea factor. The H factor presumably suppresses the rescuing factor of Medea. Using gamma irradiation, Beeman and Friesen (1999) succeeded in knocking out the gene for the cytoplasmic lethality factor while retaining the zygotic rescue activity of the M allele. The molecular basis of Medea is unknown. Medea shows all the properties of a selfish modification–rescue system with a suicidal defense against invasion of competing systems. The vast distribution of Medea over continents has inspired several models to explain its population behavior (Wade and Beeman, 1994; Hastings, 1994; Smith, 1998; Hatcher, 2000).

A mutant of the common wall cress *Arabidopsis thaliana*, which is also called MEDEA (*MEA*), leads to 50% abortion upon self-fertilization of heterozygotes with MEDEA (Grossniklaus et al., 1998). MEDEA is a mutant of the FERTILIZATION-INDEPENDENT SEED development locus *FIS1*. *FIS1*/*MEA* is imprinted and expressed only from the maternal genome. It encodes a SET-domain protein similar to *Enhancer of zeste* of *Drosophila*, a member of the Polycomb group of proteins, which in animals ensure the stable inheritance of expression pattern through cell division. Paternally inherited *MEA* alleles are transcriptionally silent in the young embryo. In *Arabidopsis*, *FIS1/MEA* is likely to repress transcription of loci of the maternally derived genome that are normally only expressed from the paternal genome. Mutations at the *DECREASE IN DNA METHYLATION 1* (*DDM1*) locus are able to rescue *MEA* by functionally reactivating paternally inherited *MEA* alleles during development. The maintenance of the genomic imprint at the *MEDEA* locus requires zygotic *ddm1* activity. Because *DDM1* encodes a putative chromatin-remodeling factor, chromatin structure is likely to be interrelated with genomic imprinting. In animals, histone deacetylases and histone lysine methyltransferases have been implicated in genomic imprinting (Vielle-Calzada et al., 1999; Sewalt et al., 2002). *MEA* leads to a paternalization of the egg, and hypomethylation in the pollen leads to maternalization; this reversal can rescue the incompatibility phenotype without the pollen contributing a wild-type *FIS1* allele (Yadegari et al., 2000; Grossniklaus et al., 2001; Spielman et al., 2001).

On the other hand, the Medea protein of *Drosophila* is not involved in CI. It is rather required for embryonic dorsal-ventral and imaginal disc patterning. Medea is a functional homolog of the human tumor-suppressor DPC4/Smad4. It mediates the signaling of the extracellular morphogen decapentaplegic to the nucleus (Das et al., 1998; Hudson et al., 1998; Wisotzkey et al., 1998).

A Medea-like killing system has been identified in mice (Peters and Barker, 1993; Hurst, 1993). Named *scat*$^+$ after its pathological manifestation, it displays the familiar pattern of a modification-and-rescue system with a cytoplasmic lethality factor and a gene-based zygotic rescue factor such as Medea. A severe combination of anemia and thrombocytopenia causes the death in *scat*$^+$ mice. The *scat*$^+$ locus has been mapped on chromosome 8. It is believed that *scat*$^+$ originated from a spontaneous mutation in a BALB/cBY colony of the Jackson Laboratory. It is not known outside the laboratory. The gene of the *scat*$^+$ locus has not been identified, and the molecular mechanism for rescue is not known either.

Cytoplasmic Male Sterility

The oldest and easily the most famous description of CI in plants that is characterized by male sterility comes from Charles Darwin (1877). The majority of flowering plants are hermaphroditic. Darwin called the coexistence of hermaphroditic individuals and individuals that are fertile females but sterile males gynodioecy. He considered this dimorphism as a transition toward dioecy, the separation of sexual types in different individuals. Gynodioecy is the second most frequent reproductive strategy after hermaphroditism in flowering plants. Mitochondrial-borne cytoplasmic male sterility (CMS) is the natural cause of gynodioecy and is characterized by the maternally inherited inability to produce functional pollen without affecting the plant otherwise. This trait is now commercially used for the production of F1 hybrid seeds in several cultivated species such as maize, sugar beets, sunflower, onions, rice, and wheat to eliminate labor-intensive emasculation by hand.

Molecularly, CMS is composed of a toxin and an antitoxin that are cotranscribed from the mitochondrial genome and translated from bicistronic mRNA. The toxin in the Texas type of maize CMS, Urf 13, is coded by a chimeric sequence composed of *atp6*, subunit 6 of the ATP synthase, the 3′-flanking region of *rrn26*, ribosomal RNA, an unidentified sequence, and a part of the coding sequence of *rrn26*. The resulting protein is hydrophobic and localizes in the internal membrane of the mitochondrion. The antitoxin, orf221, is an essential mitochondrial membrane protein that is also found in wild-type mitochondria (Priolo et al., 1993; Rhoads et al., 1998). In Japanese radish, the toxin, orf138, is made up of different radish mitochondrial sequences. The matching antitoxin, orfB, corresponds to subunit 8 of ATP synthetase (Terachi et al., 2001). In rapeseed, the toxin contains parts of NADH dehydrogenase subunit 3 and is rescued by subunit 6 of ATP synthase (L'Homme et al., 1997; Brown, 1999). The petunia toxin is composed of the 5′ end of subunit 9 of ATP synthase, parts of cytochrome *c* oxidase subunit II, and an unknown sequence, and this is compensated by subunit

3 of NADH dehydrogenase (Hanson et al., 1999). The sunflower toxin has similarities with subunit 8 of ATP synthase, and the alpha subunit of ATP synthase acts as the antidote (Horn, 2002). By comparison with bacterial toxins, the mitochondrial proteins are on the large side with some 150 amino acids, and all are membrane proteins. With the exception of CMS in the bean, the toxin–antitoxin units are constitutively expressed in all tissues of the plant without any obvious negative effects. These systems show a high degree of polymorphism. In sunflowers, 64 different systems are known. All these systems have very likely originated from genetic recombination, which must be very common in plant mitochondria (Budar and Pelletier, 2001; Knoop and Brennicke, 2002).

Jacobs proposed that subunit 8 of the mitochondrial ATP synthase (Atp8) originated from a bacterial toxin–antitoxin system. Atp8 has no known homolog in any prokaryotic or plastid ATP synthase but shows similarity to members of the Hok family of toxins mentioned earlier, some of which are involved in post-segregational killing in bacteria (Jacobs, 1991).

A good indication that these CMS systems are nothing else but selfish systems can be seen in the fact that they have always elicited a host response that tries to overcome male sterility. This also questions the role of gynodioecy in the evolution of separate sexes. The host response manifests in a variety of nuclear restorers. The large majority of these nuclear-restorer genes presumably act on the maturation of the mitochondrial mRNA of the toxin–antitoxin unit. The molecular mechanisms are not yet understood. However, one of the nuclear restorers of maize that does not act on the mRNA level has recently been identified as aldehyde dehydrogenase (ALDH). Soluble ALDH accumulates in the mitochondrial matrix. This led to the hypothesis that the toxin in Texas CMS might cause an increased influx of electrons reducing the redox components of the chain, which would lead to an overproduction of reactive oxygen such as superoxide. Superoxide will react with, among other things, polyunsaturated fatty acids in the membrane, and one of the end products is malondialdehyde, which itself is reactive and, eventually, the mitochondrion will induce the killing of the cell. ALDH in the mitochondrial matrix could prevent the accumulation of dangerous levels of malondialdehyde and CMS (Schnable and Wise, 1998; Liu et al., 2001; Møller, 2001; Schnable, 2001).

Intercellular Mechanisms

Intercellular means here that one of two initially independent cells tries to kill the other if it does not have the same toxin–antitoxin system. With the example of the bacteriocins we would like to stress again that it is an ancestral mechanism that has proven itself over time to such a degree that most bacteria have retained it. For this reason, we should expect its action in intracellular symbiotic bacteria as well, where it may contribute to quorum sensing or CI.

Bacteriocins

Bacteriocins are proteic toxins that are exported from the cell to kill bacteria of the same or related species that do not carry that toxin system. Two classes are generally recognized. Class I members are often encoded by a gene cluster on a large conjugative transposon, usually cysteine rich and highly post-translationally modified. They contain α,β–unsaturated amino acids like dehydroalanine or dehydrobutyrine and typically intramolecular thioester bridges known as β–methyl lanthionines, which are formed between the sulfhydryl group of a cysteine and the double bond of a dehydroamino acid. This has led to the name of lantibiotics. They are common in Gram-positive bacteria. They have a rather broad spectrum of antimicrobial activity, which has led to a wide range of applications in food and feed preservation. Lantibiotics act through pore formation in the cytoplasmic membrane. The antitoxin, in these cases often called the immunity factor or immunity protein, is a lipoprotein encoded by the same gene cluster. It prevents pore formation. The expression of the system is highly (auto)regulated in a way that allows quorum sensing in the bacterial population.

Class II bacteriocins are not post-translationally modified; they are mainly plasmid coded and show a much narrower host range and antimicrobial activity. Bacteria that carry a bacteriocinogenic plasmid are immune to the action of the toxin. The best known bacteriocin is colicin, which is found in many *E. coli* strains. The colicin-producing plasmid ColE1 is the ancestor of most plasmids used in molecular biology. The mode of action of various colicins is quite different. Colicin E1 permealizes the cytoplasmic membrane, colicin E2 leads to the degradation of DNA, and colicin E3 leads to the degradation of ribosomal RNA. Some bacteriocins like enterocoliticin have evolved from bacteriophages. The toxins form high-molecular-weight phage tail-like structures that in some cases still show contraction upon contact with susceptible bacteria. The R2 pyocin of *Pseudomonas aeruginosa* is related to phage P2, and the F2 pyocin is related to phage lambda (Nakayama et al., 2000; Strauch et al., 2001). The nature of the plasmid-encoded antitoxins is much less well understood. Many lactic-acid bacteria harbor more than one bacteriocin system with often complementary antimicrobial activities. Bacteriocins of both classes, initially synthesized as preproteins, are secreted by a dedicated ABC transporter system that is also encoded by the bacteriocin cluster. A few bacteriocins make use of the more general *sec*-dependent secretory system (Baba and Schneewind, 1998; Riley, 1998; McAuliffe et al., 2001; O'Connor and Shand, 2002).

Intracellular Microorganisms

The bacterial relE toxin can kill yeast and human cells, and its cognate bacterial antitoxin exhibits rescue activity in eukaryotes (Kristoffersen et al., 2000; Yamamoto et al., 2002). CMS caused by an organelle killing its eukaryotic host cell represents an independent system actively secreting components of its addiction system, and it may indicate a paradigm of how a prokaryote (or a symbiont) starts to manipulate its eukaryotic hosts. It also documents vividly the conflict between selfish systems in the cytoplasm and the nucleus. If seen from the outside, cytoplasmic inheritance leads to intragenomic conflict, which likely is the driving force behind uniparental cytoplasmic inheritance (Cosmides and Tooby, 1981; Law and Hutson, 1992).

Not surprisingly, there are a wide and steadily increasing variety of microorganisms, and in particular bacteria, that are cytoplasmic symbionts exerting drastic effects on their host cells (Chapter 12, this book; Braig et al., 2002). It seems to be evolutionarily easy to develop into a selective killing system. This is best illustrated by the many bacteria that selectively kill males very early on during embryogenesis (see Chapter 12). However, there has been so far only one bacterium identified whose host killing can manifest itself as CI. This bacterium is *W. pipientis*.

WOLBACHIA-INDUCED CI

The bacterium *Wolbachia* was first described by Hertig and Wolbach as a rickettsial infection of the gonads of *Culex pipiens* (Hertig and Wolbach, 1924; Hertig, 1936), but it was not until 1971 that Yen and Barr proposed that *Wolbachia* was the etiological factor for CI (Yen and Barr, 1971, 1973; Yen, 1972). CI in *C. pipiens* had already been observed in the late 1930s (Marshall and Staley, 1937), and Laven had recognized its potential to control mosquitoes by 1951 (Laven, 1951). By the late 1960s CI was employed as a vector-control strategy (Laven, 1967; Awahmukalah and Brooks, 1983; Braig and Yan, 2002; Chapter 13 of this book). *W. pipientis* has a long and successful relationship with arthropods and nematodes, collectively the "Ecdyozoa." Presumably, more than 15% of all known insect species carry it (Werren and O'Neill, 1997; Bourtzis and Braig, 1999; Stouthamer et al., 1999; Anderson and Karr, 2001; Stevens et al., 2001). Here we describe the CI phenomenon and its distribution as well as the current knowledge about the mechanisms involved.

Description

The genetic phenomenon of CI was originally described in the mosquito *C. pipiens* when Laven and Ghelelovitch confirmed the occurrence of reproductive isolation between different mosquito populations (Laven, 1951, 1953; Ghelelovitch, 1952). Ghelelovitch showed that males from a

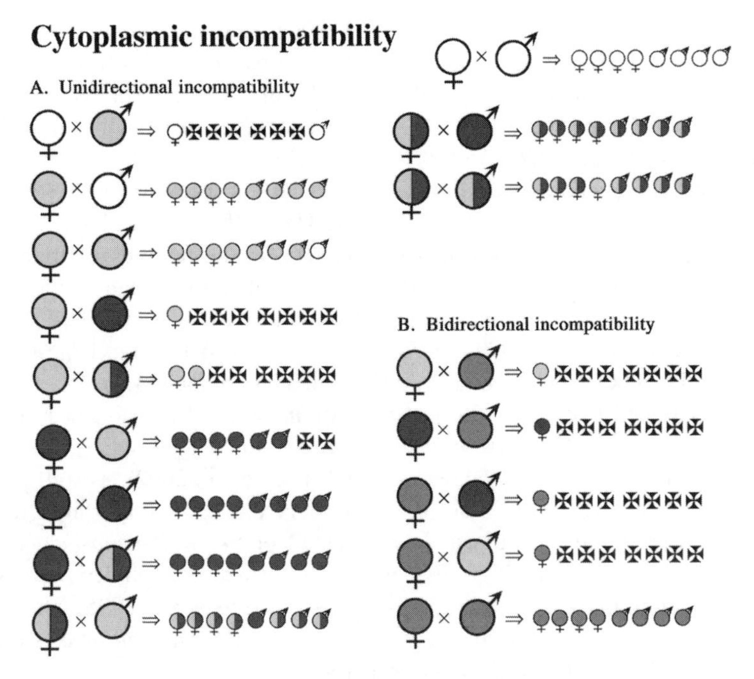

FIGURE 14.1 (Color figure follows p. 206.) Crossing types of *Wolbachia* infections. Red, blue, and violet represent different *Wolbachia* strains. All infected strains are unidirectionally incompatible with uninfected strains; this is shown only for the red strain. The red and blue strains are unidirectionally incompatible with each other. Females of the red strain cannot rescue the modification of the blue strain, whereas females of the blue strain can rescue the modification of the red strain. The violet strain is bidirectionally incompatible with both the red and the blue strain; neither strain can rescue the other's modification. Females of the double-infected red–blue strain can rescue sperm from double-infected, both single-infected and the uninfected strains.

given strain failed to produce progeny when mated with females from other strains, while the reverse cross was compatible. Laven further found that in some cases, this reproductive isolation occurred in crosses of both directions. They both concluded that a maternally inherited cytoplasmic factor was responsible for these incompatibilities. CI results in embryonic mortality in crosses between insects with different *Wolbachia* infection status (Figure 14.1). It can be either unidirectional or bidirectional. Unidirectional CI is typically expressed when an infected male is crossed with an uninfected female. The reciprocal cross (infected female and uninfected male) is fully compatible, as are crosses between infected individuals. Bidirectional CI usually occurs in crosses between infected individuals harboring different strains of *Wolbachia*. In most insects, the expression of CI is lethal to the developing embryo, but in insects with haplodiploid sex determination (Hymenoptera) the end result of CI can be a sex-ratio shift to the haploid sex, which is male. As a consequence of CI, *Wolbachia* infections can spread and remain in nature by replacing uninfected populations, because infected females can successfully mate with both infected and uninfected males, while uninfected females can only successfully mate with uninfected males. Several factors other than CI can affect invasion dynamics and are discussed elsewhere (see Chapter 13).

Distribution

Wolbachia-induced CI has been reported in almost all major insect orders (Diptera, Coleoptera, Hemiptera, Hymenoptera, Orthoptera, and Lepidoptera) as well as in Arachnida and Isopoda. As mentioned above, *Wolbachia*-induced CI was first reported in the order Diptera, in *C. pipiens* in the family Culicidae (e.g., Ghelelovitch, 1952; Yen and Barr 1973, 1974; Trpis et al., 1981;

Kambhampati et al., 1993; McGraw and O'Neill, 1999; Jamnongluk et al., 2001; Dobson et al., 2002; Kittayapong et al., 2002), and it has since been reported in the families Drosophilidae (e.g., Hoffmann et al., 1986; Hoffmann, 1988; O'Neill and Karr, 1990; Bourtzis et al., 1994, 1996; Giordano et al., 1995; Merçot et al., 1995; Werren and Jaenike, 1995; James and Ballard, 2000) and Tephritidae (Boller et al., 1976; Boller, 1989); in Coleoptera, in the families Curculionidae (Blickenstaff, 1965; Hsiao and Hsiao, 1985) and Tenebrionidae (Wade and Stevens, 1985); in Hemiptera, in the families Aleurodidae (De Barro and Hart, 2000) and Delphacidae (Noda, 1984; Rousset et al., 1992; Noda et al., 2001); in Hymenoptera, in the families Figitidae (Vavre et al., 2000, 2001, 2002), Proctotrupoidae (A. van Alphen, personal communication cited in Werren and O'Neill, 1997), ants (van Borm et al., 2001), and Pteromalidae (Saul, 1961; Breeuwer and Werren, 1990; Perrot-Minnot and Werren, 1999); in Orthoptera, in the family Grylidae (Giordano et al., 1997; Kamoda et al., 2000; Mandel et al., 2001); and in Lepidoptera, in the family Pyralidae (Brower, 1976; Sasaki and Ishikawa, 1999). *Wolbachia*-induced CI has also been reported in two families of the class Arachnida, Phytoseiidae (Johanowicz and Hoy, 1998), and Tetranychidae (Breeuwer, 1997; Gotoh et al., 1999; Vala et al., 2000; Egas et al., 2002), as well as in the family Porcellionidae of Crustacea (Legrand et al., 1986; Moret et al., 2001). It is evident that *Wolbachia*-induced CI is the most frequent and widely distributed of the *Wolbachia*-induced phenotypes. Phylogenetic analysis suggests that CI-*Wolbachia* do not form a monophyletic group with respect to the *Wolbachia* strains that cause other phenotypes (Werren et al., 1995; Zhou et al., 1998). Actually, the distribution of CI within the general phylogeny of *Wolbachia* makes it parsimonious to assume that it was an ancestral *Wolbachia* property.

The Mechanism of *Wolbachia*-Induced CI

Although the mechanism of CI has not yet been elucidated on the molecular level, several lines of evidence suggest that *Wolbachia* somehow modifies the paternal chromosomes during spermatogenesis (mature sperm do not contain the bacteria), thereby influencing their behavior during the first mitotic divisions and resulting in loss of mitotic synchrony (Breeuwer and Werren, 1990; Reed and Werren, 1995; Lassy and Karr, 1996; Callaini et al., 1997; Tram and Sullivan, 2002). Based on the available genetic and cytogenetic data, Werren (1997) proposed the so-called mod/resc model, which assumes the presence of two distinct bacterial functions. First, the modification (mod), a kind of an "imprinting" effect, which is expressed in the male germline, probably during spermatogenesis, and second, the rescue (resc), which is expressed in the egg. Sperm imprinting may be due either to secreted *Wolbachia* proteins that modify the paternal chromosomes or the removal of host proteins that are necessary for proper condensation/decondensation of the paternal chromosomal set before or during zygote formation. Similarly, the presence of the same *Wolbachia* strain in the egg may result in the production and secretion of rescue factors or alternatively the recruitment of host molecules that are capable of rescuing the sperm "imprint" in a *Wolbachia* strain-specific manner. The mod and resc functions are now characterized through a number of properties, which must be accounted for by any hypothesis regarding their as yet unknown molecular nature. Molecular studies are now in progress to identify the proteins involved in CI (Sasaki et al., 1998; Braig et al., 1998; Harris and Braig, 2001).

Charlat et al. (2001, 2002) summarized five properties:

1. The mod intensity is a variable factor. Indeed, the levels of modification, expressed as embryonic mortality in incompatible crosses, range from 0 to 100% (Poinsot et al., 1998). Experimental evidence suggests that both bacterial and host factors are responsible for the variation in mod intensity (Boyle et al., 1993; Rousset and de Stordeur, 1994; Giordano et al., 1995; Sinkins et al., 1995b; Poinsot et al., 1998; Poinsot and Merçot, 1999).
2. The mod and resc factors interact in a specific manner. The phenomenon of bidirectional CI supports this conclusion. The current data obtained by crossing experiments indicate

that any *Wolbachia* strain is only fully compatible with itself. This means that there is a specific recognition between the different mod and resc factors (O'Neill and Karr, 1990; Braig et al., 1994). Crossing patterns involving multiple infections can also be explained by specific mod/resc, complete or partial, recognitions (Merçot et al., 1995; Rousset and Solignac, 1995; Sinkins et al., 1995a; Perrot-Minnot et al., 1996; Poinsot et al., 1998; Rousset et al., 1999).

3. Mod and resc are distinct bacterial functions. The identification of a *Wolbachia* strain that was unable to induce CI but capable of rescuing CI induced by other strains strongly suggests that mod and resc are separate functions (Bourtzis et al., 1998; Merçot and Poinsot, 1998). Based on this finding, four different CI-*Wolbachia* types (strains) can exist:
   a. The mod+/resc+ type, which can induce CI and rescue its own modification. For example, three different mod+/resc+ types have been reported in *D. simulans*, *w*Ri, *w*Ha and *w*No (Hoffmann et al., 1986; O'Neill and Karr, 1990; Braig et al., 1994; Merçot et al., 1995), while only one has been reported in *D. melanogaster*, *w*Mel (Hoffmann, 1988; Hoffmann et al., 1994; Bourtzis et al., 1994, 1996, 1998; Reynolds and Hoffmann, 2002; Weeks et al., 2002).
   b. The mod–/resc– type, which is unable to induce CI itself or rescue the CI effect of other mod+ strains; such a type is, for example, present in the *w*Cof (or *w*Au) strain of *D. simulans* (Hoffmann et al., 1996).
   c. The mod+/resc– type, which can induce CI but cannot rescue its own effect. This type is suicidal and has never been detected in natural or laboratory populations (but see Charlat et al., 2001 for an alternative view).
   d. The mod–/resc+ type, which was described above. Such a type is present in the *w*Ma strain of *D. simulans* and *D. mauritiana* and can rescue the CI effect induced by the *w*No strain (Bourtzis et al., 1998; Merçot and Poinsot, 1998).
4. The mod intensity is linked to bacterial density. Breeuwer and Werren (1993) introduced the bacterial-density model to explain the different CI levels and patterns observed. Although the different CI patterns could not be resolved, several studies have suggested a relationship between *Wolbachia* density and CI levels (Hoffmann et al., 1986; Binnington and Hoffmann, 1989; Boyle et al., 1993; Bressac and Rousset, 1993; Rousset and de Stordeur, 1994; Solignac et al., 1994; Merçot et al., 1995; Sinkins et al., 1995a; Bourtzis et al., 1996, 1998; Poinsot et al., 1998; Rousset et al., 1999). In a recent study, Veneti et al. (2002) determined and compared the distribution and the density of *Wolbachia* during spermatogenesis in 16 different naturally infected and transinfected *Drosophila* hosts using a *Wolbachia*-specific antibody. They investigated the relationship between the distribution and density of *Wolbachia* within sperm cysts and the expression levels of CI by comparing the same *Wolbachia* strain in different hosts and different *Wolbachia* strains in the same host genetic background. Their experimental data show a positive correlation between the number of infected sperm cysts and CI levels.
5. The mod function alters the mitotic behavior of the paternal chromosomes after fertilization. Cytological studies have shown that *Wolbachia* disrupts the kinetics of decondensation/recondensation of the pronuclei during fertilization. Details about the cell biology of *Wolbachia*-induced CI are presented below.

Several *Wolbachia* genome projects are in progress aiming to identify the complete genome sequence of several *Wolbachia* strains (Bandi et al., 1999; O'Neill, 1999; Slatko et al., 1999; Oehler and Bourtzis, 2000). The genome sequence of the *w*Mel *Wolbachia* strain has been completed, while that of *w*No is at an advanced stage (O'Neill, 2002; Bourtzis, 2002). These two strains are bidirectionally incompatible. Coordinated efforts using comparative genomics, proteomics, and post-genomics approaches will ultimately lead to the identification and characterization of *Wolbachia* genes involved in the induction of the three phenotypes as well as host genes involved in host–*Wolbachia* interactions. Identification of these genes will be a major breakthrough in deciphering the biology of this unculturable bacterium, understanding *Wolbachia*–host symbiotic associations, and uncovering the evolution of intracellular symbiosis.

Genetics and Cell Biology of *Wolbachia*-Induced CI

Genetic Analysis of Host/Symbiosis

Although five *Wolbachia* types have been described in *D. simulans,* the exact number of distinct *Wolbachia* strains found in *D. simulans* is under question (e.g., see James et al., 2002, for a discussion of strain nomenclature). The origins, evolution, and genetic basis of these different incompatibility types are not known. However, their presence does suggest that each unique strain of *Wolbachia* differentially modifies and rescues sperm, which is almost certainly the case for bidirectional incompatibility (Braig et al., 1994; Rousset et al., 1999). In this section, the genetic basis of this host/symbiosis system will be discussed. Unfortunately, our present knowledge of *Wolbachia* genetics is extremely limited. This is due to a lack of molecular genetic analyses of the kind that has proven so extremely powerful for microorganisms such as *E. coli.* Unlike free-living facultative anaerobes such as *E. coli*, *Wolbachia* is an obligate intracellular microbe for which cell-free culture is not yet available. Therefore, essentially nothing is known of its genetic structure, although this will certainly change as whole genomic sequencing and subsequent analyses of *Wolbachia* genomes are realized.

A number of factors have been identified that effect the expression of CI in *Drosophila.* These include (1) *Wolbachia* type, (2) host strain and host age, (3) heat treatment, and (4) mating history (Boyle et al., 1993; Turelli and Hoffmann, 1995; Sinkins et al., 1995b; Hoffmann et al., 1996; Bourtzis et al., 1998; Clancy and Hoffmann, 1998; Karr et al., 1998; Poinsot et al., 1998; James and Ballard, 2000; Snook et al., 2000). Each of these identified factors are potential targets for genetic analyses of CI. Below, we provide a brief general background of previous work relevant to genetic analyses of the system and further focus on one particular area of contemporary research, the influence of host genetic structure on *Wolbachia* growth and tissue distribution in the *Drosophila* testis.

Historically, inferences about the genetics of *Wolbachia* symbiosis have been indicated by studies involving (1) comparative work on different host-strain genetic backgrounds, (2) replacement of the host genome by repeated backcrosses, (3) direct transfer of *Wolbachia* into novel or naïve hosts, or (4) genetic analyses of the effect of host mutations on *Wolbachia* biology. Each of these approaches provides a wealth of information about the overall impact of host genetics in each system studied and has generally suggested extensive genetic interactions between host and microbe. Each of these areas of study is summarized below.

Comparative and Functional Studies

A variety of *Drosophila* species have been analyzed for *Wolbachia* infection (Bourtzis et al., 1994, 1996; Giordano et al., 1995; Werren and Jaenike, 1995; Hurst et al., 2000; Lachaise et al., 2000). Taken together, only 12 out of 52 tested species have been identified as infected. From this limited data it appears that *Wolbachia* may be present in about 23% of *Drosophila* species in agreement with current estimates that these infect anywhere from 16 to 25% of insects (Werren and Windsor, 2000).

Studies on different *D. melanogaster* strains exhibited an interesting variation in the expression of CI associated with *Wolbachia* infection (Bourtzis et al., 1996; Clark et al., 2002). Presumably, the considerable variation in CI expression observed in infected *D. melanogaster* mutant and wild-type lines reflects at some level the effect of host genetic backgrounds on CI expression. For example, egg mortality rates among various *D. melanogaster* lines range from as low as 10% to as high as 50% (Bourtzis et al., 1996). This would indicate, assuming the same *Wolbachia* infects all *D. melanogaster* lines, that the variation observed was due to the differing genetic backgrounds of the host. However, the precise genetic basis for these differences is not known, and these observations suggest that further genetic tests that target specific cellular processes and traits will identify key host genetic loci involved in the symbiosis.

Because *Wolbachia* affect sperm function during fertilization, research has focused on *Wolbachia* distribution and behavior during spermatogenesis (Bressac and Rousset, 1993; Ndiaye and Mattei, 1993; Karr et al., 1998; Clark and Karr, in press; Clark et al., in press). *Wolbachia* presumably impair the male pronucleus but not the extranuclear component of the sperm (Presgraves, 2000). Recent studies have demonstrated the presence of two major groups of *Drosophila–Wolbachia* associations within infected testes: group I is characterized by the presence of high numbers of *Wolbachia*-infected sperm cysts, while in group II *Wolbachia* are absent from cyst cells and found mainly within somatic cells (Clark et al., 2002, in press; Veneti et al., in review). In all cases studied, *Wolbachia* infection of sperm cysts was tightly linked to CI expression. For example, group I infection consisted of two subgroups, *D. simulans* Coffs (*w*Cof) and *D. mauritiana* (*w*Ma), that exhibit high numbers of heavily *Wolbachia*-infected sperm cysts, but they did not express CI, while the second contained variable numbers of *Wolbachia*-infected sperm cysts and express variable levels of CI. Based on these studies, a "WIS" hypothesis (for *Wolbachia*-Infected Spermatocyte) was proposed and a model built on this hypothesis suggested. In this model, cyst infection is a necessary, but not sufficient, condition for the expression of CI. Using this definition as a starting point, genetic studies can focus on host mutations that change or otherwise alter *Wolbachia* in the developing spermatocyst.

While no concrete conclusions can be reached as to the genetic bases underlying the growth and maintenance of *Wolbachia* in the expression of a WIS phenotype, the consistency of the results strongly imply such a basis. The relative contributions of host genetic backgrounds to the expression of CI have been studied by introgression of host genetic backgrounds into a *Wolbachia*-infected maternal cytoplasm (see Genome Replacement via Introgression, below). This is a more direct approach to the study of genetic interactions, as it compares differing host genetic backgrounds in an otherwise identical *Wolbachia* strain infecting the same maternal cytoplasm.

Genome Replacement via Introgression

The contributions of host genetic backgrounds to the expression of CI in the parasitic hymenopteran *Nasonia* have been well studied (Breeuwer and Werren, 1993; Bordenstein and Werren, 1998). In this system, two species, *Nasonia vitripennis* and *N. giraulti*, harbor double *Wolbachia* infections (A and B strains). Introgressions of the *N. giraulti* genome into a doubly-infected *N. vitripennis* line did not affect the expression of bidirectional CI or unidirectional CI when compared to the parental *N. vitripennis* males, i.e., the *N. giraulti* genome did not affect this modification–rescue system. However, introgression of *N. giraulti* into a singly-infected *Wolbachia* A strain of *N. vitripennis* did affect CI expression levels compared to this same A strain in *N. vitripennis*, suggesting the differences noted were due to host genetic backgrounds.

Parallel studies in *Drosophila* confirmed and extended the basic conclusions obtained in the *Nasonia* system. Reciprocally introgressed genomes between host lines harboring different *Wolbachia* strains were measured for their ability to express CI (Clark and Karr, in press). These introgression experiments demonstrated that the *Wolbachia* type determined the ability or inability to modify sperm independent of host genetic background and confirmed the comparative studies mentioned above. Although an intrinsic *Wolbachia* factor appears to determine CI type, host genetic backgrounds were shown to clearly affect *Wolbachia* growth in the testis and, as a result, to affect the level of CI in those lines with mod+ *Wolbachia*.

Introgression experiments are limited because the original host genotype cannot be completely replaced, and epistatic interactions in the host may influence the results and obscure direct genetic interactions of host and symbiont (although epistasis in this context might reveal interesting interactions themselves). Another useful avenue of research that avoids these limitations is direct transfer of *Wolbachia* via microinjection, as described below.

Direct Inter- and Intraspecific Transfer between Species and Lines

Direct transfer of *Wolbachia* into novel hosts has been a useful tool for the study of genetic interactions in *Drosophila* (Boyle et al., 1993; Braig et al., 1994; Giordano et al., 1995; Poinsot et al., 1998; McGraw et al., 2001, 2002). This approach transfers the microbe (and small amounts of egg cytoplasm) into uninfected host eggs via direct cytoplasmic transfer by microinjection, and infected lines are established (reviewed in Karr, 1994). Once infected lines are established, CI and bacterial densities can be measured and compared to the parental line.

The first successful interspecies transinfection of *Wolbachia* (Boyle et al., 1993) from *D. simulans* into *D. melanogaster* showed a clear effect of host genetic background on CI expression. Further, the lack of CI expression was due to the presence of far fewer bacteria in the novel *D. melanogaster* host. This suggested two important points: first, CI expression was not an exclusive intrinsic property of *Wolbachia*, and, second, the host could regulate bacterial load. A linear correlation between bacterial numbers in the egg and the level of CI expression in isofemale lines selected for increased levels of CI further suggested a host genetic basis for this trait (Boyle et al., 1993; Sinkins, 1995a). A number of subsequent studies in a variety of systems have shown the efficacy and challenges of cross-taxa transfer of *Wolbachia* (Braig et al., 1994; Poinsot et al., 1998; van Meer and Stouthamer, 1999).

Recently, horizontal transfer of a virulent strain of *Wolbachia* ("popcorn" [Min and Benzer, 1997]) has revealed the importance of host–symbiont interactions in the expression, growth, and maintenance of infection, particularly the role of host genetic background (McGraw et al., 2001). In its original host, the popcorn-strain variant does not express CI (but see Reynolds and Hoffmann, 2002; Weeks et al., 2002) and does not reside in the testis (McGraw et al., 2001; Veneti et al., in review). Upon transfer to an uninfected *D. simulans* strain known to harbor a CI-inducing strain of *Wolbahcia* (the DSR strain), the popcorn strain infected the testis and expressed CI, clearly showing a recurring theme of the importance of host genetic background.

Mutant Analyses

The three approaches described above reveal general patterns of host genotype on *Wolbachia* biology. However, they cannot provide information about individual genes and their potential impact on *Wolbachia* growth, tissue distributions, or expression of CI. Genetic analyses would greatly benefit from the application of classical genetic techniques (e.g., identification of mutants that effect symbiosis parameters mentioned above and the concomitant identification of individual genes involved). As mentioned above, the lack of *in vitro* cultures precludes direct genetic analyses of *Wolbachia*, but fortunately insights into the genetics of this system can be obtained by study of host genetics relying on the extensive genetic information and genomic data available for *D. melanogaster* (Lindsley and Zimm, 1992). Although in its infancy, this approach has proven successful in identifying situations where mutant phenotypes are affected by *Wolbachia* infection, as described below. In its broader application, these initial studies strongly suggest that a dedicated screen for other genetic loci will identify host genes that directly or indirectly interact with *Wolbachia*.

The sxl^4 *mutation. Wolbachia* has been found to restore fertility to *D. melanogaster* females carrying the sxl^4 mutation (Starr and Cline, 2002). This mutation results in sterile females because of its effect on sex determination (Cline and Meyer, 1996). Prior phenotypic analyses had been performed in this mutant line without knowledge of its infection status. Interestingly, restoration of fertility by *Wolbachia* is not observed to the same extent in other germline-specific *sxl* alleles; therefore, fertility restoration most probably results from a specific interaction with *Sxl* protein rather than from a bypass of the normal germline requirement for this developmental pathway. As such, *sxl–Wolbachia* interactions indicate the importance of *Wolbachia* insinuation of host reproduction and the degree to which it has occurred.

The chico mutation. Mutations in the insulin-receptor pathway and the PTEN/MMAC signaling pathway affect body, tissue, and organ size in *D. melanogaster* (Bohni et al., 1999; Goberdhan et al., 1999; Huang et al., 1999). In a survey of stocks infected by *Wolbachia* (M. Clark and T. L. Karr, unpublished data), an allele of the insulin-receptor substrate gene *chico* was found to be infected by *Wolbachia*. Mutations in *chico* are normally homozygous viable; however, only one is viable following removal of *Wolbachia* infection, suggesting that *chico* might interact directly with the microbe. However, interactions between *Wolbachia* in this genetic background are complex and cannot be explained by the *chico* mutation alone but must involve other loci located on the second and third chromosomes. Mapping of these additional loci, which may be effectors of *chico* function, could provide valuable genetic insights into this complex genetic pathway and, as in the sxl^4 case, underpins the potential impact of *Wolbachia* infection on phenotypic analyses of known mutations in *D. melanogaster*.

Genetics of Growth Control in *Drosophila*

Coordination of bacterial growth with that of host replication is under stringent control. The presence of *Wolbachia* in a specific subset of tissues (usually the gonads) is further evidence of integrated functioning of host and parasite biology. This tissue tropism is most evident in gonadal tissues, where bacterial numbers are highest, presumably because vertical transmission of infection is transovarial. *Wolbachia* has been detected in somatic tissues in a number of taxa, most notably in isopods, where heavy infections in multiple tissues have been described (Rousset et al., 1992; Juchault et al., 1994; Bouchon et al., 1998; Dobson et al., 1999). However, reproductive tissues of germline origin are the predominant target tissue of *Wolbachia*, suggesting specific mechanisms for their deployment during development. In this regard, it is interesting to note that *Wolbachia* become incorporated within pole cells soon after fertilization in *Drosophila* (O'Neill and Karr, 1990; Hadfield and Axton, 1999).

Infections are closely integrated with host development, and the timing and location of bacterial proliferation are organized in a manner that fosters bacterial transmission with minimal adverse effects on host viability. For example, direct bacterial counts using confocal microscopy revealed that no microbe growth occurred in the fertilized egg for a substantial period of embryonic development (Kose and Karr, 1995). Taken together, these observations predict a mechanism (or mechanisms) governing bacterial replication such that replication is prevented in inappropriate cellular environments and permitted in appropriate tissues at appropriate, and specific, stages in host development. However, the mechanisms by which this is achieved are unknown.

Critical questions relevant to the prevailing model for *Wolbachia* action (the modification/rescue model) can best be formulated when the subcellular localization and behavior of *Wolbachia* during spermatogenesis are understood (Clark et al., 2002). Spermatogenesis in adult *Drosophila* has been described (Lindsley and Tokuyasu, 1980; Fuller, 1993). Sperm development proceeds within an elongated coiled tube and begins in the germinal proliferation center located at the apical hub of the testis where the gonial stem cells and cyst progenitor cells are located (Figure 14.2). A primary gonial cell buds off from the germline stem cell and is surrounded by two somatically derived cyst cells that bud off the nearby cyst progenitor cells. As the primary spermatogonial cell undergoes four rounds of mitotic division before entering meiosis, it moves down the tube of the testis away from the apical hub. Cytokinesis is not complete in either mitosis or meiosis, the result being 64 interconnected haploid cells. Both cyst cells as well as germ cells comprise one cyst. Following the completion of meiosis, axoneme growth occurs as the cyst elongates, eventually growing the entire length of the testis, with the sperm nuclei toward the seminal vesicle. Following elongation, cysts undergo individualization, the stripping away of the major mitochondrial derivative, as well as most cytoplasmic factors via the individualization complex, a network of cytoskeletal factors. The individualization complex is seen as a bulge proceeding along in the cyst, pushing the stripped-away material into the waste bag, a bulge in the tail end of the cyst. The nuclear end of the cyst becomes anchored to the terminal epithelium, followed by the coiling of tightly packed sperm tails prior to liberation into the seminal vesicle. Fully mature, fertilization-competent sperm in the

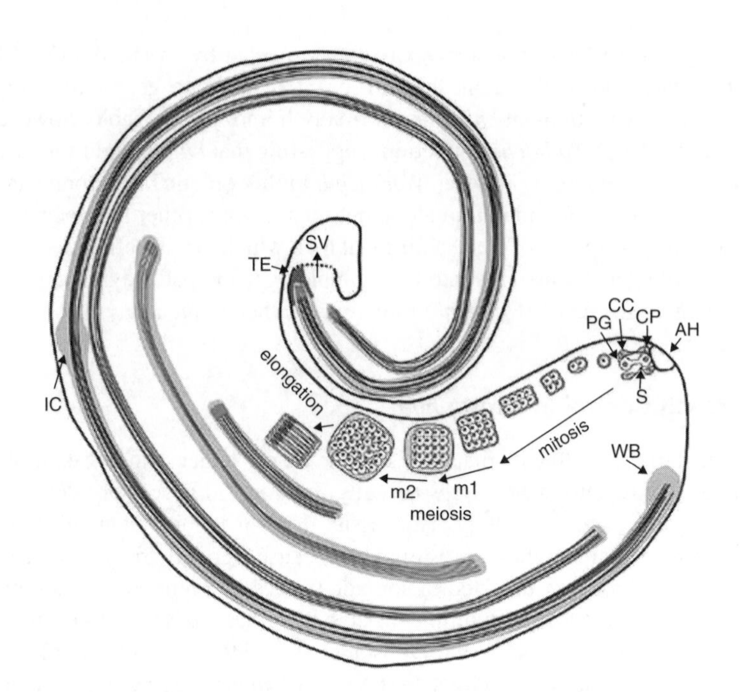

FIGURE 14.2 (Color figure follows p. 206.) An idealized version of progressive stages of cyst development and their approximate location within an adult testis. Anchored to the apical hub (AH) are the gonial stem cells (S) as well as the cyst progenitor cells, which bud off to form a primary gonial cell (PG — white with red nuclei) surrounded by two somatically derived cyst cells (CC — gray). The primary gonial cell undergoes four rounds of mitotic division before entering meiosis. Cytokinesis is not complete in either mitosis or meiosis, the result being 64 interconnected haploid cells. The cyst cells (gray, nuclei not shown) do not undergo division; rather, they expand to form a continuous layer around the germ cells. Both cyst cells and germ cells comprise one cyst. Following the completion of meiosis, axoneme growth occurs, as the cyst elongates, eventually growing the entire length of the testis, with the sperm nuclei toward the seminal vesicle (SV). Individualization is the stripping away of the major mitochondrial derivative, as well as most cytoplasmic factors via the individualization complex (IC), a network of cytoskeletal factors, which is seen as a bulge proceeding along in the cyst pushing the stripped-away material into the waste bag (WB), a bulge in the tail end of the cyst. The nuclear end of the cyst becomes anchored to the terminal epithelium (TE), followed by the coiling of tightly packed sperm tails prior to liberation into the seminal vesicle.

seminal vesicle have been shown to be devoid of *Wolbachia* (Snook et al., 2000), which is consistent with the idea that sperm modification must take place earlier during spermatogenesis.

Two lines, *D. simulans* Riverside (DSR) and *D. melanogaster* Bermuda (DMB), contain *Wolbachia* within developing spermatocytes and spermatids of young males (Clark et al., 2002). *Wolbachia* numbers increased during spermiogenesis, between the initiation of elongation and individualization, in both lines. However, comparison of these two lines revealed striking differences in the apparent timing of *Wolbachia* growth and in resulting *Wolbachia* density and distribution (Figure 14.3). In DSR, noticeable *Wolbachia* growth began during the spermatocyte growth phase, while little *Wolbachia* growth occurred at this stage in DMB, where growth was delayed until the initiation of spermatid elongation.

Elucidation of the genetic bases for these differences, both within and between closely related *Drosophila* species, has been suggested by recent work (Clark et al., 2002, in press; Clark and Karr, in press; Veneti et al., in review), and currently under way are genetic screens designed to identify loci affecting *Wolbachia* growth, maintenance, and CI expression. For example, a screen for haploinsufficiency loci using the "deficiency kit" from the Bloomington Drosophila Stock Center has identified one region of genome necessary for *Wolbachia* growth in spermatocysts (M. Clark, unpublished data). Additional screening and functional tests should reveal additional interacting loci.

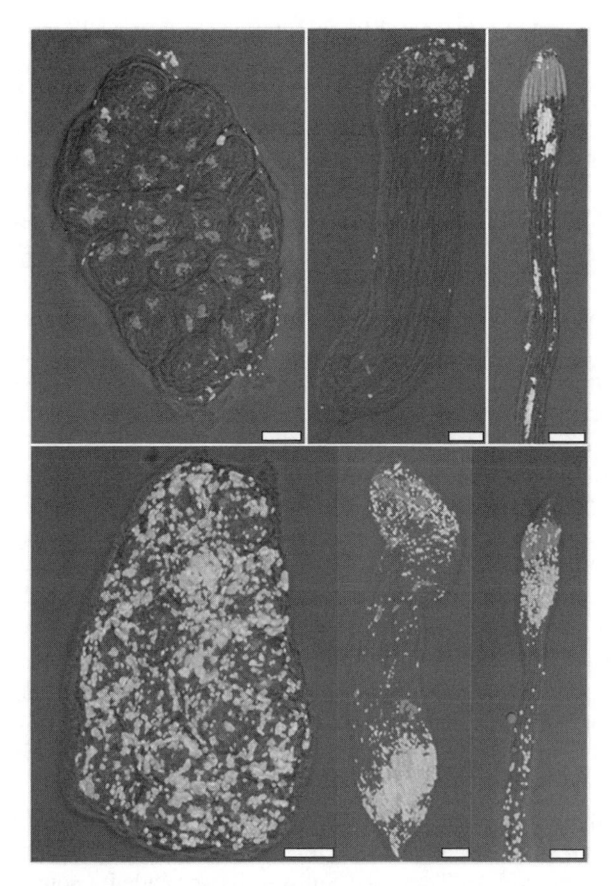

FIGURE 14.3 (Color figure follows p. 206.) *Wolbachia* distribution and proliferation within cysts during meiosis and sperm elongation from *D. melanogaster* (DMB, top row) and *D. simulans* (DSR, bottom row). Each of three images shows progressively later stages of spermatogenesis: left, 16-cell spermatocyt; middle, beginning of spermatid elongation; right, apical region of fully elongated spermatid. DNA is shown as red (propidium iodide), and *Wolbachia* are seen as small punctate dots varying in color from yellow to green, depending on the relative staining intensity of the propidium and the FITC-labeled secondary antibody. In DMB at the 16-cell spermatocyst stage (top left), essentially no *Wolbachia* are present, while DSR (bottom left) display abundant *Wolbachia* (yellow–green), which are present in numbers great enough to obscure spermatocyte nuclei (red). During early stages of spermatid elongation (middle) DSR (bottom) contain a substantial number of *Wolbachia* at either end of the spermatid, while DMB (top) spermatids contain few, if any, bacteria. Fully elongated spermatids of both (right) contain *Wolbachia*, although DSR (bottom) have substantially greater numbers than DMB (top). All scale bars = 10 μm. [Adapted from Clark, M.E., Veneti, Z., Bourtzis, K., and Karr, T.L. (2002). *Mech. Dev.* **111:** 3–15.]

Cell Biology of *Wolbachia*-Induced CI

Cytological Studies

CI results in early egg lethality following sperm entry (Jost, 1970; O'Neill and Karr, 1990; Reed and Werren, 1995; Callaini et al., 1996; Lassy and Karr, 1996). As a central element of the "modification" portion of the mod/resc model, understanding how and when *Wolbachia* affect sperm function during spermatogenesis is of primary interest. However, manifestation of the "modification" that results in sperm dysfunction occurs only during and following fertilization. One of the earliest observed defects is chromosomal abnormalities at the first division (Callaini et al., 1997). However, this early phenotype is not fully penetrant, and numerous other developmental abnormalities are present. These additional developmental anomalies include aberrant mitotic figures and chromosomal fragmentation. Indeed, Callaini et al. (1996) reported developmental anomalies at later stages of development that included segmentation defects, suggesting that CI expression could affect various stages of development. Neither the nature of these differences nor their origins during development are yet known. The analysis of CI in *Drosophila* is further complicated by the fact that CI is never 100% penetrant, with egg lethality in incompatible crosses varying from 60 to 90%. This is in stark contrast to CI in mosquitoes and *Nasonia*, where the early CI phenotype is essentially 100% penetrant (Breeuwer and Werren, 1990, 1993). The phenotype of CI in this system is almost exclusively observed as diffuse paternal chromatin and fragmentation during the first division (Reed and Werren, 1995). Although a description of all *Drosophila* CI phenotypes is beyond the scope of this chapter, suffice it to note that substantial variation in these phenotypes exists, suggesting that different molecular mechanisms may be at work in these different species groups. Although no mechanistic explanation for the variety of CI phenotyes observed in *Drosophila* has yet been put forward, they do suggest that *Wolbachia* may operate by different mechanisms in different taxa. If true, it may indicate that *Wolbachia* has evolved a variety of mechanisms in which to affect host reproduction and CI expression.

Mechanistic Studies of CI in Nasonia

A significant advance in our understanding of the underlying mechanisms has come from recent work in *Nasonia* that utilized real-time imaging and indirect immunofluorescence to visualize early developmental events leading to the expression of CI and consequent egg lethality (Figure 14.4). These results strongly supported a model whereby *Wolbachia* affects the timing of nuclear-envelope breakdown (NEB) prior to the crucial first gonomeric division (Tram and Sullivan, 2002). Additional studies by the same group further suggested that NEB delay was due to delayed activation of Cdk1, a regulatory kinase whose activity is required for entry into mitosis.

These first molecular clues to the CI mechanism are consistent with a "kinetic" model of CI that suggested *Wolbachia* somehow affected the relative timing of the events following sperm entry (Kose, 1995). The current work of Tram and Sullivan has provided a sound cell biological basis for the phenomenon. It will be of great interest to determine whether other cell-cycle components are affected by *Wolbachia* or whether similar mechanisms operate in diverse taxa.

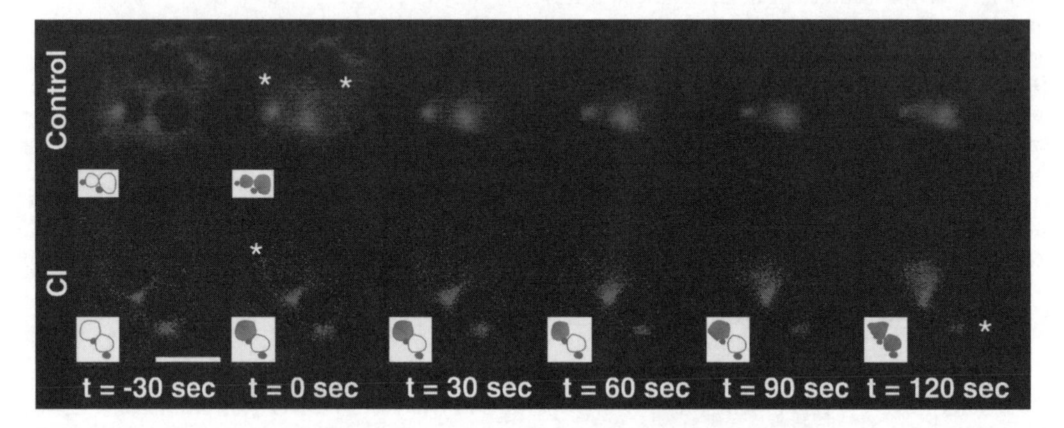

FIGURE 14.4 (Color figure follows p. 206.) Nuclear-envelope breakdown (NEB) assessed by time-lapsed confocal microscopy of 0- to 1-h embryos injected with rhodamine-tubulin. When the nuclear envelope was intact, the nucleus appeared as a black circle surrounded by a ring of red (rhodamine-tubulin). During NEB rhodamine-tubulin invaded the nucleus (compare asterisks in upper and lower panels, $t = 0$ sec). The CI cross nucleus remains relatively devoid of tubulin (indicated by dark regions of the intact nucleus), indicating delayed NEB. (Insets) Schematized interpretation of NEB. [Adapted from Tram, U. and Sullivan, W. (2002). *Science* **296:** 1124–1126.]

Finally, how do the results above relate to the "mod/resc" model for CI? A number of models have been put forward over the years, but all suffer from the same lack of detailed information about the basic molecular mechanisms affected by *Wolbachia*. In this particular case, the kinetic model proposed by Tram and Sullivan can be thought of as a specific case of mod/resc whereby "modification" includes altering the ability of sperm to enter the first mitosis and "rescue" is the ability of *Wolbachia* (in the egg) to synchronize the female pronucleus to match the delay in sperm maturation. However, in many ways the mod/resc model is akin to the original "lock-and-key" model for enzyme action in that it is so general and broad in scope as to include everything while at the same time excluding little or nothing. As we now know, the lock-and-key model of enzyme action has been replaced by dynamic models that include, for example, substrate-induced conformational change (induced fit), binding of cofactors, and, most recently, enzyme action coupled to mechanochemical force transduction in polymeric-protein systems (Dickinson and Purich, 2002). This last point is relevant to the present discussion as the model was developed to explain bacterial motility inside eukaryotic cells and underscores the extent to which prevailing models of enzyme action (and, by analogy, *Wolbachia* action) must be modified or altered to explain existing data. It will be of great interest to follow the course of events as more cellular and molecular mechanistic details of CI and *Wolbachia* action are discovered that will, it is hoped, provide refined insights into the nature of *Wolbachia*-induced incompatibility.

ACKNOWLEDGMENTS

KB and HRB acknowledge funding by the European Commission (EUWOL, QLRT-2000-01079). KB is supported by intramural funds of the University of Ioannina and by funding from the Empirikion Foundation. HRB is supported by funding from the Biotechnology and Biological Research Council (5/S11854) and the Natural Environment Research Council (GR3/13199). TLK acknowledges support from the National Science Foundation (MCB-0135166) and the Royal Society Wolfson Merit Award program. The authors thank Dr. Harriet L. Harris and Dr. Stefan Oehler for reviewing the manuscript.

REFERENCES

Ameisen, J.C. (2002). On the origin, evolution, and nature of programmed cell death: a timeline of four billion years, *Cell Death Differ.* **9:** 367–393.

Anderson, C.L. and Karr, T.L. (2001). *Wolbachia*: evolutionary novelty in a rickettsial bacteria, *BMC. Evol. Biol.* **1:** 10.

Awahmukalah, D.S. and Brooks, M.A. (1983). Reproduction of an inbred strain of *Culex pipiens* prevented by loss of *Wolbachia pipientis*. *J. Invertebr. Pathol.* **41:** 184–190.

Baba, T. and Schneewind, O. (1998). Instruments of microbial warfare: bacteriocin synthesis, toxicity and immunity. *Trends Microbiol.* **6:** 66–71.

Bandi, C., Slatko, B., and O'Neill, S.L. (1999). *Wolbachia* genomes and the many faces of symbiosis. *Parasitol. Today* **15:** 428–429.

Beeman, R.W. and Friesen, K.S. (1999). Properties and natural occurrence of maternal-effect selfish genes ('*Medea*' factors) in the red flour beetle, *Tribolium castaneum*. *Heredity* **82:** 529–534.

Beeman, R.W., Friesen, K.S., and Denell, R.E. (1992). Maternal-effect selfish genes in flour beetles. *Science* **256:** 89–92.

Binnington, K.C. and Hoffmann, A.A. (1989). *Wolbachia*-like organisms and cytoplasmic incompatibility in *Drosophila simulans*. *J. Invertebr. Pathol.* **54:** 344–352.

Blickenstaff, C.C. (1965). Partial intersterility of Eastern and Western U.S. strains of the alfalfa weevil. *Ann. Entomol. Soc. Am.* **58:** 523–526.

Bohni, R., Riesgo-Escovar, J., Oldham, S., Brogiolo, W., Stocker, H., Andruss, B.F., Beckingham, K., and Hafen, E. (1999). Autonomous control of cell and organ size by CHICO, a *Drosophila* homolog of vertebrate IRS1–4. *Cell* **97:** 865–875.

Boller, E.F., Russ, K., Vallo, V., and Bush, G.L. (1976). Incompatible races of European cherry fruit fly *Rhagoletis cerasi* (Diptera: Tephritidae) their origin and potential use in biological control. *Entomol. Exp. Appl.* **20:** 237–247.

Boller, E.F. (1989). Cytoplasmic incompatibility in *Rhagoletis cerasi*. In *Fruit Flies, Their Biology, Natural Enemies and Control, World Crop Pests 3B* (A.S. Robinson and G. Hooper, Eds.), pp. 69–74. Elsevier, Amsterdam.

Bordenstein, S.R. and Werren, J.H. (1998). Effects of A and B *Wolbachia* and host genotype on interspecies cytoplasmic incompatibility in *Nasonia*. *Genetics* **148:** 1833–1844.

Bouchon, D., Rigaud, T., and Juchault, P. (1998). Evidence for widespread *Wolbachia* infection in isopod crustaceans: molecular identification and host feminization. *Proc. R. Soc. London (B)* **265:** 1081–1090.

Bourtzis, K. (2002). The European *Wolbachia* Project. 2nd International *Wolbachia* Conference July 9–15, Crete, Greece.

Bourtzis, K. and Braig, H.R. (1999). The many faces of *Wolbachia*. In *Rickettsiae and Rickettsial Diseases at the Turn of the Third Millennium* (D. Raoult and P. Brouqui, Eds.), pp. 199–219. Elsevier, Paris.

Bourtzis, K., Nirgianaki, A., Onyango, P., and Savakis, C. (1994). A prokaryotic *dnaA* sequence in *Drosophila melanogaster*: *Wolbachia* infection and cytoplasmic incompatibility among laboratory strains. *Insect Mol. Biol.* **3:** 131–142.

Bourtzis, K., Nirgianaki, A., Markakis, G., and Savakis, C. (1996). *Wolbachia* infection and cytoplasmic incompatibility in *Drosophila* species. *Genetics* **144:** 1063–1073.

Bourtzis, K., Dobson, S.L., Braig, H.R., and O'Neill, S.L. (1998). Rescuing *Wolbachia* have been overlooked. *Nature* **391:** 852–853.

Boyle, L., O'Neill, S.L., Robertson, H.M., and Karr, T.L. (1993). Inter- and intraspecific horizontal transfer of *Wolbachia* in *Drosophila*. *Science* **260:** 1796–1799.

Braig, H. R. and Yan, G. (2002). The spread of genetic constructs in natural insect populations. In *Genetically Engineered Organisms: Assessing Environmental and Human Health Effects* (D.K. Letourneau and B.E. Burrows, Eds.), pp. 251–314. CRC Press, Boca Raton, FL.

Braig, H.R., Guzman, H., Tesh, R.B., and O'Neill, S.L. (1994). Replacement of the natural *Wolbachia* symbiont of *Drosophila simulans* with a mosquito counterpart. *Nature* **367:** 453–455.

Braig, H.R., Zhou, W., Dobson, S.L., and O'Neill, S.L. (1998). Cloning and characterization of a gene encoding the major surface protein of the bacterial endosymbiont *Wolbachia pipientis*. *J. Bacteriol.* **180:** 2373–2378.

Braig, H.R., Turner, B.D., Normark, B.B., and Stouthamer, R. (2002). Microorganism-induced parthenogenesis. In *Progress in Asexual Reproduction. Volume XI. Reproductive Biology of Invertebrates* (R.N. Hughes, Ed.), pp. 1–62. John Wiley & Sons, Chichester, U.K.

Breeuwer, J.A.J. (1997). *Wolbachia* and cytoplasmic incompatibility in the spider mite *Tetranychus urticae* and *T. turkestani*. *Heredity* **79:** 41–47.

Breeuwer, J.A.J. and Werren, J.H. (1990). Microorganisms associated with chromosome destruction and reproductive isolation between two insect species. *Nature* **346:** 558–560.

Breeuwer, J.A.J. and Werren, J.H. (1993). Cytoplasmic incompatibility and bacterial density in *Nasonia vitripennis*. *Genetics* **135:** 565–574.

Bressac, C. and Rousset, F. (1993). The reproductive incompatibility system in *Drosophila simulans*: DAPI-staining analysis of the *Wolbachia* symbionts in sperm cysts. *J. Invertebr. Pathol.* **61:** 226–230.

Brower, J.H. (1976). Cytoplasmic incompatibility: occurrence in a stored product pest, *Ephestia cautella*. *Ann. Entomol. Soc. Am.* **69:** 1011–1015.

Brown, G.G. (1999). Unique aspects of cytoplasmic male sterility and fertility restoration in *Brassica napus*. *J. Heredity* **90:** 351–356.

Budar, F. and Pelletier, G. (2001). Male sterility in plants: occurrence, determinism, significance and use. *C. R. Acad. Sci. III Vie* **324:** 543–550.

Callaini, G., Riparbelli, M.G., Giordano, R., and Dallai, R. (1996). Mitotic defects associated with cytoplasmic incompatibility in *Drosophila simulans*. *J. Invertebr. Pathol.* **67:** 55–64.

Callaini, G., Dallai, R., and Riparbelli, M.G. (1997). *Wolbachia*-induced delay of paternal chromatin condensation does not prevent maternal chromosomes from entering anaphase in incompatible crosses of *Drosophila simulans*. *J. Cell Sci.* **110:** 271–280.

Charlat, S., Calmet, C., and Merçot, H. (2001). On the mod resc model and the evolution of *Wolbachia* compatibility types. *Genetics* **159:** 1415–1422.

Charlat, S., Bourtzis, K., and Merçot, H. (2002). *Wolbachia*-induced cytoplasmic incompatibility, in *Symbiosis, Mechanisms and Model Systems, Vol. 4, Cellular Origin and Life in Extreme Habitats* (J. Seckbach, Ed.), pp. 623–644. Kluwer, Dordrecht, the Netherlands.

Clancy, D.J. and Hoffmann, A.A. (1998). Environmental effects on cytoplasmic incompatibility and bacterial load in *Wolbachia*-infected *Drosophila simulans*. *Entomol. Exp. Appl.* **86:** 13–24.

Clark, M.E. and Karr, T.L. Distribution of *Wolbachia* within *Drosophila* reproductive tissues: implications for the expression of cytoplasmic incompatibility. *Integr. Comp. Biol.* In press.

Clark, M.E., Veneti, Z., Bourtzis, K., and Karr, T.L. (2002). The distribution and proliferation of the intracellular bacteria *Wolbachia* during spermatogenesis in *Drosophila*. *Mech. Dev.* **111:** 3–15.

Clark, M.E., Veneti, Z., Bourtzis, K., and Karr, T.L. *Wolbachia* distribution and cytoplasmic incompatibility in *Drosophila*: the cyst as the basic cellular unit of CI expression. *Mech. Dev.* In press.

Cline, T.W. and Meyer, B.J. (1996). Vive la difference: males vs. females in flies vs worms. *Annu. Rev. Genet.* **30:** 637–702.

Cooper, T.F. and Heinemann, J.A. (2000). Postsegregational killing does not increase plasmid stability but acts to mediate the exclusion of competing plasmids. *Proc. Natl. Acad. Sci. U.S.A.* **97:** 12643–12648.

Cosmides, L.M. and Tooby, J. (1981). Cytoplasmic inheritance and intragenomic conflict. *J. Theor. Biol.* **89:** 83–129.

Dao-Thi, M.-H., Charlier, D., Loris, R., Maes, D., Messens, J., Wyns, L., and Backmann, J. (2002). Intricate interactions within the ccd plasmid addiction system. *J. Biol. Chem.* **277:** 3733–3742.

Darwin, C. (1877). *The Different Forms of Flowers on Plants of the Same Species*. J. Murray, London.

Das, P., Maduzia, L.L., Wang, H., Finelli, A.L., Cho, S.H., Smith, M.M., and Padgett, R.W. (1998). The *Drosophila* gene *Medea* demonstrates the requirement for different classes of Smads in dpp signaling. *Development* **125:** 1519–1528.

De Barro, P.J. and Hart, P.J. (2000). Mating interactions between two biotypes of the whitefly, *Bemisia tabaci* (Hemiptera: Aleyrodidae) in Australia. *Bull. Entomol. Res.* **90:** 103–112.

Dickinson, R.B. and Purich, D.L. (2002). Clamped-filament elongation model for actin-based motors. *Biophys. J.* **82:** 605–617.

Dobson, S.L., Bourtzis, K., Braig, H.R., Jones, B.F., Zhou, W., Rousset, F., and O'Neill, S.L. (1999). *Wolbachia* infections are distributed throughout insect somatic and germ line tissue. *Insect Biochem. Mol. Biol.* **29:** 153–160.

Dobson, S.L. Marsland, E.J., and Rattanadechakul, W. (2002). Mutualistic *Wolbachia* infection in *Aedes albopictus*: accelerating cytoplasmic drive. *Genetics* **160:** 1087–1094.

Eddy, S.R. (2001). Non-coding RNA genes and the modern RNA world. *Nat. Rev. Genet.* **2:** 919–929.

Egas, M., Vala, F., and Breeuwer, J.A.J. (2002). On the evolution of cytoplasmic incompatibility in haplodiploid species. *Evolution* **56:** 1101–1109.

Engelberg-Kulka, H. and Glaser, G. (1999). Addiction modules and programmed cell death and antideath in bacterial cultures. *Annu. Rev. Microbiol.* **53:** 43–70.

Engelberg-Kulka, H., Reches, M., Narasimhan, S., Schoulaker-Schwarz, R., Klemes, Y., Aizenman, E., and Glaser, G. (1998). rexB of bacteriophage lambda is an anti-cell death gene. *Proc. Natl. Acad. Sci. U.S.A.* **95:** 15481–15486.

Fuller, M.T. (1993). Spermatogenesis. In *The Development of Drosophila melanogaster*, Vol. I (M. Bate and A.M. Arias, Eds.), pp. 71–147. Cold Spring Harbor Laboratory Press, Cold Spring Harbor, NY.

Gazit, E. and Sauer, R.T. (1999). The Doc toxin and Phd antidote proteins of the bacteriophage P1 plasmid addiction system form a heterotrimeric complex. *J. Biol. Chem.* **274:** 16813–16818.

Gerdes, K., Gultyaev, A.P., Franch, T., Pedersen, K., and Mikkelsen, N.D. (1997). Antisense RNA-regulated programmed cell death. *Annu. Rev. Genet.* **31:** 1–31.

Ghelelovitch, S. (1952). Sur le déterminisme génétique de la stérilité dans les croisements entre différentes souches de *Culex autogenicus* Roubaud. *C. R. Acad. Sci. III Vie* **234:** 2386–2388.

Giordano, R., O'Neill, S.L., and Robertson, H.M. (1995). *Wolbachia* infections and the expression of cytoplasmic incompatibility in *Drosophila sechellia* and *D. mauritiana*. *Genetics* **140:** 1307–1317.

Giordano, R., Jackson, J.J., and Robertson, H.M. (1997). The role of *Wolbachia* bacteria in reproductive incompatibilities and hybrid zones of *Diabrotica* beetles and *Gryllus* crickets. *Proc. Natl. Acad. Sci. U.S.A.* **94:** 11439–11444.

Goberdhan, D.C.I., Paricio, N., Goodman, E.C., Mlodzik, M., and Wilson, C. (1999). *Drosophila* tumor suppressor PTEN controls cell size and number by antagonizing the Chico/PI3-kinase signaling pathway. *Genes Dev.* **13:** 3244–3258.

Gotfredsen, M. and Gerdes, K. (1998). The *Escherichia coli relBE* genes belong to a new toxin-antitoxin gene family. *Mol. Microbiol.* **29:** 1065–1076.

Gotoh, T., Gomi, K., and Nagata, T. (1999). Incompatibility and host plant differences among populations of *Tetranychus kanzawai* Kishida (Acari: Tetranychidae). *Appl. Entomol. Zool.* **34:** 551–561.

Greenfield, T.J., Franch, T., Gerdes, K., and Weaver, K.E. (2001). Antisense RNA regulation of the par post-segregational killing system: structural analysis and mechanism of binding of the antisense RNA, RNAII and its target, RNAI. *Mol. Microbiol.* **42:** 527–537.

Gronlund, H. and Gerdes, K. (1999). Toxin-antitoxin systems homologous with *relBE* of *Escherichia coli* plasmid P307 are ubiquitous in prokaryotes. *J. Mol. Biol.* **285:** 1401–1415.

Grossniklaus, U., Vielle-Calzada, J.-P., Hoeppner, M.A., and Gagliano, W.B. (1998). Maternal control of embryogenesis by *MEDEA*, a *polycomb* group gene in *Arabidopsis*. *Science* **280:** 446–450.

Grossniklaus, U., Spillane, C., Page, D.R., and Kohler, C. (2001). Genomic imprinting and seed development: endosperm formation with and without sex. *Curr. Opinion Plant Biol.* **4:** 21–27.

Hadfield, S.J. and Axton, J.M. (1999). Germ cells colonized by endosymbiotic bacteria. *Nature* **402:** 482.

Handa, N., Nakayama, Y., Sadykov, M., and Kobayashi, I. (2001). Experimental genome evolution: large-scale genome rearrangements associated with resistance to replacement of a chromosomal restriction-modification gene complex. *Mol. Microbiol.* **40:** 932–940.

Hanson, M.R., Wilson, R.K., Bentolila, S., Kohler, R.H., and Chen, H.C. (1999). Mitochondrial gene organization and expression in petunia male fertile and sterile plants. *J. Heredity* **90:** 362–368.

Hargreaves, D., Giraldo, R., Santos-Sierra, S., Boelens, R., Rice, D.W., Diaz-Orejas R., and Rafferty, J.B. (2002). Crystallization and preliminary x-ray crystallographic studies on the *parD*-encoded protein Kid from *Escherichia coli* plasmid R1. *Acta Crystallogr. D* **58:** 355–358.

Harris, H.L. and Braig, H.R. (2001). Sperm nuclear basic proteins in *Drosophila simulans* undergoing *Wolbachia*-induced cytoplasmic incompatibility. *Dev. Biol.* **235:** 212.

Hastings, I.M. (1994). Selfish DNA as a method of pest control. *Philos. Trans. R. Soc. London (B)* **344:** 313–324.

Hatcher, M.J. (2000). Persistence of selfish genetic elements: population structure and conflict. *Trends Ecol. Evol.* **15:** 271–277.

Hayes, F. (1998). A family of stability determinants in pathogenic bacteria. *J. Bacteriol.* **180:** 6415–6418.

Hazan, R., Sat, B., Reches, M., and Engelberg-Kulka, H. (2001). Postsegregational killing mediated by the P1 phage "addiction module" phd-doc requires the *Escherichia coli* programmed cell death system mazEF. *J. Bacteriol.* **183:** 2046–2050.

Hertig, M. (1936). The rickettsia, *Wolbachia pipientis* (gen. et sp. n.) and associated inclusions of the mosquito, *Culex pipiens*. *Parasitology* **28:** 453–486.

Hertig, M. and Wolbach, S.B. (1924). Studies on rickettsia-like microorganisms in insects. *J. Med. Res.* **44:** 329–374.

Hille, M., Kies, S., Gotz, F., and Peschel, A. (2001). Dual role of GdmH in producer immunity and secretion of the staphylococcal lantibiotics gallidermin and epidermin. *Appl. Environ. Microbiol.* **67:** 1380–1383.

Hoffmann, A.A. (1988). Partial cytoplasmic incompatibility between two Australian populations of *Drosophila melanogaster*. *Entomol. Exp. Appl.* **48:** 61–67.

Hoffmann, A.A., Turelli, M., and Simmons, G.M. (1986). Unidirectional incompatibility between populations of *Drosophila simulans*. *Evolution* **40:** 692–701.

Hoffmann, A.A., Clancy, D.J., and Merton, E. (1994). Cytoplasmic incompatibility in Australian populations of *Drosophila melanogaster*. *Genetics* **136:** 993–999.

Hoffmann, A.A., Clancy, D.J., and Duncan, J. (1996). Naturally-occurring *Wolbachia* infection in *Drosophila simulans* that does not cause cytoplasmic incompatibility. *Heredity* **76:** 1–8.

Horn, R. (2002). Molecular diversity of male sterility inducing and male-fertile cytoplasms in the genus *Helianthus*. *Theor. Appl. Genet.* **104:** 562–570.

Hsiao, C. and Hsiao, T.H. (1985). Rickettsia as the cause of cytoplasmic incompatibility in the alfalfa weevil, *Hypera postica*. *J. Invertebr. Pathol.* **45:** 244–246.

Huang, H., Potter, C.J., Tao, W.F., Li, D.M., Brogiolo, W., Hafen, E., Sun, H., and Xu, T.A. (1999). PTEN affects cell size, cell proliferation and apoptosis during *Drosophila* eye development. *Development* **126:** 5365–5372.

Hudson, J.B., Podos, S.D., Keith, K., Simpson, S.L., and Ferguson, E.L. (1998). The *Drosophila Medea* gene is required downstream of dpp and encodes a functional homolog of human Smad4. *Development* **125:** 1407–1420.

Hurst, L.D. (1993). *scat*+ Is a selfish gene analogous to *Medea* of *Tribolium castaneum*. *Cell* **75:** 407–408.

Hurst, G.D.D., Johnson, A.P., von der Schulenburg, J.H.G., and Fuyama, Y. (2000). Male-killing *Wolbachia* in *Drosophila*: a temperature sensitive trait with a threshold bacterial density. *Genetics* **156:** 699.

Jacobs, H.T. (1991). Structural similarities between a mitochondrially encoded polypeptide and a family of prokaryotic respiratory toxins involved in plasmid maintenance suggest a novel mechanism for the evolutionary maintenance of mitochondrial DNA. *J. Mol. Evol.* **32:** 333–339.

James, A.C. and Ballard, J.W.O. (2000). Expression of cytoplasmic incompatibility in *Drosophila simulans* and its impact on infection frequencies and distribution of *Wolbachia pipientis*. *Evolution* **54:** 1661–1672.

James, A.C., Dean, M.D., McMahon, M.E., and Ballard, J.W.O. (2002). Dymanics of double and single *Wolbachia* infections in *Drosophila simulans* from New Caledonia. *Heredity* **88:** 182–189.

Jamnongluk, W., Kittayapong, P., Baisley, K.J., and O'Neill, S.L. (2001). *Wolbachia* infection and expression of cytoplasmic incompatibility in *Armigeres subalbatus* (Diptera: Culicidae). *J. Med. Entomol.* **37:** 53–57.

Jensen, R.B. and Gerdes, K. (1995). Programmed cell death in bacteria — proteic plasmid stabilization systems. *Mol. Microbiol.* **17:** 205–210.

Jiang, Y., Pogliano, J., and Helinski, D.R. (2002). ParE toxin encoded by the broad-host-range plasmid RK2 is an inhibitor of *Escherichia coli* gyrase. *Mol. Microbiol.* **44:** 971–979.

Johanowicz, D.L. and Hoy, M.A. (1998). Experimental induction and termination of non-reciprocal reproductive incompatibilities in a parahaploid mite. *Entomol. Exp. Appl.* **87:** 51–58.

Jost, E. (1970). Genetische Untersuchungen zur Kreuzungssterilität im *Culex pipiens* complex. *Theor. Appl. Genet.* **40:** 251–256.

Juchault, P., Frelon, M., Bouchon, D., and Rigaud, T. (1994). New evidence for feminizing bacteria in terrestrial isopods: evolutionary implications. *C. R. Acad. Sci. III Vie* **317:** 225–230.

Kambhampati, S., Rai, K.S., and Burgun, S.J. (1993). Unidirectional cytoplasmic incompatibility in the mosquito, *Aedes albopictus*. *Evolution* **47:** 673–677.

Kamoda, S., Masui, S., Ishikawa, H., and Sasaki, T. (2000). *Wolbachia* infection and cytoplasmic incompatibility in the cricket *Teleogryllus taiwanemma*. *J. Exp. Biol.* **203:** 2503–2509.

Karr, T.L. (1994). Giant steps sideways. *Curr. Biol.* **4:** 537–540.

Karr, T.L., Yang, W., and Feder, M.E. (1998). Overcoming cytoplasmic incompatibility in *Drosophila*. *Proc. R. Soc. London (B)* **265:** 391–395.

Kim, W.S., Hall, R.J., and Dunn, N.W. (1998). Improving nisin production by increasing nisin immunity/resistance genes in the producer organism *Lactococcus lactis*. *Appl. Microbiol. Biotechnol.* **50:** 429–433.

Kittayapong, P., Mongkalangoon, P., Baimai, V., and O'Neill, S.L. (2002). Host age effect and expression of cytoplasmic incompatibility in field populations of *Wolbachia*-superinfected *Aedes albopictus*. *Heredity* **88:** 270–274.

Knoop, V. and Brennicke, A. (2002). Molecular biology of the plant mitochondrion. *Crit. Rev. Plant Sci.* **21:** 111–126.

Kobayashi, I. (2001). Behavior of restriction-modification systems as selfish mobile elements and their impact on genome evolution. *Nucleic Acids Res.* **29:** 3742–3756.

Kose, H. (1995). Cytological and biochemical analyses of *Wolbachia pipientis*, causative reagent of cytoplasmic incompatibility in *Drosophilia*. Ph.D. thesis, University of Illinois, Urbana-Champaign.

Kose, H. and Karr, T.L. (1995). Organization of *Wolbachia pipientis* in the *Drosophila* fertilized egg and embryo revealed by an anti-*Wolbachia* monoclonal antibody. *Mech. Dev.* **51:** 275–288.

Kristoffersen, P., Jensen, G.B., Gerdes, K., and Piskur, J. (2000). Bacterial toxin-antitoxin gene system as containment control in yeast cells. *Appl. Environ. Microbiol.* **66:** 5524–5526.

Kroemer, G. (1997). Mitochondrial implication in apoptosis: towards an endosymbiont hypothesis of apoptosis evolution. *Cell Death Differ.* **4:** 443–456.

Kulakauskas, S., Lubys, A., and Ehrlich, S.D. (1995). DNA restriction-modification systems mediate plasmid maintenance. *J. Bacteriol.* **177:** 3451–3454.

Lachaise, D., Harry, M., Solignac, M., Lemeunier, F., Bénassi, V., and Cariou, M.-L. (2000). Evolutionary novelties in islands: *Drosophila santomea*, a new *melanogaster* sister species from São Tomé. *Proc. R. Soc. London (B)* **267:** 1487–1495.

Lassy, C.W. and Karr, T.L. (1996). Cytological analysis of fertilization and early embryonic development in incompatible crosses of *Drosophila simulans*. *Mech. Dev.* **57:** 47–58.

Laven, H. (1951). Crossing experiments with *Culex* strains. *Evolution* **5:** 370–375.

Laven, H. (1953). Reziprok unterschiedliche Kreuzbarkeit von Stechmücken (Culicidae) und ihre Deutung als plasmatische Vererbung. *Z. Indukt. Abstamm. Vererbungsl.* **85:** 118–136.

Laven, H. (1967). Eradication of *Culex pipiens fatigans* through cytoplasmic incompatibility. *Nature* **216:** 383–384.

Law, R. and Hutson, V. (1992). Intracellular symbionts and the evolution of uniparental cytoplasmic inheritance. *Proc. R. Soc. London (B)* **248:** 69–77.

Legrand, J.J., Juchault, P., Moraga, D., and Legrand-Hamelin, E. (1986). Symbiotic microorganisms and speciation. *Bull. Soc. Zool. France – Évol. Zool.* **111:** 135–147.

L'Homme, Y., Stahl, R.J., Li, X.Q., Hameed, A., and Brown, G.G. (1997). *Brassica* nap cytoplasmic male sterility is associated with expression of a mtDNA region containing a chimeric gene similar to the pol CMS-associated orf224 gene. *Curr. Genet.* **31:** 325–335.

Lin, L.F., Posfai, J., Roberts, R.J., and Kong, H.M. (2001). Comparative genomics of the restriction-modification systems in *Helicobacter pylori*. *Proc. Natl. Acad. Sci. U.S.A.* **98:** 2740–2745.

Lindsley, D.L. and Tokuyasu, K.T. (1980). Spermatogenesis. In *The Genetics and Biology of* Drosophila, 2nd ed. (M. Ashburner and T.R.F. Wright, Eds.), pp. 226–294. Academic Press, London.

Lindsley, D.L. and Zimm, G.G. (1992). *The Genome of Drosophila melanogaster.* Academic Press, San Diego.

Liu, F., Cui, X.Q., Horner, H.T., Weiner, H., and Schnable, P.S. (2001). Mitochondrial aldehyde dehydrogenase activity is required for male fertility in maize. *Plant Cell* **13:** 1063–1078.

Mandel, M.J., Ross, C.L., and Harrison, R.G. (2001). Do *Wolbachia* infections play a role in unidirectional incompatibilities in a field cricket hybrid zone? *Mol. Ecol.* **10:** 703–709.

Marshall, J.F. and Staley, J. (1937). Some notes regarding the morphological and biological differentiation of *Culex pipiens* Linnaeus and *Culex molestus* Forskal (Diptera, Culicidae). *Proc. R. Entomol. Soc. London A* **12:** 17–26.

McAuliffe, O., Ross, R.P., and Hill, C. (2001). Lantibiotics: structure, biosynthesis and mode of action. *FEMS Microbiol. Rev.* **25:** 285–308.

McGraw, E.A. and O'Neill, S.L. (1999). Evolution of *Wolbachia pipientis* transmission dynamics in insects. *Trends Microbiol.* **7:** 297–302.

McGraw, E.A., Merritt, D.J., Droller, J.N., and O'Neill, S.L. (2001). *Wolbachia*-mediated sperm modification is dependent on the host genotype in *Drosophila*. *Proc. R. Soc. London (B)* **268:** 2565–2570.

McGraw, E.A., Merritt, D.J., Droller, J.N., and O'Neill, S.L. (2002). *Wolbachia* density and virulence attenuation after transfer into a novel host. *Proc. Natl. Acad. Sci. U.S.A.* **99:** 2918–2923.

Meinhart, A., Alings, C., Strater, N., Camacho, A.G., Alonso, J.C., and Saenger, W. (2001). Crystallization and preliminary x-ray diffraction studies of the epsilon zeta addiction system encoded by *Streptococcus pyogenes* plasmid pSM19035. *Acta Crystallogr. D* **57:** 745–747.

Merçot, H. and Poinsot, D. (1998). Rescuing *Wolbachia* have been overlooked and discovered on Mount Kilimanjaro. *Nature* **391:** 853.

Merçot, H., Llorente, B., Jacques, M., Atlan, M., and Montchamp-Moreau, C. (1995). Variability within the Seychelles cytoplasmic incompatibility system in *Drosophila simulans*. *Genetics* **141:** 1015–1023.

Min, K.T. and Benzer, S. (1997). *Wolbachia*, normally a symbiont of *Drosophila*, can be virulent, causing degeneration and early death. *Proc. Natl. Acad. Sci. U.S.A.* **94:** 10792–10796.

Møller, I.M. (2001). A more general mechanism of cytoplasmic male fertility? *Trends Plant Sci.* **6:** 560.

Moret, Y., Juchault, P., and Rigaud, T. (2001). *Wolbachia* endosymbiont responsible for cytoplasmic incompatibility in a terrestrial crustacean: effects in natural and foreign hosts. *Heredity* **86:** 325–332.

Murray, N.E. (2002). Immigration control of DNA in bacteria: self versus non-self. *Microbiology* **148:** 3–20.

Nagel, J.H.A., Gultyaev, A.P., Gerdes, K., and Pleij, C.W.A. (1999). Metastable structures and refolding kinetics in *hok* mRNA of plasmid R1. *RNA* **5:** 1408–1418.

Naito, T., Kusano, K., and Kobayashi, I. (1995). Selfish behavior of restriction-modification systems. *Science* **267:** 897–899.

Nakayama, K., Takashima, K., Ishihara, H., Shinomiya, T., Kageyama, M., Kanaya, S., Ohnishi, M., Murata, T., Mori, H., and Hayashi, T. (2000). The R-type pyocin of *Pseudomonas aeruginosa* is related to P2 phage, and the F-type is related to lambda phage. *Mol. Microbiol.* **38:** 213–231.

Ndiaye, M. and Mattei, X. (1993). Endosymbiotic relationship between a rickettsia-like microorganism and the male germ-cells of *Culex tigripes*. *J. Submicrosc. Cytol. Pathol.* **25:** 71–77.

Noda, H. (1984). Cytoplasmic incompatibility in allopatric field populations of the small brown planthopper, *Laodelphax striatellus*, in Japan. *Entomol. Exp. Appl.* **35:** 263–267.

Noda, H., Koizumi, Y., Zhang, Q., and Deng, K. (2001). Infection density of *Wolbachia* and incompatibility level in two planthopper species, *Laodelphax striatellus* and *Sogatella furcifera*. *Insect Biochem. Mol. Biol.* **31:** 727–737.

Oberer, M., Zangger, K., Prytulla, S., and Keller, W. (2002). The anti-toxin ParD of plasmid RK2 consists of two structurally distinct moieties and belongs to the ribbon-helix-helix family of DNA-binding proteins. *Biochem. J.* **361:** 41–47.

O'Connor, E.M. and Shand, R.F. (2002). Halocins and sulfolobicins: the emerging story of archaeal protein and peptide antibiotics. *J. Indust. Microbiol. Biotech.* **28:** 23–31.

Oehler, S. and Bourtzis, K. (2000). First international *Wolbachia* conference: *Wolbachia* 2000. *Symbiosis* **29:** 151–161.

O'Neill, S.L. (1999). *Wolbachia*: why these bacteria are important to genome research. *Microb. Comp. Genomics* **3:** 159.

O'Neill, S.L. (2002). Using genomics approaches to better understand *Wolbachia*/host interactions. Second International *Wolbachia* Conference. July 9–15, Crete, Greece.

O'Neill, S.L. and Karr, T.L. (1990). Bidirectional cytoplasmic incompatibility between conspecific populations of *Drosophila simulans*. *Nature* **348:** 178–180.

Parret, A.H.A. and De Mot, R. (2002). Bacteria killing their own kind: novel bacteriocins of pseudomonas and other gamma-proteobacteria. *Trends Microbiol.* **10:** 107–112.

Pedersen, K. and Gerdes, K. (1999). Multiple *hok* genes on the chromosome of *Escherichia coli*. *Mol. Biol.* **32:** 1090–1102.

Perrot-Minnot, M.-J. and Werren, J.H. (1999). *Wolbachia* infection and incompatibility dynamics in experimental selection lines. *J. Evol. Biol.* **12:** 272–282.

Perrot-Minnot, M.-J., Guo, L.R., and Werren, J.H. (1996). Single and double infections with *Wolbachia* in the parasitic wasp *Nasonia vitripennis*: effects on incompatibility. *Genetics* **143:** 961–972.

Peters, L.L. and Barker, J.E. (1993). Novel inheritance of the murine severe combined anemia and thrombocytopenia (*scat*) phenotype. *Cell* **74:** 135–142.

Picardeau, M., Ren, S.X., and Girons, I.S. (2001). Killing effect and antitoxic activity of the *Leptospira interrogans* toxin-antitoxin system in *Escherichia coli*. *J. Bacteriol.* **183:** 6494–6497.

Poinsot, D. and Merçot, H. (1999). *Wolbachia* can rescue from cytoplasmic incompatibility while being unable to induce it. In *From Symbiosis to Eukaryotism — Endocytobiology VII* (E. Wagner et al., Eds.), pp. 221–234. Universities of Geneva, Switzerland, and Freiburg im Breisgau, Germany.

Poinsot, D., Bourtzis, K., Markakis, G., Savakis, C., and Merçot, H. (1998). *Wolbachia* transfer from *Drosophila melanogaster* to *D. simulans*: host effect and cytoplasmic incompatibility relationships. *Genetics* **150:** 227–237.

Presgraves, D.C. (2000). A genetic test of the mechanism of *Wolbachia*-induced cytoplasmic incompatibility in *Drosophila*. *Genetics* **154:** 771–776.

Priolo, L.M., Huang, J.T., and Levings, C.S. (1993). The plant mitochondrial open reading frame *orf221* encodes a membrane-bound protein. *Plant Mol. Biol.* **23:** 287–295.

Rawlings, D.E. (1999). Proteic toxin-antitoxin: bacterial plasmid addiction systems and their evolution with special reference to the *pas* system of pTF-FC2. *FEMS Microbiol. Lett.* **176:** 269–277.

Reed, K.M. and Werren, J.H. (1995). Induction of paternal genome loss by the paternal-sex-ratio chromosome and cytoplasmic incompatibility bacteria (*Wolbachia*): a comparative study of early embryonic events. *Mol. Reprod. Dev.* **40:** 408–418.

Reynolds, K.T. and Hoffmann, A.A. (2002). Male age and the weak expression or non-expression of cytoplasmic incompatibility in *Drosophila* strains infected by maternally-transmitted *Wolbachia. Genet. Res.* In press.

Rhoads, D.M., Brunner-Neuenschwander, B.B., Levings, C.S., and Siedow, J.N. (1998). Cross-linking and disulfide bond formation of introduced cysteine residues suggest a modified model for the tertiary structure of URF13 in the pore-forming oligomers. *Arch. Biochem. Biophys.* **354:** 158–164.

Riley, M.A. (1998). Molecular mechanisms of bacteriocin evolution. *Annu. Rev. Genet.* **32:** 255–278.

Riley, M.A., Pinou, T., Wertz, J.E., Tan, Y., and Valletta, C.M. (2001). Molecular characterization of the klebicin B plasmid of *Klebsiella pneumoniae. Plasmid* **45:** 209–221.

Roberts, R.J. and Macelis, D. (2001). REBASE — restriction enzymes and methylases. *Nucleic Acids Res.* **29:** 268–269.

Rousset, F. and de Stordeur, E. (1994). Properties of *Drosophila simulans* strains experimentally infected by different clones of the bacteria *Wolbachia. Heredity* **72:** 325–331.

Rousset, F. and Solignac, M. (1995). Evolution of single and double *Wolbachia* symbioses during speciation in the *Drosophila simulans* complex. *Proc. Natl. Acad. Sci. U.S.A.* **92:** 6389–6393.

Rousset, F., Bouchon, D., Pintureau, B., Juchault, P., and Solignac, M. (1992). *Wolbachia* endosymbionts responsible for various alterations of sexuality in arthropods. *Proc. R. Soc. London (B)* **250:** 91–98.

Rousset, F., Braig, H.R., and O'Neill, S.L. (1999). A stable triple *Wolbachia* infection in *Drosophila* with nearly additive incompatibility effects. *Heredity* **82:** 620–627.

Ryan, S.L. and Saul, G.B., II. (1968). Post-fertilization effect of incompatibility factors in *Mormoniella. Mol. Gen. Genet.* **103:** 29–36.

Santos-Sierra, S., Giraldo, R., and Diaz-Orejas, R. (1997). Functional interactions between homologous conditional killer systems of plasmid and chromosomal origin. *FEMS Microbiol. Lett.* **152:** 51–56.

Santos-Sierra, S., Giraldo, R., and Diaz-Orejas, R. (1998). Functional interactions between chpB and parD, two homologous conditional killer systems found in the *Escherichia coli* chromosome and in plasmid R 1. *FEMS Microbiol. Lett.* **168:** 51–58.

Santos-Sierra, S., Pardo-Abarrio, C., Giraldo, R., and Diaz-Orejas, R. (2002). Genetic identification of two functional regions in the antitoxin of the parD killer system of plasmid R1. *FEMS Microbiol. Lett.* **206:** 115–119.

Sasaki, T. and Ishikawa, H. (1999). *Wolbachia* infections and cytoplasmic incompatibility in the almond moth and the Mediterranean flour moth. *Zool. Sci.* **16:** 739–744.

Sasaki, T., Braig, H.R., and O'Neill, S.L. (1998). Analysis of *Wolbachia* protein synthesis in *Drosophila in vivo. Insect Mol. Biol.* **7:** 101–105.

Sat, B., Hazan, R., Fisher, T., Khaner, H., Glaser, G., and Engelberg-Kulka, H. (2001). Programmed cell death in *Escherichia coli*: some antibiotics can trigger mazEF lethality. *J. Bacteriol.* **183:** 2041–2045.

Saul, G.B. (1961). An analysis of non-reciprocal cross incompatibility in *Mormoniella vitripennis* (Walker). *Z. Indukt. Abstamm. Vererbungsl.* **92:** 28–33.

Sayeed, S., Reaves, L., Radnedge, L., and Austin, S. (2000). The stability region of the large virulence plasmid of *Shigella flexneri* encodes an efficient postsegregational killing system. *J. Bacteriol.* **182:** 2416–2421.

Schnable, P. (2001). A more general mechanism of cytoplasmic male fertility? *Trends Plant Sci.* **6:** 560.

Schnable, P.S. and Wise, R.P. (1998). The molecular basis of cytoplasmic male sterility and fertility restoration. *Trends Plant Sci.* **3:** 175–180.

Sewalt, R.G.A.B., Lachner, M., Vargas, M., Hamer, K.M., den Blaauwen, J.L., Hendrix, T., Melcher, M., Schweizer, D., Jenuwein, T., and Otte, A.P. (2002). Selective interactions between vertebrate polycomb homologs and the SUV39H1 histone lysine methyltransferase suggest that histone H3-K9 methylation contributes to chromosomal targeting of Polycomb group proteins. *Mol. Cell. Biol.* **22:** 5539–5553.

Sinkins, S.P., Braig, H.R., and O'Neill, S.L. (1995a). *Wolbachia* superinfections and the expression of cytoplasmic incompatibility. *Proc. R. Soc. London (B)* **261:** 325–330.

Sinkins, S.P., Braig, H.R., and O'Neill, S.L. (1995b). *Wolbachia pipientis*: bacterial densities and unidirectional cytoplasmic incompatibility between infected populations of *Aedes albopictus. Exp. Parasitol.* **81:** 284–291.

Slatko, B.E., O'Neill, S.L., Scott, A.L., Werren, J.H., Blaxter, M.L., and the Wolbachia Genome Consortium. (1999). Meeting summary: The *Wolbachia* Genome Consortium. *Microb. Comp. Genomics* **3:** 161–165.

Smith, N.G.C. (1998). The dynamics of maternal-effect selfish genetic elements. *J. Theor. Biol.* **191:** 173–180.

Snook, R.R., Cleland, S.Y., Wolfner, M.F., and Karr, T.L. (2000). Offsetting effects of *Wolbachia* infection and heat shock on sperm production in *Drosophila simulans*: comparative analyses of fecundity, fertility and accessory gland proteins. *Genetics* **155:** 167–178.

Solignac, M., Vautrin, D., and Rousset, F. (1994). Widespread occurence of the proteobacteria *Wolbachia* and partial cytoplasmic incompatibility in *Drosophila melanogaster. C. R. Acad. Sci. III Vie.* **317:** 461–470.

Spielman, M., Vinkenoog, R., Dickinson, H.G., and Scott, R.J. (2001). The epigenetic basis of gender in flowering plants and mammals. *Trends Genet.* **17:** 705–711.

Starr, D.J. and Cline, T. W. (2002). A host-parasite interaction rescues *Drosophila* oogenesis defects. *Nature* **418:** 76–79.

Stein, T., Borchert, S., Kiesau, P., Heinzmann, S., Kloss, S., Klein, C., Helfrich, M., and Entian, K.D. (2002). Dual control of subtilin biosynthesis and immunity in *Bacillus subtilis. Mol. Microbiol.* **44:** 403–416.

Stevens, L., Giordano, R., and Fialho, R.F. (2001). Male-killing, nematode infections, bacteriophage infection, and virulence of cytoplasmic bacteria in the genus *Wolbachia. Annu. Rev. Ecol. Syst.* **32:** 519–545.

Stouthamer, R., Breeuwer, J.A.J., and Hurst, G.D.D. (1999). *Wolbachia pipientis*: microbial manipulator of arthropod reproduction. *Annu. Rev. Microbiol.* **53:** 71–102.

Strauch, E., Kaspar, H., Schaudinn, C., Dersch, P., Madela, K., Gewinner, C., Hertwig, S., Wecke, J., and Appel, B. (2001). Characterization of enterocoliticin, a phage tail-like bacteriocin, and its effect on pathogenic *Yersinia enterocolitica* strains. *Appl. Environ. Microbiol.* **67:** 5634–5642.

Terachi, T., Yamaguchi, K., and Yamagishi, H. (2001). Sequence analysis on the mitochondrial *orfB* locus in normal and Ogura male-sterile cytoplasms from wild and cultivated radishes. *Curr. Genet.* **40:** 276–281.

Thomson, M.S. and Beeman, R.W. (1999). Assisted suicide of a selfish gene. *J. Heredity* **90:** 191–194.

Thomson, M.S., Friesen, K.S., Denell, R.E., and Beeman, R.W. (1995). A hybrid incompatibility factor in *Tribolium castaneum. J. Heredity* **86:** 6–11.

Tian, Q.B., Ohnishi, M., Murata, T., Nakayama, K., Terawaki, Y., and Hayashi, T. (2001). Specific protein-DNA and protein-protein interaction in the hig gene system, a plasmid-borne proteic killer gene system of plasmid Rts1. *Plasmid* **45:** 63–74.

Tram, U. and Sullivan, W. (2002). Role of delayed nuclear envelope breakdown and mitosis in *Wolbachia*-induced cytoplasmic incompatibility. *Science* **296:** 1124–1126.

Trpis, M., Perrone, J.B., Reissig, M., and Parker, K.L. (1981). Control of cytoplasmic incompatibility in the *Aedes scutellaris* complex. *J. Heredity* **72:** 313–317.

Turelli, M. and Hoffmann, A.A. (1995). Cytoplasmic incompatibility in *Drosophila simulans*: dynamics and parameter estimates from natural populations. *Genetics* **140:** 1319–1338.

Vala, F., Breeuwer, J.A.J., and Sabelis, M.W. (2000). *Wolbachia*-induced "hybrid breakdown" in the two-spotted spider mite *Tetranychus urticae* Koch. *Proc. R. Soc. London (B)* **267:** 1931–1937.

Van Borm, S., Wenseleers, T., Billen, J., and Boomsma, J.J. (2001). *Wolbachia* in leafcutter ants: a widespread symbiont that may induce male killing or incompatible matings. *J. Evol. Biol.* **14:** 805–814.

Van Meer, M.M. and Stouthamer, R. (1999). Cross-order transfer of *Wolbachia* from *Muscidifurax uniraptor* (Hymenoptera: Pteromalidae) to *Drosophila simulans* (Diptera: Drosophilidae). *Heredity* **82:** 163–169.

Van Melderen, L. (2002). Molecular interactions of the CcdB poison with its bacterial target, the DNA gyrase. *Int. J. Med. Microbiol.* **291:** 537–544.

Vavre, F., Fleury, F., Varaldi, J., Fouillet, P., and Boulétreau, M. (2000). Evidence for female mortality in *Wolbachia*-mediated cytoplasmic incompatibility in haplodiploid insects: epidemiologic and evolutionary consequences. *Evolution* **54:** 191–200.

Vavre, F., Dedeine, F., Quillon, M., Fouillet, P., Fleury, F., and Boulétreau, M. (2001). Within-species diversity of *Wolbachia*-induced cytoplasmic incompatibility in haplodiploid insects. *Evolution* **55:** 1710–1714.

Vavre, F., Fleury, F., Varaldi, J., Fouillet, P., and Boulétreau, M. (2002). Infection polymorphism and cytoplasmic incompatibility in Hymenoptera-*Wolbachia* associations. *Heredity* **88:** 361–365.

Veneti, Z., Clark, M.E., Zabalou, S., Savakis, C., Karr, T.L., and Bourtzis, K. Cytoplasmic incompatibility and *Wolbachia* infection during spermatogenesis in different *Drosophila*-bacterial associations. *Genetics*. In review.

Vielle-Calzada, J.P., Thomas, J., Spillane, C., Coluccio, A., Hoeppner, M.A., and Grossniklaus, U. (1999). Maintenance of genomic imprinting at the *Arabidopsis medea* locus requires zygotic *DDM1* activity. *Genes Dev.* **13:** 2971–2982.

Wade, M.J. and Beeman, R.W. (1994). The population dynamics of maternal-effect selfish genes. *Genetics* **138:** 1309–1314.

Wade, M.J. and Stevens, L. (1985). Microorganism mediated reproductive isolation in flour beetles (genus *Tribolium*). *Science* **227:** 527–528.

Wagner, E.G.H. and Flärdh, K. (2002). Antisense RNAs everywhere? *Trends Genet.* **18:** 223–226.

Wagner, E.G.H. and Simons, R.W. (1994). Antisense RNA control in bacteria, phages, and plasmids. *Annu. Rev. Microbiol.* **48:** 713–724.

Weeks, A.R., Reynolds, K.T., and Hoffmann, A.A. (2002). *Wolbachia* dynamics and host effects: what has (and has not) been demonstrated? *Trends Ecol Evol.* **17:** 257–262.

Werren, J.H. (1997). Biology of *Wolbachia. Annu. Rev. Entomol.* **42:** 587–609.

Werren, J.H. and Jaenike, J. (1995). *Wolbachia* and cytoplasmic incompatibility in mycophagous *Drosophila* and their relatives. *Heredity* **75:** 320–326.

Werren, J.H. and O'Neill, S.L. (1997). The evolution of heritable symbionts. In *Influential Passengers: Inherited Microorganisms and Arthropod Reproduction* (S.L. O'Neill, A.A. Hoffman, and J.H. Werren, Eds.), pp. 1–41. Oxford University Press, Oxford, U.K.

Werren, J.H. and Windsor, D.M. (2000). *Wolbachia* infection frequencies in insects: evidence of a global equilibrium? *Proc. R. Soc. London (B)* **267:** 1277–1285.

Werren, J.H., Zhang, W., and Guo, L.R. (1995). Evolution and phylogeny of *Wolbachia*-reproductive parasites of arthropods. *Proc. R. Soc. London (B)* **261:** 55–63.

Wisotzkey, R.G., Mehra, A., Sutherland, D.J., Dobens, L.L., Liu, X.Q., Dohrmann, C., Attisano, L., and Raftery, L.A. (1998). Medea is a *Drosophila* Smad4 homolog that is differentially required to potentiate DPP responses. *Development* **125:** 1433–1445.

Yadegari, R., Kinoshita, T., Lotan, O., Cohen, G., Katz, A., Choi, Y., Katz, A., Nakashima, K., Harada, J.J., Goldberg, R.B., Fischer, R.L., and Ohad, N. (2000). Mutations in the *FIE* and *MEA* genes that encode interacting polycomb proteins cause parent-of-origin effects on seed development by distinct mechanisms. *Plant Cell* **12:** 2367–2381.

Yamamoto, T.-A.M., Gerdes, K., and Tunnacliffe, A. (2002). Bacterial toxin RelE induces apoptosis in human cells. *FEBS Lett.* **519:** 191–194.

Yen, J.H. (1972). The microorganismal basis of cytoplasmic incompatibility in the *Culex pipiens* complex. Ph.D. thesis, University of California, Los Angeles.

Yen, J.H. and Barr, A.R. (1971). New hypothesis of the cause of cytoplasmic incompatibility in *Culex pipiens* L. *Nature* **232:** 657–658.

Yen, J.H. and Barr, A.R. (1973). The etiological agent of cytoplasmic incompatibility in *Culex pipiens. J. Invertebr. Pathol.* **22:** 242–250.

Yen, J.H. and Barr, A.R. (1974). Incompatibility in *Culex pipiens.* In *The Use of Genetics in Insect Control* (R. Pal and M.J. Whitten, Eds.), pp. 97–118. Elsevier-North Holland, Amsterdam.

Zhou, W., Rousset, F., and O'Neill, S.L. (1998). Phylogeny and PCR-based classification of *Wolbachia* strains using *wsp* gene sequences. *Proc. R. Soc. London (B)* **265:** 509–515.

15 Parthenogenesis Associated with *Wolbachia*

Martinus E. Huigens and Richard Stouthamer

CONTENTS

PARTHENOGENESIS MEDIATED BY MICROBIAL INFECTIONS

Since the 19th century biologists have been puzzled by the fact that sexual reproduction is so common and that only some, mostly lower, organisms have parthenogenetic reproduction. Parthenogenetic reproduction has advantages due to its simplicity, efficiency, effectiveness, and low cost (Crow, 1994). For an individual, many disadvantages are associated with sex, such as finding a mate, competition for mates, and the chance of obtaining sexually transmitted diseases. In one of the many reviews on the advantages of sexual reproduction, Crow (1994) mentions two of the most plausible explanations. The "oldest" explanation is that greater genetic variability in a population through sex can keep the population level at pace with changes in the environment. Second, harmful mutations can be eliminated more easily through recombination and do not accumulate according to the Muller's ratchet (Felsenstein, 1974), as happens in parthenogenetic populations (Muller, 1964). Most studies on the advantages of sexual reproduction are theoretical, and hypotheses remain very difficult to test experimentally.

Recent studies on different modes of reproduction have forced us to change our perspective because the evolutionary forces behind parthenogenetic reproduction are often not the genes of the organism itself but those of their bacterial symbionts (Stouthamer et al., 1990a). In this chapter, we will focus on microbe-induced parthenogenesis.

Studies on extraordinary sex ratios by Hamilton (1967, 1979) and Cosmides and Tooby (1981) suggested that the differences in inheritance of cytoplasmic factors and nuclear genes could result in a conflict over offspring sex ratios. While cytoplasmic factors that are maternally inherited benefit from an offspring sex ratio of 100% females, nuclear genes located on autosomal chromosomes favor an optimal sex ratio with at least some male offspring in populations that are able to reproduce

0-8493-1286-8/03/$0.00+$1.50

sexually. Many cytoplasmically inherited factors cause female-biased sex ratios (Werren et al., 1988; Wrensch and Ebbert, 1993), but the extreme bias of 100% females and the induction of parthenogenesis were shown for the first time by Stouthamer et al. (1990a). In many cases, microbe-induced parthenogenesis may not be optimal for the organism (host) but is forced upon them by the fitness advantages their symbionts derive from their host's parthenogenetic reproduction.

These parthenogenesis-inducing (PI) symbionts inhabit the cytoplasm and can therefore be transmitted only vertically through egg cells. Males are considered a dead end for the symbionts. By inducing parthenogenesis the symbionts enhance their own transmission despite the cost they may inflict on their host. Two different microbes have been described as being associated with parthenogenesis in insects, and a potential third case was found in nematodes.

At present the most common symbiotic microbe found to induce parthenogenesis is the bacterium *Wolbachia pipientis* (Chapter 13). It was first identified as the microbe inducing parthenogenesis in parasitoid wasps of the genus *Trichogramma* (Rousset et al., 1992; Stouthamer et al., 1993). In general, *Wolbachia* was thought to be unique in its ability to induce parthenogenesis, but recently Zchori-Fein et al. (2001) found an undescribed vertically transmitted bacterium also associated with parthenogenesis. The microbe was found in six parthenogenetic *Encarsia* species. *Encarsia* is a genus of parasitoids that use whiteflies as their host. The bacterium belongs to the Cytophaga-Flexibacter-Bacteroid (CFB) group, thus being unrelated to *Wolbachia*. Bacteria belonging to the CFB group are also capable of inducing feminization (Weeks et al., 2001). Vandekerckhove et al. (2000) discovered a potential third case of microbial involvement in parthenogenesis. They detected a verrucomicrobial species that is maternally transmitted and seems to be associated with parthenogenesis in the nematode species *Xiphinema americanum*. These bacteria are the first endosymbionts found among the *Verrucomicrobia*. More experimental evidence is needed to show that these bacteria do indeed induce parthenogenesis.

This chapter describes current knowledge of PI-*Wolbachia* in insects. Because studies on the involvement of the CFB bacterium and the verrucomicrobial species in parthenogenesis have just started, these endosymbionts will only be mentioned briefly (but see Chapter 11 of this book).

WOLBACHIA-INDUCED PARTHENOGENESIS IN HAPLODIPLOIDS

The main requirement for parthenogenesis induction by *Wolbachia* seems to be a haplodiploid mode of reproduction of the host species. Haplodiploidy is known from the arthropod groups Hymenoptera, Acari, Thysanoptera, and a few genera in the Coleoptera (Wrensch and Ebbert, 1993). However, there is no fundamental reason why species with a diplodiploid sex-determination system should not be vulnerable to microbial induction of parthenogenesis.

Haplodiploidy, or arrhenotoky, is a mixture of parthenogenetic and sexual reproduction; males develop from unfertilized (haploid) eggs, whereas females develop from fertilized (diploid) eggs (Hartl and Brown, 1970; White, 1973). A completely parthenogenetic mode of reproduction is thelytoky. In thelytoky, all eggs, fertilized or not, develop into females. In arrhenotoky, only a single barrier must be overcome for the induction of parthenogenesis, namely, diploidization of the unfertilized egg. An additional barrier to complete parthenogenesis exists in diplodiploid species: egg development must also be induced, which is normally initiated by sperm penetration in sexual species. The terminology surrounding parthenogenesis in haplodiploids is complicated and clearly in need of revision (Luck et al., 1992; Stouthamer, 1997); in this chapter, we call arrhenotoky and thelytoky, respectively, sexual and parthenogenetic reproduction.

The first indications for microbial involvement in parthenogenesis were found through exposure of the parthenogenetic parasitoid wasp *Habrolepis rouxi* to elevated rearing temperatures (Flanders, 1945). Females reared at 26.6°C or less produced only daughters, whereas females reared at 32.2°C produced sons and daughters or only sons. Several other studies confirmed male production by thelytokous wasps at higher temperatures (Wilson and Woolcock, 1960a,b; Flanders, 1965; Bowen and Stern, 1966; Orphanides and Gonzalez, 1970; Laraichi, 1978; Jardak et al., 1979; Cabello and Vargas, 1985).

To determine the genetic basis of parthenogenesis, Stouthamer et al. (1990b) backcrossed the nuclear genome of a temperature-treated parthenogenetic line of *T. pretiosum* into a sexual line of the same species. If the parthenogenesis trait had been inherited through genes on the chromosomes, the expectation was that the backcrossed females, when kept at low rearing temperatures, would reproduce by parthenogenesis. After nine generations of backcrossing, unmated *T. pretiosum* females still produced only male offspring. Therefore, they concluded that parthenogenesis was not caused by a simple Mendelian trait and might be caused by a cytoplasmic factor. Strong evidence for microbial involvement came from feeding antibiotics to parthenogenetic *Trichogramma* wasps (Stouthamer et al., 1990a). Antibiotic treatment (tetracycline hydrochloride, sulfamethoxazole, and rifampicin) caused male offspring production and reverted females from parthenogenetic to sexual reproduction. Temperature treatment had the same effect. Three years later, Stouthamer and Werren (1993) showed the presence of microbes in eggs of parthenogenetic *Trichogramma* females. They were absent in eggs from lines cured by antibiotic treatment and in field-collected sexual lines. The microbes were identified as *Wolbachia* (Rousset et al., 1992; Stouthamer et al., 1993). Grenier et al. (1998) provided definitive proof that *Wolbachia* was the causal agent of parthenogenesis. They infected eggs of the sexual species *T. dendrolimi* with PI-*Wolbachia* from a *T. pretiosum* line through microinjection. After several generations the *Wolbachia* was still present in *T. dendrolimi*, and a low level of parthenogenesis was induced. Pintureau et al. (2000b) tested the dynamics of infection in two transfected *T. dendrolimi* lines and found the frequency of infected females to decrease dramatically from 52.9% in generation 44 to 3.7% in generation 60 in one line and from 75.5% in generation 32 to 4.5% in generation 48 in the other line.

Recently we found evidence for horizontal transfer of PI-*Wolbachia* under natural conditions, followed by complete expression of parthenogenesis (Huigens et al., 2000). When infected and originally uninfected *T. kaykai* larvae share the same food source, a butterfly egg, approximately 40% of the female offspring of the uninfected line acquire the infection and produce some daughters from unfertilized eggs. In subsequent generations, perfect (100%) transmission of, and PI by, *Wolbachia* was observed. This study, together with the work of Grenier et al. (1998), showed that *Wolbachia* was the causal agent of parthenogenesis in *Trichogramma*.

It still remains unclear how common *Wolbachia*-induced parthenogenesis is. So far, it has been detected mainly in Hymenoptera because the reproduction in these wasps has been studied intensively for their application in biological control (Stouthamer, 1997). A total of 66 hymenopteran species have been reported as being most likely infected with a PI microbe. *Wolbachia* was detected in 46 of these species (Table 15.1), 14 cases are unknown (Table 15.1), and 6 *Encarsia* species are infected with the CFB bacterium.

In some parthenogenetic species, there are strong indications of PI by microbes other than *Wolbachia*, as is the case in *Galeopsomyia fausta*, where evidence for microbial involvement was found through antibiotic treatment, but *Wolbachia* could not be detected (Argov et al., 2000).

Outside the Hymenoptera, Werren et al. (1995) detected *Wolbachia* in a parthenogenetic beetle, *Naupactus tesselatus*, but it remains uncertain if *Wolbachia* causes parthenogenesis in this species. Pintureau et al. (1999) found *Wolbachia* in two parthenogenetic thrips species, *Heliothrips haemorrhoidalis* and *Hercinothrips femoralis*. Antibiotic or heat treatment should show *Wolbachia*'s involvement in their parthenogenesis. Arakaki et al. (2001a) showed the first strong evidence of PI by *Wolbachia* in the predatory thrips *Franklinothrips vespiformis*. In this case, a population fixed for the infection and completely parthenogenetic is found on a Japanese island. Sexual populations occur in Central and South America.

In Acari, the only "noninsect" order with haplodiploidy, parthenogenesis is widely distributed. In oribatid mites, whole families reproduce parthenogenetically (Norton et al., 1993). Perrot-Minnot and Norton (1997) tested eight oribatid species for the presence of *Wolbachia* but could not find the symbiont. Weeks and Breeuwer (2001) found *Wolbachia* infection associated with

TABLE 15.1A
Cases of Parthenogenetic Reproduction in Which Evidence Exists for *Wolbachia* Involvement

Taxon	h	a	w	p	c[a]	Ref.
			Insecta			
			Hymenoptera			
Pteromalidae						
Muscidifurax uniraptor	+	+	+	f	–	Legner, 1985a,b; Stouthamer et al., 1993, 1994
Spalangia fuscipes	?	?	+	?	?	Werren et al., 1995; Van Meer et al., 1999
Aphelinidae						
Aphytis chilensis	?	?	+	?	?	Gottlieb et al., 1998
A. chrysomphali	?	?	+	?	?	Gottlieb et al., 1998
A. diaspidis	?	?	+	?	?	Zchori-Fein et al., 1994, 1995
A. lingnanensis	?	+	+	?	+[b]	Zchori-Fein et al., 1994, 1995
A. yanonensis	?	+	+	?	?	H. Nadel, pers. commun.; Werren et al., 1995
Encarsia formosa	?	+	+	f	–	Zchori-Fein et al., 1992; Van Meer et al., 1995; Werren et al., 1995
Eretmocerus staufferi	?	?	+	?	?	Van Meer et al., 1999
E. mundus	?	+	+	f	+[b]	De Barro and Hart, 2001
Platygasteridae						
Amitus fuscipennis	?	?	+	?	?	Van Meer et al., 1999; Manzano et al., 2000
Encyrtidae						
Apoanagyrus diversicornis	+	+	+	f	+[c]	Pijls et al., 1996; Van Meer, 1999
Coxxidoxenoides peregrinus	+	?	+	f	+	Flanders, 1965; Van Meer et al., 1999
Scelionidae						
Telonomus nawai	+	+	+	f	+[b]	Arakaki et al., 2000
Trichogrammatidae						
Trichogramma brevicapillum	+	+	+	m	+	Stouthamer et al., 1990a,b; Werren et al., 1995
T. chilonis	+	+	+	m	+	Stouthamer et al., 1990a,b; Chen et al., 1992; Schilthuizen and Stouthamer, 1997
T. cordubensis	+	+	+	f	+	Cabello and Vargas, 1985; Stouthamer et al., 1990b, 1993; Silva and Stouthamer, 1996
T. deion	+	+	+	m	+	Stouthamer et al., 1990a,b, 1993
T. embryophagum	+	+	+	?	+	Birova, 1970; Stouthamer et al., 1990b; R.P. De Almeida, pers. commun.
T. evanescens (rhenana)	+	+	?	?	+	Stouthamer et al., 1990b; Van Oosten, H., pers. commun.
T. kaykai	?	+	+	m	+	Stouthamer and Kazmer, 1994; Schilthuizen and Stouthamer, 1997, 1998
T. nubilalae	?	?	+	?	?	Schilthuizen and Stouthamer, 1997; Van Meer et al., 1999
T. oleae	+	+	+	?	+	Stouthamer et al., 1990b; Rousset et al., 1992
T. platneri	+	+	+	m	+	Stouthamer et al., 1990a; Schilthuizen and Stouthamer, 1997
T. pretiosum	+	+	+	m	+	Orphanides and Gonzalez, 1970; Stouthamer et al., 1990a,b

TABLE 15.1A (CONTINUED)
Cases of Parthenogenetic Reproduction in Which Evidence Exists for *Wolbachia* Involvement

Taxon	h	a	w	p	c[a]	Ref.
T. sibericum	+	?	+	?	?	Schilthuizen and Stouthamer, 1997; Van Meer et al., 1999
T. atopovirilla	?	+	+	?	?	Ciociola et al., 2001; R.P. De Almeida, pers. commun.
T. semblidis	?	?	+	?	?	Pintureau et al., 2000a
Eucoilidae						
Gronotoma micromorpha	?	+	+	f	+[b]	Arakaki et al., 2001b
Figitidae						
Leptopilina australis	?	?	+	?	?	Werren et al., 1995
L. clavipes	?	?	+	?	?	Werren et al., 1995
Cynipidae						
Diplolepis rosae	?	?	+	f, m	–	Stille and Dävring, 1980; Van Meer et al., 1995; Plantard et al., 1999
D. spinosissima	?	?	+	f, m	?	Plantard et al., 1998, 1999
D. mayri	?	?	+	f	?	Plantard et al., 1999
D. fructuum	?	?	+	m	?	Plantard et al., 1999
D. eglanteriae	?	?	+	f	?	Plantard et al., 1999
D. bicolor	?	?	+	f	?	Plantard et al., 1999
D. californica	?	?	+	f	?	Plantard et al., 1999
D. nodulosa	?	?	+	f	?	Plantard et al., 1999
D. polita	?	?	+	f	?	Plantard et al., 1999
D. radicum	?	?	+	f	?	Plantard et al., 1999
D. spinosa	?	?	+	f	?	Plantard et al., 1999
Liposthenes glechomae	?	?	+	f	?	Plantard et al., 1999
Timaspis lampsanae	?	?	+	f	?	Plantard et al., 1999
Phanacis hypochaeridis	?	?	+	f	?	Plantard et al., 1999
P. centaureae	?	?	+	f	?	Plantard et al., 1999
				Coleoptera		
Curculionidae						
Naupactus tesselatus	?	?	+	?	?	Werren et al., 1995
				Thysanoptera		
Thripidae						
Heliothrips haemorrhoidalis	?	?	+	?	?	Pintureau et al., 1999
Hercinothrips femoralis	?	?	+	?	?	Pintureau et al., 1999
Aeolothripidae						
Franklinothrips vespiformis	+	+	+	f	+[b]	Arakaki et al., 2001a
				Arachnida		
				Acari		
Tetranychidae						
Bryobia kissophila	?	?	+	f	?	Weeks and Breeuwer, 2001
B. praetiosa	?	+	+	f	+	Weeks and Breeuwer, 2001
B. graminum	?	?	+	f	?	Weeks and Breeuwer, 2001

(*continued*)

TABLE 15.1A (CONTINUED)
Cases of Parthenogenetic Reproduction in Which Evidence Exists for *Wolbachia* Involvement

Taxon	h	a	w	p	c[a]	Ref.
B. rubrioculus	?	?	+	f	?	Weeks and Breeuwer, 2001
B. neopraetisosa	?	?	+	f	?	Weeks and Breeuwer, 2001
Bryobia sp. *x*	?	+	+	f	+	Weeks and Breeuwer, 2001

Note: The evidence is classified as males following heat treatment (h), males following antibiotic treatment (a), and molecular evidence for *Wolbachia* presence (w). In addition, information is given if the parthenogenetic forms are found in populations where parthenogenesis is fixed in the population or if it occurs mixed with sexual reproduction (p) and if the males and females are capable of successful copulations (c). + = Evidence exists, ? = information not available, f = parthenogenesis fixed in population, m = parthenogenesis and sexual reproduction occur in populations.

[a] Copulations are successful (+) or not (–).
[b] Mating and sperm transfer take place, but no successful fertilization of eggs
[c] Mating of males of parthenogenetic lines is successful with closely related sexual females, but not with parthenogenetic females.

parthenogenesis in six species within the phytophagous mite genus *Bryobia.* Through antibiotic treatment they showed that in two of those species, *B. praetiosa* and an unidentified species, the *Wolbachia* infection was strictly associated with parthenogenesis.

We expect the incidence of PI-*Wolbachia* among species to be much higher than reported so far. In particular, low infection frequencies with PI-*Wolbachia* have most likely been underestimated. Low frequencies of infection with PI-*Wolbachia* in field populations have only been found in *Trichogramma* sp. (Stouthamer, 1997). This may simply be a consequence of the way populations are sampled. In general, laboratory colonies of species are initiated by pooling large numbers of field-collected individuals. In this way, low levels of parthenogenesis are easily overlooked because partially parthenogenetic populations cannot be distinguished from sexual populations; in both cases, the colony consists of both males and females. *Trichogramma* sp. stands out because the field populations have been sampled by establishing isofemale lines from wasps emerging from field-collected host eggs. All-female isofemale lines (= infected lines) are easily detected using this protocol (Pinto et al., 1991).

CYTOGENETICS: PARTHENOGENESIS NOT ONLY THROUGH GAMETE DUPLICATION

Cytogenetic processes involving diploidization can be divided into meiotic and postmeiotic modifications (Stouthamer, 1997). A review of parthenogenesis in insects shows that most belong to the first group (Suomalainen et al., 1987). In most species, the meiosis is entirely suppressed and the division has a mitotic character. The process of diploidization in parasitoid wasps due to *Wolbachia* infection, however, seems to be mostly a postmeiotic modification. Stouthamer and Kazmer (1994) described in detail the restoration of diploidy for several infected *Trichogramma* species. They showed that the meiosis was the same for infected and uninfected eggs. At first, in both infected and uninfected eggs, the haploid number of chromosomes is doubled in the prophase of the first mitotic division. In the normal anaphase, each haploid set of chromosomes is pulled to a different pole. In infected eggs, however, the two haploid sets of chromosomes do not separate during the anaphase and result in a single nucleus containing the two identical sets of haploid chromosomes. The following mitotic divisions are the same in infected and uninfected eggs. An

TABLE 15.1B
Cases of Parthenogenetic Reproduction in Which Evidence Exists for Microbial Involvement

Taxon	h	a	w	p	c[a]	Ref.
				Insecta		
				Hymenoptera		
Tenthredinoidae						
Pristiphora erichsonii	+	?	?	?	?	Smith, 1955
Aphelinidae						
Aphelinus asynchus	+	?	?	?	?	Schlinger and Hall, 1959
Aphytis mytilaspidis	?	?	?	m	+	Rössler and DeBach, 1973
Signiforidae						
Signiphora borinquensis	+	?	?	?	+	Quezada et al., 1973
Encyrtidae						
Pauridia peregrina	+	?	?	f	+	Flanders, 1965
Ooencyrtus submetallicus	+	?	?	?	–	Wilson and Woolcock, 1960a,b; Wilson, 1962
O. fecundus	+	?	?	?	?	Laraichi, 1978
Plagiomerus diaspidis	+	?	?	?	–	Gordh and Lacey, 1976
Trechnites psyllae	?	+	?	?	?	T.R. Unruh, pers. commun.
Habrolepis rouxi	+	?	?	?	?	Flanders, 1945
Trichogrammatidae						
Trichogramma sp.	+	?	?	?	?	Bowen and Stern, 1966
T. telengai	+	?	?	?	?	Sorakina, 1987
Eulophidae						
Galeopsomyia fausta	?	+	?	?	+[b]	Argov et al., 2000
Cynipidea						
Hexicola sp. *near websteri*	+	?	?	?	?	Eskafi and Legner, 1974

Note: The evidence is classified as males following heat treatment (h), males following antibiotic treatment (a), and molecular evidence for *Wolbachia* presence (w). In addition, information is given if the parthenogenetic forms are found in populations where parthenogenesis is fixed in the population or if it occurs mixed with sexual reproduction (p) and if the males and females are capable of successful copulations (c). + = Evidence exists, ? = information not available, f = parthenogenesis fixed in population, m = parthenogenesis and sexual reproduction occur in populations.

[a] Copulations are successful (+) or not (–).

[b] Mating and sperm transfer take place, but no successful fertilization of eggs.

unfertilized infected egg develops into a diploid female, homozygous at all loci, and an unfertilized uninfected egg into a haploid male. This diploidization process, which results in a fusion of two identical sets of chromosomes, is called gamete duplication. Allozyme analysis showed that infected females from the three *Trichogramma* species studied — *T. pretiosum*, *T. deion*, and *T. kaykai* — still fertilized their eggs with sperm from conspecific males. The heterozygous F1 virgin females produced only homozygous offspring. This confirmed the gamete duplication. When infected eggs are fertilized, sperm prevents the diploidization during the first mitotic division caused by *Wolbachia* infection and heterozygous infected females develop.

Besides *Trichogramma* spp., two additional cases of gamete duplication due to *Wolbachia* infection have been reported. Stille and Dävring (1980) and Gottlieb et al. (in press) observed a slightly different cytogenetic process in the parthenogenetic gall wasp *Diplolepis rosae* and the pteromalid wasp *Muscidifurax uniraptor*, respectively. Diploidy restoration is not achieved because of an aberrant anaphase but following the completion of the first mitotic division. At that time, the products of the two mitotic nuclei fuse. These studies show that either different forms of *Wolbachia*-induced gamete duplication exist or the observations of Stouthamer and Kazmer (1994) are incorrect.

The gamete duplication in some *Trichogramma* species appears to be far from perfect. Tagami et al. (2001) found in lines of two infected *Trichogramma* spp. a higher prepupal mortality of infected eggs compared to uninfected eggs. Cytological analysis of the developmental stage of 6- to 48-h-old eggs and larvae showed that in up to 35% of unfertilized infected eggs the embryonic development was arrested in the early mitotic stages. This arrestment was not found in eggs of the sexual forms or in a *T. cacoeciae* line where parthenogenesis was not associated with *Wolbachia*. The researchers concluded that the high prepupal mortality of infected eggs was a result of these failures in gamete duplication.

Several studies on the cytogenetic process of cytoplasmic incompatibility (CI) induced by *Wolbachia* have also been carried out (Chapter 14). There is some similarity in this process to gamete duplication. Incompatibility in the mosquito *Culex pipiens* showed that the paternal chromosome set did not faithfully fuse with the female chromosomes after fertilization (Jost, 1970). Reed and Werren (1995) observed the same phenomenon in the parasitoid wasp *Nasonia vitripennis*. Both cytogenetic processes of gamete duplication and incompatibility due to *Wolbachia* show defects in the early mitotic divisions. Lassay and Karr (1996) state that the cytogenetic process of incompatibility is pleiotropic and can be classified into several categories. One defect, in *Drosophila simulans* for example, seems to occur in the anaphase of the first mitotic division (Callaini et al., 1996; Callaini et al., 1997) as with gamete duplication. In this fruit fly, the maternal and paternal chromosomes did not condense synchronically when the father was infected and the mother uninfected. The maternal chromosomes normally enter the anaphase of the first mitotic division and migrate to the two poles. The paternal chromosomes are delayed and stay in the mid-zone of the spindle. This results in embryos with aneuploid or haploid nuclei that eventually die (Callaini et al., 1997). In general, we can say that *Wolbachia* acts in the early mitotic divisions. The main difference between the cytogenetic processes of diploidization and incompatibility induced by *Wolbachia* is of course that gamete duplication is prevented by sperm, whereas incompatibility occurs after fertilization.

From the cytogenetic studies done in Hymenoptera, one might conclude that the most common diploidization process in PI-*Wolbachia*-infected eggs is gamete duplication. However, recent work by Weeks and Breeuwer (2001) shows that another mechanism of *Wolbachia*-induced parthenogenesis is found in some mites, i.e., a meiotic modification in eggs infected with PI-*Wolbachia*. Microsatellite loci in six infected mite species of the genus *Bryobia* indicate the mechanism of parthenogenesis to be most likely a meiotic modification, with progeny being identical to their infected heterozygous mother. It is clear that *Wolbachia* has evolved different mechanisms to induce parthenogenetic development.

DISTRIBUTION AND DENSITY OF *WOLBACHIA* IN PARTHENOGENETIC WASPS

Transmission of *Wolbachia* from mother to daughter is through the cytoplasm of the eggs. A study on two *Aphytis* species suggested that, inside the ovaries, the microorganisms multiply inside the nurse cells and then move, together with all other maternal substances, into the developing oocytes through cytoplasmic bridges (Zchori-Fein et al., 1998). Then, they move to the posterior pole where they are found in freshly laid eggs, as was also shown in *Trichogramma* spp. (Figure 15.1; Stouthamer and Werren, 1993). In later stages, they migrate to the center of the eggs surrounding the nuclei throughout the embryo, although this does not appear to be the case in *Trichogramma* spp. eggs (Stouthamer and Werren, 1993). Nothing is known about the distribution of PI-*Wolbachia* throughout adult host tissues other than the reproductive tissue.

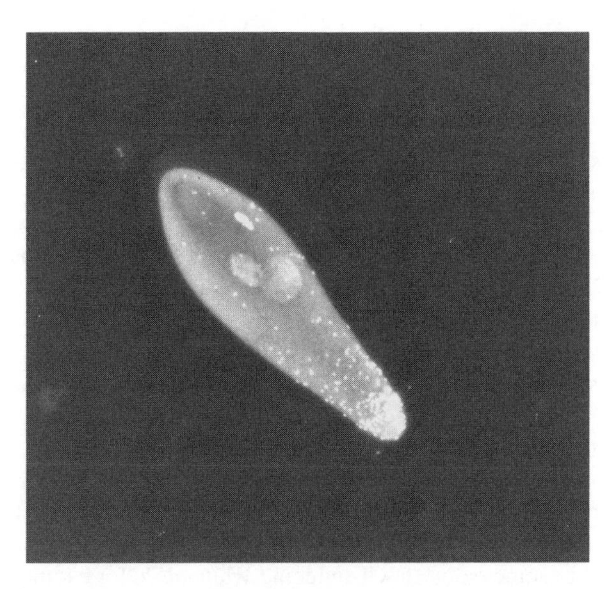

FIGURE 15.1 (Color figure follows p. 206.) *Wolbachia* located at the posterior pole of a freshly laid *Trichogramma kaykai* egg. (Photo courtesy of Merijn Salverda.)

PI is most likely related to the density of *Wolbachia*. As in CI it seems to be a titer effect. In *M. uniraptor*, male production increased with increasing antibiotic dose (Zchori-Fein et al., 2000). In *Trichogramma* spp., infected females start producing more and more males after 5 or 6 d of being able to parasitize a nonlimiting number of host eggs (R. Stouthamer, unpublished results). The titer of *Wolbachia* in the ovaries most likely becomes too low to infect all the oocytes after a few days of oviposition. The uninfected oocytes then develop into males. This might also be the case in *T. kaykai* females that acquired *Wolbachia* through horizontal transfer (Huigens et al., 2000). Newly infected virgin females always produce both sons and daughters.

PHYLOGENETICS: NO CLUSTERING OF PI-*WOLBACHIA*

PI-*Wolbachia* do not form a monophyletic group within the phylogeny of *Wolbachia* based on genes used thus far for reconstructing its phylogeny, i.e., *ftsZ* (Holden et al., 1993; Werren et al., 1995; von der Schulenburg et al., 2000), *wsp* (Braig et al., 1998; Zhou et al., 1998; von der Schulenburg et al., 2000), 16S rDNA (O'Neill et al., 1992; Stouthamer et al., 1993), 23S rDNA (Rousset et al. 1992), and SR2 and 5S rDNA regions (Fialho and Stevens, 1997; Van Meer et al., 1999).

Wolbachia phenotypes are intermixed in the phylogeny of *Wolbachia* (Chapter 16). Several hypotheses exist to explain the scattered distribution of the *Wolbachia* phenotype over the phylogeny (Stouthamer, 1997). According to one, the induction of parthenogenesis undergoes multiple evolution. A second hypothesis is that specific host effects allow certain *Wolbachia* phenotypes to express themselves in a particular host. Curing a host that shows phenotype A and infecting it with a *Wolbachia* that induces phenotype B in its original host may prove this. Infecting cured parasitoid wasps that previously showed CI with *Wolbachia* associated with parthenogenesis could provide support for this hypothesis. Van Meer and Stouthamer (1999) microinjected uninfected *D. simulans* embryos with PI-*Wolbachia* from *M. uniraptor*. Infection could be detected in the new host, but no effect was expressed, and the infection persisted for only seven generations. A third hypothesis states that the genes used for phylogenies are not linked to the genes associated with the effect. Movable genetic elements within the genome of *Wolbachia*, such as the recently discovered bacteriophage (Masui et al., 2000), may induce the effect and not *Wolbachia* itself. Masui et al. (2000) show evidence for frequent horizontal transmission of this phage among different *Wolbachia*.

Many PI-*Wolbachia* are distributed in the B group, but a few also occur in the A group (Werren et al., 1995; Gottlieb et al., 1998; Zhou et al., 1998; Plantard et al., 1999; Vavre et al., 1999; van Meer et al., 1999; von der Schulenburg et al., 2000). When we review these studies and calculate the percentage of PI-*Wolbachia* in both groups, using only those *Wolbachia* cases where the phenotype has been established, we find PI-*Wolbachia* in 28% (10 of 36) of the *Wolbachia* in group A and 54% (20 in 37) in group B.

Only in *Trichogramma* spp. do PI-*Wolbachia* show a clear phylogenetic pattern. Independent of the studied gene, *wsp*, *ftsZ*, or 16S rDNA, these *Wolbachia* always form a separate cluster. Schilthuizen and Stouthamer (1997) studied the phylogenetic relationships between PI-*Wolbachia* and their *Trichogramma* hosts in detail. A comparison of the phylogenetic tree of the host species with the tree of *Wolbachia* in these species showed them to be incongruent. The major reason for this incongruence is most likely horizontal transfer of the *Wolbachia* among different *Trichogramma* species.

Do PI-*Wolbachia* differ in this respect from *Wolbachia* inducing other effects? At present we cannot make strong statements concerning this pattern because our sample size is too small. For parasitoids of the genus *Aphytis* several PI-*Wolbachia*-infected lines have been sequenced using the *ftsZ* gene (Gottlieb et al., 1998). In this genus, the *ftsZ* sequence of the *Wolbachia* is identical, suggesting that here too we find a high similarity of PI-*Wolbachia* within the genus. Similar patterns are also found to some extent in CI-inducing *Wolbachia*, for example, the sibling species of *Nasonia* spp. carry similar *Wolbachia*. The same applies to *Culex* spp. (Zhou et al., 1998). *Wolbachia* in *Diplolepis*, most likely associated with PI, show an extraordinary pattern: the *Wolbachia* in five *Diplolepis* species from France cluster together in the B group, whereas five other PI-*Wolbachia* from North American species are distributed over the A and the B group (Plantard et al., 1999). This distribution would be consistent with the hypothesis that the peculiarities of the host induce the PI effect. The *Leptopilina* spp. are a very interesting case because here CI- and PI-*Wolbachia* exist within the same genus. The pattern we observe is that all CI-*Wolbachia* in *Leptopilina* are distributed over the A group, mostly associated with the *Wolbachia* of their host species (Vavre et al., 1999), but the PI-*Wolbachia* in *L. clavipes* and *L. australis* occur in a separate cluster in the B group.

Besides Schilthuizen and Stouthamer (1997), many other phylogenetic studies detected indirect evidence of horizontal transfer (O'Neill et al., 1992; Stouthamer et al., 1993; Rousset and Solignac, 1995; Werren et al., 1995; Vavre et al., 1999). Host-to-parasitoid and parasitoid-to-parasitoid transfer of *Wolbachia* has been considered mainly as potential transmission routes. Horizontal transmission was shown for PI-*Wolbachia* by Huigens et al. (2000): horizontal transfer of PI-*Wolbachia* between *T. kaykai* parasitoids occurs when they share the same food source.

HOST FITNESS: HIGHER COSTS OF CARRYING *WOLBACHIA* IN MIXED POPULATIONS

The effects of *Wolbachia* on the fitness of its host are influenced by the following factors: transmission route, presence of a genomic conflict between *Wolbachia* and host, and the physiological cost of carrying bacteria. The relationship between *Wolbachia* and its host can range from an adversarial relationship in populations where only a fraction of the females is infected with the PI-*Wolbachia* (under these circumstances we expect a genomic conflict in which the nuclear genes of the host may try to suppress the *Wolbachia* or its effect) to a completely mutualistic relationship in which the infection with *Wolbachia* in a host population has gone to fixation and all females of a population are infected with *Wolbachia*.

Transmission Route

Since the early part of the 20th century, the evolution of parasite virulence has received much attention. Although PI-*Wolbachia* are not true parasites but nonobligatory symbionts, theories on the evolution of virulence are applicable to PI-*Wolbachia*. In general, the mode of parasite transmission, vertical or horizontal, is considered to be the key selective force behind the evolution of virulence (Ewald, 1994). The two modes of transmission are expected to select for opposing virulence levels.

Vertical transmission should select for lower parasite "virulence" to the host. Parasites depend on the host and benefit from a high reproductive success of their host. Hosts with less virulent parasites should produce more offspring than hosts infected with more virulent forms (Lipsitch et al., 1996). On the other hand, horizontal transmission is expected to select for higher virulence because parasites are not dependent on the reproductive capacity of the host. When only horizontally transmitted, parasites obtain the highest fitness through a tradeoff between the negative effect of the parasites' multiplication (virulence) on the longevity of the infected host. Too many parasites could make the host less capable of transmitting parasites (Messenger et al., 1999).

Some parasites have two modes of transmission. They provide ideal models for studying the evolution of virulence. Bull et al. (1991) studied phage virulence evolution in *Escherichia coli*. These phages can be both vertically and horizontally transmitted. Two forms of the phage occur in natural populations — a mild form and a more virulent form that decreases the growth speed of the host. In situations where only vertical transmission was possible, the mild form prevailed. This was not the case when horizontal transmission was also allowed.

In *T. kaykai*, PI-*Wolbachia* are also transmitted in two ways. Vertical transmission is the main mode, but at high parasitoid densities horizontal transfer is expected to be frequent in the field (Huigens et al., 2000). The presence of both transmission forms in the PI-*Wolbachia* of *T. kaykai* will allow us to study the tradeoffs between virulence and transmission of *Wolbachia*.

Nuclear-Cytoplasmic Conflict

The theory of parasite virulence evolution cannot be wholly applied to PI-*Wolbachia* because they are not parasites in the true meaning of the word. They are reproductive parasites that drive through a host population by manipulating the sex ratio of the host. Therefore, when *Wolbachia* are only vertically transmitted, they can induce relatively high fitness costs and still spread through the host population. For PI-*Wolbachia*, several aspects of a host population have implications for the fitness costs we can expect (Table 15.2).

First, fitness costs of carrying *Wolbachia* are expected to be different among host populations fixed for the infection and populations where infected and uninfected individuals coexist (Stouthamer, 1997). (A) Initial infections that spread fastest through a population are the ones with

TABLE 15.2
Fitness Costs for Hosts Carrying PI-*Wolbachia*, Ranked from Lowest (1) to Highest (4), Expected in Several Situations

Infection Status	Horizontal Transmission	Fitness Cost
Fixed	–	1
Fixed	+	2
Mixed	–	3
Mixed	+	4

Costs for hosts from populations fixed or mixed for the infection and from populations where horizontal transmission of *Wolbachia* does or does not occur are mentioned.

little or no effect on the host fitness. Once an infection has spread throughout the entire host population the selective pressures that act on both host and symbionts are assumed to minimize the costs to the host of carrying these symbionts. (B) In cases where the infection has not reached fixation in populations, a nuclear cytoplasmic conflict between the *Wolbachia* and the nuclear genome of the host is expected. *Wolbachia* favors a 100% female bias, whereas the nuclear genes favor a sex ratio involving at least some males. In these situations, an arms race between the nuclear genes and those of the *Wolbachia* may result in nuclear genes trying to suppress the *Wolbachia* or its effect. Consequently, a much higher physiological cost of being infected is expected in mixed host populations where infected and uninfected individuals co-occur.

In addition to the contribution of the genomic conflict to the potential fitness cost of the PI-*Wolbachia* in mixed populations, the opportunities for horizontal transmission in such populations may also select for additional costs of being infected. In fixed populations, all individuals are already infected, and we expect the horizontal transmission to be selected against.

Physiological Cost

It is generally assumed that the physiological costs are minimized in those populations where the infection has gone to fixation. But this does not imply there is no cost at all because energy-consuming bacteria are still present inside the host.

The best way to determine the effect of *Wolbachia* infection on offspring production is by "curing" a line of its infection. Such cured lines should be maintained for several generations; then the offspring production of two genetically identical lines that differ in their infection status can be compared. Unfortunately, this has been impossible to do for those cases where sexual lines cannot be established because males and females are no longer capable of sexual reproduction. In this case it is only possible to compare the offspring production of infected females with cured (antibiotic-treated) females. When such comparisons are made, infected females fed antibiotics are compared with those fed honey. It is important in these comparisons to avoid interference of toxic effects that the antibiotics may have on the females. Therefore, the lowest possible effective dose should be applied. Stouthamer and Mak (2002) showed that at a concentration of 50 mg/ml honey, the antibiotic tetracycline was toxic to *E. formosa* females, while at 5 mg/ml the treated females produced significantly fewer offspring than controls. At a concentration of 1 mg/ml no difference in offspring production was found between treated and control females. Both the 5 mg/ml and the 1 mg/ml concentrations cured the treated females of their infection.

Fitness cost for infected hosts from mixed populations — a rare situation for *Wolbachia*-induced parthenogenesis — has been studied only in *Trichogramma* spp. In *T. pretiosum*, *T. deion*, and *T. kaykai*, *Wolbachia* has a negative influence on lifetime offspring production under laboratory conditions (Stouthamer and Luck, 1993; van Meer, 1999; Silva, 1999). Infected wasps produce fewer offspring than genetically identical wasps cured from their infection when given an abundance of host eggs every day during their entire lifespan (Stouthamer and Luck, 1993). Sometimes the number of daughters produced was even lower for infected mothers (Stouthamer et al., 1990b; Stouthamer and Luck, 1993). These results show that the host can suffer severe fitness costs from being infected, but lifetime offspring production cannot be extrapolated very well from the lab to the field because females are expected to be host-limited in the field (Stouthamer and Luck, 1993). Some parameters that may be important for host fitness in the field were also studied. Van Meer (1999) determined preadult survival and showed it was reduced for infected *T. kaykai* and *T. deion* females. We compared several infected lines with several naturally uninfected lines of *T. kaykai* and found that offspring of infected females had a lower preadult survival and survived even dramatically less when both forms competed for the same food source (M.E. Huigens, unpublished results). Under high parasitoid densities, sharing of a food source may occur frequently in the field. The high mortality rates of unfertilized eggs in infected *T. kaykai* and *T. deion* lines found by Tagami et al. (2001) also confirm the general pattern of high fitness costs in mixed populations.

When the *Wolbachia* infections have gone to fixation in populations, the infections do not seem to have a negative impact on offspring production. In two *Trichogramma* species with fixed populations, *T. cordubensis* and *T. oleae*, sexual lines could be established after curing (Silva, 1999; van Meer, 1999). Infected females and sexual females from cured lines had the same offspring production in both species. In all other studies, the effects on host fitness of PI-*Wolbachia* have been studied only through comparison of untreated infected females with infected females treated with antibiotics of heat. Infected wasps of *M. uniraptor* produced as many offspring as antibiotic-fed wasps (Horjus and Stouthamer, 1995). Another study on the same species confirmed the absence of a negative effect of *Wolbachia* on fecundity, longevity, and offspring survival (Zchori-Fein et al., 2000).

In *Eretmocerus mundus,* fecundity was unaffected by *Wolbachia*, but offspring of infected females had much higher survival than offspring from antibiotic-treated females (De Barro and Hart, 2001). This low survival may be caused by the very high antibiotic dose of 29.8 mg/ml Rifampicin that was used. Two cases of *Wolbachia*-induced parthenogenesis outside the Hymenoptera, with the population fixed for the infection, also do not show a negative effect of *Wolbachia* on host fitness. In the parthenogenetic mite species *B. praetiosa* and *B.* sp. *x*, antibiotic-treated females (0.15% tetracycline) produced fewer offspring than infected females (Weeks and Breeuwer, 2001). Arakaki et al. (2000) studied the predatory thrips species *F. vespiformis* and found the same longevity and offspring production for infected and treated females. Pijls et al. (1996) compared parthenogenetic *Apoanagyrus diversicornis* females with sexual uninfected conspecific females, which were allopatric, and studied the effect of antibiotic treatment in both forms. They found a lower offspring survival for treated originally parthenogenetic females. For *A. diversicornis* the results are, however, slightly ambiguous because allopatric females were compared. At the time of their study *Wolbachia* had not been detected yet in *A. diversicornis*, but this was later confirmed by van Meer et al. (1999). In general, infection induces a fitness cost in mixed populations but not in populations fixed for the infection. For some aspects of host fitness, infection might have evolved to neutrality or even benevolence in fixed populations.

DYNAMICS OF PI-*WOLBACHIA* AND THE EVOLUTION OF "VIRGINITY" MUTANTS

The dynamics of PI-*Wolbachia* infections are in general very simple. A PI-*Wolbachia* infects a female in a population, and as long as the infected female produces more infected daughters than an uninfected female produces uninfected daughters the infection will spread in the population, barring stochastic events. The infection frequency among females that such an infection can reach is a function of a number of variables: (1) the fitness cost of the infection, in terms of offspring production, (2) the transmission fidelity of the PI-*Wolbachia* infection, and (3) the egg-fertilization frequency (Stouthamer, 1997). Models show that the transmission fidelity of the infection is one of the most important variables that determine the equilibrium infection frequency. Once the infection has reached equilibrium it may stay at that level for a long time, unless suppressor alleles, i.e., traits that either kill off the *Wolbachia* or suppress the *Wolbachia*-induced phenotype, evolve and spread (Stouthamer et al., 2001). Alternatively, *Wolbachia* may also evolve to a higher level of transmission and eventually reach fixation. However, another option is very likely to be the most common. In this model, the infection reaches fixation in the population because a mutation that causes females not to fertilize their eggs any longer has an enormous fitness advantage when a substantial part of the population has become infected with the PI-*Wolbachia*.

Imagine a population where half of the females are infected and produce mainly infected daughters as offspring, and assume 10% of the offspring of infected females is uninfected because of inefficient transmission of the *Wolbachia*. Let us also assume that all infected females mate. The uninfected females produce male and female offspring with a sex ratio that is optimal for the species. The male offspring of these uninfected females have a high fitness compared to both uninfected and infected females because they can mate with several females. Assume that a mutation takes place in an uninfected female and that this mutation stops her either from mating or from fertilizing her eggs so that now she produces exclusively male offspring. This "virginity" mutation has a large fitness advantage because of the female-biased sex ratio in the population. When the mutant male mates with an infected female, part of her offspring will be heterozygous for the mutant, and some of the offspring (50% of the unfertilized eggs) of these infected females will become homozygote for the "virginity" mutation and no longer mate with males. These infected mutant females will produce some uninfected eggs because of the inefficient transmission. These eggs will be unfertilized and thus become males that are carriers of the mutation. These males also experience the large fitness benefit from being male in a population consisting largely of females. If these males mate with uninfected females, they will increase the frequency of the mutant allele in the uninfected population, thereby reducing the number of uninfected females produced. In the infected population, an increasingly larger number of females will carry the mutation and become unavailable for mating with males, while the infected females not yet homozygous for the virginity mutation will still be willing to mate.

Simulations (R. Stouthamer and F. Vavre, unpublished results) have shown that even with a large negative impact on the fitness of their carriers in terms of offspring production this "virginity" mutation will spread through the population and result in a rapid fixation of the *Wolbachia* infection. In addition, the resulting all-female population will now consist entirely of mutant females that are no longer able or willing to mate. However, if males are induced by feeding the infected females antibiotics, these males are still capable of mating successfully with nonmutant females. This is indeed the pattern that is observed in a number of populations where both completely infected and uninfected populations exist. Pijls et al. (1996) and Arakaki et al. (2000) studied species where both completely sexual and completely infected populations existed allopatrically. When males were produced by antibiotic treatment from the infected population, they did not mate successfully with females of the infected line, while successful matings took place with females from the sexual line. The females from the infected line did not mate successfully with either males from the infected or from the uninfected line. The lack of successful matings between cured males and infected females is a common phenomenon in populations where the infection has gone to fixation.

For example, in parthenogenetic parasitoid species such as *E. formosa* (Zchori-Fein et al., 1992), *M. uniraptor* (Stouthamer et al., 1993, 1994; Zchori-Fein et al. 2000), *A. diversicornis* (Pijls et al., 1996), *Aphytis* spp. (Zchori-Fein et al., 1995), *Telenomus nawaii* (Arakaki et al., 2000), and *E. mundus* (de Barro and Hart, 2001), females do not produce offspring from fertilized eggs, when allowed to mate with conspecific males derived by antibiotic treatment. In addition, in the parthenogenetic mite species *B. praetiosa*, copulation took place between cured males and females, but sperm was not used by the females (Weeks and Breeuwer, 2001). In cured females of the thrips *F. vespiformis*, sperm was found after they mated with "parthenogenetic" males (Arakaki et al., 2001a) but also not used.

In mixed populations, mating behavior of infected females does not seem to be different from uninfected females. Within several species — *T. kaykai*, *T. pretiosum*, and *T. deion* — infected females successfully mate with males from cured or uninfected forms (Stouthamer et al., 1990b; Stouthamer and Luck, 1993; van Meer, 1999; Stouthamer et al., 2001). In all mixed populations, the virginity mutation could evolve and should lead to fixation of the infection in the population. The cases where mixed populations persist are an indication of the presence of some suppressing factor. Suppressor genes as such have not yet been found for PI-*Wolbachia*. In *T. kaykai*, detailed work on the prolonged persistence of infected and uninfected individuals within a population did not find any evidence for the presence of suppressors but brought to light the presence of a second sex-ratio distorter (paternal sex ratio, PSR) in the population. About 10% of the males carry this chromosome (Stouthamer et al., 2001). Such a PSR chromosome was until then found only in the parasitoid wasp *N. vitripennis* (Werren, 1991). When a female mates with a PSR-carrying male, the paternal chromosomes in fertilized eggs are functionally destroyed and only the maternal chromosomes and the PSR factor itself remain. Such fertilized eggs develop into sons that carry a haploid set of chromosomes from the mother and the PSR chromosome from their father. PSR therefore causes uninfected and infected females to produce sons from their fertilized eggs. The mating structure of the uninfected *T. kaykai* is such that a large fraction of the uninfected females mate with their brothers (54 to 65%, Stouthamer and Kazmer, 1994). In contrast, the infected females mate with males from the population, and 10% of these are carriers of PSR. As a consequence, approximately 10% of the infected females mate with PSR males and only 5.5 to 6.4% of the uninfected females. The consequence of this asymmetry is that the higher daughter production of the infected population is suppressed more than that of the uninfected population, allowing for a stable coexistence of both forms within the population (Stouthamer et al., 2001).

CONCLUSION

Since the beginning of the last decade, when the association between *Wolbachia* and parthenogenesis was first identified, our knowledge of this association has grown extensively. After the first evidence of bacterial involvement in parthenogenesis through antibiotic treatment and the identification of *Wolbachia*, the recent success of finding intraspecific (Huigens et al., 2000) and interspecific (Grenier et al., 1998) horizontal transfer delivered the final proof that this symbiont was a causal agent of parthenogenesis. Since the last review it has also become clear that PI-*Wolbachia* may influence cytogenetic events in different ways to cause parthenogenesis, even through meiotic modifications (Weeks and Breeuwer, 2001), and obviously more cytogenetic studies are needed to determine if yet unknown mechanisms of PI exist. One of the main questions that remains to be answered is: Which *Wolbachia* genes cause the traits expressed in their hosts? And similarly, how important are host effects in expressing the *Wolbachia* phenotype?

The different genome projects that are under way should result in the sequence of several CI-*Wolbachia*, a PI-*Wolbachia*, and a feminizing *Wolbachia*. Cooperative studies on these sequences may help in solving this problem. It is hoped that we will then also be able to explain why PI-*Wolbachia* are scattered throughout the phylogenetic trees based on several genes.

The study on the CFB bacterium that causes parthenogenesis in several *Encarsia* wasps showed that the oviposition behavior of the infected females was modified by the infection (Zchori-Fein et al., 2001). We do not know if PI-*Wolbachia* also modify their host's behavior.

On the population level, in at least one species we can now explain why infected and uninfected forms coexist sympatrically. The presence of a suppressor in the form of the *psr* chromosome in *T. kaykai* is able to keep the infection in the population at a low level. A *psr*-like factor could occur in other species as well (Stouthamer et al., 2001).

More work remains to be done on the *Wolbachia*–host association in fixed populations. Studies on genetic variation in these populations can tell us something about how *Wolbachia* went to fixation. Furthermore, the genetics of the "virginity" mutation remains to be studied.

Little has been done on applied aspects of the PI-*Wolbachia*. One greenhouse study by Silva et al. (2000) showed that in inundative biocontrol the use of PI-infected *Trichogramma* wasps is more economic than the use of sexual forms. However, more studies are needed to determine if this is a general pattern or only a specific example. In addition, similar studies should be done with wasps that are used in classical biological control. The use of PI-infected wasps for augmenting native sexual populations should be monitored closely because it may lead to the replacement of native sexual forms with the released infected form.

Rendering wasps parthenogenetic through microinjection of PI-*Wolbachia* may only be effective when the inoculum originates from mixed populations. *Wolbachia* from fixed populations are expected to have coevolved to some extent with their hosts, which may have led to the loss of genes needed to function in other host species

Recently, other symbionts involved in parthenogenesis have been discovered, i.e., the CFB bacterium in the parasitoid wasp *Encarsia* and the verrucomicrobial species in nematodes. The latter case shows that not only haplodiploids should be looked upon as possible hosts for PI microbes; in addition, the fact that PI has evolved outside *Wolbachia* should stimulate the search for additional parthenogenesis-causing symbionts.

REFERENCES

Arakaki, N., Noda, H., and Yamagishi, K. (2000). *Wolbachia*-induced parthenogenesis in the egg parasitoid *Telonomus nawai*. *Entomol. Exp. Appl.* **96:** 177–184.

Arakaki, N., Miyoshi, T., and Noda, H. (2001a). *Wolbachia*-mediated parthenogenesis in the predatory thrips *Franklinothrips vespiformis* (Thysanoptera: Insecta). *Proc. R. Soc. London (B)* **268:** 1011–1016.

Arakaki, N., Oishi, T., and Noda, H. (2001b). Parthenogenesis induced by *Wolbachia* in *Gronotoma micromorpha* (Hymenoptera: Eucolidae). *Entomol. Sci.* **4:** 9–15.

Argov, Y., Gottlieb, Y., Amin, S.S., and Zchori-Fein, E. (2000). Possible symbiont-induced thelytoky in *Galeopsomyia fausta*, a parasitoid of the citrus leafminer *Phyllocnistis citrella*. *Phytoparasitica* **28:** 212–218.

Birova, H. (1970). A contribution to the knowledge of the reproduction of *Trichogramma embryophagum*. *Acta Entomol. Bohemoslovakia* **67:** 70–82.

Bowen, W.R. and Stern, V.M. (1966). Effect of temperature on the production of males and sexual mosaics in uniparental race of *Trichogramma semifumatum*. *Ann. Entomol. Soc. Am.* **59:** 823–834.

Braig, H.R., Zhou, W., Dobson, S., and O'Neill, S.L. (1998). Cloning and characterization of a gene encoding the major surface protein of the bacterial endosymbiont *Wolbachia*. *J. Bacteriol.* **180:** 2373–2378.

Bull, J.J., Molineux, I.J., and Rice, W.R. (1991). Selection of benevolence in a host-parasite system. *Evolution*, **45:** 875–882.

Cabello, T. and Vargas, P. (1985). Temperature as a factor influencing the form of reproduction of *Trichogramma cordubensis*. *Z. Angew. Entomol.* **100:** 434–441.

Callaini, G., Dallai, R., and Riparbelli, M.G. (1997). *Wolbachia*-induced delay of paternal chromatin condensation does not prevent maternal chromosomes from entering anaphase in incompatible crosses of *Drosophila simulans*. *J. Cell Sci.* **110:** 271–280.

Callaini, G., Riparbelli, M.G., Giordano, R., and Dallai, R. (1996). Mitotic defects associated with cytoplasmic incompatibility in *Drosophila simulans*. *J. Invertebr. Pathol.* **67:** 55–64.

Chen, B.H., Kfir, R., and Chen, C.N. (1992). The thelytokous *Trichogramma chilonis* in Taiwan. *Entomol. Exp. Appl.* **65:** 187–194.

Ciociola, A.I., Jr., De Almeida, R.P., Zucchi, R.A., and Stouthamer, R. (2001). Detacção de *Wolbachia* em uma população Telítoca de *Trichogramma atopovirrilia* Oatman and Platner (Hymenoptera: Trichogrammatidae) via PCR com o primer específico wsp. *Neotrop. Entomol.* **30:** 489–491.

Cosmides, L.M. and Tooby, J. (1981). Cytoplasmatic inheritance and intragenomic conflict. *J. Theor. Biol.* **89:** 83–129.

Crow, J.F. (1994). Advantages of sexual reproduction. *Dev. Genet.* **15:** 205–213.

De Barro, P.J. and Hart, P.J. (2001). Antibiotic curing of parthenogenesis in *Eretmocerus mundus* (Australian parthenogenetic form). *Entomol. Exp. Appl.* **99:** 225–230.

Eskafi, F.M. and Legner, E.F. (1974). Parthenogenetic reproduction in *Hexacola* sp. near *websteri*, a parasite of *Hippelates* eye gnats. *Ann. Entomol. Soc. Am.* **67:** 767–768.

Ewald, P.W. (1994) *Evolution of Infectious Disease*. Oxford University Press, Oxford, U.K.

Felsenstein, J. (1974). The evolutionary advantage of recombination. *Genetics* **78:** 737–756.

Fialho, R.F. and Stevens, L. (1997). Molecular evidence for single *Wolbachia* infections among geographic strains of the flour beetle *Tribolium confusum*. *Proc. R. Soc. London (B)* **264:** 1065–1068.

Flanders, S.A. (1945). The bisexuality of uniparental hymenoptera, a function of the environment. *Am. Nat.* **79:** 122–141.

Flanders, S.E. (1965). On the sexuality and sex ratios of hymenopterous populations. *Am. Nat.* **99:** 489–494.

Gordh, G. and Lacey, L. (1976). Biological studies of *Plagiomerus diaspidis*, a primary internal parasite of diaspidid scale insects. *Proc. Entomol. Soc. Wash.* **78:** 132–144.

Gottlieb, Y., Zchori-Fein, E., Faktor, O., and Rosen, D. (1998). Phylogenetic analysis of parthenogenesis-inducing *Wolbachia* in the genus *Aphytis* (Hymenoptera: Aphelinidae). *Insect Mol. Biol.* **7:** 393–396.

Gottlieb, Y., Zchori-Fein, E., Werren, J.H., and Karr, T.L. Diploidy restoration in *Muscidifurax uniraptor* (Hymenopetra: Pteromalidae). *J. Invertebr. Pathol.* In press.

Grenier, S, Pintureau, B., Heddi, A., Lassablière, F., Jager, C., Louis, C., and Khatchadourian, C. (1998) Successful horizontal transfer of *Wolbachia* symbionts between *Trichogramma* wasps. *Proc. R. Soc. London (B)* **265:** 1441–1445.

Hamilton, W.D. (1967). Extraordinary sex ratios. *Science* **156:** 477–488.

Hamilton, W.D. (1979). Wingless and fighting males in fig wasps and other insects. In *Sexual Selection and Reproductive Competition in Insects* (M.S. Blum and N.A. Blum, Eds.), pp. 167–220. Academic Press, New York.

Hartl, D.L. and Brown, S.W. (1970). The origin of male haploid genetic systems and their expected sex ratio. *Theor. Population Biol.* **1:** 165–190.

Holden, P.R., Brookfield, J.F., and Jones, P. (1993). Cloning and characterization of a ftsZ homologue from a bacterial symbiont of *Drosophila melanogaster. Mol. Gen. Genet.* **240:** 213–220.

Horjus, M. and Stouthamer, R. (1995). Does infection with thelytoky-causing *Wolbachia* in the pre-adult and adult life stages influence the adult fecundity of *Trichogramma deion* and *Muscidifurax raptor. Proc. Sect. Exp. Appl. Entomol. Neth. Entomol. Soc.* **6:** 35–40.

Huigens, M.E., Luck, R.F., Klaassen, R.H.G., Maas, M.F.P.M., Timmermans, M.J.T.N., and Stouthamer, R. (2000). Infectious parthenogenesis. *Nature* **405:** 178–179.

Hurst, G.D.D., Hammerton, T.C., Bandi, C., Majerus, T.M.O., Bertrand, D., and Majerus, M.E.N. (1997). The diversity of inherited parasites of insects: the male-killing agent of the ladybird beetle *Coleomegilla maculata* is a member of the Flavobacteria. *Genet. Res.* **70:** 1–6.

Jardak, T., Pintureau, B., and Voegele, J. (1979). Mise en evidence d'une nouvelle espece de *Trichogramme.* Phenomene d'intersexualite, etude enzymatique. *Ann. Soc. Entomol. Fr.* **15:** 635–642.

Jost, E. (1970). Untersuchungen zur Inkompatibilität im *Culex pipiens* Komplex. *Wilhelm Roux' Archiv* **166:** 173–188.

Laraichi, M. (1978). L'effect de hautes temperatures sur le taux sexuel de *Ooencyrtus fecundus. Entomol. Exp. Appl.* **23:** 237–242.

Lassay, C.W. and Karr, T.L. (1996). Cytological analysis of fertilzation and early embryonic development in incompatible crosses of *Drosophila simulans. Mech. Dev.* **57:** 47–58.

Legner, E.F. (1985a). Effects of scheduled high temperature on male production in thelytokous *Muscidifurax uniraptor. Can. Entomologist* **117:** 383–389.

Legner, E.F. (1985b). Natural and induced sex ratio changes in populations of thelytokous *Muscidifurax uniraptor. Ann. Entomol. Soc. Am.* **78:** 398–402.

Lipsitch, M., Siller, S., and Nowak, M.A. (1996). The evolution of virulence in pathogens with vertical and horizontal transmission. *Evolution* **50:** 1729–1741.

Luck, R.F., Stouthamer, R., and Nunney, L. (1992). Sex determination and sex ratio patterns in parasitic hymenoptera. In *Evolution and Diversity of Sex Ratio in Haplodiploid Insects and Mites* (D.L. Wrensch and M.A. Ebbert, Eds.), pp. 442–476. Chapman & Hall, New York.

Manzano, M.R., van Lenteren, J.C., Cardona, C., and Drost, Y. (2000). Developmental time, sex ratio, and longevity of *Amitus fuscipennis* MacGown and Nebeker (Hymenoptera: Platygasteridae) on the greenhouse whitefly. *Biol. Control* **18:** 94–100.

Masui, S., Kamoda, S., Sasaki, T., and Ishikawa, H. (2000). Distribution and evolution of bacteriophage WO in *Wolbachia*, the endosymbiont causing sexual alterations in arthropods. *J. Mol. Evol.* **51:** 491–497.

Messenger, S.L., Molineux, I.J., and Bull, J.J. (1999). Virulence evolution in a virus obeys a trade-off. *Proc. R. Soc. London (B)* **266:** 397–404.

Muller, H.J. (1964). The relation of recombination to mutational advantage. *Mutation Res.* **1:** 2–9.

Norton, R.A., Kethley, J.B., Johnston, D.E., and O'Connor, B.M. (1993). Phylogenetic perspectives on genetic systems and reproductive modes of mites. In *Evolution and Diversity of Sex Ratio in Insects and Mites* (D.L. Wrensch and M.A. Ebbert, Eds.), pp 8–99. Chapman & Hall, New York.

O'Neill, S.L., Giordano, R., Colbert, A.M.E., Karr, T.L., and Robertson, H.M. (1992). 16S RNA phylogenetic analysis of the bacterial endosymbionts associated with cytoplasmic incompatibility in insects. *Proc. Natl. Acad. Sci. U.S.A.* **89:** 2699–2702.

Orphanides, G.M. and Gonzalez, D. (1970). Identity of a uniparental race of *Trichogramma pretiosum*. *Ann. Entomol. Soc. Am.* **63:** 1784–1786.

Perrot-Minnot, M.J. and Norton, R.A. (1997). Obligate thelytoky in oribatid mites: no evidence for *Wolbachia* inducement. *Can. Entomol.* **129:** 691–698.

Pijls, J.W.A.M., van Steenbergen, J.J., and van Alphen, J.J.M. (1996). Asexuality cured: the relations and differences between sexual and asexual *Apoanagyrus diversicornis*. *Heredity* **76:** 506–513.

Pinto, J.D., Stouthamer, R., Platner, G.R., and Oatman, E.R. (1991). Variation in reproductive compatibility in *Trichogramma* and its taxonomic significance. *Ann. Entomol. Soc. Am.* **84:** 37–46.

Pintureau, B., Lassablière, F., Khatchadourian, C., and Daumal, J. (1999). Parasitoïdes oophages et symbiotes de deux thrips europeens. *Ann. Soc. Entomol. Fr.* **35:** 416–420.

Pintureau, B., Chaudier, S., Lassablière, F., Charles, H., and Grenier, S. (2000a). Addition of *wsp* sequences to the *Wolbachia* phylogenetic tree and stability of the classification. *J. Mol. Evol.* **51:** 374–377.

Pintureau, B., Grenier, S., Boleat, B., Lassablière, F., Heddi, A., and Khatchadourian, C. (2000b). Dynamics of *Wolbachia* populations in transfected lines of *Trichogramma*. *J. Invertebr. Pathol.* **76:** 20–25.

Plantard, O., Rasplus, J.Y., Mondor, G., Le Clainche, I., and Solignac, M. (1998). *Wolbachia*-induced thelytoky in the rose gall-wasp *Diplolepis spinosissimae* (Giraud) (Hymenoptera: Cynipidae), and its consequences on the genetic structure of its host. *Proc. R. Soc. London (B)* **265:** 1075–1080.

Plantard, O., Rasplus, J.Y., Mondor, G., Le Clainche, I., and Solignac, M. (1999). Distribution and phylogeny of *Wolbachia* inducing thelytoky in Rhoditini and 'Aylacini' (Hymenoptera: Cynipidae). *Insect Mol. Biol.* **8:** 185–191.

Quezada, J.R., Debach, P., and Rosen, D. (1973). Biological and taxonomical studies of *Signophora borinquensis*, new species, (Hym: Signiphoridae), a primary parasite of diaspine scales. *Hilgardia* **41:** 543–604.

Reed, K.M. and Werren, J.H. (1995). Induction of paternal genome loss by the paternal-sex-ratio chromosome and cytoplasmic incompatibility bacteria (*Wolbachia*): a comparative study of early embryonic events. *Mol. Reprod. Dev.* **40:** 408–418.

Rössler, Y. and DeBach, P. (1973). Genetic variability in the thelytokous form of *Aphytis mytilaspidis*. *Hilgardia* **42:** 149–175.

Rousset, F. and Solignac, M. (1995). Evolution of single and double *Wolbachia* symbioses during speciation in the *Drosophila simulans* complex. *Proc. Natl. Acad. Sci. U.S.A.* **92:** 6389–6393.

Rousset, F., Bouchon, D., Pintureau, B., Juchault, P., and Solignac, M. (1992). *Wolbachia* endosymbionts responsible for various alterations of sexuality in arthropods. *Proc. R. Soc. London (B)* **250:** 91–98.

Schilthuizen, M. and Stouthamer, R. (1997). Horizontal transmission of parthenogenesis-inducing microbes in *Trichogramma* wasps. *Proc. R. Soc. London (B)* **264:** 361–366.

Schilthuizen, M., Honda, J., and Stouthamer, R. (1998). Parthenogenesis-inducing *Wolbachia* in *Trichogramma kaykai* (Hymenoptera: Trichogrammatidae) originates from a single infection. *Ann. Entomol. Soc. Am.* **91:** 410–414.

Schlinger, E.I. and Hall, J.C. (1959). A synopsis of the biologies of three imported parasites of the spotted alfalfa aphid. *J. Econ. Entomol.* **52:** 154–157.

Silva, I.M.M.S. (1999). Identification and evaluation of *Trichogramma* parasitoids for biological pest control. Ph.D. thesis, Wageningen University, the Netherlands.

Silva, I.M.M.S. and Stouthamer, R. (1996). Can the parthenogenesis *Wolbachia* lead to unusual courtship behavior in *Trichogramma*? *Proc. Sect. Exp. Appl. Entomol. Neth. Entomol. Soc.* **7:** 27–31.

Silva, I.M.M.S., van Meer, M.M.M., Roskam, M.M., Hoogenboom, A., Gort, G., and Stouthamer, R. (2000). Biological control potential of *Wolbachia*-infected versus uninfected wasps: laboratory and greenhouse evaluation of *Trichogramma cordubensis* and *T. deion* strains. *Biocontrol Sci. Technol.* **10:** 223–228.

Smith, S.G. (1955). Cytogenetics of obligatory parthenogenesis. *Can. Entomol.* **87:** 131–135.

Sorakina, A.P. (1987). Biological and morphological substantiation of the specific distinctness of *Trichogramma telengai* sp. n. *Entomol. Rev.* **66:** 20–34.

Stille, B. and Dävring, L. (1980). Meiosis and reproductive stategy in the parthenogenetic gall wasp *Diplolepis rosae*. *Heriditas* **92:** 353–362.

Stouthamer, R. (1997). *Wolbachia*-induced parthenogenesis. In *Influential Passengers: Inherited Microorganisms and Arthropod Reproduction* (S.L. O'Neill, A.A. Hoffmann, and J.H. Werren, Eds.), pp. 102–122. Oxford University Press, Oxford, U.K.

Stouthamer, R. and Kazmer, D.J. (1994). Cytogenetics of microbe-associated parthenogenesis and its consequences for gene flow in *Trichogramma* wasps. *Heredity* **73:** 317–327.

Stouthamer, R. and Luck, R.F. (1993). Influence of microbe-associated parthenogenesis on the fecundity of *Trichogramma deion* and *T. pretiosum*. *Entomol. Exp. Appl.* **67:** 183–192.

Stouthamer, R. and Mak, F. Influence of antibiotics on the offspring production of the *Wolbachia*-infected parthenogenetic parasitoid *Encarsia formosa*. *J. Invertebr. Pathol.* **80:** 41–45.

Stouthamer, R. and Werren, J.H. (1993). Microorganisms associated with parthenogenesis in wasps of the genus *Trichogramma*. *J. Invertebr. Pathol.* **61:** 6–9.

Stouthamer, R., Luck, R.F., and Hamilton, W.D. (1990a). Antibiotics cause parthenogenetic *Trichogramma* to revert to sex. *Proc. Natl. Acad. Sci. U.S.A.* **87:** 2424–2427.

Stouthamer, R., Pinto, J.D., Platner, G.R., and Luck, R.F. (1990b). Taxonomic status of thelytokous forms of *Trichogramma*. *Ann. Entomol. Soc. Am.* **83:** 475–481.

Stouthamer, R., Breeuwer, J.A.J., Luck, R.F., and Werren, J.H. (1993). Molecular identification of microorganisms associated with parthenogenesis. *Nature* **361:** 66–68.

Stouthamer, R., Luko, S., and Mak, F. (1994). Influence of parthenogenesis *Wolbachia* on host fitness. *Norwegian J. Agric. Sci.* **16:** 117–122.

Stouthamer, R., Breeuwer, J.A.J, and Hurst, G.D.D. (1999). *Wolbachia pipientis*: microbial manipulator of arthropod reproduction. *Annu. Rev. Microbiol.* **53:** 71–102.

Stouthamer R., van Tilborg, M., De Jong, H., Nunney, L., and Luck, R.F. (2001). Selfish element maintains sex in natural populations of a parasitoid wasp. *Proc. R. Soc. London (B)* **268:** 617–622.

Suomalainen, E., Saura, A., and Lokki, J. (1987). *Cytology and Evolution in Parthenogenesis*. CRC Press, Boca Raton, FL.

Tagami, Y., Miura, K., and Stouthamer, R. (2001). How does infection with parthenogenesis-inducing *Wolbachia* reduce the fitness of Trichogramma? *J. Invertebr. Pathol.* **78:** 267–271.

Vandekerckhove, T.M., Willems, A., Gillis, M., and Coomans, A. (2000). Occurrence of a novel verrumicrobial species, endosymbiotic and associated with parthenogenesis in *Xiphinema americanum*-group species (Nematoda, Longidoridae). *Int. J. Syst. Evol. Microbiol.* **50:** 2197–2205.

Van Meer, M.M.M. (1999). Phylogeny and host-symbiont interactions of thelytoky inducing *Wolbachia* in Hymenoptera. Ph.D. thesis, Wageningen University, Wageningen, the Netherlands.

Van Meer, M.M.M. and Stouthamer, R. (1999). Cross-order transfer of *Wolbachia* from *Muscidifurax uniraptor* (Hymenoptera: Pteromalidae) to *Drosophila simulans* (Diptera: Drosophilidae). *Heredity* **82:** 163–169.

Van Meer, M.M.M., van Kan, F.J.M.P., Breeuwer, J.A.J., and Stouthamer, R. (1995). Identification of symbionts associated with parthenogenesis in *Encarsia formosa* and *Diplolepis rosae*. *Proc. Sect. Exp. Appl. Entomol. Neth. Entomol. Soc.* **6:** 81–86.

Van Meer, M.M.M., Witteveldt, J., and Stouthamer, R. (1999). Phylogeny of the arthropod endosymbiont *Wolbachia* based on the *wsp* gene. *Insect Mol. Biol.* **8:** 399–408.

Vavre, F., Fleury, F., Lepetit, D., Fouillet, P., and Boulétreau, M. (1999). Phylogenetic evidence for horizontal transmission of *Wolbachia* in host–parasitoid associations. *Mol. Biol. Evol.* **16:** 1711–1723.

von der Schulenburg, J.H.G., Hurst, G.D.D., Huigens, M.E., van Meer, M.M.M., Jiggins, F.M., and Majerus, M.E.N. (2000). Molecular evolution and phylogenetic utility of *Wolbachia* ftsZ and wsp gene sequences with special reference to the origin of male-killing. *Mol. Biol. Evol.* **17:** 584–600.

Weeks, A.R. and Breeuwer, J.A.J. (2001). *Wolbachia*-induced parthenogenesis in a genus of phytophagous mites. *Proc. R. Soc. London (B)* **268:** 2245–2251.

Weeks, A.R., Marec, F., and Breeuwer, J.A.J. (2001). A mite species that consists entirely of haploid females. *Science* **292:** 2479–2482.

Werren, J.H. (1991). The paternal sex ratio chromosome of *Nasonia*. *Am. Nat.* **137:** 392–402.

Werren, J.H., Nur, U., and Wu, C. (1988). Selfish genetic elements. *Trends Ecol. Evol.* **11:** 297–302.

Werren, J.H., Z.W., and Guo, L.R. (1995). Phylogeny of *Wolbachia* bacteria: reproductive parasites of arthropods. *Proc. R. Soc. London (B)* **250:** 91–98.

White, M.J.D. (1973). *Animal Cytology and Evolution*. Cambridge University Press, Cambridge, U.K.

Wilson, F. (1962). Sex determination and gynandromorph production in aberrant and normal strains of *Ooencyrtus submetallicus*. *Aust. J. Zool.* **10:** 349–359.

Wilson, F. and Woolcock, L.T. (1960a). Environmental determination of sex in a parthenogenetic parasite. *Nature* **186:** 99–100.

Wilson, F. and Woolcock, L.T. (1960b). Temperature determination of sex in a parthenogenetic parasite, *Ooencyrtus submetallicus. Aust. J. Zool.* **8:** 153–169.

Wrensch, D.L. and Ebbert, M.A., Eds. (1993). *Evolution and Diversity of Sex Ratio in Insects and Mites.* Chapman & Hall, New York.

Zchori-Fein, E., Roush, R.T., and Hunter, M.S. (1992). Male production induced by antibiotic treatment in *Encarsia formosa*, an asexual species. *Experientia* **48:** 102–105.

Zchori-Fein, E., Rosen, D., and Roush, R.T. (1994). Microorganisms associated with thelytoky in *Aphytis lingnanensis. Int. J. Insect Morphol. Embryol.* **23:** 169–172.

Zchori-Fein, E., Faktor, O., Zeidan, M., Gottlieb, Y., Czosnek, H., and Rosen, D. (1995). Parthenogenesis-inducing microorganisms in *Aphytis. Insect Mol. Biol.* **4:** 173–178.

Zchori-Fein, E., Roush, R.T., and Rosen, D. (1998). Distribution of Parthenogenesis-inducing symbionts in ovaries and eggs of *Aphytis* (Hymenoptera: Aphelinidae). *Curr. Microbiol.* **36:** 1–8.

Zchori-Fein, E., Gottlieb, Y., and Coll, M. (2000). *Wolbachia* density and host fitness components in *Muscifdifurax uniraptor* (Hymenoptera: Pteromalidae). *J. Invertebr. Pathol.* **75:** 267–272.

Zchori-Fein, E., Gottlieb, Y., Kelly, S.E., Brown, J.K., Wilson, J.M., Karr, T.L., and Hunter, M.S. (2001). A newly discovered bacterium associated with parthenogenesis and a change in host selection behavior in parasitoid wasps. *Proc. Natl. Acad. Sci. U.S.A.* **98:** 12555–12560.

Zhou, W., Rousset, F., and O'Neill, S. (1998). Phylogeny and PCR-based classification of *Wolbachia* strains using wsp gene sequences. *Proc. R. Soc. London (B)* **265:** 509–515.

16 Insights into *Wolbachia* Obligatory Symbiosis

Franck Dedeine, Claudio Bandi, Michel Boulétreau, and Laura H. Kramer

CONTENTS

INTRODUCTION

It is now widely accepted that maternal transmission of symbiotic microorganisms (from host mother to offspring) can lead to the establishment of two different types of host–symbiont relationships: (1) mutualism, where the symbiont increases its own fitness by increasing the fitness of host females, and (2) reproductive parasitism, which regroups the associations where the symbiont manipulates the reproduction of its host in ways that enhance its spread in the population (Werren and O'Neill, 1997; Bandi et al., 2001a). In mutualism, although males are not involved in the transmission of the symbiont, they often also benefit from the infection. The main difference between mutualism and reproductive parasitism is the consequences on host evolution. In mutualism, selective pressures acting on the host and the symbiont are convergent and can lead to stable

0-8493-1286-8/03/$0.00+$1.50

infection, coadaptation, and reciprocal dependence between partners (i.e., obligatory symbiosis) (Lipsitch et al., 1995). Inversely, in reproductive parasitism the symbiont can spread, even though it has a detrimental effect on infected females. Moreover, the "manipulations" performed by reproductive parasites are detrimental to those individuals that do not transmit them; this obviously implies a decrease of the fitness of individuals in the host population. For these reasons, selective pressures acting on each partner can be antagonistic and lead to conflict, unstable association, and sometimes loss of infection (Hurst and McVean, 1996; Rigaud, 1997; Vavre et al., in press).

Interestingly, the strictly intracellular *Wolbachia* bacteria have adopted both strategies in their invertebrate hosts. In filarial nematodes, the hosts seem to be dependent on the infection. Indeed, the removal of bacteria has detrimental effects on development, survival, and reproduction (Bandi et al., 2001b). Inversely, in arthropods, *Wolbachia* infections most often induce reproductive alterations, including cytoplasmic incompatibility (CI), male killing, feminization, and parthenogenesis (Werren and O'Neill, 1997). Assuming that *Wolbachia* in nematodes and arthropods share a common symbiotic ancestor, the evolutionary interpretation of such diversity in *Wolbachia* lifestyles is not clearly understood. Moreover, obligatory symbiosis was recently found in insects, suggesting that host differences between arthropods and nematodes are not sufficient to explain this diversity. The goal of this chapter is to underline those differences between obligatory and facultative *Wolbachia* infections to try to answer the following two questions: (1) What are the main conditions required for a *Wolbachia* infection to evolve toward obligatory symbiosis? (2) What are the evolutionary mechanisms capable of generating such evolutionary transitions?

In the first part of the chapter, we summarize basic information and focus on those aspects of symbiosis between *Wolbachia* and filarial nematodes that could help understand how an association with these bacteria could evolve to obligatory symbiosis.

In the second part, we present those cases in insects where *Wolbachia* infection is either beneficial (but facultative) or obligatory for host reproduction. Examples of these associations are few and poorly documented, but they raise important evolutionary questions.

In the final part, we propose possible hypotheses on the conditions required for a *Wolbachia* infection to evolve in an obligatory symbiosis. In particular, we focus on the roles of horizontal transfer, multiple *Wolbachia* infections, and bacterial virulence in this evolution. Finally, we speculate on the significance of *Wolbachia* obligatory symbiosis in the mutualism concept.

OBLIGATORY *WOLBACHIA* IN FILARIAL NEMATODES

Several recent reviews have been published on the *Wolbachia* endosymbionts of filarial nematodes (Taylor and Hoerauf, 1999; Bandi et al., 1999, 2001b). Owing to space constraints, not all the original work up to the beginning of 2001 will be cited, and the reader is referred to the above reviews.

A SUMMARY OF CURRENT INFORMATION

Six supergroups of *Wolbachia* (A–F) have thus far been described on the basis of branching and clustering patterns in unrooted phylogenetic trees derived from 16S rDNA and *ftsZ* gene sequences (Lo et al., 2002). The majority of nematode *Wolbachia* have been assigned to supergroups C and D. There is, however, one species of filaria (*Mansonella ozzardi*) whose *Wolbachia* has not been precisely assigned to any of the six supergroups (Lo et al., 2002). Supergroups A, B, E, and F encompass the *Wolbachia* of arthropods. The branching order of these six supergroups of *Wolbachia* is not clear, and some of these groups might be paraphyletic. It is, however, clear that all the *Wolbachia* thus far found in nematodes are phylogenetically very distant from any of the *Wolbachia* thus far found in arthropods. There is therefore no evidence for current transmission of *Wolbachia* from arthropods to nematodes, or vice versa.

Wolbachia is present in several species of filariae, including the main agents of human and animal filariasis. Examples of infected filariae are *Onchocerca volvulus* (the agent of river blindness), *Wuchereria bancrofti* and *Brugia malayi* (agents of tropical elephantiasis), and *Dirofilaria immitis* (the agent of canine and feline heartworm disease). In species harboring *Wolbachia*, the prevalence of infection appears to be 100%. Moreover, the infection appears stable along evolutionary times: main branches of filarial evolution are composed of species harboring *Wolbachia* (Casiraghi et al., 2001). In adult nematodes, *Wolbachia* is present in the hypodermal cells of the lateral chords; the cytoplasm of some of these cells is filled with *Wolbachia* and resembles insect bacteriocytes. While in females *Wolbachia* is also present in the ovaries, *Wolbachia* has not been demonstrated in the male reproductive apparatus (Sacchi et al., 2002). The bacterium is vertically transmitted through the cytoplasm of the egg, and there is no evidence of horizontal transmission or paternal transmission. Indeed, the phylogeny of *Wolbachia* matches that of the host filariae (Casiraghi et al., 2001). It is also noteworthy that there are no indications for multiple infection of a single host or for the presence of different "types" of *Wolbachia* in a given filarial species. There is thus overall evidence that the infection is stable and species-specific.

Current information on the distribution and phylogeny of *Wolbachia* in filarial nematodes thus suggests that *Wolbachia* is needed by the host nematode (100% prevalence in infected species, whole branches of filaria evolution composed by infected species, consistency of host and symbiont phylogenies). The results of experiments with tetracycline (and derivates) on animal hosts infected by filariae are in agreement with the idea that the *Wolbachia*–filaria association is obligatory. Tetracyclines have indeed been shown to (1) inhibit the development from the infective larva (L3) to the adult, (2) inhibit embryogenesis and microfilaria (L1) production, (3) interfere with the long-term survival of adult nematodes, and (4) interfere with L1–L3 development of the filaria in mosquito hosts (reviewed in Bandi et al., 2001b). It must be emphasized that some of the results above have been obtained on more than one filarial species (i.e., 1 and 2), while in other cases the results are relevant to a single species (3 and 4). There is, however, overall agreement among the results of treatment experiments using tetracycline and derivates, showing that these antibiotics, in reducing the amount of *Wolbachia* in nematodes from treated hosts, also have detrimental effects on the filarial nematode. Similar results have been obtained in *in vitro* experiments, and other antibiotics have been shown to cause attrition on filariae harboring *Wolbachia*.

It still must be demonstrated that there is a cause-and-effect relationship between the activity of antibiotics on *Wolbachia* and the deleterious effects on the nematodes. It is, however, suggestive that in a filarial species that is free of *Wolbachia* (*Achantocheilonema viteae*), tetracycline treatments do not interfere with L3-adult development and with microfilaria production. The fact that tetracycline and other antibiotics are detrimental to filarial species harboring *Wolbachia* has opened the way to new therapeutic strategies for the control of filariases. For example, the combination of a "traditional" antifilarial drug (ivermectin) associated with doxycycline (a derivate of tetracycline) appears very promising for the therapy of human onchocerciasis (Taylor et al., 2000; Hoerauf et al., 2001).

Evidence for Single Infection in Filarial Nematode Ancestor

There is an important difference between arthropod and nematode *Wolbachia*. In arthropods, *Wolbachia* infects a wide taxonomic range of hosts, particularly insects, but also other classes (Werren, 1997; Stouthamer et al., 1999; Jeyaprakash and Hoy, 2000). In nematodes, *Wolbachia* has so far been observed only in filariae (which represent a single family, the Onchocercidae). Other groups of nematodes may harbor *Wolbachia*, and screenings for *Wolbachia* in nematodes are in progress in various laboratories (S. Bordenstein, personal communication). We must, however, keep in mind that the *Wolbachia* we presently know in nematodes is restricted to the well-defined and quite restricted taxonomic group of the filariae. We may hypothesize that *Wolbachia* infection was (or is?) a more widespread characteristic in nematodes, retained by filarial nematodes. But we do have evidence that suggests that this is not the case. First, the phylogeny of *Wolbachia* matches that of filarial nematodes (Casiraghi et al., 2001). This indicates that *Wolbachia* cospeciated with the host during the evolutionary radiation of filarial nematodes. Second, arthropod and nematode *Wolbachia* appear to have originated through a star-like radiation: in the global *Wolbachia* tree, the internal branches are quite short compared to the length of the subtrees of arthropod and nematode *Wolbachia* (Lo et al., 2002). This suggests that the time elapsed between the split of arthropod and nematode *Wolbachia* and the start of the evolutionary radiation of *Wolbachia* in filarial nematodes was quite short. This suggests that there has not been enough time for nematode *Wolbachia* to have a significant evolutionary radiation in nematodes other than the filariae after its separation from arthropod *Wolbachia*. In other words, based on the genes thus far used for global phylogenetics of *Wolbachia* (*ftsZ* and 16S rDNA), the six supergroups of *Wolbachia* appear approximately equidistant at the nucleotide level. If we assume that the rate of molecular evolution has been even roughly constant in *Wolbachia*, we should conclude that *Wolbachia* moved from an arthropod to a close ancestor of the filariae or, conversely, from this hypothetical close ancestor of the filariae to some ancestral arthropod. The widespread presence in arthropods of bacteria phylogenetically related to *Wolbachia* (i.e., the various alpha-proteobacteria) makes more realistic the hypothesis that *Wolbachia* was originally present in arthropods and then moved to filarial nematodes. That genome size of arthropod *Wolbachia* is around 1.5 megabases (Mb), compared to around 1 Mb in the nematode *Wolbachia* (Sun et al., 2001), also points to the ancestral status of the former to the latter. In summary, there is overall evidence that suggests that the "character" *Wolbachia* was acquired during the first steps of the evolutionary radiation of filarial nematodes. Consistent with this hypothesis is the absence of *Wolbachia* in nematodes closely related to the filariae, the Thelazidae (Bandi, unpublished observation). Furthermore, *Wolbachia* is absent in those filarial species that appear to represent deep branches in the evolution of the filariae (Casiraghi et al., 2001; C. Bandi, unpublished observation). Should *Wolbachia* be found in nematodes phylogenetically distant from the filariae, this could represent the result of an infection event independent of the one that led to filarial infection or from horizontal transmission from a filaria to another nematode (but, of course, future phylogenetic evidence could prove otherwise).

Vertical vs. Horizontal Transmission; Sex vs. No Sex

There is no evidence for horizontal transmission of *Wolbachia* in filarial nematodes. There is no evidence for polymorphism in the presence/absence of *Wolbachia* in a given species of filarial (see above and Bandi et al., 2001b; Casiraghi et al., 2001). Lack of evidence for something does not, of course, rule it out. However, based on the biology of filariae, it would be difficult to hypothesize how horizontal transmission of *Wolbachia* could occur in these nematodes. In arthropods are found a wealth of parasites, parasitoids, and predators of other arthropods, members of different species living in close contact, with chances for individuals to be mechanically injured (which perhaps made the transmission of microorganisms through the hemolymph possible), and detritivory and coprophagy. In summary, among arthropods can be found all the kinds of "dangerous liaisons" and "unsanitary behaviors" that provide a "road network" for the transmission of microorganisms including the intracellular ones like *Wolbachia*. In filarial nematodes, life is probably more "clean and protected": localization of different species of filariae is usually quite different in the vertebrate host; chances for mechanical injuries are low; in the case of injuries within the host, fragments of the nematode (including *Wolbachia*) are likely to be quickly removed by the host immune system; parasites or parasitoids of filarial nematodes that could transmit microorganisms among individuals are not known. Microorganisms are thus unlikely to have many chances for horizontal transmission in filarial nematodes, in particular among members of different species.

The lack of evidence for recombination in filarial *Wolbachia* further argues against frequent occurrence of horizontal transmission. In arthropod *Wolbachia*, disagreement among phylogenies based on different genes (or on different portions of the same gene) provides evidence for the occurrence of genetic recombination among strains (Jiggins et al., 2001; Werren and Bartos, 2001). In nematode *Wolbachia*, phylogenies based on three different genes are congruent (Casiraghi et al., 2001). For a detectable recombination this must occur between strains showing differences in their gene sequences; this requires that an individual host become infected with two or more strains. Where transmission is strictly vertical, and where the bacteria that are vertically transmitted form a clone, recombination, if it occurs, will not be detectable through the comparison of phylogenies based on different genes. There are thus three possibilities for nematode *Wolbachia*: (1) recombination occurs, but we do not detect it because of the exchange of genetic material among a vertically transmitted, clonal population of bacteria; (2) the capacity of undergoing recombination has been lost because of weaker selection for maintaining this capacity among members of a clone; or (3) recombination simply does not occur, even where chances for exchanging genetic material among not-too-closely-related bacteria exist. The first two options imply that the chances for multiple *Wolbachia* infection of a single nematode host (and thus of horizontal, in addition to vertical, transmission to occur) are (or have been) very low. The third option appears the most unlikely: why should nematode *Wolbachia* have lost the capacity to exchange genetic material if this can occur among not-too-closely-related bacteria? We should emphasize that sequence analysis of cloned polymerase chain reaction (PCR) products of *Wolbachia* genes, obtained from individual nematode hosts, has not provided any evidence so far for genetic variation (Bandi et al., 1999; C. Bandi, unpublished observation).

In conclusion, based on the available information (consistency of host and symbiont phylogenies, no evidence for genetic recombination, and no evidence for *Wolbachia* polymorphism in individual nematodes), it is reasonable to assume that horizontal transmission in nematode *Wolbachia* does not occur or, if it does, involves only conspecific nematodes, harboring very closely related *Wolbachia* strains.

BENEFICIAL AND OBLIGATORY *WOLBACHIA* IN INSECTS

Beneficial and obligatory *Wolbachia* infections in insects are poorly documented and have never been reviewed. Here we present all the cases, to our knowledge, where *Wolbachia* have either a positive effect on host fecundity or an obligatory role for insect reproduction.

FACULTATIVE *WOLBACHIA* INFECTIONS THAT ENHANCE INSECT FECUNDITY

The first beneficial *Wolbachia* was described in a wasp species of the genus *Trichogramma* (Hymenoptera: Pteromalidae). *Trichogramma* are common parasitoids that infest numerous Lepidoptera species, including several important pests. Females oviposit into the eggs of their hosts, within which parasitic larvae feed and develop into adult wasps. Comparing offspring production in two Moroccan strains of *T. bourarachae*, Mimouni (1991) established that the higher fecundity of the "high" strain (61.3 ± 1.4 per female during 5 days) was maternally inherited when crossed and backcrossed with the "low" strain, which produced only 25.8 ± 1.8 descent. Based on their demonstration that antibiotic treatments decreased fecundity of "H" females and did not affect "L" ones and supporting their hypothesis with microscopic observations, Girin and Boulétreau (1995) showed that a maternally inherited microbe infecting the H strain could account for its phenotype. Using PCR with *Wolbachia*-specific primers, Vavre et al. (1999a) demonstrated the role of a single particular *Wolbachia* located in the A supergroup, which thus appeared as a facultative fecundity enhancer that did not induce CI.

Another case of beneficial *Wolbachia* infection was recently described in mosquitoes. Mutualistic *Wolbachia* infection was suspected in *Culex pipiens* (Awahmukalah and Brooks, 1985) and *Aedes albopictus* (Dobson et al., 2001; Dobson and Rattanadechkul, 2001) and has been clearly demonstrated in *A. albopictus* by Dobson et al. (2002). The authors demonstrated reduced performances in an aposymbiotic line derived from a naturally infected one (from Houston, TX). The

Wolbachia-free status determines in females a net decrease (17.7%) in realized fecundity, which combines offspring production and longevity of females and also reduces longevity in males (Dobson et al., 2002). All individuals of the naturally infected line were infected with two distinct *Wolbachia* strains, which also induce CI (Sinkins et al., 1995). Beneficial *Wolbachia* strains are here also able to induce CI, which is considered a reproductive alteration, and for the first time *Wolbachia* infection was shown to be involved in both mutualism and reproductive parasitism. This case perfectly illustrates the hypothesis that microorganisms simultaneously exhibiting both strategies can be expected. However, we do not know so far whether the two co-occurring *Wolbachia* strains are each specialized or are equally involved in both effects.

Other less clear or controversial examples of facultative beneficial *Wolbachia* have been reported. In the pteromalid wasp *Nasonia vitripennis*, removal of infection presumably reduced fecundity of uninfected females, suggesting a positive effect of infection on host fitness (Stolk and Stouthamer, 1996). However, the significance of this result was later refuted due to possible confusion of infection effects with differences in genetic background of strains (Bordenstein and Werren, 2000). In *Drosophila simulans*, a decrease in female fecundity following elimination of *Wolbachia* was equally observed by Poinsot and Merçot (1997), but the effect was not permanent in the course of generations, making conclusions difficult. Finally, two intriguing cases were independently found where *Wolbachia* infection increased fertility in males. In the flour beetle *Tribolium confusum*, *Wolbachia* enhanced male fertility at the expense of other fitness components, including viability and female fecundity (Wade and Chang, 1995). Because *Wolbachia* also induces CI in this species, the male fertility effect was interpreted as a way of accelerating *Wolbachia* spread through the host population. Such an advantage for male fertility was also reported by Hariri et al. (1998) in the fly *Sphyracephala beccarii* (Diopsidae), but in this case *Wolbachia* did not induce CI, and interpretations are unclear.

Obligatory *Wolbachia* in Insects

Obligatory *Wolbachia*-Induced Parthenogenesis

Among reproductive manipulations caused by *Wolbachia*, thelytokous parthenogenesis has been observed in certain haplodiploid parasitic wasps (Stouthamer, 1997; Huigens and Stouthamer, Chapter 15 of this book). In these species, infection permits females to produce all-female offspring without being fertilized. Because *Wolbachia* are maternally inherited, elimination of males procures a high selective advantage for *Wolbachia*. In some species, this effect is so strong that males do not exist in natural populations and can be induced only by antibiotic treatments that remove *Wolbachia* infection (Stouthamer et al., 1990). Except for species of the genus *Trichogramma*, these males fail to produce offspring with conspecific females and consequently cannot transmit their genes. Gottlieb and Zchori-Fein (2001) reviewed different reproductive barriers that could occur in crosses between antibiotic-induced males with conspecific females. Disorders of sexual functionality vary among species. In *Encarsia formosa* (Aphelinidae), antibiotic-induced males produce sperm and sometimes mate with conspecific females, but insemination fails (Zchori-Fein et al., 1992). A more dramatically degenerate state occurs in *Muscidifurax uniraptor* (Pteromalidae), where males are incapable of producing sperm and females are not sexually receptive. Moreover, females are totally devoid of a muscle in the spermathecae, which plays a major role both in drawing sperm into the spermathecal reservoir after copulation and in egg fertilization (Gottlieb and Zchori-Fein, 2001).

In all cases, the reproductive barrier between sexes is complete and irreversible, and establishment of stable sexual lines from asexual ones is impossible. Thus, *Wolbachia*-induced parthenogenesis is the only possible mode of reproduction, making *Wolbachia* an obligatory symbiotic partner. Such host dependence could have arisen as follows: in infected wasps, parthenogenesis has made sexual traits useless, and the absence of selective pressures on them has permitted accumulation of random deleterious mutations in their genetic determinants, leading to their loss in the course of evolution.

Wolbachia Dependence for Oogenesis

Another case of obligatory *Wolbachia* infection in insects was reported in the wasp *Asobara tabida* (Dedeine et al., 2001). However, involvement of *Wolbachia* in this obligatory association suggests different types of interaction that could follow different evolutionary origins. For this reason, we will review what is known about this biological model.

A. tabida (Braconidae) is a common solitary larval endoparasitoid of *Drosophila* species, which live in fermenting fruits and sap fluxes. Females oviposit in *Drosophila* larvae in which parasitoid feed and develop. Its large distribution in Europe and its easy rearing make *A. tabida* an appropriate model to study several aspects of host–parasitoid interactions (Kraaijeveld and Godfray, 1997). *Wolbachia* in *A. tabida* was first detected by Werren et al. (1995) and confirmed by Vavre et al. (1999b). In the latter study, the authors described a case of triple *Wolbachia* infection in which individuals can be simultaneously infected by three different *Wolbachia* strains, each characterized by its partial sequence of the bacterial *wsp* gene. Based on sequences of the same gene, phylogenetic reconstruction showed that these *Wolbachia* strains were well differentiated even if they all belonged to the A *Wolbachia* supergroup, which has been detected only in arthropods. Using diagnostic PCR, it was shown that all individuals (males and females) from 13 European strains were all simultaneously infected by the three *Wolbachia* strains (F. Dedeine, unpublished data). This total infection in laboratory strains suggests a high triple-infection prevalence in populations and a very strong efficient transmission of bacteria through the germ line, at least under laboratory conditions. This high transmission rate could be explained in part by the particular localization of *Wolbachia* in germ cells. Indeed, microscopy has localized *Wolbachia* in the posterior pole of mature oocytes. Because this region of the cytoplasm contains germ-cell determinants in insects, the presence of *Wolbachia* in this region could be interpreted as an adaptation to enhance bacterial transmission to host progeny, as already suggested in other cases of intracellular symbiosis (Breeuwer and Werren, 1990; Stouthamer and Werren, 1993). However, *Wolbachia* were also detected in isolated thoraxes, demonstrating the presence of bacteria in tissues other than the germline (Dedeine et al., 2001).

A study on the impact of *Wolbachia* infection on *A. tabida* reproduction and performance had an astonishingly unexpected result. Treated with antibiotics, aposymbiotic individuals (cured of the infection) had a normal overall physiological state and normal length; males were fertile and females had an apparently normal oviposition behavior. However, females were completely incapable of producing offspring, while each control (untreated) female produced around 200 offspring (Dedeine et al., 2001).

Why does antibiotic treatment completely and specifically sterilize females? The explanation is very simple: aposymbiotic females do not produce any eggs and consequently cannot have any offspring. In this species, vitellogenesis is almost completed at the time of emergence, taking place during the pupal stage at the expense of larval reserves (proovogenic oogenesis). Thus, the mature oocytes, which are localized at the basal region of ovarioles, can be easily counted in each female. After curative antibiotic treatments, results obtained were strongly demonstrative: while symbiotic females produce more than approximately 240 mature oocytes 5 days after emergence, ovarioles from aposymbiotic females are simply empty of oocytes. Emptiness of ovaries means that preimaginal oogenesis is inhibited, and no further oogenesis occurs in the early adult stage after treatment has ceased. Moderate antibiotic doses that do not totally eliminate *Wolbachia* are compatible with limited egg production, which shows a clear relationship with bacterial density. The possibility that inhibition of oocyte production is caused directly by antibiotics, or indirectly through the release of toxins from decaying *Wolbachia* bacteria, can be discarded (Dedeine et al., 2001). There is thus overall evidence that egg production in *A. tabida* is wholly dependent on *Wolbachia* infection.

Moreover, a recent unpublished study demonstrated that the possible involvement of bacteria other than *Wolbachia* is also highly unlikely. The three *Wolbachia* strains in *A. tabida* have been named *w*At1, *w*At2, and *w*At3. The goal of the study was to determine the respective roles of each strain in oogenesis (Dedeine et al., in preparation). By using moderate antibiotic treatments followed by a selection of isofemale lines, the three *Wolbachia* could be separated and stabilized into different lines. During the five-generation selection, a strong relationship was found between individual offspring production and the presence of one strain, *w*At3. All the females that lost this strain were unable to produce progeny. Conversely, most of the females in which *w*At3 was detected produced offspring. These results have two consequences: first, they are not in agreement with the presence of an unknown bacterium. Indeed, there is no reason *a priori* that such an unknown agent would be strictly associated with *w*At3 only. Second, these results strongly suggest that only *w*At3 is obligatory for oogenesis.

That *w*At3 is needed for oogenesis was confirmed at the end of the selection procedure. Indeed, four lines were finally established and stabilized (of the eight combinations normally possible with three bacterial strains): a line harboring the three *Wolbachia* strains "Pi(123)," the two bi-infected lines "Pi(13)" and "Pi(23)," and the monoinfected one "Pi(3)," which is infected only by *w*At3. Fecundities of these lines were the same as in untreated control lines, demonstrating that only the presence of *w*At3 is obligatory for oogenesis (Dedeine et al., in preparation). Moreover, intercrossing these lines demonstrated that variants *w*At1 and *w*At2 induced bidirectional CI resulting in high female mortality among offspring (≈75%) compared to the compatible crosses (Dedeine et al., in preparation). Female mortality was already described as FM CI-phenotype (for "Female Mortality") in haplodiploid species (Breeuwer, 1997; Vavre et al., 2000; Vavre et al., 2002).

In conclusion, triple *Wolbachia* infection in *A. tabida* is associated with different relationships with the host. In each individual, one *Wolbachia* (*w*At3), which is obligatory for oogenesis, coexists with two reproductive parasitic ones (*w*At1 and *w*At2), which induce bidirectional CI. We know that *Wolbachia* can have a wide range of effects on hosts, but the respective contributions of host and bacteria in this diversity are generally unknown. In the case of *A. tabida*, the coinfection by an obligatory strain and two reproductive parasitic strains in the same host individual (i.e., in the same host genotype) demonstrates the decisive role of bacterial genotype in induction of each phenotype observed. Thus, different *Wolbachia* can be genetically specialized in different relationships within the same host.

Two questions have arisen from the studies on *A. tabida*. First, what mechanisms are responsible for the dependence on *Wolbachia* for oogenesis? And what evolutionary scenarios account for it?

A Novel Reproductive Manipulation by Wolbachia*?*

Sterility of cured females could result from differences in their own infection status compared to that of their mothers, leading to some kind of intergeneration incompatibility. Its principle could be summarized as follows: when *Wolbachia* is present in the mother, it must also be present in the daughters, or these will be completely sterilized. Such an effect must consider two independent products of *Wolbachia*. First, *Wolbachia* could act by producing a sort of "time-bomb toxin" in infected mothers. After maternal transmission to offspring, this toxin could be inactivated in infected daughters by a second bacterial product ("defusing molecules") but not in uninfected daughters that are lacking this protective factor and that thus suffer deleterious effects of the toxin that made them sterile. These modes of action imply three functional constraints. First, the effect of the toxin must act specifically in oogenesis. Second, the toxin must still be active when the defusing molecule is absent (the two active molecules could have different turnover kinetics). Third, the effect of the toxin must be maintained after it is lost from the insect (the toxin could induce epigenetic modifications that persist in the insect). This kind of phenotype has never been shown to be associated with the presence of symbiotic microorganisms. However, a single nuclear gene has been reported to induce a similar effect in the beetle *T. castaneum* (Beeman et al., 1992), showing that this hypothesis is not unreasonable. In this case, when the selfish gene is present in the mother, it must also be present in the zygote or the zygote will die. In reference to *Medea*, the Greek mythological mother who killed her own offspring, this phenotype was called the *Medea* phenotype.

Although a certain selective advantage is evident for a nuclear gene to induce such a phenotype, certain conditions are necessary for a *Medea*-inducing gene when localized in the genome of a maternally inherited symbiont. In the case of *A. tabida*, *Wolbachia* will only increase in frequency when the sterilization of uninfected daughters increases the fitness of the infected ones. For example, selective pressure could occur when high resource competition exists between the offspring of closely related females. Such competition is likely to occur in *A. tabida*, where progeny can be laid in the same local host resource and where cases of superparasitism (i.e., more than one parasitic egg is deposited in each *Drosophila* larvae by several conspecific females) have been described (van Alphen and Nell, 1982).

Because this *Medea* effect resembles the modification-rescue model of CI, it can easily be compared to *Wolbachia*–induced CI, and an evolutionary switch from CI to *Medea* induction has recently been proposed to explain the case of *A. tabida* (Charlat and Merçot, 2001).

Direct Involvement of Bacterial Factors in Insect Oogenesis?

A second hypothesis that may explain how *Wolbachia* is involved in the reproductive physiology of females is that bacterial factors actively participate in some metabolic pathways necessary to oogenesis.

As recently reviewed by McFall-Ngai (2002), many bacteria are involved in the development, survival, and reproduction of their hosts, forming obligatory symbiosis, where both the host and the symbiont depend on the association. In insects, such microorganisms are common, and they form distinctive associations called "primary symbiosis" (Moran and Baumann, 2000) or "bacteriocyte symbiosis" (Douglas, 1998). This type of symbiosis groups all cases where strictly intracellular microorganisms are restricted to one host cell type (called bacteriocyte), which is specialized in housing them. In such primary symbiosis, it is widely accepted that host dependence has a nutritional base, making exploitation of nutritionally poor or unbalanced diets possible through providing essential nutrients such as amino acids, vitamins, or lipids (Douglas, 1994, 1998). In these intimate associations, host and symbiont phylogenies are most often congruent, demonstrating stable infection, cospeciation, and long coevolutionary history of partners (Moran and Baumann, 2000). This primary symbiosis opposes "secondary" (or facultative) symbiosis, where the association is sporadic and bacteria occupy various host tissues. Effects on the host are usually not known, and in all cases symbionts are not obligatory to host survival or reproduction.

Because infection is obligatory for *A. tabida* reproduction, we could consider their *Wolbachia* primary symbionts. However, they also share several features with secondary symbionts. First, since they can be detected in various body parts (Dedeine et al., 2001), they are not restricted to a precise cell type. Second, closely related *Asobara* species (*A. citri*, *A. persimilis*, and another undescribed species from the United States) are naturally free of *Wolbachia* (F. Dedeine, unpublished data), suggesting recent common evolutionary history of partners. Moreover, we would expect a major difference in the evolutionary origin of bacterial dependence. In primary symbiosis, host dependence probably originated in acquisition of beneficial bacterial metabolic functions that permitted the insect to exploit poor diets and consequently to extend (or change) their previous ecological niche. In this evolutionary scenario (the "evolutionary novelty"), symbionts would have become obligatory for the host by providing additional functions that are necessary for exploiting specific habitats. This scenario could apply to a number of symbiosis cases (Margulis and Fester, 1991) but certainly not to the *Wolbachia–A. tabida* association, since autonomous egg production by insects obviously preexisted *Wolbachia* acquisition.

We shall now consider a hypothetical "substitution" evolutionary scenario. In an initial step, the ancestral lineage of *A. tabida* would have become associated with a *Wolbachia* strain (*w*At3) that spread into host populations thanks to CI induction. This *Wolbachia* may have had a particular genetic determinant encoding for some factor directly involved in oogenesis of the wasp. In this situation, both host and bacterial genes would have acted in the same insect function, and such nucleocytoplasmic redundancy could therefore have resulted in the loss of nuclear determinants by the insect, making the bacteria totally obligatory for reproduction. A similar substitution process may have occurred in the endosymbiotic association between the intracellular x-bacteria and the unicellular eukaryote *Amoeba proteus*. In this case, the initially harmful host–bacteria association evolved to a beneficial state after ≈200 generations in culture, and experiments demonstrated that the host nucleus had become dependent on the infective organisms for its own functioning (Jeon, 1972).

WOLBACHIA SPECIALIZATIONS AND EVOLUTIONARY TRANSITIONS

Vertical transmission can lead to the establishment of different kinds of host–symbiont relationships or specializations (Werren and O'Neill, 1997; Bandi et al., 2001a). An obvious pathway is toward mutualistic symbiosis: the symbiont will increase its own fitness by increasing the fitness of the host that is involved in its transmission. Another possible outcome is to become a reproductive parasite: by manipulating host reproduction, the symbiont can reduce the fitness of those members of the host species that are not involved in its transmission. There is, however, no intrinsic conflict between mutualistic symbiosis and reproductive parasitism. For example, a maternally inherited microorganism could be beneficial toward females (the host sex responsible for transmission to the offspring) while being detrimental toward males (which are not involved in transmission). In *Wolbachia* symbiosis, the only example of reproductive parasitism and mutualism coexisting is observed in *A. albopictus*, where a CI-inducing *Wolbachia* strain also increases female fecundity (Dobson et al., 2002). Indeed, in all other cases, *Wolbachia* seems to be specialized either in obligatory mutualism in filarial nematodes or reproductive parasitism in the majority of arthropod species.

Reasons for Specialization in *Wolbachia* Symbiosis

We believe that four main evolutionary keys may explain specialization in the two host phyla: (1) the different rate of horizontal transmission, (2) the evolutionary stability of association, (3) the efficacy of vertical transmission, and (4) multiple infection. In filarial nematodes, we assume that horizontal transmission and multiple infection do not occur (or occur involving closely related *Wolbachia* strains). Moreover, the prevalence of infection appears complete. Conversely, as stated previously, *Wolbachia* in arthropods likely has chances for horizontal transmission and multiple infections. In addition, *Wolbachia* is haphazardly distributed within arthropod taxa and populations: not all populations within a species appear to harbor *Wolbachia* and not all species within a genus appear to harbor it (and so on along the taxonomic hierarchy). This pattern is consistent with theoretical models on the population biology of bacteria that induce CI: both loss of infection and invasion by new compatibility types are expected to occur in the host populations. Once *Wolbachia* has become fixed within a population, selective pressures for maintaining the ability to modify sperm should decrease. This could be followed by the loss of the sterilizing trait and, finally, by the loss of the infection in the population (Hurst and McVean, 1996; Vavre et al., in press). Both empirical and recent experimental studies are consistent with this conclusion (Werren and Windsor, 2000; Vavre et al., 2002). From these differences, we will now speculate on two general questions that address the difference between the general picture of *Wolbachia* associations in arthropods and nematodes.

Why Has the Association Filarial Nematode/*Wolbachia* Led to Obligatory Symbiosis Instead of Reproductive Parasitism?

If we assume that CI-inducing *Wolbachia* are systematically lost in the course of the host–symbiont evolution, the ability to be horizontally transferred appears necessary for a CI-*Wolbachia* strain to perpetuate through evolutionary times. In absence of horizontal transfer in nematodes, CI-inducing *Wolbachia* certainly cannot be retained in these species, even though we hypothesize that a CI phenotype could be expressed in these hosts. Therefore, the remaining strategy for a maternally transmitted bacterium such as *Wolbachia* is to become beneficial and obligatory for the host. On the contrary, mutualism and reproductive parasitism can be retained in arthropods. Since reproductive parasitism does not require coadaptation, it should spread more easily into new host species through horizontal transmission and thus be overrepresented. Following this hypothesis, the few cases of mutualistic *Wolbachia* associations in arthropods could be explained by a sampling bias.

Why Has the Association Arthropod/*Wolbachia* Not Led More Frequently to Obligatory Symbiosis?

First, it is possible that instability of *Wolbachia* associations in arthropods, either through replacement by other CI-inducing variants or through infection loss, does not allow long coevolutionary times and consequently species-specific coadaptations.

Another possibility is related to vertical transmission, which requires that a compromise be reached between maximal transmission efficacy to the eggs and minimal attrition on the host (i.e., minimal effect on host fitness). Transmission efficacy is obviously related to *Wolbachia* replication rate and *Wolbachia* density in the eggs. Replication rate of a microorganism is, however, usually related to its virulence. In addition, high density of microorganisms in eggs could result in developmental arrest. It is thus generally thought that the control of the replication rate of microorganisms is crucial for establishing a well-integrated symbiotic system, with reduction of replication rate leading to virulence reduction (in some cases, there might be a decoupling in the replication rate of a symbiont in different host tissues, with a delay of the pathogenetic effects; see McGraw et al., 2002). It has recently been shown that significant virulence attenuation can actually occur in *Wolbachia* in an insect host, in terms of a reduction of bacterial replication and density in the ovaries, with reduction of detrimental effects on host fecundity (McGraw et al., 2002).

In the presence of strict vertical transmission and clonality, reduction of *Wolbachia* replication rate during generations could possibly occur through kin-selection phenomena. In the presence of a high degree of horizontal transmission (and, thus, of infections with more than one strain), competition among unrelated bacteria could possibly hamper selection for reduced rate of replication. The presence of microorganisms that are unable to minimize their effects on host fitness through the optimization of their replication rate will probably urge the host to develop countermeasures to control the microorganism population. In these conditions, the arms race between host and symbionts is not likely to allow a definitive virulence reduction and the development of a stable, well-regulated, and highly reproducible system of transmission. However, let us assume that the degree of horizontal transmission is not high and that virulence reduction occurs in strains of CI-inducing *Wolbachia* through reduction of replication rate. Infection with these strains could become unsuccessful in the long term: slow-replicating strains could be overtaken by invading fast-replicating strains. Even a relatively limited degree of horizontal transmission of CI *Wolbachia* could thus hamper the establishment of obligatory symbiosis, by reducing the duration of associations with low-virulence, slow-replicating strains.

In summary, we have offered two hypotheses that could explain the dichotomous situation between arthropod and nematode *Wolbachia*. In nematodes (1) the lack of horizontal transmission does not permit the maintenance of a CI-inducing *Wolbachia*, thus making obligatory mutualism the only strategy for nematode *Wolbachia*, while in arthropods (2) horizontal transmission of CI-inducing *Wolbachia* could hamper the development of obligatory symbiosis in two ways, depending on its rate: by making unlikely evolution of strains toward reduced virulence/slow replication rate or by rendering strains with reduced virulence/slow replication rate (compared to the virulent, fast-replicating ones) unsuccessful.

Evolutionary Transitions between Symbiotic Lifestyles

Coadaptations Require (Quite) Long Evolutionary Times

We will now consider a hypothetical system where transmission of a CI-inducing *Wolbachia* is strictly vertical. If we assume the existence of polymorphism for the presence of *Wolbachia* in the host population, this will produce a strong selective advantage in being able to induce CI. Once *Wolbachia* has become fixed in the population, the CI-inducing capacity will lose its selective advantage. If this *Wolbachia* is not capable of horizontal transmission to another population (and thus has no chances to exploit CI induction for its spreading), there will be no strain in which preservation of CI capacity will be under strong selective pressure. It is, of course, possible to hypothesize that CI-inducing capacity could be maintained, at least in the short term, in the presence of incomplete transovarial transmission of *Wolbachia*. In any case, the lack of invasion of new compatibility types (determined by the absence of horizontal transmission) and the clonal structure of the vertically transmitted bacterial population could possibly provide conditions for virulence reduction through kin-selection phenomena. We can thus expect a reduction of the arms race between *Wolbachia* and its host, with optimization of *Wolbachia* replication and transmission. In the presence of perfect vertical transmission, in the long term we should also expect the development of some coadaptation to the reciprocal presence and possibly the establishment of some forms of mutualistic interactions. If *Wolbachia* becomes in some way needed for embryonic development, pressure for preservation of the CI-inducing capacity should disappear (but there could be a phase in which CI becomes a system by which to avoid nonuseful partial development of embryos if *Wolbachia* is needed by the host in successive phases of development). In the long term, if the presence of *Wolbachia* is in some way needed by the host, we should expect development of host mechanisms to ensure its transmission and definitive loss of pressure for maintaining CI-inducing capacity.

The establishment of *Wolbachia* symbiosis in filarial nematodes could have followed a similar pathway, but it is also possible that the initial *Wolbachia* phenotype was different. A CI-inducing *Wolbachia* is, however, attractive as a hypothetical starting point for the development of obligatory symbiosis: it promotes its own survival and diffusion while not being too costly to the host when it nears fixation.

In other forms of reproductive parasitism, fitness costs to the host remain high even at fixation (e.g., in male killing phenotypes), and the arms races is likely to be constant between *Wolbachia* and its hosts. Thus, where the behavior of a given kind of reproductive parasite is not incompatible with obligatory symbiosis, we do not expect this kind of association to develop if the effect on host fitness is above the threshold that will determine counter-responses by the host (i.e., the establishment of an arms race).

Rapid Evolutionary Mechanisms for Transition

The above hypothesis concludes that evolution from a facultative to an obligatory symbiosis requires long coevolution to obtain species-specific coadaptation between partners and reduction of virulence. However, the two examples in insects where *Wolbachia* infection is obligatory for reproduction suggest another transitive mechanism, which could occur more rapidly through the evolutionary time. Indeed, in both the thelytokous wasps and *A. tabida*, host dependence has probably resulted from the accumulation of deleterious mutations in the insect genome. In thelytokous wasps, because sexual functions are no longer under selective pressure, random mutations in genes related to this function can fixate and accumulate, making *Wolbachia*-induced parthenogenesis the only possible mode of reproduction. In *A. tabida*, we have proposed two different hypotheses that could account for the total *Wolbachia* dependence of females in producing eggs. The first one postulates a novel reproductive *Wolbachia* manipulation (i.e., the *Medea* phenotype) and will not be discussed here. The alternative hypothesis postulates that certain bacterial factors are directly involved in the oogenesis of the insect. In this case, we have speculated on an evolutionary scenario that includes a step in which both host and bacterial genomes would have a genetic determinant encoding for a factor involved in and necessary for oogenesis. As in parthenogenetic wasps, we can therefore expect that obligatory symbiosis would have been acquired rapidly by accumulation of deleterious mutations in the genetic determinant of the insect, making those of *Wolbachia* the only functional genetic determinants for insect reproduction.

We believe that these cases could occur rapidly in the course of evolution, without the need for long coevolution and sophisticated coadaptation between partners. Indeed, this mode of evolution is based on mutational events. Moreover, gene loss and genome degradation have often been evoked to explain the total host dependence of several strictly intracellular bacteria. Indeed, genome-size reduction was recently recognized to be a general phenomenon observed in many obligate intracellular bacteria. The strict adaptation to a constant environment, such as that provided by the cytoplasm of a eukaryotic cell, may allow the bacteria to abolish many of the adaptive responses that are required by free-living bacteria and may also allow them to reduce their anabolic capacity if they succeed in recruiting metabolic precursors from the host cell's metabolism. As a consequence of the loss of essential characteristics, bacteria therefore become wholly dependent on their hosts (Andersson and Kurland, 1998; Andersson and Andersson, 1999; Ochman and Moran, 2001). Following the same principle, we believe that the reverse situation has occurred in parthenogenetic wasps and *A. tabida*, where the hosts themselves have lost an essential function for their own reproduction, thereby becoming wholly dependent on *Wolbachia*.

Wolbachia Obligatory Symbiosis: Mutualism or Trickery?

Is obligatory symbiosis mutualistic? This question has previously been underlined by Douglas and Smith (1989). These authors do not believe that two interdependent symbiotic partners necessarily form a mutualistic association, in the sense that each partner does not necessarily benefit from the other. In numerous cases, it is accepted that the host benefits from the association by utilizing products of its symbiont metabolism (essential nutrients in insects, photosynthetic activity in some invertebrates, cellulose degradation in ruminants, etc.). However, there is little evidence that the symbiont also benefits from its host. Some authors even argue that in certain cases the symbiont could be considered "a slave," which produces some beneficial products for its "proslaver" host. Why, therefore, are most symbionts strictly symbiotic and unable to survive in a free-living state? From an evolutionary point of view, we can speculate that, at the beginning of the association, the host would have developed some features to capture some "interesting" free-living bacteria. In the course of evolution, adaptations would have allowed the hosts to conserve their symbionts during life and would also have acquired the ability to transmit them to offspring. In such a situation, where the symbiont is now in a permanent relationship with its host, we could expect it to lose metabolic capabilities, as mentioned in the above paragraph, becoming wholly dependent on its host habitat. In this way, each partner is entirely dependent on the other. However, can we consider this obligatory symbiosis mutualistic? The answer is evident if we consider the association at the present time: each partner requires the presence of the other to survive. However, we can also consider that only the host actually benefits from the association in the course of evolution. In other words, Douglas and Smith (1989) described these associations as "slavery," where the host captures bacteria for its own benefit.

Can we expect such "slavery" associations in obligatory *Wolbachia* symbiosis? This question is not easy to answer. However, we can expect a possible unbalanced benefit between the two interdependent protagonists, at least in insects. Indeed, while *Wolbachia* is believed to benefit from the association, the existence of such benefit for the host remains uncertain. In parthenogenetic wasps, the effect induced procures an evolutionary advantage for *Wolbachia* by accelerating the invasion of host populations (Stouthamer, 1997). However, what is the advantage for the host to be obligatorily linked with such bacteria? Does the host benefit from its asexual mode of reproduction? In *A. tabida*, the total dependence of host reproduction assures the maintenance of *Wolbachia* infection through evolutionary times. However, what is the interest for females to be wholly *Wolbachia* dependent to produce their eggs? Does this dependence confer better egg production, associated with a more effective fecundity? Of course, we cannot exclude the possibility that *Wolbachia* provides other metabolic capabilities that have thus far eluded detection. However, the highly specific inhibition of oogenesis in aposymbiotic *A. tabida* females makes this hypothesis unlikely.

In conclusion, we think obligatory *Wolbachia* symbiosis in insects may represent a "slavery" association, where the bacteria, for their own advantage, have captured their host. In this sense, as considered for the other reproductive alterations induced by *Wolbachia*, obligatory *Wolbachia* could also be qualified as parasites that have manipulated host reproduction to the point of rendering the host incapable of independence. These new cases of reproductive parasitism could therefore be qualified as "obligatory parasitism."

ACKNOWLEDGMENTS

The authors thank Fabrice Vavre for stimulating discussion on the manuscript.

REFERENCES

Andersson, J.O. and Andersson, S.G.E. (1999). Insights into the evolutionary process of genome degradation. *Curr. Opinion Genet. Dev.* **9:** 664–671.

Andersson, S.G.E. and Kurland, C.G. (1998). Reductive evolution of resident genomes. *Trends Microbiol.* **7:** 263–268.

Awahmukalah, D.S.T. and Brooks, M.A. (1985). Viability of *Culex pipiens* eggs affected by nutrition and aposymbiosis. *J. Invertebrate Pathol.* **45:** 225–230.

Bandi, C., McCall, J.W., Genchi, C., Corona, S., Venco, L., and Sacchi, L. (1999). Effects of tetracycline on the filarial worms *Brugia pahangi* and *Dirofilaria immitis* and their bacterial endosymbionts *Wolbachia. Int. J. Parasitol.* **29:** 357–364.

Bandi, C., Dunn, A.M., Hurst, G.D.D., and Rigaud, T. (2001a). Inherited microorganisms, sex-specific virulence and reproductive parasitism. *Trends Parasitol.* **17:** 88–94.

Bandi, C., Trees, A.J., and Brattig, N.W. (2001b). *Wolbachia* in filarial nematodes: evolutionary aspects and implications for the pathogenesis and treatment of filarial diseases. *Vet. Parasitol.* **98:** 215–238.

Beeman, R.W., Friesen, K.S., and Denell, R.E. (1992). Maternal-effect selfish genes in flour beetles. *Science* **256:** 89–92.

Bordenstein, S.R. and Werren, J.H. (2000). Do *Wolbachia* influence fecundity in *Nasonia vitripennis*? *Heredity* **84:** 54–62.

Breeuwer, J.A.J. (1997). *Wolbachia* and cytoplasmic incompatibility in the spider mites, *Tetranychus urticae* and *T. turkestani. Heredity* **79:** 41–47.

Breeuwer, J.A.J. and Werren, J.H. (1990). Microorganisms associated with chromosome destruction and reproductive isolation between two insect species. *Nature* **346:** 558–560.

Casiraghi, M., Anderson, T.J.C., Bandi, C., Bazzocchi, C., and Genchi, C. (2001). A phylogenetic analysis of filarial nematodes: comparison with the phylogeny of *Wolbachia* endosymbionts. *Parasitology* **122:** 93–103.

Charlat, S. and Merçot, H. (2001). Did *Wolbachia* cross the border? *Trends Ecol. Evol.* **16:** 540–541.

Dedeine, F., Vavre, F., Fleury, F., Loppin, B., Hochberg, M.E., and Boulétreau, M. (2001). Removing symbiotic *Wolbachia* bacteria specifically inhibits oogenesis in a parasitic wasp. *Proc. Natl. Acad. Sci. U.S.A.* **98:** 6247–6252.

Dedeine, F., Vavre, F., Fouillet, P., and Boulétreau, M. Lifestyle evolution in symbiotic bacteria: evidence for intra-individual coexistence of obligatory and facultative *Wolbachia* bacteria in the parasitic wasp *Asobara* tabida. In preparation.

Dobson, S.L. and Rattanadechakul, W. (2001). A novel technique for removing *Wolbachia* infections from *Aedes albopictus* (Diptera: Culicidae). *J. Mediterranean Entomol.* **38:** 844–849.

Dobson, S.L., Marsland, E.J., and Rattanadechakul, W. (2001). *Wolbachia*-induced cytoplasmic incompatibility in single and superinfected *Aedes albopictus* (Diptera: Culicidae). *J. Mediterranean Entomol.* **38:** 382–387.

Dobson, S.L., Marsland, E.J., and Rattanadechakul, W. (2002). Mutualistic *Wolbachia* infection in *Aedes albopictus*: accelerating cytoplasmic drive. *Genetics* **160:** 1087–1094.

Douglas, A.E. (1994). *Symbiotic Interactions*. Oxford University Press, New York.

Douglas, A.E. (1998). Nutritional interactions in insect-microbial symbioses: aphids and their symbiotic bacteria *Buchnera*. *Annu. Rev. Entomol.* **43:** 17–37.

Douglas, A.E. and Smith, D.C. (1989). Are endosymbiosis mutualistic? *Trends Ecol. Evol.* **4:** 350–352.

Girin, C. and Boulétreau, M. (1995). Microorganism-associated variations in host infestation efficiency in a parasitoid wasp, *Trichogramma bourarachae* (Hymenoptera: Trichogrammatidae). *Experientia* **51:** 398–401.

Gottlieb, Y. and Zchori-Fein, E. (2001). Irreversible thelytokous reproduction in *Muscidifurax uniraptor*. *Entomol. Exp. Appl.* **100:** 271–278.

Hariri, A.R., Werren, J.H., and Wilkinson, G.S. (1998). Distribution and reproductive effects of *Wolbachia* in stalk-eyed flies (Diptera: Diopsidae). *Heredity* **81:** 254–260.

Hoerauf, A., Mand, S., Adjei, O., Fleischer, B., and Buttner, D.W. (2001). Depletion of *Wolbachia* in *Onchocerca volvulus* by doxycycline and microfilaridermia after ivermectin treatment. *Lancet* **357:** 1415–1416.

Hurst, G.D.D. and McVean, G.T. (1996). Clade selection, reversible evolution and the persistence of selfish elements: the evolutionary dynamics of cytoplasmic incompatibility. *Proc. R. Soc. London (B)* **263:** 97–104.

Jeon, K.W. (1972). Development of cellular dependence on infective organisms: micrurgical studies in *Amoebas*. *Science* **176:** 1122–1123.

Jeyaprakash, A. and Hoy, M.A. (2000). Long PCR improves *Wolbachia* DNA amplification: wsp sequences found in 76% of sixty-three arthropod species. *Insect Mol. Biol.* **9:** 393–405.

Jiggins, F.M., von der Schulenberg, J.H., Hurst, G.D.D., and Majerus, M.E.N. (2001). Recombination confounds interpretations of *Wolbachia* evolution. *Proc. R. Soc. London (B)* **268:** 1423–1427.

Kraaijeveld, A.R. and Godfray, H.C.J. (1997). Trade-off between parasitoid resistance and larval competitive ability in *Drosophila melanogaster*. *Nature* **389:** 278–280.

McFall-Ngai, M.J. (2002). Unseen forces: the influence of bacteria on animal development. *Dev. Biol.* **242:** 1–14.

Lipsitch, M., Nowak, M.A., Ebert, D., and May, R.M. (1995). The population dynamics of vertically and horizontally transmitted parasites. *Proc. R. Soc. London (B)* **260:** 321–327.

Lo, N., Casiraghi, M., Salati, E., Bazzocchi, C., and Bandi, C. (2002). How many *Wolbachia* supergroups exist? *Mol. Biol. Evol.* **19:** 341–346.

Margulis, L. and Fester, R. (1991). *Symbiosis as a Source of Evolutionary Innovation*. MIT Press, Cambridge, MA.

McGraw, E.A., Merritt, D.J., Droller, J.N., and O'Neill, S.L. (2002). *Wolbachia* density and virulence attenuation after transfer into a novel host. *Proc. Natl. Acad. Sci. U.S.A.* **99:** 2918–2923.

Moran, A.N. and Baumann, P. (2000). Bacterial endosymbionts in animals. *Curr. Opinion Microbiol.* **3:** 270–275.

Mimouni, F. (1991). Microorganism-associated variation in host infestation efficiency in a parasitoid wasp, *Trichogramma bourarachae* (Hymenoptera: Trichogrammatidae) *Redia* **124:** 393–400.

Ochman, H. and Moran, N.A. (2001). Gene lost and genes found: evolution of bacterial pathogenesis and symbiosis. *Science* **292:** 1096–1098.

Poinsot, D. and Merçot, H. (1997). *Wolbachia* infection in *Drosophila simulans*: does the female host bear a physiological cost? *Evolution* **5:** 180–186.

Rigaud, T. (1997). Inherited microorganisms and sex determination of arthropod hosts. In *Influential Passengers: Inherited Microorganisms and Arthropod Reproduction* (S.L. O'Neill, J.H. Werren, and A.A. Hoffmann, Eds.), pp. 81–101. Oxford University Press, New York.

Sacchi, L., Corona, S., Casiraghi, M., and Bandi, C. (2002). Does fertilization in the filarial nematode *Dirofilaria immitis* occur through endocytosis of spermatozoa? *Parasitology* **124:** 87–95.

Sinkins, S.P., Braig, H.R., and O'Neill, S.L. (1995). *Wolbachia* superinfections and the expression of cytoplasmic incompatibility. *Proc. R. Soc. London (B)* **261:** 325–330.

Stolk, C. and Stouthamer, R. (1996). Influence of a cytoplasmic incompatibility-inducing *Wolbachia* on the fitness of the parasitoid wasp *Nasonia vitripennis*. *Proc. Sect. Exp. Appl. Entomol. Neth. Entomol. Soc.* **7:** 33–37.

Stouthamer, R. (1997). *Wolbachia*-induced parthenogenesis. In *Influential Passengers: Inherited Microorganisms and Arthropod Reproduction* (S.L. O'Neill, J.H. Werren, and A.A. Hoffmann, Eds.), pp. 102–124. Oxford University Press, New York.

Stouthamer, R. and Werren, J.H. (1993). Microorganisms associated with parthenogenesis in wasps of the genus *Trichogramma. J. Invertebrate Pathol.* **61:** 6–9.

Stouthamer, R., Luck, R.F., and Hamilton, W.D. (1990). Antibiotics cause parthenogenesis *Trichogramma* to revert to sex. *Proc. Natl. Acad. Sci. U.S.A.* **87:** 2424–2427.

Stouthamer, R., Breeuwer, J.A., and Hurst, G.D. (1999). *Wolbachia pipientis*: microbial manipulator of arthropod reproduction. *Annu. Rev. Microbiol.* **53:** 71–102.

Sun, L.V., Foster, J.M., Tzertzinis, G., Ono, M., Bandi C., Slatko, B., and O'Neill, S.L. (2001). Determination of *Wolbachia* genome size by pulsed-field gel electrophoresis. *J. Bacteriol.* **183:** 2219–2225.

Taylor, M.J. and Hoerauf, A.M. (1999). *Wolbachia* bacteria of filarial nematodes. *Parasitol. Today* **15:** 437–442.

Taylor, M.J., Bandi, C., Hoerauf, A.M., and Lazdins, J. (2000). *Wolbachia* bacteria of filarial nematodes: a target for control? *Parasitol. Today* **16:** 179–180.

Van Alphen, J.J.M. and Nell, H.W. (1982). Superparasitism and host discrimination by *Asobara tabida* Nees (Braconidae: Alysiinae), larval parasitoid of Drosophilidae. *Neth. J. Zool.* **32:** 232–260.

Vavre, F., Girin, C., and Boulétreau, M. (1999a). Phylogenetic status of a fecundity-enhancing *Wolbachia* that does not induce thelytoky in *Trichogramma. Insect Mol. Biol.* **8:** 67–72.

Vavre, F., Fleury, F., Lepetit, D., Fouillet, P., and Boulétreau, M. (1999b). Phylogenetic evidence for horizontal transmission of *Wolbachia* in host-parasitoid associations. *Mol. Biol. Evol.* **16:** 1711–1723.

Vavre, F., Fleury, F., Varaldi, J., Fouillet, P., and Boulétreau, M. (2000). Evidence for female mortality in *Wolbachia*-mediated cytoplasmic incompatibility in haplodiploid insects, epidemiologic and evolutionary consequences. *Evolution* **54:** 191–200.

Vavre, F., Fleury, F., Varaldi J., Fouillet, P., and Boulétreau, M. (2002). Infection polymorphism and cytoplasmic incompatibility in Hymenoptera-*Wolbachia* associations. *Heredity* **88:** 361–365.

Vavre, F., Fouillet, P., and Fleury, F. Between and within host species selection in CI-inducing *Wolbachia* in haplodiploids. Reply to Egas et al. (2002). In press.

Wade, M.J. and Chang, N.W. (1995). Increased male fertility in *Tribolium confusum* beetles after infection with the intracellular parasite *Wolbachia. Nature* **373:** 72–74.

Werren, J.H. (1997). Biology of *Wolbachia. Annu. Rev. Entomol.* **42:** 587–609.

Werren, J.H. and Bartos, J.D. (2001). Recombination in *Wolbachia. Curr. Biol.* **11:** 431–435.

Werren, J.H. and O'Neill, S.L. (1997). The evolution of heritable symbionts. In *Influential Passengers: Inherited Microorganisms and Arthropod Reproduction* (S.L. O'Neill, J.H. Werren, and A.A. Hoffmann, Eds.), pp. 1–41. Oxford University Press, New York.

Werren, J.H. and Windsor, D.M. (2000). *Wolbachia* infection in insects: evidence of a global equilibrium? *Proc. R. Soc. London (B)* **267:** 1277–1285.

Werren, J.H., Zhang, W., and Guo, L.R. (1995). Evolution and phylogeny of *Wolbachia*, a reproductive parasite of arthropods. *Proc. R. Soc. London (B)* **262:** 197–204.

Zchori-Fein, E., Roush, R.T., and Hunter, M.S. (1992). Male production induced by antibiotic treatment in *Encarsia formosa* (Hymenoptera: Aphelinidae), an asexual species. *Experientia* **48:** 102–105.

17 Symbiosis and the Origin of Species

Seth R. Bordenstein

CONTENTS

It is a rather startling proposal that bacteria, the organisms which are popularly associated with disease, may represent the fundamental causative factor in the origin of species.

Ivan E. Wallin, 1927

INTRODUCTION

In making his case for the bacterial nature of mitochondria, Ivan E. Wallin (1927) became the first serious advocate of symbiont-induced speciation — the process by which symbiotic organisms split one host species into two. In his book *Symbionticism and the Origin of Species*, Wallin reasoned that the universality of bacterial-derived mitochondria reflected the importance of bacterial symbionts as building blocks of evolutionary change and ultimately new species. To Wallin, bacteria were not just pathogenic agents but also heritable units that could generate large phenotypic changes leading to new species. His views on microbial symbiosis were received as plausible at the time, though they were quickly overshadowed by T. H. Morgan's discovery of chromosomal genes and

0-8493-1286-8/03/$0.00+$1.50

the emergence of population genetics in the 1930s (Sapp, 1990). During this time, chromosomal genes quickly replaced symbionts as what were considered the main building blocks of evolutionary change. The rise of the modern synthesis led to a fall in symbiosis research, and Wallin's thesis was ultimately characterized by skeptical remarks, such as "it is within the range of possibility that they (symbionts) may some day call for more serious attention" (Wilson, 1925).

Some 75 years later, Wallin's day may have finally arrived. The last decade has brought renewed interest and even some agreement on the role that endosymbionts play in evolutionary processes. There are clear footprints of endosymbiosis in major evolutionary transitions (Gray et al., 1999) as well as microevolutionary processes including adaptation and speciation (Margulis and Fester, 1991; Douglas, 1998). Although endosymbionts are certainly not the major causal factor in species formation, as Wallin argued, they are now viewed as at least one of the factors that can promote the speciation process. Nowhere is this more clear than in studies of *Wolbachia*, a group of α-proteobacteria that are among the most abundant endosymbionts in the world. Werren (1998) last reviewed the topic, but significant advances have been made since then that have shifted the controversy over *Wolbachia*-associated speciation from whether it is plausible to how important it is.

Inferences about how symbionts can generate evolutionary novelty and new species have lurked in the background of speciation research for quite some time (Hoyt and Osborne, 1971; Williamson et al., 1971; Howard et al., 1985; Nardon and Grenier, 1991; Margulis, 1993; Adams and Douglas, 1997). Indeed, there is an immense body of work on the host cellular, biochemical, developmental, and evolutionary changes caused by symbiotic miroorganisms (Margulis and Fester, 1991; Chapter 1, this book). But these studies are rarely taken as serious evidence of symbiont-induced speciation because it is unclear which and how often these intraspecies changes affect the process of speciation. In this regard, the cytoplasmically inherited bacterium *Wolbachia* has emerged as the poster child for studies of speciation by means of symbionts. The reason is simple — these bacteria alter reproduction and therefore can directly affect compatibility between populations or species.

My objectives here are to briefly introduce the reader to *Wolbachia*-associated speciation and to highlight the advances that have been made in the field since the last review of the topic. I will also evaluate the criticisms of *Wolbachia*-associated speciation and emphasize those areas where more research is badly needed. Finally, I will discuss some alternative systems in which symbionts may drive host speciation. Throughout this chapter, it is important to keep in mind that more than 60 years of speciation research have fortified the nuclear gene as a dominant agent of species formation (Coyne and Orr, 1998); but it is this very fact that makes the possibility of symbiont-induced speciation still so "startling."

DEFINING A SPECIES AND SYMBIONT

Defining a species is a controversial issue (Harrison, 1998; Hey, 2001; Noor, 2002), especially when considering a symbiotic basis of speciation. I will avoid lengthy discussions on this topic, as this is not the focus of the chapter. I simply adhere to the widely accepted Biological Species Concept (Mayr, 1963), in which species are "reproductively isolated" groups comprised of potentially interbreeding individuals. Reproductive isolation simply refers to those mechanisms that prevent or reduce interbreeding between such groups. There are two general forms of reproductive isolation: (1) postmating isolation refers to those mechanisms that hinder the flow of genes after mating takes places (e.g., hybrid sterility or inviability), and (2) premating isolation refers to those mechanisms that affect interbreeding before mating takes places (e.g., mate discrimination, habitat differences). Both pre- and postmating isolation ultimately reduce gene exchange between groups and therefore allow different species or diverging populations to evolve independently of each other. By adhering to this species concept, speciation can be equated to the evolution of reproductive isolation.

A symbiont is one of the organisms involved in an intimate association between two organisms. Symbionts are sometimes strictly considered to be mutualists, but I will use the more general and historical view that symbionts can take on any relationship with the other organism, including parasitism and commensalism (de Bary, 1879). The text will focus on endosymbionts — those symbionts that exist inside (e.g., intracellularly) their hosts.

WOLBACHIA AS A MODEL SYSTEM FOR SYMBIONT-INDUCED SPECIATION

Before considering the advances in the frontier of *Wolbachia*-associated speciation, it is important to summarize some key biological features of *Wolbachia*. These features, which have been discovered only within the last decade, form the conceptual landscape for why *Wolbachia* stand apart from other symbionts implicated in promoting speciation (Margulis and Fester, 1991). For those readers unfamiliar with *Wolbachia* biology, this section will also serve as a brief introduction to this fascinating bacterium. There are four important features:

ABUNDANCE

Wolbachia are among the most abundant endosymbiotic bacteria on the planet, due in part to their unparalleled host range. First discovered in the mosquito *Culex pipiens* (Hertig and Wolbach, 1924), *Wolbachia* are estimated to occur in 20 to 75% of all insect species (Werren et al., 1995a; Jeyaprakash and Hoy, 2000), 35% of terrestrial isopods (Bouchon et al., 1998), 43% of mites (Breeuwer and Jacobs, 1996), and almost all filarial nematodes (Bandi et al., 2001). Thus, *Wolbachia* infect at least two animal phyla (Arthropoda and Nematoda) and are at high frequencies within two of the most speciose groups of animals — insects and mites. Extrapolating these various infection frequencies to the estimated number of species in these taxa places *Wolbachia* in several million host species. These numbers speak for themselves and have obvious implications for the potential importance of these symbionts in host speciation. Limits to the host range (e.g., vertebrates or other invertebrate groups) are currently not known.

There are at least four major subgroups of *Wolbachia*, labeled A through D. Subgroups A and B diverged ~60 million years ago (Werren et al., 1995b) and occur strictly in arthropods. Subgroups C and D are specific to filarial nematodes and diverged from the common ancestor of A and B ~100 million years ago (Bandi et al., 1998). Arthropod species can be either singly or multiply infected with A and B *Wolbachia* (Werren et al., 1995a). Based on phylogenetic work and the occurrence of double infections, it is inferred that horizontal transmission of the A and B *Wolbachia* must occur at some level (O'Neill et al., 1992; Werren et al., 1995a; Stouthamer et al., 1999).

REPRODUCTIVE ALTERATIONS

Unlike other symbionts that spread through host populations by enhancing the fitness of their host, *Wolbachia* can spread by reducing the fitness of their host. In arthropods, *Wolbachia* parasitize host reproductive strategies in four basic ways — male killing, feminization, parthenogenesis, and cytoplasmic incompatibility (CI) (Werren, 1997). Because these bacteria are inherited through egg cytoplasm, they are selected to increase the number of infected females (i.e., the transmitting sex) in a population, even at the expense of males. Such examples illustrate the ongoing cytonuclear conflict over sex determination and sex ratios, which can in turn play an important role in rapid evolutionary changes and subsequent genetic divergence among populations.

Briefly, male killing occurs when infected male embryos die such as in the ladybird *Adalia bipunctata*, the butterfly *Acraea encedon*, and *Drosophila bifasciata* (Hurst et al., 1999; Hurst et al., 2000). This effect imparts a fitness advantage to infected female siblings, perhaps through reducing the fitness cost of competition with siblings. Feminization occurs when infected genetic males are converted to phenotypic females who are able to transmit the bacteria (Rousset et al., 1992). Parthenogenesis induction (PI) typically occurs in haplodiploid wasps in which infected virgin females

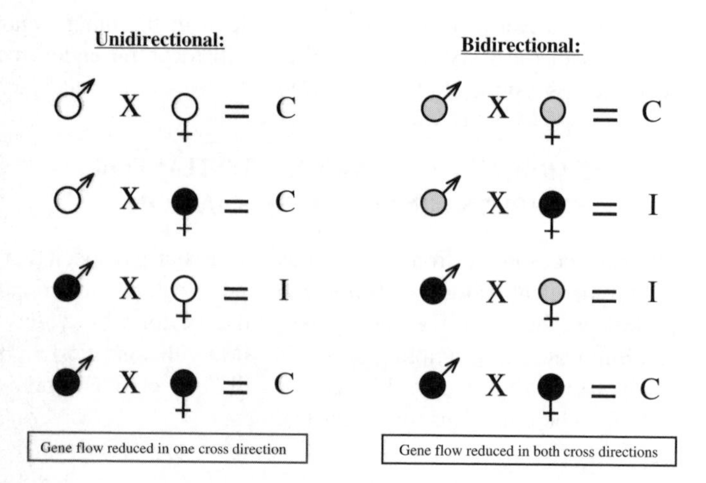

FIGURE 17.1 The dynamics of *Wolbachia*-induced CI. C and I denote compatible and incompatible crosses, respectively. Unidirectional CI occurs when sperm from infected males (black fill) fertilize eggs from uninfected females (no fill). Bidirectional CI occurs in both cross directions when sperm from a male infected with one *Wolbachia* variant (black fill) fertilize infected eggs with a different *Wolbachia* variant (gray fill).

asexually produce all female offspring (Stouthamer et al., 1990, 1993). Unlike male-killing *Wolbachia*, PI-*Wolbachia* can give rise to all female populations that can persist. Finally, CI is the most common alteration and occurs in all the major insect orders as well as in mites and isopods (Hoffmann and Turelli, 1997). This effect is typically characterized by a sperm modification that leads to abnormalities in post-fertilization paternal chromosome behavior and, ultimately, embryonic mortality (O'Neill and Karr, 1990; Reed and Werren, 1995; Presgraves, 2000). It is typically expressed in crosses between an infected male and uninfected female, thereby reducing the fitness of uninfected females (Figure 17.1).

The key point here is that *Wolbachia* are in various ways in the business of modifying reproduction, the central element of speciation. If reproduction is barred between two populations by whatever means, then reproductive isolation has evolved and speciation is under way. This feature makes *Wolbachia* a more likely symbiont for promoting speciation than other symbionts that, for example, impart a novel biochemical ability to their hosts.

Reproductive Isolation

Some of the reproductive alterations induced by *Wolbachia* within species can quite easily be associated with post- and premating isolation among species (Werren, 1998). For example, CI can play a direct role in postmating isolation by causing F1 hybrid inviability among populations infected with different CI-*Wolbachia* strains. It is worth noting here that, like an F1-dominant genetic incompatibility, CI could have a severe effect on gene-flow reduction, more so than typical recessive genetic incompatibilities. Such recessive incompatibilities are often expressed in F2 hybrids, backcross hybrids, or the heterogametic sex (Orr, 1997), thereby allowing gene flow through certain hybrid combinations. In addition to postmating isolation, *Wolbachia* can also be associated with premating isolation through the induction of parthenogenesis. Gene flow between infected parthenogenetic and uninfected sexual populations could be reduced due to these differences in reproductive strategies. Thus, it is not farfetched to imagine cases where *Wolbachia* could accelerate the speciation process alone or in conjunction with other genetically based reproductive isolation barriers.

Rapid Speciation

Mendelian nuclear genes and selfish genetic elements may promote speciation at different rates. This possibility is important to consider because *Wolbachia* fall into the latter class of elements.

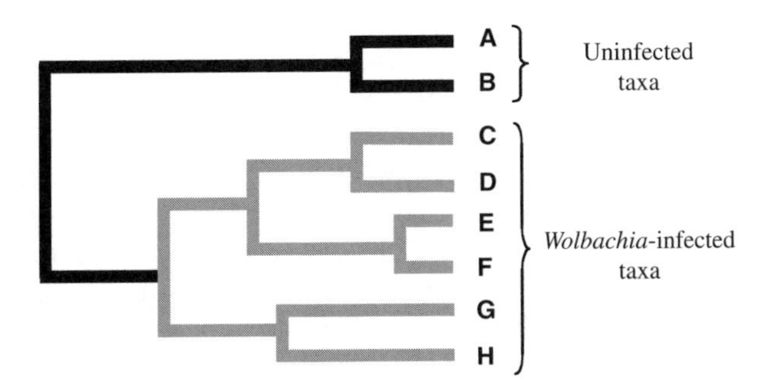

FIGURE 17.2 The hypothesized effect of *Wolbachia* on species phylogeny. If *Wolbachia* accelerate the speciation process, then groups of organisms infected with *Wolbachia* (gray branches) may be more speciose than related, uninfected groups (black branches).

Wolbachia manipulate host sex ratios and reproductive strategies to gain a transmission advantage into the next generation. By doing so, they may spread more rapidly through populations than does the average incompatibility gene. Therefore, reproductive isolation associated with a selfish genetic element like *Wolbachia* may require fewer generations on average to evolve than genetically based reproductive isolation, which is thought to arise through the gradual accumulation of several isolating barriers (Coyne and Orr, 1997). The rapidity with which reproductive isolation evolves is important when considering rates of speciation and the amount of time two allopatric populations have before coming back into contact in sympatry, where they could potentially interbreed. If *Wolbachia* indeed accelerate the speciation process, then groups of organisms harboring *Wolbachia* are predicted to be more speciose than closely related uninfected groups (Werren, 1998; Figure 17.2). Data are currently not complete enough to address this prediction.

CYTOPLASMIC INCOMPATIBILITY AND SPECIATION

Most of the recent progress on *Wolbachia* and speciation has centered on CI-*Wolbachia*. CI is a post-fertilization incompatibility that typically leads to F1 inviability between infected males and uninfected females, or females harboring a strain of *Wolbachia* different than that of the male (Figure 17.1). Because CI reduces or eliminates the production of F1 hybrids, it can hinder gene flow among hybridizing populations. The phenomenon was first described, without knowledge of its infectious nature, in the mosquito *C. pipiens* by the German scientist Hans Laven (1951). He fortuitously came upon CI in intraspecific crosses among geographic races (some only miles apart) that showed complex incompatibility relationships. In fact, he found approximately 15 different "crossing types" within the *C. pipiens* complex (Laven, 1959). He viewed these crossing types as evidence that species could arise without morphological divergence. Laven also found that the causal factor of this incompatibility was not nuclear but cytoplasmic, suggesting that the cytoplasm may play a more important role in animal speciation than previously thought. Thus, the first work to link CI and speciation coincided with the discovery of the CI phenomenon. It would only be realized some 12 years later that the causative agent of CI was *Wolbachia* (Yen and Barr, 1971).

Two Forms of CI

In his early experiments, Laven observed the two major forms of CI: unidirectional and bidirectional (Figure 17.1). Unidirectional CI results in crosses between an infected male and uninfected female. Although the precise molecular mechanisms underlying CI are unknown, the cytological effects of CI are clear and occur in various insect species. For example, in both *Drosophila* and the parasitic wasp *Nasonia*, sperm from an infected male fertilize the uninfected egg, but the paternal chromosomes do not undergo proper condensation during early mitotic divisions of the egg and are not used by the developing embryo (O'Neill and Karr, 1990; Reed and Werren, 1995; Presgraves, 2000). This alteration in paternal chromosome behavior ultimately leads to embryonic death. The outcome of CI is a fitness decrease for uninfected females when they mate to infected males. Because infected females (the transmitting sex) are compatible with either infected or uninfected males, they do not suffer this fitness reduction, and unidirectional CI can therefore rapidly spread *Wolbachia* through host populations (Turelli and Hoffmann, 1991; Turelli, 1994). This effect will work for any cytoplasmically inherited element; thus, we should bear in mind that while all cases of CI so far described are due to *Wolbachia*, it is possible that other cytoplasmically inherited elements also induce CI and achieve the same rapid spread through host populations.

Bidirectional CI is the second form of CI (Figure 17.1) that presumably arises as an incidental byproduct of divergence in CI components (Charlat et al., 2001). In this case, males and females that harbor genetically different strains of *Wolbachia* are reciprocally incompatible. Despite the fact that the egg is infected in these crosses, the same cytological defects occur as in the unidirectional incompatible crosses. Unless the egg has the same strain of *Wolbachia* as that in the male, the paternal chromosome "modification" cannot be "rescued" and F1 embryonic inviability results (Werren, 1997). However, when the male and female harbor the same strain of *Wolbachia*, the paternal modification is "rescued" in the egg and normal embryogenesis is restored.

Four Models of CI-Assisted Speciation

There are at least four ways in which CI can contribute to speciation. For a more detailed treatment of them, the reader is referred to Werren (1998). The take-home message from recent work is that significant progress in these areas cannot be made alone by long-term studies within a single species complex. The most effective way to advance our understanding of the plausibility and importance of *Wolbachia*-induced speciation is to broaden the number of taxa that we study. It is also clear that theoretical insights into the models discussed below are seriously needed (e.g., Telschow et al., 2002a,b). In doing so, we may one day determine both the probability of speciation when *Wolbachia* enter a new host and which of the four models below are important in CI-assisted speciation.

Model 1 (CI Alone)

Consider two genetically identical populations, each fixed for a CI-*Wolbachia* infection. If the infections are the same, then the two populations would be compatible upon contact because there has not been enough divergence for genetic or cytoplasmic incompatibilities to arise. If the infections are genetically distinct, however, the two populations may not be compatible because bidirectional CI is a byproduct of genetic divergence in the components that underlie unidirectional CI (Werren, 1997; Charlat et al., 2001). Therefore, when two populations with different CI-*Wolbachia* come into contact, their hybrids will be inviable due to CI, regardless of divergence in the nuclear genome. Indeed, this prediction turns out to be true when populations or young species with unrelated *Wolbachia* strains are brought together in the lab and CI is carefully measured (Table 17.1; O'Neill and Karr, 1990; Bordenstein et al., 2001).

This phenomenon has obvious implications for the evolution of reproductive isolation. Based on the Biological Species Concept, populations with identical genetic backgrounds could be considered different species if they are isolated by bidirectional CI. Thus, presence or absence of endosymbiont-based reproductive isolation can form the basis of a species diagnosis. This nonnuclear view of species should not be trivialized. Speciation geneticists often ask how many nuclear genes are required in the origin of species (Coyne, 1992). That the answer may sometimes be zero is generally unexpected. For this reason, speciation via bidirectional CI is perhaps the most intriguing of the various models on *Wolbachia*-assisted speciation models — it is the only one in which species could arise in the absence of nuclear divergence.

TABLE 17.1
Some Host Systems Where the Characterization of *Wolbachia*-Induced Bidirectional CI Has Been Conducted or Seems Imminent

Host System	Number of CI Types	Number of CI Types Acquired by Horizontal Transfer/ Number Assayed	Ref.
Culex pipiens mosquitoes	15	0/4[a]	Laven, 1959; Guillemaud et al., 1997
Nasonia wasps	6	5/6[b]	Breeuwer and Werren, 1990; Werren et al., 1995b; Bordenstein and Werren, 1998; Bordenstein et al., 2001
Drosophila simulans flies	5	5/5	O'Neill and Karr, 1990; Merçot and Poinsot, 1998; James and Ballard, 2000
Coleomegilla maculata beetles	2	2/2	Jeyaprakash and Hoy, 2000; Perez and Hoy, 2002
Trichopria drosophilae wasps	2	2/2	Werren et al., in review
Total	30	14/19	

The number of CI types are phenotypically determined by the number of host strains that are uni- or bidirectionally incompatible with one another. The number of CI types acquired by horizontal transfer is determined by the DNA-sequence relationships of the *Wolbachia* infections. For example, CI types that are not closely related, based on one to three different *Wolbachia* gene sequences, are classified into the horizontal transfer group. *Wolbachia* sequences of CI types that are identical (= a) or share their most recent common ancestor (= b) are classified into the alternative group.

The importance of bidirectional CI to speciation depends critically on how often species or populations actually harbor multiple incompatibility types (i.e., CI-*Wolbachia* strains that are reciprocally incompatible). Data are limited in this regard (Table 17.1), but there are two good reasons to think that the number of host systems that harbor mutually incompatible *Wolbachia* has been grossly underestimated. First, much more is known about overall infection frequencies throughout major taxonomic groups than intraspecies variation in *Wolbachia* strains (Jeyaprakash and Hoy, 2000; Werren and Windsor, 2000; Jiggins et al., 2001). Table 17.1 shows those systems where bidirectional CI has been characterized or where the characterization of bidirectional CI seems imminent based on indirect evidence. Some of these host systems appear quite vulnerable to harboring several infections. Second, the systems known to harbor multiple *Wolbachia* infections are especially well studied, including the *Nasonia* species complex (*N. vitripennis*, *N. giraulti*, and *N. longicornis*), *D. simulans*, and *C. pipiens*. The number of systems with bidirectional incompatibility will therefore likely increase with the number of systems studied. Given that estimates of infection frequencies across arthropod species run upwards of 75% (Jeyaprakash and Hoy, 2000), it appears that we have only scratched the surface of a large mountain of data.

The well-studied system of *Nasonia* stands out in this regard since it single-handedly revived interest in Laven's original ideas about CI and speciation in mosquitoes (Laven, 1959). Some 30 years after Laven's work, Breeuwer and Werren (1990) found complete bidirectional CI between two closely related species of parasitic wasps, *N. vitripennis* and *N. giraulti*. It was the first such study to show that bacterial microbes played a direct role in interspecific reproductive isolation and has remained a hallmark case for *Wolbachia*-based reproductive isolation. Shortly after, a second study in *D. simulans* showed the same effect within species — bidirectional CI between strains isolated from California and Hawaii (O'Neill and Karr, 1990). Despite the findings, both

TABLE 17.2
The Presence (+) or Absence (–) of Various Types of Reproductive Isolation in *Nasonia*

Isolating Barrier	Older Species Pair (*N. giraulti/N. vitripennis*)	Younger Species Pair (*N. giraulti/N. longicornis*)
Wolbachia-induced bidirectional CI	+	+
F1 hybrid inviability	+	–
F1 hybrid infertility	–	–
F2 hybrid inviability	+	–
F2 hybrid behavioral infertility	+	–
Sexual isolation	+	+

Data are based on laboratory characterizations of reproductive isolation by Breeuwer and Werren (1990, 1995), Bordenstein et al. (2001), as well as unpublished data. Divergence times for the older and younger species pair are approximately 0.800 and 0.250 million years ago, respectively (Campbell et al., 1993). The isolation profile in the younger species pair indicates that *Wolbachia*-induced bidirectional CI and sexual isolation can evolve early in speciation prior to the evolution of other isolating barriers.

cases were greeted with some skepticism. In the former case, it became apparent that CI was not the only mechanism preventing gene flow between *N. vitripennis* and *N. giraulti*. Severe levels of hybrid inviability, hybrid sterility, and sexual isolation were also found (Breeuwer and Werren, 1995; Bordenstein and Werren, 1998; Drapeau and Werren, 1999; Table 17.2). Which form of reproductive isolation played the primary causal role in speciation was a mystery. By not knowing the temporal order in which these isolating barriers evolved (a difficult task in practice for any speciation researcher), one could argue that *Wolbachia* came into these two species after speciation was complete. In the *D. simulans* case, infections were still spreading through fly populations, and uninfected individuals were found to be regenerated each generation because of incomplete transmission of the *Wolbachia* through females (Turelli and Hoffmann, 1995). Therefore, gene flow could still be fluid through uninfected individuals in the two populations. Taken together, these criticisms dampened the perceived plausibility of speciation via bidirectional CI.

The most recent study on speciation via bidirectional CI (Bordenstein et al., 2001) has to some extent overcome these criticisms. Using the youngest *Nasonia* species pair, *N. giraulti* and *N. longicornis*, the authors found that, in addition to different *Wolbachia* strains being fixed in their respective species, bidirectional CI was the primary form of reproductive isolation: no other postmating isolation was found, and premating isolation was weak and asymmetric (Table 17.2). Therefore, *Wolbachia*-induced bidirectional CI in this hybridization preceded the evolution of other forms of isolation. This study, as well as the *D. simulans* study, also show that multiple CI strains can easily occur within a population or sister species. It is therefore plausible for *Wolbachia* to play a causal role in speciation and act as the wedge that splits one species into two. If *Wolbachia* are speciation agents, we expect to find similar cases in other arthropod systems.

One caveat is that the species studied by Bordenstein et al. (2001) are not known to occur sympatrically. They inhabit the eastern and western regions of North America (Darling and Werren, 1990). Therefore, bidirectional CI here will play no causal role in speciation, unless the species come into contact or are found to already be hybridizing in geographic areas not surveyed. This caveat outlines a task for future studies of CI and speciation — to discover systems of sympatric, hybridizing species that are isolated because of CI. Here the conditions that restrict or promote CI-assisted speciation in nature could be thoroughly investigated. However, such hybrid zones may be transient and go unsampled, which in turn highlights the additional need for controlled, population-cage experiments of these phenomena as well. Work by Bordenstein and Werren (unpublished) indicates that CI alone can have strong effects on gene flow in experimental populations of *N. vitripennis*. Using decay in linkage disequilibria between visible mutants as an estimate of gene flow, they showed that in cages with no CI or unidirectional CI the decay proceeded rapidly; in contrast, in cages with bidirectional CI there was little to no decay in linkage disequilibria. Thus, bidirectional CI can limit gene flow and maintain genetic divergence between populations. Recent theoretical work by Telschow et al. (2002b) also shows that bidirectional CI can accelerate genetic divergence among populations even when bacterial transmission is inefficient (i.e., infected females lay some uninfected eggs) and CI levels are incomplete (i.e., not all offspring die in incompatible crosses). Thus, CI may assist speciation over a broad range of biologically realistic conditions.

Any attempt to understand how bidirectional CI can promote speciation is confronted with the question of how bidirectional CI arises, that is, how can two or more different CI-*Wolbachia* make it into the same host species? Briefly, there are four paths to the evolution of reciprocal CI. The first is independent acquisition, in which two allopatric populations acquire different *Wolbachia* through independent horizontal-transfer events. The phylogenetic evidence largely favors this path (Table 17.1), and clear examples can be found in *D. simulans* (Clancy and Hoffmann, 1996) and *Nasonia* (Werren et al., 1995b; Bordenstein and Werren, 1998; Werren and Bartos, 2001). The second path is codivergence, in which a single, ancestral *Wolbachia* diverges into two CI-*Wolbachia* types within its host's species. There is phylogenetic evidence of codivergence at the genetic level in the species pair *N. giraulti* and *N. longicornis* (Werren et al., 1995b; Werren and Bartos, 2001). However, both of these species are infected with strains of A and B *Wolbachia*, and only the B *Wolbachia* has undergone codivergence. Measuring bidirectional CI between the two B strains will require isolation of single B-infected *N. giraulti* and *N. longicornis*, an effort that has so far proven difficult (S.R. Bordenstein and J.H. Werren, unpublished). Third, segregation of a double infection (e.g., A and B) can lead to individuals harboring single A and B *Wolbachia* that are bidirectionally incompatible. However, the establishment of single A- and B-infected populations would be unlikely because double infections can spread easily against single infections (Sinkins et al., 1995; Perrot-Minnot et al., 1996). Finally, different host genetic influences on the same *Wolbachia* variant could possibly lead to bidirectional CI between populations or species. The diversity of incompatibility types in *C. pipiens*, but lack of *Wolbachia* sequence variation among these CI types, is consistent with this model (Guillemaud et al., 1997). However, only one *Wolbachia* gene has been surveyed for genetic variation in this system. Taken together, results thus far indicate that bidirectional CI typically evolves through independent acquisition: 14 of 19 cases show evidence of independent acquisition via horizontal transfer (Table 17.1).

Model 2 (CI Coupled with Genetically Based Isolation)

Wolbachia-induced CI will probably play its most significant role in speciation when it is coupled with additional isolating barriers. This view is consistent with at least three lines of evidence. First, the likelihood that *Wolbachia* will be the only cause of speciation is reduced due to the typically incomplete levels of CI (Boyle et al., 1993; Breeuwer and Werren, 1993), inefficient bacterial transmission (Turelli and Hoffmann, 1995), and any reproductive isolation that has evolved before *Wolbachia* enter a host system. Second, unidirectional CI between allopatric populations is common, probably more so than bidirectional CI. Because unidirectional CI is a one-way cross incompatibility between infected and uninfected populations, gene flow can still be fluid through populations via the compatible cross direction (uninfected male × infected female). And third, there is a growing consensus that speciation rarely occurs due to a single form of reproductive isolation. Rather, as populations diverge and begin to sustain independent evolutionary fates, reproductive isolation will gradually evolve due to the accumulation of several (incomplete) isolation barriers (Coyne and Orr, 1997; Sasa et al., 1998; Presgraves, 2002). However, as we will see, *Wolbachia* can still play an essential role in promoting species formation even when CI is coupled with other isolating barriers.

Consider two allopatric populations, one fixed for a *Wolbachia* infection and the other uninfected. What would happen if infected individuals migrated into the range of the uninfected population or vice versa? Would these two populations fuse back into one? Would the *Wolbachia* infection now sweep into the susceptible uninfected population? Or would individuals from the two populations not exchange genes, at least in part due to unidirectional CI, where an infected male and uninfected female fail to produce hybrids? The answers partly depend on the amount of genetically based reproductive isolation that has accrued since the split of these populations. For example, considerable gene flow will clearly not allow stable coexistence of infected and uninfected populations because the infected cytoplasm will sweep through (Turelli, 1994). Additionally, it also depends on how strong natural selection (e.g., disruptive selection) opposes migration and the spread of "foreign" genes (Telschow et al., 2002a). While theoretical and experimental evolution studies are poised to significantly enhance our understanding of these issues, empirical studies of natural systems are now motivating the questions and offering us a glimpse into how unidirectional CI can promote the origin of new species.

One recent study is largely responsible for this new outlook on CI-assisted speciation. Shoemaker et al. (1999) characterized pre- and postmating isolation in the mushroom-feeding species pair *D. recens* and *D. subquinaria. D. recens* is infected with *Wolbachia*, while the closely related *D. subquinaria* is uninfected. Levels of mitochondrial diversity in *D. recens* are reduced but consistent with a sweep of *Wolbachia* through this species in the distant evolutionary past. What is most intriguing about these species is that their ranges are likely to overlap or have overlapped in the past in the north-central part of the United States or central Canada. Areas of overlap create opportunities for interspecific hybridization, and genetic evidence indeed suggests that hybridization has occurred between these species (Shoemaker et al., 1999, and unpublished). But as Table 17.3 shows, interspecific gene flow could be severely reduced due to the complementary action of three main isolating barriers: unidirectional CI, sexual isolation, and hybrid male sterility, the latter two being genetically based. One key feature in this species pair is that unidirectional CI and sexual isolation act asymmetrically but in opposite directions. Thus, instead of gene flow being confined to one cross direction, as would be expected with the existence of just one of these isolating barriers (or if they both operated in the same direction), gene flow is limited in both directions. This pattern provides just the right fit for a more stable coexistence of the infected and uninfected species. And if that is not enough, hybrid male sterility awaits the surviving males produced in the F1 generation. The assortment of symbiotic and genetically based isolating barriers provides clear evidence that *Wolbachia* can act in concert with other barriers and still be essential to the speciation process. Whether it is essential in this *Drosophila*

TABLE 17.3
The Percent of Fit Hybrids between *Drosophila recens* and *D. subquinaria* Is Successively Reduced When Multiple Isolating Barriers Are Considered

Cross (male x female)	No Isolation (%)	Sexual Isolation (%)	Sexual Isolation + Unidirectional CI (%)	Sexual Isolation + Unidirectional CI + Hybrid Male Sterility (%)
D. recens × *D. subquinaria*	100.0	69.3	10.1	5.1
D. subquinaria × *D. recens*	100.0	28.9	28.9	14.5

Data are modified from laboratory measurements of reproductive isolation in Shoemaker et al. (1999). Percent fit hybrids is estimated by multiplying the strength of each isolating barrier considered (sexual isolation is calculated by dividing the interspecific mating frequency by the intraspecific mating frequency with relation to the same species female; unidirectional CI between an infected *D. recens* male and uninfected *D. subquinaria* female reduces hybrid production to 14.6%; hybrid male sterility is complete and therefore reduces the percent of surviving, fertile F1 offspring by 50%, assuming a 1:1 sex ratio).

speciation event has not been fully resolved. The extent of interspecific gene flow and reproductive isolation in the zone of contact awaits full characterization (Rokas, 2000).

Other systems where unidirectional CI may have evolved between allopatric or parapatric populations/species include *Tetranychus* mites (Navajas et al., 2000; Vala et al., 2000), *Diabrotica* beetles (Giordano et al., 1997), *Gryllus* crickets (Giordano et al., 1997; Mandel et al., 2001), *Tribolium confusum* beetles (Wade et al., 1995), and *Solenopsis* fire ants (Shoemaker et al., 2000).

In *Tetranychus urticae* mites, unidirectional CI between populations on different host plants is unexpectedly associated with two types of postmating reproductive isolation (Vala et al., 2000). The first is the typical expression of CI. The second is a phenotype that closely resembles F2 hybrid breakdown in which F2 male mortality is elevated in incompatible crosses but not in compatible ones. This effect is presumably a consequence of paternal genome fragmentation and subsequent aneuploidy generated from CI in parental crosses. These mites have a holokinetic chromosome structure (e.g., microtubules can attach anywhere on chromosome or chromosomal fragments), which may make them more susceptible to aneuploidy in incompatible crosses. The frequency of F2 problems associated with CI among arthropods is currently not known. CI effects are rarely measured past the F1 generation.

The *Gryllus* crickets have served as a model system for studies of hybrid zones and speciation for many years (Harrison, 1983; Harrison, 1986). Only recently, though, have *Wolbachia* been implicated in the one-way cross incompatibility between *G. pennsylvanicus* males and *G. firmus* females. This suggestion was made by Giordano et al. (1997), who found that *G. pennsylvanicus* was infected with *Wolbachia*, while *G. firmus* was not. Thus, the one-way incompatibility could seemingly be explained by *Wolbachia*-induced unidirectional CI (Giordano et al., 1997). However, a more recent study of infection patterns and species identity by Mandel et al. (2001) has disputed this conclusion. They found more complicated patterns of infections, which precluded a clear interpretation of CI's role in the incompatibility. Unfortunately, neither study has tested for the presence of CI (with infected and uninfected individuals) in controlled laboratory crosses, so the jury is still out on whether CI is involved in at least some of the reproductive isolation in this hybrid zone.

The flow of recent empirical work emphasizes that *Wolbachia*-induced CI will probably play its most significant species-forming role in association with other forms of reproductive isolation. In the case of unidirectional CI, it is clear that other factors that restrict gene flow (such as disruptive selection and other isolating barriers) are necessary to complete speciation (Turelli, 1994; Telschow et al., 2002a). Even in cases of bidirectional CI, levels of incompatibility are sometimes not complete (O'Neill and Karr, 1990; Bordenstein et al., 2001), though CI can still help accelerate genetic divergence (Telschow et al., 2002b). Thus, *Wolbachia*-induced CI is more likely to facilitate speciation when it is one of several "steps" in the evolution of complete reproductive isolation. This view is entirely consistent with how speciation is thought to proceed — through the accumulation of multiple isolating barriers. In addition, the recent work also highlights the importance of characterizing natural systems where CI is associated with reproductive isolation in sympatry. As yet, the *Drosophila* system remains the most likely system in which this could be the case. Finally, it is not just a coincidence that *Wolbachia* are now being considered as an isolating barrier in the *Gryllus* hybrid zone model. The impressive infection frequency of *Wolbachia* in insects (Werren et al., 1995a; Jeyaprakash and Hoy, 2000) will surely place them in other insect hybrid zones as well.

Model 3 (Host Accommodation)

CI will not always act as a direct cause of reproductive isolation between species. It may indirectly be associated with reproductive isolation through a process termed host accommodation. As with any species interaction, *Wolbachia* and their hosts have the potential to coevolve and affect each other's evolutionary fate. With the onset of endosymbiosis, conflict and cooperation between the *Wolbachia* and host genomes can spawn genetic interactions between the two parties, ultimately leading to the evolution of host genotypic influences on *Wolbachia* or compensatory changes in the host to restore *Wolbachia*-altered phenotypes. In particular, the host-selection pressure to accommodate or control the presence of *Wolbachia* may accelerate host genetic substitutions and lead to genetic divergence between populations and potentially the evolution of genetically based isolation. This outcome may be especially true for a bacterium like *Wolbachia*, which can alter important aspects of host fitness, including gametogenesis, mitosis, sex determination, sex ratios, and cytonuclear genetic interactions. The genes underlying these host accommodations within species could end up causing postmating incompatibilities between species.

This model makes the prediction that *Wolbachia* will leave a genetic footprint of its presence through host genetic substitutions in the nuclear or mitochondrial genomes, some of which may be maladaptive in hybrids. Testing this prediction is not simple. Virtually nothing is known about genetic substitutions (or actual genes, for that matter) that interact with *Wolbachia*. However, there has been a recent burst of work describing host–*Wolbachia* interactions in insects. These studies can be seen as part of the first phase in characterizing the nuclear genes involved in host accommodation. In these studies, *Wolbachia* are typically "moved" from the resident species background into a naïve or foreign genetic background, either by microinjection or backcrossing methods. *Wolbachia*-induced phenotypes are then characterized in this new genetic background. The most common effect described in these studies is a change in CI level and bacterial densities when the *Wolbachia* are in a foreign genetic background (Boyle et al., 1993; Bordenstein and Werren, 1998; McGraw et al., 2001; Poinsot and Merçot, 2001). Other *Wolbachia*–host genotypic interactions include rescue of a *Drosophila* lethal mutation (Karr, 2000) and a phenotypic switch from feminization to male killing when a *Wolbachia* strain is experimentally transferred from one Lepidopteran species to another (Fujii et al., 2001).

Some of these interactions between the *Wolbachia* and host genomes may be due to selection on the host to modify *Wolbachia*. For example, in *Nasonia*, CI type differs between *N. vitripennis* and the sister species *N. giraulti* and *N. longicornis*, and this difference is genetically controlled (Bordenstein et al., in press). CI type has apparently been modified by the *N. vitripennis* nuclear genome to increase host fitness. The nuclear genetic divergence underlying this difference could possibly contribute to the genetically based postmating incompatibilities observed between these young species, though this notion remains speculative.

Taken together, the collection of studies across diverse insects provides strong support for ongoing host–*Wolbachia* coevolutionary interactions. Because developmental processes such as gametogenesis and embyrogenesis are likely to be altered by *Wolbachia* and subsequently accommodated by the host, it seems probable that evidence will eventually mount in favor of this model. More detailed genetic experiments are now needed to dissect the host genes involved and their potential association with hybrid incompatibilities.

Model 4 (Reinforcement)

The expression of CI can select for additional forms of reproductive isolation through reinforcement — the process by which postmating isolation acts as a direct selective pressure for the evolution of premating isolation in areas of sympatry (Dobzhansky, 1937; Noor, 1999). Reinforcement is historically viewed as a means by which speciation will be completed (e.g., premating isolation seals off any remaining gene flow from incomplete postmating isolation). Premating isolation is selected for because postmating isolation is a "wasteland" for parental gametes: since hybrid offspring are dead or sterile, they cannot pass on genes themselves. Selection will consequently favor parents that mate preferentially with homospecific mates, thereby maximizing their potential to pass genes into future generations. Despite the elegance of this reasoning, the reinforcement hypothesis has been the subject of some controversy (Coyne and Orr, 1998; Noor, 1999). How does *Wolbachia* fit into all this and perhaps provide new insight into reinforcement?

Consider two allopatric populations that have accrued independent genetic substitutions and harbor different CI-*Wolbachia* strains. What would the fate of these incipient species be if they were to make contact and hybridize? Hybridization would lead to high levels of F1 hybrid inviability due to CI, and parents would be selected to mate discriminately with compatible individuals. How strong must CI be to select for mate-discrimination genes? What levels of migration will restrict or promote CI-assisted reinforcement? Will the required linkage disequilibrium between cytotype and genotype be maintained to complete reinforcement, and will the incipient species just fuse or will one displace the other? These questions await answers from theoretical studies as well as natural and experimental systems where the conditions that affect reinforcement can be directly tested. Circumstantial evidence for this process comes from *Nasonia*. Two species (*N. giraulti* and *N. longicornis*) are microsympatrically embedded within the range of the third species, *N. vitripennis*. All three species are bidirectionally incompatible, and, consistent with reinforcement, *N. giraulti and N. longicornis* females show stronger discrimination against *N. vitripennis* males than the reciprocal cross (Bordenstein and Werren, 1998; Drapeau and Werren, 1999; Bordenstein et al., 2000). However, several postzygotic barriers besides CI exist between these species (Table 17.2), any of which may facilitate reinforcement.

Is there any reason to think that *Wolbachia*-induced CI may have an unusual effect on reinforcement? It seems likely that reinforcement is more probable when driven by *Wolbachia*-induced CI rather than intrinsic postzygotic genetically based incompatibilities. Let us assume that the two scenarios are identical except for the basis of the postmating isolation between the populations. Thus, for example, the amount of migration and intensity of the postmating isolation are the same. However, hybrid fitness is reduced due to *Wolbachia* in one scenario and a simple two-locus genetic incompatibility in the other. Now why might *Wolbachia* have a higher likelihood of driving reinforcement? Because CI halts gene flow at the F1 generation, whereas most genes involved in early genetic incompatibilities are recessive and limit gene flow in some F2 genotypes or the heterogametic F1 genotype (in accordance with Haldane's rule). The upshot of this difference is twofold. First, the F1 isolation caused by CI reduces more gene flow by eliminating hybrids irrespective of their sex or genotype; second, F1 isolation prevents recombination from slashing the required linkage disequilibria between the incompatibility locus and the mate-discrimination locus (Felsenstein, 1981; Kirkpatrick and Servedio, 1999). Recessive incompatibilities do not share this luxury because more fit hybrids will be produced and recombination in the previous generations can break down the required linkage disequilibria. Theoretical treatments of these issues are needed to evaluate this prediction and the conditions associated with CI-assisted reinforcement.

ASEXUALITY, SEXUAL DEGRADATION, AND THE ORIGIN OF SPECIES

The process by which an asexual population splits from a sexual population is a form of cladogenesis that can be termed asexual speciation. This process falls neatly under the Biological Species Concept because it is concerned with the severing of gene flow and the evolution of reproductive isolation between sexual and asexual populations. But first it is necessary to distinguish between two forms of asexual reproduction — arrhenotoky and thelytoky. The former is the typical mode of reproduction in haplodiploid insects where males are produced from unfertilized (haploid) eggs and females are produced from fertilized (diploid) eggs. The latter, in which all unfertilized eggs become (diploid) females, is the one we are concerned with here and will generally be referred to as asexuality in the text.

Asexuality may be under symbiotic or genetic control, though there is good reason to believe that symbiotic bacteria are more often than not the causative agent, at least in haplodiploid organisms. Over 30 cases within the Hymenoptera have been documented by Luck et al. (1992), and within the last few years there has been a modest burst of work revealing a bacterial basis of asexuality. Every case deals with a cytoplasmically inherited bacterium, and while not all the bacteria have been identified, *Wolbachia* have historically been the most common culprit (Stouthamer et al., 1993; Stouthamer, 1997). The mechanism of *Wolbachia*-induced parthenogenesis is usually gamete duplication (Stouthamer, 1997) in which haploid eggs do not complete the first mitotic division and diploidy is restored. Because of this mechanism, it is typically thought that PI *Wolbachia* would be restricted to the Hymenoptera and other haplodiploids (e.g., thrips and mites), where infected virgin females would lay all female (diploid) offspring. The finding that PI *Wolbachia* can also do their business through apomictic parthenogenesis (Weeks and Breeuwer, 2001) suggests that diplodiploid systems are also susceptible to these phenomena. Apomictic asexuality is actually the most common form of parthenogenesis in diplodiploid arthropods (Suomalainen et al., 1987).

What are the consequences of PI bacteria on species formation? The onset of asexuality does not directly prohibit gene flow between the asexual and sexual population because asexuals can still exchange genes with sexual mates, as shown in *Trichogramma* wasps (Stouthamer and Kazmer, 1994). For example, asexual females still retain the ability to mate with a sexual male and may do so unless no mates can be found. Additional isolating barriers must therefore accompany the shift in reproductive strategy to complete the speciation event.

Premating isolation may be easily achieved since the onset of asexuality can indirectly lead to the degradation of fitness characters involved in sexual reproduction, including male and female mating behavior, male fertility, secondary sexual characteristics, fertilization processes, oviposition behaviors, and developmental requirements. For example, an asexual female may lose her ability to accept a sexual male through mutational degradation of genes required for mating behavior.

Both genetic drift and selection can drive the evolution of mutations involved in sexual degradation, and these forces may show sex-specific patterns (Pijls et al., 1996). For example, deleterious mutations in male fitness are expected to accumulate neutrally because males are neither produced nor needed in asexual races, assuming that male–female sexually antagonistic alleles are rare. In contrast, mutations in genes encoding female sexual traits may be strongly selected for due to antagonistic pleiotropy. For example, mutations that erode sexual traits in females (e.g., mating behavior or sperm usage) could pleiotropically cause a fitness increase in asexual females. Female sexual degradation may therefore proceed rapidly via natural selection. This potential difference in decay rate for male and female sexual traits leads to the prediction that female traits may degrade before male traits, at least in large populations where the efficacy of selection is strong and drift is low.

This asymmetry actually strengthens the possibility for asexual speciation because decay of female function may fortify reproductive isolation more so than decay of male function. If asexuality is complete and males are never (or rarely) produced, then male sexual decay is not necessary for the completion of asexual speciation. In contrast, female sexual decay is necessary because females make up the bulk of an asexual population and have the potential to mate with sexual males. Thus, the simplest mode of asexual speciation is the acquisition of a completely penetrant PI *Wolbachia*, followed by the fixation of a single mutation that inhibits mating between sexual males (uninfected) and asexual females (infected). Table 17.4 shows a list of studies that have examined the kinds of sexual degradation

TABLE 17.4
Asexuality in the Hymenoptera Frequently Leads to Sexual Degradation in Female Traits, but Not in Male Traits

System	Female Trait	Male Trait	Ref.
Trichogramma	—	—	Stouthamer et al., 1990
Apoanagyrus diversicornis	Mating behavior	—	Pijls et al., 1996
Encarsia hispida	Oviposition behavior	—	Hunter, 1999
Galeopsomyia fausta	Mating behavior	—	Argov et al., 1999
Telenomus nawai	Fertilization	—	Arakaki et al., 2000
Encarsia pergandiella	Oviposition behavior	—	Zchori-Fein et al., 2001
Muscidifurax uniraptor	Mating behavior and anatomy	Sperm production	Gottlieb and Zchori-Fein, 2001

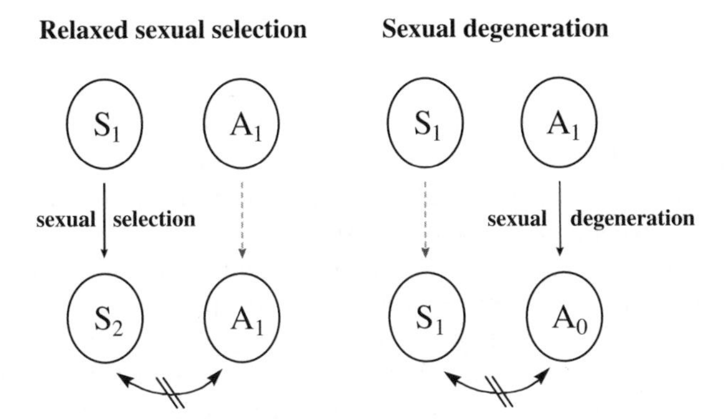

FIGURE 17.3 Two models of asexual speciation. S and A represent the sexual and asexual population, respectively. When the sexual and asexual population have the same subscript number (subscript 1), they can interbreed. However, as evolutionary time runs downward, the sexual population can evolve newly derived components necessary for mating through sexual selection (subscript 2), or the asexual population can lose sexual traits through sexual degradation (subscript 0). The end result under both models is premating isolation between the two populations.

associated with bacterial-induced parthenogenesis and the sex specificity of the trait affected. The key result here is that when one sex suffers from sexual decay, it is the female sex (five out of five cases). Female traits are indeed the more frequent target of sexual degradation. There is only a single conclusive case of male degradation, but this system also shows severe female degradation. Taken together, the findings are consistent with selection as the evolutionary force driving sexual degradation. The consistency of the pattern across several Hymenopteran families suggests that there is a general cost associated with female sexual traits in an asexual population and that male sexual degradation evolves more slowly. The fact that females are the preferred target of sexual degradation also supports the possibility of asexual speciation via parasitic bacteria such as *Wolbachia*. Such populations may easily become "locked" into a parthenogenetic mode of reproduction, which in evolutionary terms would complete the speciation event between sexual and asexual races (Werren, 1998).

Before concluding this section, it is important to point out that some of the apparent effects of sexual degradation can also be explained by an alternative hypothesis, which will be termed "relaxed sexual selection" (Figure 17.3). Here, isolation can evolve between sexuals and asexuals because sexual selection is ongoing and drives changes in the sexual population but is relaxed and stagnant in the asexual population. Thus, new adaptations in the sexual population can bring about isolation with the asexual population. For clarity, consider the imagery of an adaptive landscape where sexual and asexual populations are initially on top of the same peak for traits involved in sexual selection. In contrast to sexual degradation, where the asexual population descends down the peak and ends up in a valley due to mutational decay and pleiotropic selection, relaxed sexual selection is a process by which the sexual population moves to a new or higher peak and the asexual population stays put, thereby leading to divergence in sexually selected traits. Since sexual selection is typically thought to be an ongoing evolutionary process, where the fitness optimum is constantly moving for sexual traits, this process may easily drive divergence between asexuals and sexuals. It could also explain the apparent sexual degradation in female mating behavior and fertilization shown in Table 17.4. Put more simply, these characters perhaps did not degrade at all; they are just no longer sufficient to conduct a successful mating with the newly derived characters in the sexual population.

SKEPTICAL VIEWS ON *WOLBACHIA*-ASSISTED SPECIATION

Any new or unexpected concept will be confronted with criticisms and doubts, and *Wolbachia*-assisted speciation is no exception in this regard (Hurst and Schilthuizen, 1998; Coyne and Orr,

1998; Rokas, 2000; Ballard et al., 2002). Three main remarks have been leveled against the role of *Wolbachia* in speciation.

The first is the least formidable but the most often cited. It concerns the fact that *Wolbachia* have not been shown to be a "cause" of speciation. Specifically, it is not known if *Wolbachia* actively prevent gene flow between any hybridizing species in nature. This criticism, while true, fails to distinguish *Wolbachia*-induced isolation from most other forms of reproductive isolation. Rarely do speciation workers know whether the particular isolating barrier they are investigating is or was a cause of speciation. This is a more general and practical problem to the study of speciation, as it takes a long time. Whether or not *Wolbachia* prove to be a causative agent now is not necessarily an indicator of whether *Wolbachia* have played a role in speciation in the past or will play a role in the future. Additionally, there is no reason to think that an isolating mechanism that is not a cause of speciation in one system will not be a cause of speciation in another system. Most reproductive isolation likely evolves in allopatry as an incidental byproduct of divergence. Therefore, the barriers that end up playing a causal role in speciation are just the ones that existed when the two species became sympatric. There is no reason to expect *Wolbachia* to preferentially evolve early or late in speciation. Simply demonstrating that *Wolbachia* at least evolve early enough in the speciation process to potentially assist speciation may be the closest we get to this issue in real time (Bordenstein et al., 2001).

The second criticism, offered by Coyne and Orr (1998), is that *Wolbachia*-assisted speciation cannot explain Haldane's rule — the common pattern that when one sex is sterile or inviable, it is most often the heterogametic sex (Haldane, 1922). Haldane's rule is of paramount significance to understanding speciation for two reasons. First, it is obeyed in every animal phylum in which it has been looked for, and second, it has been shown to frequently evolve early in the speciation process of various *Drosophila* and Lepidopteran hybridizations (Wu and Davis, 1993; Presgraves, 2002). Thus, its generality and impact on incipient speciation suggest that it may often be part of the path to forming a new species. The fact that *Wolbachia*-induced isolation cannot explain Haldane's rule indicates that *Wolbachia* are probably not part of this shared genetic process that underlies speciation in various animal species. This criticism is fair, and any claims that suggest *Wolbachia* is the major player in animal speciation would be erroneous. The fact that they are players at all, though, is largely unexpected; and whether they are important players in specific groups, such as insects and mites, remains a current research issue for the field.

The final category of skeptical views concerns the level of CI and severity of the gene-flow reduction that *Wolbachia* can cause (Hurst and Schilthuizen, 1998; Ballard et al., 2002). CI is often incomplete due to imperfect transmission of *Wolbachia* or reduced expression of CI, though not always (see Breeuwer and Werren, 1990). When CI is incomplete, gene flow may be fluid between populations. Like the first criticism, this one also fails to distinguish *Wolbachia* from other isolating barriers because genetically based pre- or postmating isolation are typically incomplete early in speciation. Indeed, Haldane's rule at best can account for only a 50% reduction in the number of F1 hybrids produced. Because speciation will frequently proceed through successive steps that increase the types and amount of isolation, speciation workers should be more concerned with the sum of the isolating barriers that complement each other and together seal off gene flow.

One possibly unique feature of CI biology is that selection at both the level of the host and *Wolbachia* is thought to favor a decrease in the penetrance of CI, which may cause CI levels to wane over time. The theory that motivated this remark actually shows that selection operates to reduce the assumed negative correlation between fitness cost to fecundity (e.g., costly levels of bacterial densities) and CI (Turelli, 1994). Thus, CI levels are not directly selected on per se. While bacterial densities are often positively correlated with CI level in practice, it is unclear how many generations this waning takes and whether it would occur before or after subsequent genetic divergence and reproductive isolation. In addition, some systems apparently do not obey this correlation (Bourtzis et al., 1996; Bordenstein and Werren, 2000).

FROM WALLIN TO *WOLBACHIA* AND BEYOND

Debates about the types of heritable elements that promote speciation began as early as the 1920s and 1930s among evolutionary biologists and geneticists. The two extremes of the debates are best represented by two books that were published within the span of 10 years — Ivan Wallin's *Symbionticism and the Origin of Species* (1927) and Theodosius Dobzhansky's *Genetics and the Origin of Species* (1937). Despite Wallin's effort to put symbiosis into the mainstream of evolutionary biology, only Dobzhansky's book would have a lasting influence through the century. From the Biological Species Concept to the Dobzhansky–Muller model of postzygotic isolation, Dobzhansky laid a solid foundation for future study of the genetics of speciation (Orr, 1996).

But now with the progress in the *Wolbachia* field and the advent of molecular biology techniques that make the identification of bacterial endosymbionts simple, perhaps a reassessment of Wallin's ideas on the symbiotic origin of species is needed. The emergence of *Wolbachia* in topical discussions of invertebrate speciation is just a start to reviving Wallin's silenced ideas on the role of endosymbionts in speciation. A full treatment of their role will go beyond *Wolbachia* and reveal the diverse ways in which other symbionts contribute to speciation. In closing, I briefly mention a few of these alternative systems below.

Many organisms complete their entire life cycle on a single host, and thus divergence in host specificity may be an important engine of speciation (Bush, 1994). Analyses of reproductive isolation associated with host shifts have focused largely on the nuclear-genetic basis of host specificity (Hawthorne and Via, 2001). However, there is ever increasing evidence that, in addition to nuclear genes, endosymbionts can also play a crucial role in host nutrition and adaptive radiations onto new resources (Margulis and Fester, 1991). For example, the weevil genus *Sitophilus* is the only genus of the Rhynochophorinae family that lives on cereal grains; the other genera live at the junction of the roots and stems of monocotyledons (Nardon and Grenier, 1991). This difference in host specificity is due to the nutrients provided by cytoplasmic bacteria that infect these weevils. *Buchnera* endosymbionts are nutritionally important to more than 4400 aphid species, many of which occur sympatrically and differ in host use (Guildemond and Mackenzie, 1994; Adams and Douglas, 1997). Perhaps divergence in *Buchnera* can also drive divergence in aphid plant use. Even host shifts in the *Rhagoletis* fruit fly genus may be influenced by their predominant symbiotic bacteria, *Klebsiella oxytoca* (Girolami, 1973; Howard et al., 1985). These kinds of symbiont-based adaptations warrant further investigation of their role in ecological shifts and host race/species formation.

Sex attractants or pheromones can also be key players in the evolution of premating isolation and speciation (Cobb and Jallon, 1990; Coyne et al., 1994). What might surprise some is the fact that symbiotic bacteria can secrete products that act as sex attractants. Males of the grass grub beetle *Costelytra zealandica* are attracted to products generated from bacteria located in the colleterial glands of this species (Hoyt and Osborne, 1971). These glands are represented as an outpocketing of the vagina and are well supplied with tracheae. Microbially induced sex attractants have been generally ignored but may prove significant in speciation given the extensive distribution of normal bacterial flora in animals.

Finally, cytoplasmic bacteria have also been implicated in more classic isolating mechanisms including hybrid male sterility in *Drosophila* species (Williamson et al., 1971; Powell, 1982) and in *Heliothis* moths (Krueger et al., 1993).

The diversity of host organisms and traits affected by symbionts suggests that our awareness of symbiont-induced speciation may be limited. While *Wolbachia* currently stand out as a model system for studies of symbiont-induced speciation, there is the more exciting prospect that in the near future we will have a collection of diverse bacteria that can act as species splitters.

ACKNOWLEDGMENTS

I am grateful to John Jaenike, Daven Presgraves, DeWayne Shoemaker, and Jack Werren, whose suggestions, corrections, and discussions greatly improved this chapter. I wrote this work during the final months of my graduate career, a time of reflection and anticipation. I am indebted to my thesis committee — Tom Eickbush, Rick Harrison, Allen Orr, and Jack Werren. Collectively and individually, they have been a steady source of help, insight, and wisdom. I especially thank Jack Werren for his consistent encouragement, critical advice, and the opportunity to pursue this chapter.

REFERENCES

Adams, D. and Douglas, A.E. (1997). How symbiotic bacteria influence plant utilization by the polyphagous aphid, *Aphis fabae*. *Oecologia* **110:** 528–532.

Argov, Y., Gottlieb, Y., Amin-Spector, S., and Zchori-Fein, E. (2000). Possible symbiont-induced thelytoky in *Galeopsomyia fausta*, a parasitoid of the citrus leafminer *Phyllocnistis citrella*. *Phytoparasitica* **28:** 212–218.

Arakaki, N., Hiroaki, N., and Yamagishi, K. (2000). *Wolbachia*-induced parthenogenesis in the egg parasitoid *Telenomus nawai*. *Entomol. Exp. Appl.* **96:** 177–184.

Ballard, J.W.O., Chernoff, B., and James, A.C. (2002). Divergence of mitochondrial DNA is not corroborated by nuclear DNA, morphology, or behavior in *Drosophila simulans*. *Evolution* **56:** 527–545.

Bandi, C., Anderson, T.J.C., Genchi, C., and Blaxter, M.L. (1998). Phylogeny of *Wolbachia*-like bacteria in filarial nematodes. *Proc. R. Soc. London (B)* **265:** 2407–2413.

Bandi, C., Trees, A.J., and Brattig, N.W. (2001). *Wolbachia* in filarial nematodes: evolutionary aspects and implications for the pathogenesis and treatment of filirial diseases. *Vet. Parasitol.* **98:** 215–238.

Bordenstein, S.R. and Werren, J.H. (1998). Effects of A and B *Wolbachia* and host genotype on interspecies cytoplasmic incompatibility in *Nasonia*. *Genetics* **148:** 1833–1844.

Bordenstein, S.R. and Werren, J.H. (2000). Do *Wolbachia* influence fecundity in *Nasonia vitripennis*? *Heredity* **84:** 54–62.

Bordenstein, S.R., Drapeau, M.D., and Werren, J.H. (2000). Intraspecific variation in sexual isolation in the jewel wasp *Nasonia*. *Evolution* **54:** 567–573.

Bordenstein, S.R., O'Hara, F.P., and Werren, J.H. (2001). *Wolbachia*-induced incompatibility precedes other hybrid incompatibilities in *Nasonia*. *Nature* **409:** 707–710.

Bordenstein, S.R., Uy, J.J., and Werren, J.H. Host genotype determines cytoplasmic incompatibility type in *Nasonia*. *Genetics*. In press.

Bouchon, D., Rigaud, T., and Juchault, P. (1998). Evidence for widespread *Wolbachia* infection in isopod crustaceans: molecular identification and host feminization. *Proc. R. Soc. London (B)* **265:** 1081–1090.

Bourtzis, K., Nirgianaki, A., Markakis, G., and Savakis, C. (1996). *Wolbachia* infection and cytoplasmic incompatibility in *Drosophila* species. *Genetics* **144:** 1063–1073.

Boyle, L., O'Neill, S.L., Robertson, H.M., and Karr, T.L. (1993). Interspecific and intraspecific horizontal transfer of *Wolbachia* in *Drosophila*. *Science* **260:** 1796–1799.

Breeuwer, J.A.J. and Jacobs, G. (1996). *Wolbachia*: intracellular manipulators of mite reproduction. *Exp. Appl. Acarol.* **20:** 421–434.

Breeuwer, J.A.J. and Werren, J.H. (1990). Microorganisms associated with chromosome destruction and reproductive isolation between two insect species. *Nature* **346:** 558–560.

Breeuwer, J.A.J. and Werren, J.H. (1993). Cytoplasmic incompatibility and bacterial density in *Nasonia vitripennis*. *Genetics* **135:** 565–574.

Breeuwer, J.A.J. and Werren, J.H. (1995). Hybrid breakdown between two haplodiploid species — the role of nuclear and cytoplasmic genes. *Evolution* **49:** 705–717.

Bush, G.L. (1994). Sympatric speciation in animals — new wine in old bottles. *Trends Ecol. Evol.* **9:** 285–288.

Campbell, B.C., Steffen-Campbell, J.D., and Werren, J.H. (1993). Phylogeny of the *Nasonia* species complex (Hymenoptera: Pteromalidae) inferred from an rDNA internal transcribed spacer (ITS2). *Insect Mol. Biol.* **2:** 255–257.

Charlat, S., Calmet, C., and Merçot, H. (2001). On the mod resc model and the evolution of *Wolbachia* compatibility types. *Genetics* **159:** 1415–1422.

Clancy, D.J. and Hoffmann, A.A. (1996). Cytoplasmic incompatibility in *Drosophila simulans*: evolving complexity. *Trends Ecol. Evol.* **11:** 145–146.

Cobb, M. and Jallon, J.M. (1990). Pheromones, mate recognition, and courtship stimulation in the *Drosophila melanogaster* species subgroup. *J. Insect Behav.* **2:** 63–89.

Coyne, J.A. (1992). Genetics and speciation. *Nature* **355:** 511–515.
Coyne, J.A. and Orr, H.A. (1997). Patterns of speciation in *Drosophila* revisited. *Evolution* **51:** 295–303.
Coyne, J.A. and Orr, H.A. (1998). The evolutionary genetics of speciation. *Philos. Trans. R. Soc. London (B)* **353:** 287–305.
Coyne, J.A., Crittenden, A.P., and Mah, K. (1994). Genetics of a pheromonal difference contributing to reproductive isolation in *Drosophila. Science* **265:** 1461–1464.
Darling, D.C. and Werren, J.H. (1990). Biosystematics of two new species of *Nasonia* (Hymenoptera: Pteromalidae) reared from birds' nests in North America. *Ann. Entomol. Soc. Am.* **83:** 352–370.
De Bary, A. (1879). *Die Erscheinung der Symbiose.* Verlag von Karl J. Trubner, Strasbourg, France.
Dobzhansky, T. (1937). *Genetics and the Origin of Species.* Columbia University Press, New York.
Douglas, A.E. (1998). Nutritional interactions in insect-microbial symbioses: aphids and their symbiotic bacteria *Buchnera. Ann. Rev. Entomol.* **43:** 17–37.
Drapeau, M.D. and Werren, J.H. (1999). Differences in mating behavior and sex ratio between three sibling species of *Nasonia. Evol. Ecol. Res.* **1:** 223–234.
Felsenstein, J. (1981). Skepticism towards Santa Rosalia, or why are there so few kinds of animals? *Evolution* **35:** 124–138.
Fujii, Y., Kageyama, D., Hoshizaki, S., Ishikawa, H., and Sasaki, T. (2001). Transfection of *Wolbachia* in Lepidoptera: the feminizer of the adzuki bean borer *Ostrinia scapulalis* causes male killing in the Mediterranean flour moth *Ephestia kuehniella. Proc. R. Soc. London (B)* **268:** 855–859.
Giordano, R., Jackson, J.J., and Robertson, H.M. (1997). The role of *Wolbachia* bacteria in reproductive incompatibilities and hybrid zones of *Diabrotica* beetles and *Gryllus* crickets. *Proc. Natl. Acad. Sci. U.S.A.* **94:** 11439–11444.
Girolami, V. (1973). Reperti morfo-istologici sulle batteriosimbiosis del *Dacus oleae* Gmelin e di altri tripetidi, in natura e negli allevamenti su substrati artificiciali. *Estratto da Redia* **54:** 26294.
Gottlieb, Y. and Zchori-Fein, E. (2001). Irreversible thelytokous reproduction in *Muscidifurax uniraptor. Entomol. Exp. Appl.* **100:** 271–278.
Gray, M.W., Burger, G., and Lang, B.F. (1999). Mitochondrial evolution. *Science* **283:** 1476–1481.
Guildemond, J.A. and Mackenzie, A. (1994). Sympatric speciation in aphids I. In *Individuals, Populations, and Patterns in Ecology* (S.R. Leather, A.D. Watt, N.J. Mills, and K. Walters, Eds.), pp. 396–397. Intercept, Andover, MA.
Guillemaud, T., Pasteur, N., and Rousset, F. (1997). Contrasting levels of variability between cytoplasmic genomes and incompatibility types in the mosquito *Culex pipiens. Proc. R. Soc. London (B)* **264:** 245–251.
Haldane, J.B.S. (1922). Sex ratio and unisexual sterility in animal hybrids. *J. Genet.* **12:** 101–109.
Harrison, R.G. (1983). Barriers to gene exchange between closely related cricket species. I. Laboratory hybridization studies. *Evolution* **37:** 245–251.
Harrison, R.G. (1986). Pattern and process in a narrow hybrid zone. *Heredity* **56:** 337–349.
Harrison, R.G. (1998). Linking evolutionary pattern and process. In *Endless Forms: Species and Speciation* (D.J. Howard and S.H. Berlocher, Eds.), pp. 19–31. Oxford University Press, New York.
Hawthorne, D.J. and Via, S. (2001). Genetic linkage of ecological specialization and reproductive isolation in pea aphids. *Nature* **412:** 904–907.
Hertig, M. and Wolbach, S.B. (1924). Studies on rickettsia-like microorganisms in insects. *J. Med. Res.* **44:** 329–374.
Hey, J. (2001). *Genes, Categories, and Species: The Evolutionary and Cognitive Causes of the Species Problem.* Oxford University Press, New York.
Hoffmann, A.A. and Turelli, M. (1997). Cytoplasmic incompatibility in insects. In *Influential Passengers: Inherited Microorganisms and Arthropod Reproduction* (S.L. O'Neill, A.A. Hoffmann, and J.H. Werren, Eds.), pp. 42–80. Oxford University Press, New York.
Howard, D.J., Bush, G.L., and Breznak, J.A. (1985). The evolutionary significance of bacteria associated with *Rhagolettis. Evolution* **39:** 405–417.
Hoyt, C.P. and Osborne, G.O. (1971). Production of an insect sex attractant by symbiotic bacteria. *Nature* **230:** 472–473.
Hurst, G.D.D. and Schilthuizen, M. (1998). Selfish genetic elements and speciation. *Heredity* **80:** 2–8.
Hurst, G.D.D., Jiggins, F.M., von der Schulenburg, J.H.G., Bertrand, D., West, S.A., Goriacheva, I.I., Zakharov, I.A., Werren, J.H., Stouthamer, R., and Majerus, M.E.N. (1999). Male-killing *Wolbachia* in two species of insect. *Proc. R. Soc. London (B)* **266:** 735–740.

Hurst, G.D.D., Johnson, A.P., von der Schulenburg, J.H.G., and Fuyama, Y. (2000). Male-killing *Wolbachia* in *Drosophila*: a temperature sensitive trait with a threshold bacterial density. *Genetics* **156:** 699–709.

Hunter, M.S. (1999). The influence of parthenogenesis-inducing on the oviposition behavior and sex-specific developmental requirements of autoparasitoid wasps. *J. Evol. Biol.* **12:** 765–741.

James, A.C. and Ballard, J.W.O. (2000). Expression of cytoplasmic incompatibility in *Drosophila simulans* and its impact on infection frequencies and distribution of *Wolbachia pipientis*. *Evolution* **54:** 1661–1672.

Jeyaprakash, A. and Hoy, M.A. (2000). Long PCR improves *Wolbachia* DNA amplification: *wsp* sequences found in 76% of sixty-three arthropod species. *Insect Mol. Biol.* **9:** 393–405.

Jiggins, F.M., Bentley, J.K., Majerus, M.E.N., and Hurst, G.D.D. (2001). How many species are infected with *Wolbachia*? Cryptic sex ratio distorters revealed to be common by intensive sampling. *Proc. R. Soc. London (B)* **268:** 1123–1126.

Karr, T.L. (2000). The microbe that roared: *Wolbachia* rescue of a lethal *Drosophila mutation*. *Am. Zool.* **40:** 1082–1083.

Kirkpatrick, M. and Servedio, M.R. (1999). The reinforcement of mating preferences on an island. *Genetics* **151:** 865–884.

Krueger, C.M., Degreugillier, M.E., and Narang, S.K. (1993). Size difference among 16S rRNA genes from endosymbiotic bacteria found in testes of *Heliothis virescens*, *H. subflexa*, and backcross sterile male moths. *Fla. Entomol.* **76:** 382–390.

Laven, H. (1951). Crossing experiments with *Culex* strains. *Evolution* **5:** 370–375.

Laven, H. (1959). Speciation by cytoplasmic isolation in the *Culex pipiens* complex. *Cold Spring Harbor Symp. Quant. Biol.* **24:** 166–173.

Luck, R.F., Stouthamer, R., and Nunney, L. (1992). Sex determination and sex ratio patterns in parasitic hymenoptera. In *Evolution and Diversity of Sex Ratio in Haplodiploid Insects and Mites* (D.L. Wrensch and M.A. Ebert, Eds.), pp. 442–476. Chapman & Hall, New York.

Mandel, M.J., Ross, C.L., and Harrison, R.G. (2001). Do *Wolbachia* infections play a role in unidirectional incompatibilities in a field cricket hybrid zone? *Mol. Ecol.* **10:** 703–709.

Margulis, L. (1993). Origins of species: acquired genomes and individuality. *Biosystems* **31:** 121–125.

Margulis, L. and Fester, R. (1991). *Symbiosis as a Source of Evolutionary Innovation*. MIT Press, Cambridge, MA.

Mayr, E. (1963). *Animal Species and Evolution*. Belknap Press, Cambridge, MA.

McGraw, E.A., Merritt, D.J., Droller, J.N., and O'Neill, S.L. (2001). *Wolbachia*-mediated sperm modification is dependent on the host genotype in *Drosophila*. *Proc. R. Soc. London (B)* **268:** 2565–2570.

Merçot, H. and Poinsot, D. (1998). *Wolbachia* transmission in a naturally bi-infected *Drosophila simulans* strain from New Caldonia. *Entomol. Exp. Appl.* **86:** 97–103.

Nardon, P. and Grenier, A. (1991). Serial endosymbiosis theory and weevil evolution: the role of symbiosis. In *Symbiosis as a Source of Evolutionary Innovation* (L. Margulis and R. Fester, Eds.), MIT Press, Cambridge, MA.

Navajas, M., Tsagkarakov, A., Lagnel, J., and Perrot-Minnot, M.J. (2000). Genetic differentiation in *Tetranychus urticae*: polymorphism, host races or sibling species? *Exp. Appl. Acarol.* **24:** 365–376.

Noor, M.A.F. (1999). Reinforcement and other consequences of sympatry. *Heredity* **83:** 503–508.

Noor, M.A.F. (2002). Is the biological species concept showing its age? *Trends Ecol. Evol.* **17:** 153–154.

O'Neill, S.L. and Karr, T.L. (1990). Bidirectional incompatibility between conspecific populations of *Drosophila simulans*. *Nature* **348:** 178–180.

O'Neill, S.L., Giordano, R., Colbert, A.M.E., Karr, T.L., and Robertson, H.M. (1992). 16S rRNA phylogenetic analysis of the bacterial endosymbionts associated with the cytoplasmic incompatibility in insects. *Proc. Natl. Acad. Sci. U.S.A.* **89:** 2699–2702.

Orr, H.A. (1996). Dobzhansky, Bateson, and the genetics of speciation. *Genetics* **144:** 1331–1335.

Orr, H.A. (1997). Haldane's rule. *Annu. Rev. Ecol. Syst.* **28:** 195–218.

Perez, O.G. and Hoy, M.A. (2002). Reproductive incompatibility between two subspecies of *Coleomegilla maculata* (Coleoptera: Coccinellidae). *Fla. Entomol.* **85:** 203–207.

Perrot-Minnot, M.J., Guo, L.R., and Werren, J.H. (1996). Single and double infections with *Wolbachia* in the parasitic wasp *Nasonia vitripennis*: effects on compatibility. *Genetics* **143:** 961–972.

Pijls, J.W.A.M., van Steenbergen, J.H., and van Alphen, J.J.M. (1996). Asexuality cured: the relations and differences between sexual and asexual *Apoanagyrus diversicornis*. *Heredity* **76:** 506–513.

Poinsot, D. and Merçot, H. (2001). *Wolbachia* injection from usual to naïve host in *Drosophila simulans. Eur. J. Entomol.* **98:** 25–30.
Powell, J.R. (1982). Genetic and nongenetic mechanisms of speciation. In *Mechanisms of Speciation* (C. Barigozzi, Ed.), pp. 67–74. Alan R. Liss, New York.
Presgraves, D.C. (2000). A genetic test of the mechanism of *Wolbachia*-induced cytoplasmic incompatibility in *Drosophila. Genetics* **154:** 771–776.
Presgraves, D.C. (2002). Patterns of postzygotic isolation in Lepidoptera. *Evolution* **56:** 1168–1183.
Reed, K.M. and Werren, J.H. (1995). Induction of paternal genome loss by the paternal-sex-ratio chromosome and cytoplasmic incompatibility bacteria (*Wolbachia*) — a comparative study of early embryonic events. *Mol. Reprod. Dev.* **40:** 408–418.
Rokas, A. (2000). *Wolbachia* as a speciation agent. *Trends Ecol. Evol.* **15:** 44–45.
Rousset, F., Bouchon, D., Pintureau, B., Juchault, P., and Solignac, M. (1992). *Wolbachia* endosymbionts responsible for various alterations of sexuality in arthropods. *Proc. R. Soc. London (B)* **250:** 91–98.
Sapp, J. (1990). Symbiosis in evolution: an origin story. *Endocytobiosis Cell Res.* **7:** 5–36.
Sasa, M.M., Chippindale, P.T., and Johnson, N.A. (1998). Patterns of postzygotic isolation in frogs. *Evolution* **52:** 1811–1820.
Shoemaker, D.D., Katju, V., and Jaenike, J. (1999). *Wolbachia* and the evolution of reproductive isolation between *Drosophila recens* and *Drosophila subquinaria. Evolution* **53:** 1157–1164.
Shoemaker, D.D., Ross, K.G., Keller, L., Vargo, E.L., and Werren, J.H. (2000). *Wolbachia* infections in native and introduced populations of fire ants (*Solenopsis* spp.). *Insect Mol. Biol.* **9:** 661–673.
Sinkins, S.P., Braig, H.R., and O'Neill, S.L. (1995). *Wolbachia* superinfections and the expression of cytoplasmic incompatibility. *Proc. R. Soc. London (B)* **261:** 325–330.
Stouthamer, R. (1997). *Wolbachia* induced parthenogenesis. In *Influential Passengers: Inherited Microorganisms and Arthropod Reproduction* (S.L. O'Neill, A.A. Hoffmann, and J.H. Werren, Eds.). Oxford University Press, Oxford, U.K.
Stouthamer, R. and Kazmer, D.J. (1994). Cytogenetics of microbe-associated parthenogenesis and its consequences for gene flow in *Trichogramma* wasps. *Heredity* **73:** 317–327.
Stouthamer, R., Luck, R.F., and Hamilton, W.D. (1990). Antibiotics cause parthenogenetic *Trichogramma* to revert to sex. *Proc. Natl. Acad. Sci. U.S.A.* **87:** 2424–2427.
Stouthamer, R., Breeuwer, J.A.J., Luck, R.F., and Werren, J.H. (1993). Molecular identification of microorganisms associated with parthenogenesis. *Nature* **361:** 66–68.
Stouthamer, R., Breeuwer, J.A.J., and Hurst, G.D.D. (1999). *Wolbachia pipientis*: microbial manipulator of arthropod reproduction. *Annu. Rev. Microbiol.* **53:** 71–102.
Suomalainen, E., Saura, A., and Lokki, J. (1987). *Cytology and Evolution in Parthenogenesis.* CRC Press, Boca Raton, FL.
Telschow, A., Hammerstein, P., and Werren, J.H. (2002a). Effects of *Wolbachia* on genetic divergence between populations: mainland-island model. *Integrative Comp. Biol.* **42:** 340–356.
Telschow, A., Hammerstein, P., and Werren, J.H. (2002b). Effect of *Wolbachia* on genetic divergence between populations: models with two way migration. *Am. Nat.* **160:** 554–566.
Turelli, M. (1994). Evolution of incompatibility-inducing microbes and their hosts. *Evolution* **48:** 1500–1513.
Turelli, M. and Hoffmann, A.A. (1991). Rapid spread of an inherited incompatibility factor in California *Drosophila. Nature* **353:** 440–442.
Turelli, M. and Hoffmann, A.A. (1995). Cytoplasmic incompatibility in *Drosophila simulans*: dynamics and parameter estimates from natural populations. *Genetics* **140:** 1319–1338.
Vala, F., Breeuwer, J.A.J., and Sabelis, M.W. (2000). *Wolbachia*-induced hybrid breakdown in the two-spotted mite *Tetranychus urticae* Koch. *Proc. R. Soc. London (B)* **267:** 1931–1937.
Wade, M.J., Chang, N.W., and McNaughton, M. (1995). Incipient speciation in the flour beetle *Tribolium confusum*: partial reproductive isolation between populations. *Heredity* **75:** 453–459.
Wallin, I.E. (1927). *Symbionticism and the Origin of Species.* Williams & Wilkins, Baltimore, MD.
Weeks, A.R. and Breeuwer, J.A.J. (2001). *Wolbachia*-induced parthenogenesis in a genus of phytophagous mites. *Proc. R. Soc. London (B)* **268:** 2245–2251.
Werren, J.H. (1997). Biology of *Wolbachia. Ann. Rev. Entomol.* **42:** 587–609.
Werren, J.H. (1998). *Wolbachia* and speciation. In *Endless Forms: Species and Speciation* (D.J. Howard and S.H. Berlocher, Eds.), pp. 245–260. Oxford University Press, New York.
Werren, J.H. and Bartos, J.D. (2001). Recombination in *Wolbachia. Curr. Biol.* **11:** 431–435.

Werren, J.H. and Windsor, D.M. (2000). *Wolbachia* infection frequencies in insects: evidence of a global equilibrium? *Proc. R. Soc. London (B)* **267:** 1277–1285.

Werren, J.H., Windsor, D., and Guo, L. (1995a). Distribution of *Wolbachia* mong neotropical arthropods. *Proc. R. Soc. London (B)* **262:** 197–204.

Werren, J.H., Zhang, W., and Guo, L.R. (1995b). Evolution and phylogeny of *Wolbachia* — reproductive parasites of arthropods. *Proc. R. Soc. London (B)* **261:** 55–63.

Werren, J.H., Calhoun, V., and van Alphen, J.A.J. *Wolbachia* associated with recent speciation in *Trichopria* wasps. In review.

Williamson, D.L., Ehrman, L., and Kernaghan, R.P. (1971). Induction of sterility in *Drosophila paulistorum*: effect of cytoplasmic factors. *Proc. Natl. Acad. Sci. U.S.A.* **68:** 2158–2160.

Wilson, E.B. (1925). *The Cell in Development and Heredity*, p. 730. Macmillan, New York.

Wu, C.-I. and Davis, A.W. (1993). Evolution of postmating reproductive isolation: the composite nature of Haldane's Rule and its genetic bases. *Am. Nat.* **142:** 187–212.

Yen, J.H. and Barr, A.R. (1971). New hypothesis of the cause of cytoplasmic incompatibility in *Culex pipiens*. *Nature* **232:** 657–658.

Zchori-Fein, E., Gottlieb, Y., Kelly, S.E., Brown, J.K., Wilson, J.M., Karr, T.L., and Hunter, M.S. (2001). A newly discovered bacterium associated with parthenogenesis and a change in host selection behavior in parasitoid wasps. *Proc. Natl. Acad. Sci. U.S.A.* **98:** 12555–12560.

18 Discovery of Symbiont–Host Horizontal Genome Transfer: A Beetle Carrying Two Bacterial and One Chromosomal *Wolbachia* Endosymbionts

Takema Fukatsu, Natsuko Kondo, Nobuyuki Ijichi, and Naruo Nikoh

CONTENTS

0-8493-1286-8/03/$0.00+$1.50

ENDOSYMBIOTIC BACTERIA OF THE GENUS *WOLBACHIA*

Members of the genus *Wolbachia* constitute a group of rickettsia-like endocellular bacteria in the α subdivision of the proteobacteria. Infection with *Wolbachia* is universally found in insects and less frequently in mites, spiders, crustaceans, and nematodes. *Wolbachia* endosymbionts have attracted a great deal of attention from biologists because they often cause a wide range of effects on the reproduction and physiology of arthropod hosts such as cytoplasmic incompatibility (CI), parthenogenesis, feminization, male killing, and others. Because *Wolbachia* are inherited solely through the maternal lineage of the host by vertical transmission, these reproductive phenotypes are regarded as the selfish strategies whereby *Wolbachia* increase the frequency of infected females in host populations often at the expense of the host fitness (O'Neill et al., 1997; Werren, 1997; Stouthamer et al., 1999).

THE ADZUKI BEAN BEETLE, *CALLOSOBRUCHUS CHINENSIS*

Callosobruchus chinensis (Coleoptera: Bruchidae) is known for infesting stored beans such as *Vigna* spp. (Figure 18.1). In addition to its importance as a pest insect, the beetle has been widely

FIGURE 18.1 Female adults of *Callosobruchus chinensis* ovipositing on adzuki beans *Vigna angularis*.

used as a model organism in population biology because its life history parameters related to population dynamics can conveniently be measured and manipulated under controlled laboratory conditions. For example, a strain of *C. chinensis*, jC, has been maintained on adzuki beans in petri dish for a variety of uses since 1936 when the original insects were collected in Kyoto, Japan. Therefore, *Wolbachia* infection in *C. chinensis* can be an excellent model system for understanding the interactions and dynamics between *Wolbachia* and its host insect.

DETECTION OF *WOLBACHIA* FROM *C. CHINENSIS*

In an attempt to survey endosymbiotic bacteria in bruchid beetles, 12 laboratory strains from seven species — *Zabrotes subfasciatus*, *C. chinensis*, *C. maculatus*, *C. rhodesianus*, *Acanthoselides obtetus*, *Bruchidius dorsalis*, and *Kytorhinus sharpianus* — were subjected to a polymerase-chain-reaction (PCR) assay using primers for two *Wolbachia* genes, *ftsZ* and *wsp* (Figure 18.2). Positive signals were detected only in strains of *C. chinensis*, whereas the other six species, including congeneric *C. maculatus* and *C. rhodesianus,* were all *Wolbachia*-negative.

To investigate the infection frequency in natural populations, 288 males and 334 females of *C. chinensis* collected from nine localities in Japan were subjected to the PCR assay. Surprisingly, all the individuals examined were *Wolbachia*-positive, indicating that *Wolbachia* infection is probably fixed in Japanese populations of *C. chinensis*.

SUPERINFECTION WITH THREE DISTINCT STRAINS OF *WOLBACHIA* IN *C. CHINENSIS*

To characterize the *Wolbachia* in *C. chinensis*, we amplified, cloned, and sequenced the *wsp* gene from individual insects. Unexpectedly, three distinct *wsp* sequences were reproducibly obtained from all the individuals examined. Molecular phylogenetic analysis of these sequences unequivocally indicated that *C. chinensis* was superinfected with three phylogenetically distinct *Wolbachia* strains (Figure 18.3). We designated these strains wBruCon, wBruOri, and wBruAus.

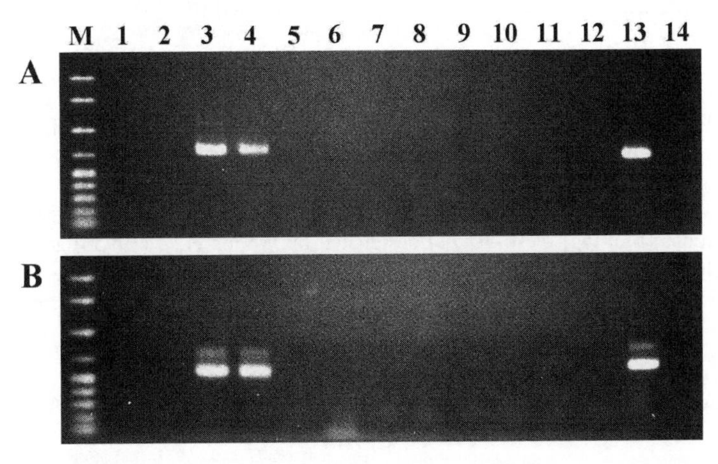

FIGURE 18.2 Diagnostic PCR detection of *Wolbachia* from laboratory strains of bruchid beetles. (A) *ftsZ* gene; (B) *wsp* gene. Lane 1, *Z. subfasciatus* C100 strain; lane 2, *Z. subfasciatus* US strain; lane 3, *C. chinensis* jC strain; lane 4, *C. chinensis* mrC97 strain; lane 5, *C. maculatus* hQ strain; lane 6, *C. maculatus* iQ strain; lane 7, *C. rhodesianus*; lane 8, *A. obtectus*; lane 9, *B. dorsalis* Harataima strain; lane 10, *B. dorsalis* Tatsuno strain; lane 11, *K. sharpianus* Mitsuma strain; lane 12, *K. sharpianus* Yoneyama strain; lane 13, *Ephestia kuehniella* infected with *Wolbachia* sp. (positive control); lane 14, without template (negative control); lane M, DNA size markers (2000, 1500, 1000, 700, 500, 400, 300, 200 and 100 bps from top to bottom). [From Kondo, N., Shimada, M., and Fukatsu, T. (1999). *Zool. Sci.* **16:** 955–962. With permission.]

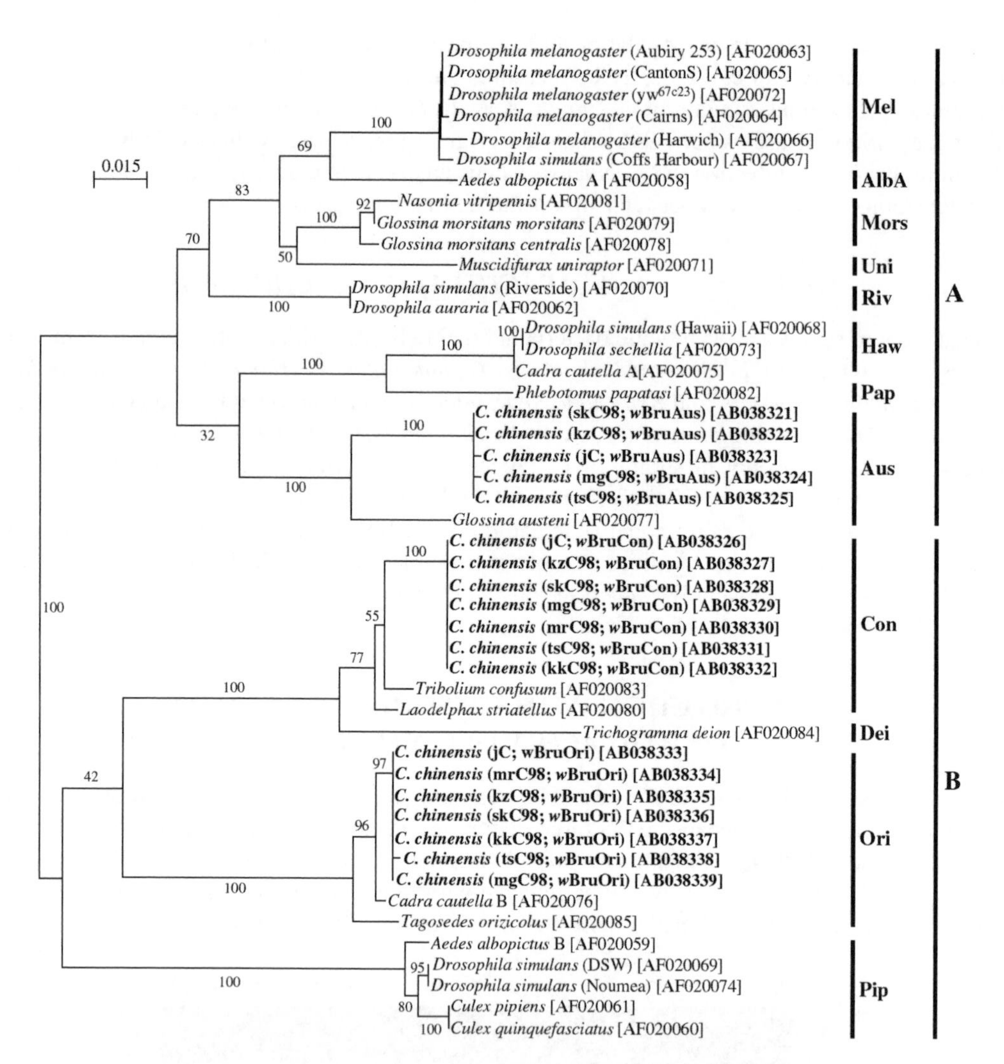

FIGURE 18.3 Molecular phylogenetic analysis of three types of *wsp* sequences identified from a laboratory strain and six Japanese populations of *C. chinensis*. The *wsp* sequences were analyzed together with the sequences investigated by Zhou et al. (1998). A total of 523 unambiguously aligned nucleotide sites were subjected to the analysis. A neighbor-joining phylogeny is shown. On the right side are shown supergroups (A and B) and groups (Mel-Pip) of *Wolbachia* designated by Werren et al. (1995b) and Zhou et al. (1998). The bootstrap values in percentage obtained with 1,000 resamplings are shown at the nodes, though values smaller than 50% are not shown. The numbers in brackets are accession numbers. [From Kondo, N., Ijichi, N., Shimada, M., and Fukatsu, T. (2002a). *Mol. Ecol.* 11: 167–180. With permission]

PREVAILING TRIPLE INFECTION IN NATURAL POPULATIONS OF *C. CHINENSIS*

Infection frequencies with wBruCon, wBruOri, and wBruAus in Japanese populations of *C. chinensis* were examined by specific PCR detection. Of 410 insects from nine Japanese local populations, wBruCon, wBruOri, and wBruAus were detected in 410 (100%), 395 (96.3%), and 398 (97.1%), respectively.

With the three *Wolbachia* strains, there are eight possible infection states: a triple infection (infection state: COA), three double infections (CO, CA, and OA), three single infections (C, O, and A), and no infection. Among these infection states, triple infection (93.7%) was highly

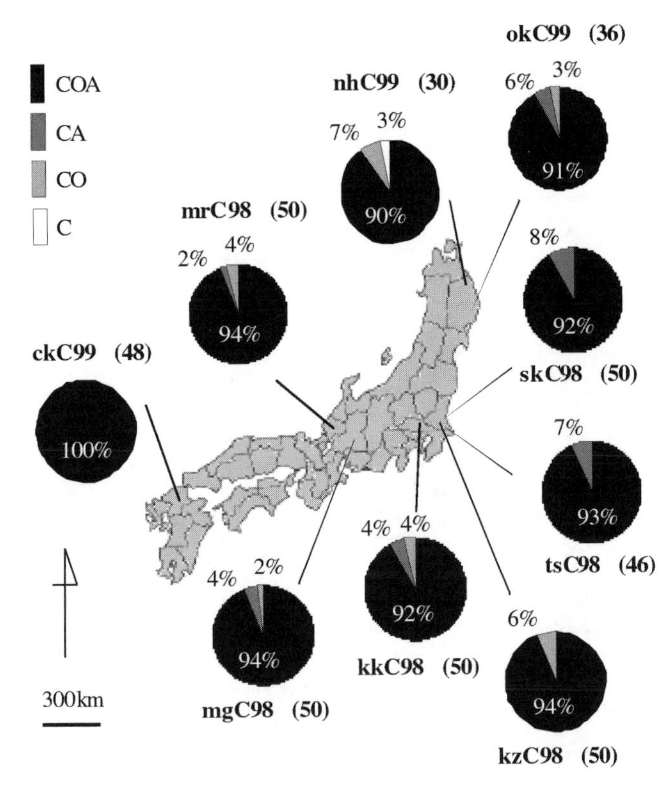

FIGURE 18.4 Infection states with the three *Wolbachia* strains in Japanese local populations of *C. chinensis*. Infections with wBruCon, wBruOri, and wBruAus, respectively, were detected by PCR using highly specific primers for their *wsp* genes. Numbers in parentheses after population names indicate the number of samples examined, including both males and females. [From Kondo, N., Ijichi, N., Shimada, M., and Fukatsu, T. (2002a). *Mol. Ecol.* **11:** 167–180. With permission.]

dominant, whereas double infection (6.1%) and single infection (0.2%) were minor. Only two types of double infection, CO and CA, and one type of single infection, C, were identified. The predominance of the triple infection was consistently observed in all nine Japanese populations examined (Figure 18.4).

WHAT ARE THE THREE *WOLBACHIA* STRAINS DOING IN THE SAME HOST INSECT?

The discovery of universal and stable infection with three distinct strains of *Wolbachia* in *C. chinensis* prompted the following questions: What effects do the respective strains of *Wolbachia* have on the host? Is there any functional differentiation between them? What reproductive and transmission strategies do they adopt in the same host? How do the three *Wolbachia* localize in the same host body? Is there any interaction/competition/conflict among them? To answer these questions we investigated various biological aspects of the *Wolbachia* triple infection in *C. chinensis*.

DIFFERENT LEVELS OF CI CAUSED BY THE THREE *WOLBACHIA* STRAINS

Using *C. chinensis* strains of different infection states, mating experiments were conducted to examine whether or not wBruCon, wBruOri, and wBruAus cause CI (Table 18.1). In crosses

TABLE 18.1
Mating Experiments to Detect CI Caused by the Respective Three Wolbachia Strains

Wolbachia Type Examined for CI	Infection State			Eggs		Hatch Rate	SD
	♂	♀	Pairs[a]	Total	Hatched		
wBruCon	**CA**	**A**	**19**	**556**	**0**	**0.000**	**0.000**
	A	CA	13	243	224	0.922	0.076
	CA	CA	15	359	322	0.897	0.076
	A	A	18	532	495	0.930	0.058
wBruOri	**COA**	**CA**	**16**	**335**	**206**	**0.615**	**0.213**
	CA	COA	15	436	352	0.807	0.263
	COA	COA	20	587	534	0.910	0.115
	CA	CA	15	359	322	0.897	0.076
wBruAus	**COA**	**CO**	**21**	**477**	**402**	**0.843**	**0.108**
	CO	COA	11	258	221	0.857	0.266
	COA	COA	20	587	534	0.910	0.115
	CO	CO	33	861	640	0.743	0.183

[a] Matings were conducted using a COA strain, three CO strains, seven CA strains, and an A strain. Bold type indicates crosses expected to be incompatible. SD = standard deviation.

between CA and A strains, A♂ × CA♀ was compatible, whereas CA♂ × A♀ showed complete incompatibility. The difference in hatch rate was unequivocal (Mann-Whitney *U*-test, $p < 0.001$). In crosses between COA and CA strains, CA♂ × COA♀ showed a normal hatch rate, whereas COA♂ × CA♀ was partially incompatible. The difference in hatch rate was statistically significant (Mann-Whitney *U*-test, $p < 0.05$). In crosses between COA and CO strains, no significant difference in hatch rate was observed between COA♂ × CO♀ and CO♂ × COA♀ (Mann-Whitney *U*-test, $p > 0.05$). These results indicated that in the same host insect, wBruCon caused complete CI, wBruOri induced moderate CI, and wBruAus showed no CI or very weak CI at an undetectable level.

QUANTIFICATION OF *WOLBACHIA* DENSITY IN HOST INSECT

Using a quantitative PCR technique, densities of wBruCon, wBruOri, and wBruAus in adult insects of *C. chinensis* were quantified in terms of *wsp* gene copies per dry body weight (Figure 18.5). Among the three *Wolbachia* strains, wBruAus exhibited remarkably lower density (around 10^7 *wsp* copies/mg) than wBruCon and wBruOri (around 10^8 *wsp* copies/mg). This pattern — lower density of wBruAus in comparison with wBruCon and wBruOri — was consistently found in samples from different local populations in Japan (data not shown). These results suggested that in the same host body, the densities of the respective three *Wolbachia* strains might be controlled to be at different levels.

POPULATION DYNAMICS OF *WOLBACHIA* THROUGH HOST DEVELOPMENT

Population dynamics of wBruCon, wBruOri, and wBruAus were monitored throughout the development of *C. chinensis* using quantitative PCR (Figure 18.6). The populations of the three *Wolbachia* strains consistently increased as development of the host proceeded, although they exhibited different population-growth patterns. In early eggs, wBruOri was the most abundant (around 4×10^5 *wsp* copies equivalent per insect), wBruCon was intermediate in amount (around 4×10^4 *wsp* copies equivalent per insect), and wBruAus was the least abundant (around 1×10^2 *wsp* copies

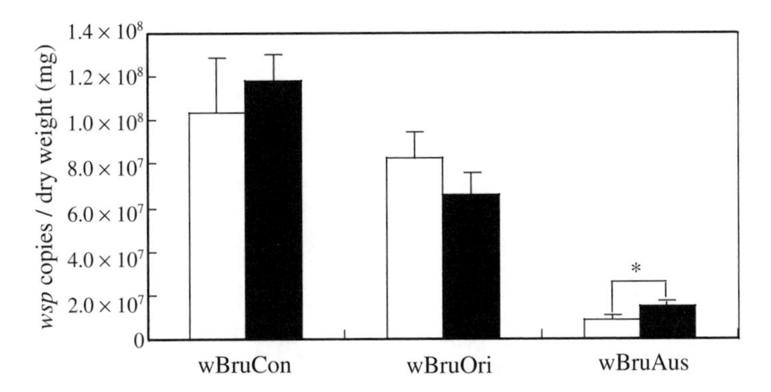

FIGURE 18.5 Comparison of the population density of wBruCon, wBruOri, and wBruAus between female and male of newly emerged adults of *C. chinensis* in terms of *wsp* copies per dry weight (mg) of insect (n = 10; error bars show standard deviation). Asterisk shows statistically significant difference (Mann-Whitney *U*-test, $p < 0.01$). [From Ijichi, N., Kondo, N., Matsumoto, R., Shimada, M., Ishikawa, H., and Fukatsu, T. (2002). *Appl. Environ. Microbiol.* **68:** 4074–4080.]

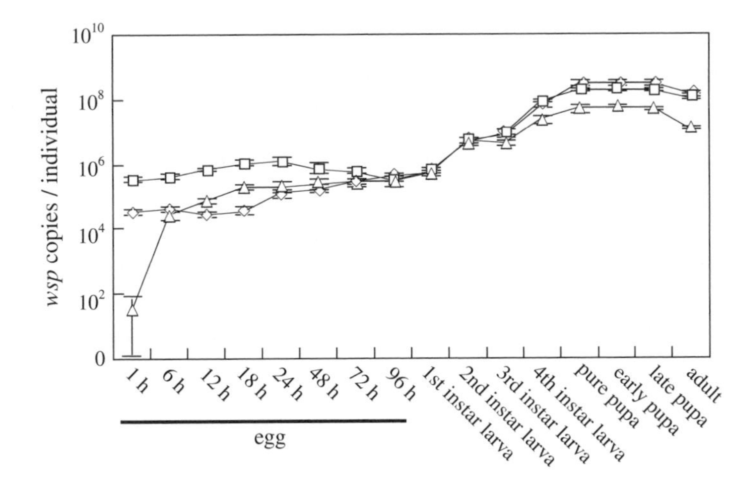

FIGURE 18.6 Infection dynamics of wBruCon, wBruOri, and wBruAus throughout the developmental course of *C. chinensis* in terms of *wsp* copies per individual insect. Diamonds, wBruCon; squares, wBruOri; triangles, wBruAus (n = 10; bars show SD). Pupae and adults were females. Sex was not distinguished at the other stages. [From Ijichi, N., Kondo, N., Matsumoto, R., Shimada, M., Ishikawa, H., and Fukatsu, T. (2002). *Appl. Environ. Microbiol.* **68:** 4074–4080. With permission.]

equivalent per insect). In late eggs and first-instar larvae, the populations of wBruCon, wBruOri, and wBruAus were at almost the same level, around 6×10^5 *wsp* copies equivalent per insect. At second larval instar and later stages, the populations of wBruCon and wBruOri were at comparable levels, whereas the populations of wBruAus were consistently and significantly smaller. In prepupae and pupae, the populations of the *Wolbachia* strains reached a plateau: wBruCon, wBruOri, and wBruAus were around 3×10^8, 2×10^8, and 5×10^7 *wsp* copies equivalent per insect, respectively.

WOLBACHIA DENSITY IN TISSUES OF *C. CHINENSIS* ADULTS

To quantitatively investigate the localization of wBruCon, wBruOri, and wBruAus in *C. chinensis* tissues, dissected tissues and organs from adult insects were subjected to quantitative PCR analysis. To make a comparison between different tissues and organs, the *wsp* gene copy number of *Wolbachia*

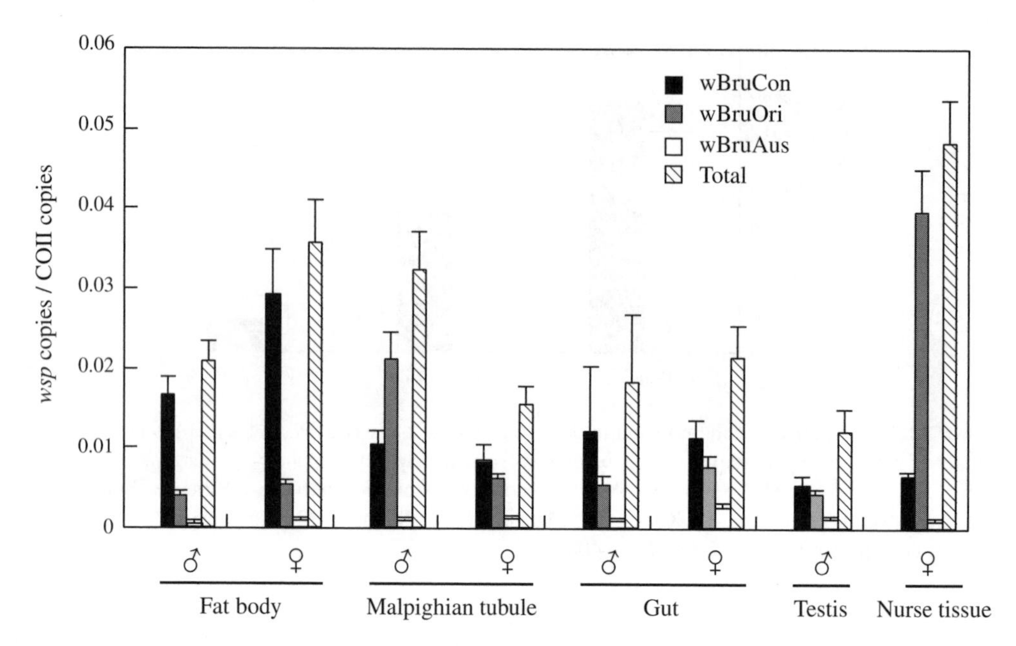

FIGURE 18.7 Infection density of wBruCon, wBruOri, and wBruAus in different tissues and organs of *C. chinensis* in terms of *wsp* copies per COII copies. The mean and standard deviation for five samples are shown for each tissue or organ. Newly emerged adults were subjected to the analysis. [From Ijichi, N., Kondo, N., Matsumoto, R., Shimada, M., Ishikawa, H., and Fukatsu, T. (2002). *Appl. Environ. Microbiol.* **68:** 4074–4080. With permission.]

was standardized by copy number of a host gene, mitochondrial COII. It was demonstrated that both the density and the composition of the three *Wolbachia* strains were specific to tissues and organs of *C. chinensis*.

Total densities of the three *Wolbachia* strains exhibited significant differences between tissues and organs (Figure 18.7). Among the tissues and organs examined, for example, higher levels of *Wolbachia* infection were detected in the nurse tissue and fat body than in the gut and testis. The total densities of the *Wolbachia* also showed sex-related differences. For example, in fat body the density of *Wolbachia* was higher in females than in males, while in Malpighian tubules the density of *Wolbachia* was lower in females than in males.

The compositions of the three *Wolbachia* strains also showed significant differences between tissues and organs (Figure 18.8). In fat body, wBruCon accounted for around 80% of the total *Wolbachia* population, wBruOri about 15 to 18%, and wBruAus a small fraction. In gut and testis, wBruCon comprised about 50 to 70% of the total *Wolbachia* population, wBruOri around 30 to 40%, and wBruAus around 10% or less. In Malpighian tubule, a sex-related difference was detected; in males, wBruOri was the predominant strain, whereas in females wBruCon was the major component. Strikingly, in nurse tissue and oocyte, wBruOri accounted for more than 80% of the total *Wolbachia* population, whereas wBruCon and wBruAus constituted only small fractions.

DIFFERENTIAL TISSUE TROPISM OF THE THREE *WOLBACHIA* STRAINS IN THE SAME HOST INSECT

Based on these results, it was demonstrated that interesting patterns of differential tissue tropism and population dynamics of wBruCon, wBruOri, and wBruAus existed in the same host insect throughout the developmental course. Although very poorly understood at present, the different patterns in tissue localization and proliferation of the three *Wolbachia* strains must reflect various biological aspects of host–symbiont and symbiont–symbiont interactions. The mechanisms controlling the tissue tropism

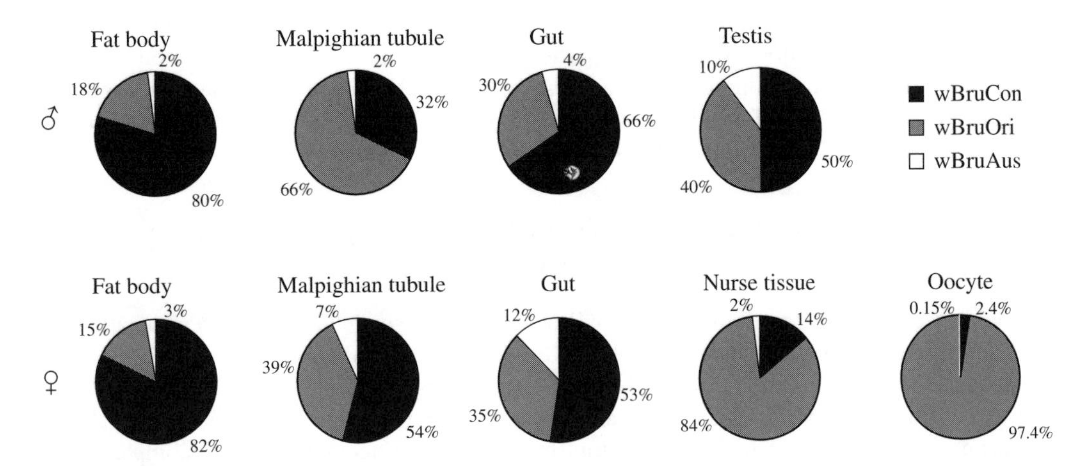

FIGURE 18.8 Relative amount of wBruCon, wBruOri, and wBruAus in different tissues and organs of *C. chinensis*. The mean values in Figure 18.7 were used to construct the pie graphs. [From Ijichi, N., Kondo, N., Matsumoto, R., Shimada, M., Ishikawa, H., and Fukatsu, T. (2002). *Appl. Environ. Microbiol.* **68:** 4074–4080. With permission.]

are unknown. However, there are two possible components that may contribute to the formation of the tissue-dependent differences in density and composition of these three *Wolbachia* strains. One component is tissue tropism inherent in the *Wolbachia* strains. The other component is interactions between coinfecting *Wolbachia* strains such as competition and niche division in the same tissue. These two components would be sorted out by examining the density and localization of the *Wolbachia* strains in singly infected and doubly infected lines of *C. chinensis*.

PECULIAR *WOLBACHIA* COMPOSITION IN NURSE TISSUE AND OOCYTE: IMPLICATION FOR VERTICAL-TRANSMISSION PROCESS

In general, intracellular symbiotic bacteria of arthropods, including *Wolbachia,* are inherited by transovarial transmission to developing oocytes in the ovarioles in the maternal body. Histologically, it has been observed that such symbionts either actively find their way to eggs or are transferred there by specialized cells or via other structures (Buchner, 1965). However, the detailed processes of vertical transmission for *Wolbachia* have been poorly described. Among the tissues examined in this study, only nurse tissue and oocyte exhibited a peculiar *Wolbachia* composition in which wBruOri was highly predominant (Figure 18.8). In the process of insect egg formation, cytoplasmic components, such as proteins, lipids, RNAs, and others, are actively synthesized by nurse cells and transported to developing oocytes through nutritive cord with trophic flow (Blackman, 1987; Foldi, 1990). One plausible explanation for the similar *Wolbachia* compositions of nurse tissue and oocyte is that the *Wolbachia* first infect and proliferate in nurse tissue and are subsequently transported to oocytes by way of the nutritive cord. The peculiar *Wolbachia* composition in nurse tissue and oocyte also suggests that wBruOri may have some adaptation for transmission to and proliferation in the female germ tissues.

FACTORS RESPONSIBLE FOR STABLE MAINTENANCE OF *WOLBACHIA* INFECTION IN HOST POPULATIONS

In natural populations of *C. chinensis*, all three *Wolbachia* strains — wBruCon, wBruOri, and wBruAus — consistently exhibited very high infection frequencies (Figure 18.4). How the three different *Wolbachia* strains are stably maintained in the host populations poses an interesting problem. Theoretically, the following factors favor the maintenance of maternally inherited endosymbionts in host populations: (1) efficient vertical transmission, (2) positive fitness effect on the host, (3) reproductive manipulation such as CI, and (4) mechanism for horizontal transmission (Caspari and Watson, 1959; Fine, 1978; Turelli, 1994). Which of these factors can be responsible for the maintenance of wBruCon, wBruOri, and wBruAus in populations of *C. chinensis*?

STRATEGIES OF wBruCon AND wBruOri TO PERSIST IN HOST POPULATIONS

We could speculate as to how wBruCon and wBruOri strains are maintained in the host populations. Considering that wBruCon caused complete CI and that its bacterial titer in eggs was small, maintenance of wBruCon may principally be realized through an efficient mechanism for CI. In contrast, wBruOri caused only partial CI, and its bacterial titer in eggs was very high, suggesting that a sophisticated mechanism for vertical transmission may compensate for the incomplete CI of wBruOri. On the other hand, it was puzzling why wBruAus, which caused no CI and accounted for a very small fraction of the bacteria in eggs, can be stably maintained in host populations.

PECULIAR PROPERTIES OF wBruAus

Among the three *Wolbachia* strains in *C. chinensis*, wBruAus exhibited peculiar properties. First, the titer of wBruAus (around 10^7 *wsp* copies equivalent per adult insect) was smaller by an order of magnitude than that of wBruCon and wBruOri (around 10^8 *wsp* copies equivalent per adult insect) (Figure 18.5). Second, infection with wBruAus showed no detectable reproductive symptoms of the host insect, whereas infections with wBruCon and wBruOri caused significant levels of CI (Table 18.1). Third, although the titer of wBruAus in oocytes and unfertilized eggs was extremely low (less than 10^2 *wsp* copies equivalent per egg) (Figure 18.6), the *Wolbachia* was stably inherited through host generations in the laboratory, and the infection rate in natural populations was over 97% on average (Figure 18.4). In addition to them, we identified further unusual properties of wBruAus atypical of endosymbiotic bacteria.

ANTIBIOTIC RESISTANCE OF wBruAus

In an attempt to obtain *Wolbachia*-free *C. chinensis*, a triple-infected strain, jC, was reared with artificial beans containing either tetracycline or rifampicin. Tetracycline treatment of only one generation was sufficient to eliminate wBruCon and wBruOri. On the other hand, wBruAus persisted throughout five generations of tetracycline treatment and could not be eliminated. After the treatment, the insects were transferred to and maintained on normal beans. The insects, named strain jC^{Aus}, contained wBruAus, but it was completely free of wBruCon and wBruOri. The same patterns of tetracycline sensitivity, resistant wBruAus, and sensitive wBruCon and wBruOri were found in other strains of *C. chinensis* (data not shown). The same results were obtained from rifampicin treatments (data not shown). Thus, wBruAus was shown to be exceptionally resistant to tetracycline and rifampicin, although *Wolbachia* and other endosymbiotic bacteria are, in general, susceptible to these antibiotics.

SEX-LINKED INHERITANCE OF wBruAus

In mating experiments between *C. chinensis* strains of different infection status, we observed the inheritance of the three *Wolbachia* strains. As expected, wBruCon and wBruOri were maternally inherited to the offspring. Unexpectedly, however, we found that wBruAus was passed to the offspring not only maternally but also paternally. The crosses between COA fathers and CO mothers produced all CO males and, surprisingly, all COA females. When these COA females were mated

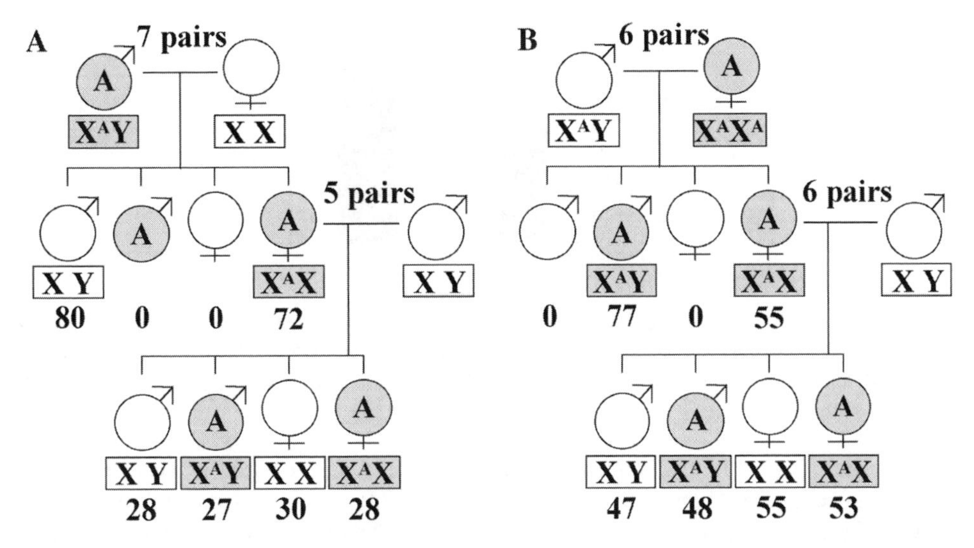

FIGURE 18.9 Inheritance of wBruAus. (A) Cross between wBruAus-bearing (COA) males and wBruAus-free (CO) females. (B) Cross between wBruAus-free (CO) males and wBruAus-bearing (COA) females. Shade indicates presence of wBruAus. The inheritance patterns are in agreement with X-linked inheritance. Sex chromosome types deduced are shown in rectangles. X^A means the X chromosome carrying wBruAus. Numbers beneath the rectangles are the number of offspring obtained. [From Kondo, N., Nikoh, N., Ijichi, N., Shimada, M., and Fukatsu, T. (2002b). *Proc. Natl. Acad. Sci. U.S.A.* **99:** 14280–14285.]

with CO males, wBruAus exhibited 1:1 segregation in both males and females (Figure 18.9A). The crosses between CO fathers and COA mothers produced all COA offspring, superficially showing typical maternal inheritance. However, when the COA female offspring were mated with CO males, wBruAus did not exhibit maternal inheritance but 1:1 segregation in both males and females (Figure 18.9B). Notably, all these patterns were perfectly explained by the sex-linked inheritance in male-heterozygotic organisms (Figure 18.9), which agreed with the karyotype of *C. chinensis* ($2n = 20$, XY) (Takenouchi, 1955; Figure 18.10).

SEX-DEPENDENT DENSITY DIFFERENCE OF wBruAus

Based on the antibiotic insensitivity and the sex-linked inheritance, it was suggested that perhaps wBruAus had no microbial entity but was a bacterial genome fragment associated with X chromo-

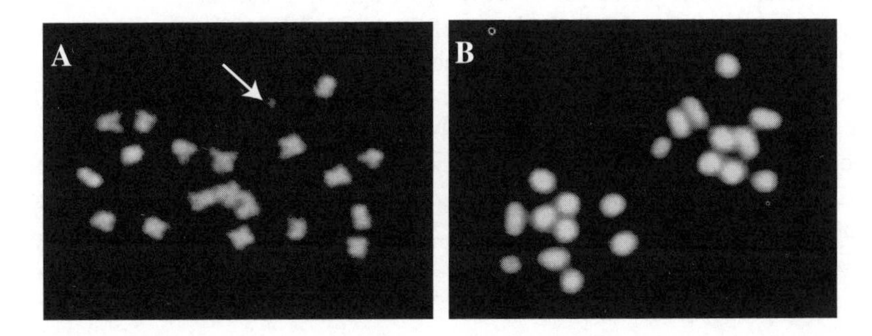

FIGURE 18.10 Chromosome specimens of *C. chinensis* prepared from testes of adult males. (A) Metaphase of first meiotic division. Condensed chromosomes ($2n = 20$) are seen. Arrow indicates the highly condensed Y chromosome. (B) Second meiotic division. Two segregating haploid chromosome sets ($n = 10$ for each) are seen. Takenouchi (1955) reported the karyotype of *C. chinensis* as $2n = 20$, XO. However, our fluorescent imaging with DAPI revealed a small Y chromosome in males, indicating that the karyotype was $2n = 20$, XY.

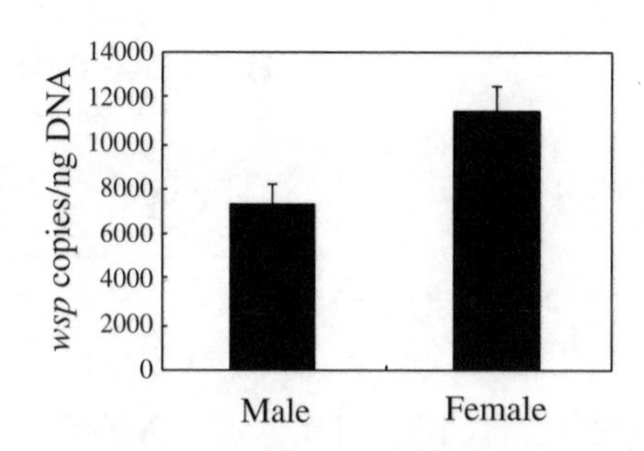

FIGURE 18.11 Sex-linked difference in titer of wBruAus. Titers of wBruAus in adult males and females were quantified in terms of *wsp* gene copies per nanogram of total insect DNA. The difference was statistically significant (Mann-Whitney *U*-test, $p < 0.001$). [From Kondo, N., Nikoh, N., Ijichi, N., Shimada, M., and Fukatsu, T. (2002b). *Proc. Natl. Acad. Sci. U.S.A.* **99:** 14280–14285.]

some of the host insect. If this were true, female insects should contain twice as much titer of wBruAus as male insects. Quantitative PCR analysis certainly demonstrated that females contained about twice as much *wsp* copies per nanogram of total DNA as males (Figure 18.11), which confirmed the idea of the X chromosome linkage.

DETECTION OF *WOLBACHIA* GENES IN CURED STRAIN OF *C. CHINENSIS*

Using specific PCR and Southern blotting, we attempted to detect *Wolbachia* genes in the cured strain jC^{Aus} that contained only wBruAus. The following 17 genes were targeted: 16S rDNA, *atpD, cytB, ftsZ, gltA, glyA, groE, gyrA, IswI, lipA, polA, rpoB, sdhA, sucD, thrA, tufA*, and *wsp*. While all the genes were detected from the triple-infected strain jC, only three genes — *ftsZ, gyrA*, and *wsp* — were detected from the strain jC^{Aus} (data not shown). These results indicated that the strain jC^{Aus} contained only a fraction of the *Wolbachia* gene repertoire, which agrees with the idea that the wBruAus strain is a genome fragment of *Wolbachia* located on the host chromosome.

STRUCTURE OF *WOLBACHIA* GENOME FRAGMENTS OBTAINED FROM CURED STRAIN OF *C. CHINENSIS*

We cloned and sequenced the amplified fragments of *wsp*, *ftsZ*, and *gyrA* genes from the total DNA of the strain jC^{Aus}. In addition, we successfully isolated large DNA fragments flanking *wsp* and *ftsZ* genes using an inverse PCR technique. A 4141-bps fragment containing *wsp* (Figure 18.12A), an 11,400-bps fragment containing *ftsZ* (Figure 18.12B), and a 373-bps fragment of *gyrA* were obtained. These DNA fragments encoded bacterial genes that showed high sequence similarity to those of the α-proteobacteria such as *Wolbachia, Rickettsia*, and *Brucella*. In the 4.1-kb fragment, four bacterial open reading frames (ORFs) were identified: *dnaX*, *wsp*, and two ORFs homologous to RP866 and RP416 in the genome of *Rickettsia prowazekii*. In the 11.4-kb fragment, seven bacterial ORFs were detected: *ftsZ, sdhB, pgpA, g3pdh, czcR*, an ORF homologous to RP741 in the genome of *R. prowazekii,* and an ORF homologous to BMEI0172 in the genome of *B. melitensis*. One ORF showed a partial weak homology to acetyltransferase genes from various bacteria, archaea, and plants.

NON-LTR RETROTRANSPOSON-LIKE SEQUENCE IN THE *WOLBACHIA* GENOME FRAGMENT

Notably, a non-long terminal repeat (non-LTR) retrotransposon-like sequence, which showed a significant similarity to *I* and *You* elements from *Drosophila* species (Berezikov et al., 2000) and

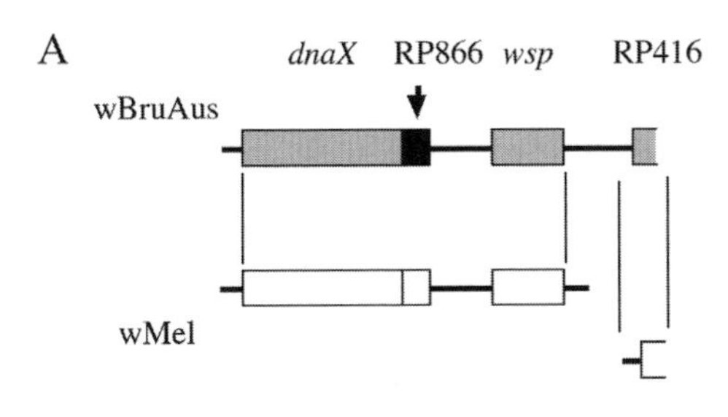

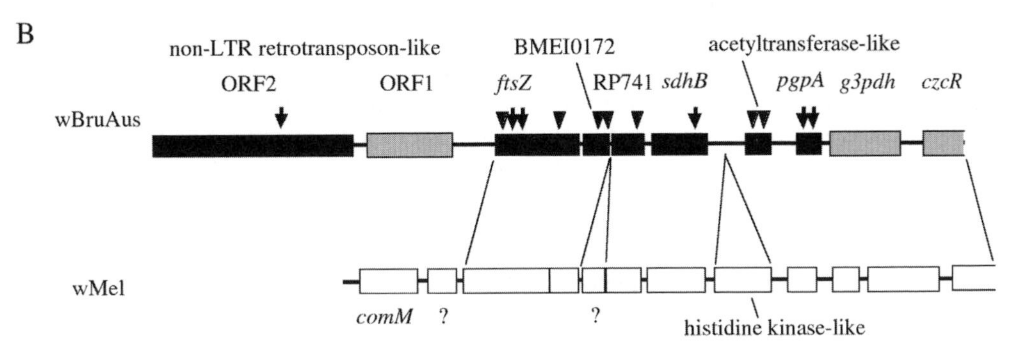

FIGURE 18.12 Structure of the genome fragments of wBruAus obtained by inverse PCR, aligned with genome sequences of wMel, a strain of *Wolbachia* from *D. melanogaster.* (A) A 4.1-kb fragment containing *wsp* gene. (B) An 11.4-kb fragment containing *ftsZ* gene. Arrows indicate the position of intermittent stop codons. Arrowheads show the position of frame shift substitutions. Filled ORFs contain either stop codons or frame shift substitutions, whereas shaded ORFs are structually intact. A non-LTR retrotransposon-like sequence was located at the upstream of *ftsZ* gene. All ORFs on wMel genome fragments are structurally intact. Question marks indicate unidentified ORFs. [From Kondo, N., Nikoh, N., Ijichi, N., Shimada, M., and Fukatsu, T. (2002b). *Proc. Natl. Acad. Sci. U.S.A.* **99:** 14280–14285.]

MosquI elements from *Aedes aegypti* (Tu and Hill, 1999), was identified in the upstream region of *ftsZ* (Figure 18.12B). The sequence contained two ORFs typical of non-LTR retrotransposons, ORF1 encoding a protein of unknown function, and ORF2 encoding a protein with endonuclease and reverse transcriptase activities. The *I* element is known to be involved in the *I–R* system of hybrid dysgenesis in *D. melanogaster* (Bucheton et al., 1984). A number of *I* and *You* elements, which constitute a distinct clade in the non-LTR retrotransposon phylogeny (Berezikov et al., 2000), have been identified in the genomes of *D. melanogaster* and other fly species. Since non-LTR retrotransposons of this family have been found in insect genomes but not in bacterial genomes, this finding may favor the idea that the *Wolbachia* genome fragment is located on a chromosome of the host insect.

COMPARISON WITH THE *WOLBACHIA* GENOME FROM *D. MELANOGASTER*

Complete genome sequencing of wMel, a strain of *Wolbachia* from *D. melanogaster*, is now in progress by the Institute for Genomic Research. Since the preliminary genome sequences are available through the Web site (http://www.tigr.org), we compared the genome fragments of wBruAus with those of wMel (Figure 18.12). Gene arrangements on the genome fragments were, in general, conserved between wBruAus and wMel, although several differences were identified. In wBruAus, *wsp* was next to RP416, whereas in wMel these genes were not on the same genome fragment. In wMel only, a histidine kinase-like sequence was found between *sdhB* and acetyltransferase-like sequence, and an unidentified ORF was located between BMEI0172 and RP741. In the upstream region of *ftsZ*, where a non-LTR retrotransposon-like sequence was identified in wBruAus, *comM* and an unidentified ORF were found in wMel. These results indicated that the genome fragments of wBruAus certainly retain the *Wolbachia* genome structure.

THE NATURE OF wBRUAUS AS "CHROMOSOMAL *WOLBACHIA*"

Taken together, these results lead to the conclusion that wBruAus, a strain of *Wolbachia* identified from *C. chinensis*, has no bacterial entity but is a genome fragment of *Wolbachia* on the genome of the host insect. The *Wolbachia* genome fragment is at least 11 kb in size (excluding the non-LTR retrotransposon region), and probably much larger, contains a number of bacterial ORFs, and is located on the X chromosome of the host insect. Based on the *wsp* sequence analyses (cf. Figure 18.3), we first thought that *C. chinensis* was infected with three distinct strains of endosymbiotic bacteria of the genus *Wolbachia*. Actually, however, it turned out that the beetle is harboring two "bacterial" and one "chromosomal" *Wolbachia* strains.

UNIVERSAL OCCURRENCE OF HORIZONTAL GENE TRANSFERS BETWEEN PROKARYOTES

For a long time, horizontal gene transfers between genomes of different organisms were believed to be rather exceptional phenomena. However, recent accumulation of microbial genome data has revealed an exciting view of evolution according to which horizontal gene transfers have commonly taken place between unrelated prokaryotic lineages in a dynamic manner (Ochman et al., 2000; Koonin et al., 2001; Bushman, 2002). Horizontal gene transfer is now widely accepted as an important and universal pathway for bacteria to reorganize their genome and to quickly acquire novel features such as drug resistance, pathogenicity, metabolic properties, and others (Mazel and Davies, 1999; Hacker and Kaper, 2000; Bushman, 2002).

HORIZONTAL GENE TRANSFERS BETWEEN PROKARYOTE AND EUKARYOTE

On the other hand, horizontal gene transfers between prokaryote and eukaryote are still regarded as unusual, except for those derived from mitochondria and chloroplasts (Martin et al., 1998; Gray et al., 1999). A number of reports have described putative prokaryote–eukaryote gene transfers based on identification of prokaryote-like genes on eukaryote genomes (Bushman, 2002). However, many, if not all, of them are considered to be dubious or inconclusive for the following reasons: (1) since the prokaryote-like gene on the eukaryote genome is fairly divergent from related prokaryotic genes, the transfer is likely to be a quite ancient event; (2) at present, complete genome-sequence data from a wide variety of eukaryotes are not available; (3) therefore, it is difficult to exclude the possibility that the prokaryote-like gene was not horizontally acquired but has been maintained in the lineage from the common ancestor of prokaryotes and eukaryotes, whereas most eukaryotic lineages have lost the gene over evolutionary time.

The most impressive case of "dubious prokaryote–eukaryote horizontal gene transfers" recently came from the human genome project. In the monumental article of the draft human genome sequence (International Human Genome Sequencing Consortium, 2001), it was reported that the human genome contained more than a hundred genes putatively transferred from bacteria. Many biologists were briefly very surprised at the abundance of alien genes in our genome, but the excitement soon vanished. By careful and intensive reanalyses of the genome data from human and other eukaryotes, Salzberg et al. (2001) and Stanhope et al. (2001) independently came to the conclusion that descent through common ancestry and gene loss provided a biologically more plausible explanation than horizontal gene transfer for the origin of "the human genes putatively from bacteria."

MECHANISMS OF PROKARYOTE–EUKARYOTE HORIZONTAL GENE TRANSFER: "FOOD HYPOTHESIS" AND "ENDOSYMBIONT HYPOTHESIS"

Mechanisms of horizontal gene transfer between prokaryotes, such as transformation, transduction, and conjugation, are relatively well understood (Ochman et al., 2000; Bushman, 2002). In contrast, mechanisms underlying prokaryote–eukaryote gene transfer are unknown, although several hypothetical models have been considered. One model is the "food hypothesis," which applies to prokaryotic genes identified in phagocytic unicellular eukaryotes like *Trichomonas* and *Entamoeba* (Köning et al., 2000; Field et al., 2000; Andersson et al., 2001). Unicellular eukaryotes often live close to prokaryotes and frequently use them as food, which means that they are constantly exposed to prokaryotic DNA (Doolittle, 1998; Berg and Kurland, 2000). However, this model does not appear applicable to multicellular eukaryotes like animals because foreign prokaryotes are not easily accessible to the germ line cells due to germ-soma separation and absence of active phagocytosis. An alternative model, the "endosymbiont hypothesis," which rests on the assumption of permanent contact between eukaryotic host cells and inhabiting microbial associates, may apply to these cases. In fact, mitochondria and chloroplasts have experienced gene transfer to the nucleus accompanied by drastic reduction in their genome size at early stages of their endosymbiotic evolution (Martin et al., 1998; Gray et al., 1999). In plants and animals, transfer of mitochondrial genes to the nuclear genome has been shown to be a currently ongoing process (Palmer et al., 2000; Bensasson et al., 2001). It appears meaningful that obligate endosymbiotic bacteria tend to exhibit a remarkably reduced genome size in comparison with their free-living relatives (Andersson and Kurland, 1998; Andersson et al., 1998; Shigenobu et al., 2000; Ochman and Moran, 2001). However, no case of horizontal gene transfer from prokaryotic endosymbiont to eukaryotic host has been described to date.

HORIZONTAL GENOME TRANSFER FROM *WOLBACHIA* ENDOSYMBIONT TO X CHROMOSOME OF *C. CHINENSIS*

In this study, we have provided unequivocal evidence of horizontal gene transfer from an endosymbiotic bacterium of the genus *Wolbachia* to a bruchid beetle. This finding is the first authentic case of horizontal gene transfer between prokaryotic endosymbiont and multicellular eukaryotic host. The horizontal transfer event we discovered is unprecedented in that (1) the transfer must be a recent event, (2) the donor bacterium is unequivocally identified as *Wolbachia*, (3) a large fragment of bacterial genome was transferred, (4) the structure of the transferred bacterial genome is highly preserved, and (5) the location of the transferred genome fragment has been identified as the X chromosome.

INACTIVATED *WOLBACHIA* GENES ON HOST CHROMOSOME

Of the 12 ORFs identified on the *Wolbachia* genome fragment (excluding the non-LTR retrotransposon region), more than half (7 ORFs) contained stop codons or frame-shift substitutions, although these ORFs in the wMel genome were structurally intact (Figure 18.12). Preliminary RT-PCR analyses showed that *wsp* and *ftsZ* were not transcribed in the strain jC^{Aus} (data not shown). These results indicate that most of the *Wolbachia* genes became pseudogenes upon or after the horizontal transfer and are no longer functional. However, it is unknown whether all the transferred *Wolbachia* genes are inactivated or whether some of them survive on the host chromosome. It will be necessary to clone the whole *Wolbachia* genome fragment, to determine its structure, and to examine the expression of all ORFs on the fragment.

PROCESS OF SYMBIONT–HOST HORIZONTAL GENE TRANSFER

At present, neither the mechanism nor the process involved in the *Wolbachia*-insect horizontal gene transfer is understood. Through generations of host insects, *Wolbachia* are maternally inherited via infection of developing oocytes. In eggs, *Wolbachia* cells localize at the posterior pole, where germ cells develop (Hadfield and Axton, 1999), and are closely associated with astral microtubules of the host cells during mitosis (Kose and Karr, 1995). Therefore, it is conceivable that these intimate associations of *Wolbachia* with proliferating germ cells have provided a favorable condition for the observed horizontal gene transfer.

EVOLUTIONARY ORIGIN

Absence of *Wolbachia* in other *Callosobruchus* species (Figure 18.2) suggests that wBruAus was acquired by the ancestor of *C. chinensis* through horizontal transmission from an unrelated host. In the DNA databases, a *wsp* sequence from the tephritid fruit fly *Dacus destillatoria* (accession No. AF295344) shows a very high similarity (98.9%) to the *wsp* gene of wBruAus, although a biological connection between the fruit fly and the bruchid beetle is obscure. The acquisition of bacterial wBruAus probably preceded the transfer of its genome fragment to the host chromosome. If so, a worldwide survey of *C. chinensis* populations might lead to the discovery of relic bacterial wBruAus, which would provide further insights into the evolutionary origin and process of the horizontal gene transfer.

MAINTENANCE MECHANISM IN HOST POPULATIONS

In all local populations of *C. chinensis* examined in Japan, the frequency of wBruAus was consistently more than 90% (Figure 18.4). At present, the mechanism whereby the *Wolbachia* genome fragment on the host chromosome prevails in populations is a mystery. Fixation by chance through drift cannot be ruled out but appears unlikely on account of the consistent prevalence in many local populations. It appears conceivable that the common ancestor of Japanese *C. chinensis* carried wBruAus on the X chromosome, which has been passed to present local populations. The chromosomal *Wolbachia* might be able to increase its frequency by hitchhiking with coexisting bacterial *Wolbachia* that cause CI. The chromosomal *Wolbachia* might be tightly linked to genes that confer a positive fitness effect to the insect or genes that enhance its own transmission in a selfish manner like meiotic drive genes (Lyttle, 1993). Alternatively, the chromosomal *Wolbachia* itself might behave as a selfish genetic element like *Medea*, a maternal-effect chromosomal factor known from flour beetles (Beeman et al., 1992). The last possibility is intriguing in that the *Medea* phenotype and *Wolbachia*-induced CI can be explained by the same "poison–antidote" or "modification–rescue" mechanism (Bull et al., 1992; Hurst and McVean, 1996).

OTHER SYMBIONT–HOST HORIZONTAL GENE TRANSFERS TO BE FOUND

Earlier studies reported that genetic recombination could occur between coinfecting *Wolbachia* strains (Werren and Bartos, 2001; Jiggins et al., 2001). In this study, we first demonstrated that genetic materials could be exchanged between *Wolbachia* and host insect. It is unknown whether the *Wolbachia*–host gene transfer is an orphan exception or whether other cases are to be found. It should be noted that in previous extensive surveys of infection and diversity of *Wolbachia* (Werren et al., 1995a; Werren and Windsor, 2000; Jeyaprakash and Hoy, 2000), only PCR detection was conducted without examining the inheritance pattern of the genes. Given that insects are the most diverse eukaryotic group in the terrestrial ecosystem (Wilson, 1989) and that infection frequency of *Wolbachia* in natural insect populations reaches around 20 to 70% worldwide (Werren and Windsor, 2000; Jeyaprakash and Hoy, 2000), it will not be surprising if future careful studies reveal other cases of *Wolbachia*-host gene transfer. Other endosymbiotic systems, such as *Buchnera* in aphids, in which the obligate symbiont genome exhibits remarkable degeneration and reduction (Shigenobu et al., 2000), might conceal similar horizontal gene transfer events. Genome sequencing of *Arabidopsis thaliana* revealed a continuous stretch of nearly 75% of mitochondrial genome located on chromosome 2 of the plant (Lin et al., 1999). Similarly, other cases of symbiont–host gene transfer might come from genome sequencing projects of various eukaryotic organisms now in progress.

EVOLUTIONARY IMPLICATIONS

It has been pointed out that *Wolbachia*-induced CI can promote reproductive isolation of host insects (Hurst and Schilthuizen, 1998; Bordenstein et al., 2001). If genes of *Wolbachia* responsible for CI are transferred to the host genome in a functional form, this would also reinforce the reproductive isolation and ultimately lead to speciation. It has been suggested that endosymbiotic associations with microorganisms act as a source of evolutionary innovations for their hosts (Margulis and Fester, 1991). Implications of the currently ongoing genome transfer between symbiont and host can be far-reaching in this context.

PERSPECTIVE

Our studies on the *Wolbachia* triple infection system in *C. chinensis* have led to the exciting discovery of the symbiont–host horizontal genome transfer. We now have a number of subjects to be investigated further.

1. Structure of the transferred genome fragment: The size and structure of the fragment, the location on the host X-chromosome, and the structure of the junction between the symbiont and host genomes must be determined for understanding the evolutionary process and molecular mechanism of the horizontal genome transfer. To clone the full length of the *Wolbachia* genome fragment, we are constructing a cosmid genomic library from the total DNA of the *C. chinensis* strain jC^{Aus}.
2. Gene expression on the transferred genome fragment: Once the full sequence of the transferred genome fragment is determined, all structurally intact ORFs on the fragment will be subjected to an expression assay using RT-PCR. If *Wolbachia* genes expressed on the host genome are identified, they should provide an important clue to understanding the biological function of the chromosomal *Wolbachia*.
3. Biological effects of the transferred genome fragment: Fitness parameters of the host insect, population and localization of the bacterial *Wolbachia*, and other aspects will be quantitatively examined in the presence and absence of the chromosomal *Wolbachia*, which would provide us with insights into the biological effects of the transferred genome fragment.
4. Diversity of chromosomal *Wolbachia* in natural populations: In Japanese populations of *C. chinensis*, diagnostic PCR analysis showed that wBruAus was prevalent at an average frequency of 97%. However, the result was based only on PCR detection of *wsp* gene. The possibility that part of the 3% negative insects are not free of the chromosomal *Wolbachia* but simply *wsp*-negative cannot be ruled out. It is conceivable that there should be variants of chromosomal *Wolbachia* in the populations with different degrees of deletions and rearrangements. To access the diversity of the chromosomal *Wolbachia*,

Southern blot analysis using many short probes that consecutively cover the full length of the genome fragment will be effective.

5. Reconstruction of the evolutionary process of the horizontal genome transfer: in addition to the Japanese populations, a survey of foreign populations of *C. chinensis* would lead to interesting discoveries such as populations without wBruAus or populations infected with bacterial wBruAus. One of our goals is to reconstruct the evolutionary process of the horizontal genome transfer through comparative and phylogenetic analyses of diverse types of wBruAus found in natural host populations.

ACKNOWLEDGMENTS

The authors thank A. Sugimura, S. Kumagai, and K. Sato for technical and secretarial assistance and M. Shimada and H. Ishikawa for encouragement. Preliminary sequence data of *Wolbachia* sp. from *Drosophila melanogaster,* which was accomplished with support from the National Institutes of Health, were obtained from the Institute for Genomic Research through the Web site at http://www.tigr.org. These studies were supported by the Program for Promotion of Basic Research Activities for Innovation Biosciences (ProBRAIN) of the Bio-Oriented Technology Research Advancement Institution.

REFERENCES

Andersson, J.O., Doolittle, W.F., and Nesbo, C.L. (2001). Are there bugs in our genome? *Science* **292:** 1848–1850.

Andersson, S.G.E. and Kurland, C.G. (1998). Reductive evolution of resident genomes. *Trends Microbiol.* **6:** 263–268.

Andersson, S.G.E., Zomorodipour, A., Andersson, J.O., Sicheritz-Pontén, T., Alsmark, U.C.M., Podowski, R.M., Näslund, A.K., Eriksson, A.S., Winkler, H.H., and Kurland, C.G. (1998). The genome sequence of *Rickettsia prowazekii* and the origin of mitochondria. *Nature* **396:** 133–140.

Beeman, R.W., Friesen, K.S., and Denell, R.E. (1992). Maternal effect selfish genes in flour beetles. *Science* **256:** 89–92.

Bensasson, D., Zhang, D.X., Hartl, D.L., and Hewitt, G.M. (2001). Mitochondrial pseudogenes: evolution's misplaced witnesses. *Trends Ecol. Evol.* **16:** 314–321.

Berezikov, E., Bucheton, A., and Busseau, I. (2000). A search for reverse transcriptase-coding sequences reveals new non-LTR retrotransposons in the genome of *Drosophila melanogaster. Genome Biol.* **1:** 0011.1–0011.15.

Berg, O.G. and Kurland, C.G. (2000). Why mitochondrial genes are most often found in nuclei. *Mol. Biol. Evol.* **17:** 951–961.

Blackman, R.L. (1987). Reproduction, cytogenetics and development. In *Aphids: Their Biology, Natural Enemies and Control,* Vol. 2A (A.K. Minks and P. Harrewijn, Eds.), pp. 163–195. Elsevier, Amsterdam.

Bordenstein, S.R., O'Hara, F.P., and Werren, J.H. (2001). *Wolbachia*-induced incompatibility precedes other hybrid incompatibilities in *Nasonia. Nature* **409:** 707–710.

Bucheton, A., Paro, R., Sang, H.M., Pelisson, A., and Finnegan, D.J. (1984). The molecular basis of I-R hybrid dysgenesis in *Drosophila melanogaster*: identification, cloning, and properties of I factor. *Cell* **38:** 153–163.

Buchner, P. (1965). *Endosymbiosis of Animals with Plant Microorganisms.* Interscience, New York.

Bull, J.J., Molineux, I.J., and Werren, J.H. (1992). Selfish genes. *Science* **256:** 65.

Bushman, F. (2002). *Lateral DNA Transfer: Mechanisms and Consequences.* Cold Spring Harbor Laboratory Press, Cold Spring Harbor, NY.

Caspari, E. and Watson, G.S. (1959). On the evolutionary importance of cytoplsmic sterility in mosquitoes. *Evolution* **13:** 568–570.

Doolittle, W.F. (1998). You are what you eat: a gene transfer ratchet could account for bacterial genes in eukaryotic nuclear genomes. *Trends Genet.* **14:** 307–311.

Field, J., Rosenthal, B., and Samuelson, J. (2000). Early lateral transfer of genes encoding malic enzyme, acetyl-CoA synthase and alcohol dehydrogenases from anaerobic prokaryotes to *Entamoeba histolytica*. *Mol. Microbiol.* **38:** 446–455.

Fine, P.E. (1978). On the dynamics of symbiote-dependent cytoplasmic incompatibility in culicine mosquitoes. *J. Invertebrate Pathol.* **30:** 10–18.

Foldi, I. (1990). Internal anatomy. In *Armored Scale Insects: Their Biology, Natural Enemies and Control,* Vol. 4A (D. Rosen, Ed.), pp. 65–84. Elsevier, Amsterdam.

Gray, M.W., Burger, G., and Lang, B.F. (1999). Mitochondrial evolution. *Science* **283:** 1476–1481.

Hacker, J. and Kaper, J.B. (2000). Pathogenicity islands and the evolution of microbes. *Annu. Rev. Microbiol.* **54:** 641–679.

Hadfield, S.J. and Axton, J.M. (1999). Germ cells colonized by endosymbiotic bacteria. *Nature* **402:** 482.

Hurst, G.D.D. and Schilthuizen, M. (1998). Selfish genetic elements and speciation. *Heredity* **80:** 2–8.

Hurst, L.D. and McVean, G.T. (1996). Clade selection, reversible evolution and the persistence of selfish elements: the evolutionary dynamics of cytoplasmic incompatibility. *Proc. R. Soc. London (B)* **263:** 97–104.

Ijichi, N., Kondo, N., Matsumoto, R., Shimada, M., Ishikawa, H., and Fukatsu, T. (2002). Internal spatiotemporal population dynamics of triple infection with *Wolbachia* strains in the adzuki bean beetle, *Callosobruchus chinensis* (Coleoptera: Bruchidae). *Appl. Environ. Microbiol.* **68:** 4074–4080.

International Human Genome Sequencing Consortium (2001). Initial sequencing and analysis of the human genome. *Nature* **409:** 860–921.

Jeyaprakash, A. and Hoy, M.A. (2000). Long PCR improves *Wolbachia* DNA amplification: *wsp* sequences found in 76% of sixty-three arthropod species. *Insect Mol. Biol.* **9:** 393–405.

Jiggins, F.M., von der Schulenburg, J.H.G., Hurst, G.D.D., and Majerus, M.E.N. (2001). Recombination confounds interpretations of *Wolbachia* evolution. *Proc. R. Soc. London (B)* **268:** 1423–1427.

Kondo, N., Shimada, M., and Fukatsu, T. (1999). High prevalence of *Wolbachia* in the azuki bean beetle *Callosobruchus chinensis* (Coleoptera, Bruchidae). *Zool. Sci.* **16:** 955–962.

Kondo, N., Ijichi, N., Shimada, M., and Fukatsu, T. (2002a). Prevailing triple infection with *Wolbachia* in *Callosobruchus chinensis* (Coleoptera: Bruchidae). *Mol. Ecol.* **11:** 167–180.

Kondo, N., Nikoh, N., Ijichi, N., Shimada, M., and Fukatsu, T. (2002b). Genome fragment of *Wolbachia* endosymbiont transferred to X chromosome of host insect. *Proc. Natl. Acad. Sci. U.S.A.* **99:** 14280–14285.

Köning, A.P., Brinkman, F.S.L., Jones, S.J.M., and Keeling, P.J. (2000). Lateral gene transfer and metabolic adaptation in the human parasite *Trichomonas vaginalis*. *Mol. Biol. Evol.* **17:** 1769–1773.

Koonin, E.V., Makarova, K.S., and Aravind, L. (2001). Horizontal gene transfer in prokaryotes: quantification and classification. *Annu. Rev. Microbiol.* **55:** 709–742.

Kose, H. and Karr, T.L. (1995). Organization of *Wolbachia pipientis* in the *Drosophila* fertilized egg and embryo revealed by an anti-*Wolbachia* monoclonal antibody. *Mech. Dev.* **51:** 275–288.

Lin, X.Y., Kaul, S.S., Rounsley, S., Shea, T.P., Benito, M.I., and 32 others (1999). Sequence and analysis of chromosome 2 of the plant *Arabidopsis thaliana*. *Nature* **402:** 761–768.

Lyttle, T.W. (1993). Cheaters sometimes prosper: distortion of Mendelian segregation by meiotic drive. *Trends Genet.* **9:** 205–210.

Margulis, L. and Fester, R. (1991). *Symbiosis As a Source of Evolutionary Innovation.* MIT Press, Cambridge, MA.

Martin, W., Stoebe, B., Goremykin, V., Hansmann, S., Hasegawa, M., and Kowallik, K.V. (1998). Gene transfer to the nucleus and the evolution of chloroplasts. *Nature* **393:** 162–165.

Mazel, D. and Davies, J. (1999). Antibiotic resistance in microbes. *Cell. Mol. Life Sci.* **56:** 742–754.

Ochman, H. and Moran, N.A. (2001). Genes lost and genes found: evolution of bacterial pathogenesis and symbiosis. *Science* **292:** 1096–1098.

Ochman, H., Lawrence, J.G., and Groisman, E.A. (2000). Lateral gene transfer and the nature of bacterial innovation. *Nature* **405:** 299–304.

O'Neill, S.L., Hoffmann, A.A., and Werren, J.H. (1997). *Influential Passengers: Inherited Microorganisms and Arthropod Reproduction.* Oxford University Press, Oxford, U.K.

Palmer, J.D., Adams, K.L., Cho, Y., Parkinson, C.L., Qiu, Y.L., and Song, K. (2000). Dynamic evolution of plant mitochondrial genomes: Mobile genes and introns and highly variable mutation rates. *Proc. Natl. Acad. Sci. U.S.A.* **97:** 6960–6966.

Salzberg, S.L., White, W., Peterson, J., and Eisen, J.A. (2001). Microbial genes in the human genome: lateral transfer or gene loss? *Science* **292:** 1903–1906.

Shigenobu, S., Watanabe, H., Hattori, M., Sakaki, Y., and Ishikawa, H. (2000). Genome sequence of the endocellular bacterial symbiont of aphids *Buchnera* sp. APS. *Nature* **407:** 81–86.

Stanhope, M.J., Lupas, A., Italia, M.J., Koretke, K.K., Volker, C., and Brown, J.R. (2001). Phylogenetic analyses do not support horizontal gene transfers from bacteria to vertebrates. *Nature* **411:** 940–944.

Stouthamer, R., Breeuwer, J.A.J., and Hurst, G.D.D. (1999). *Wolbachia pipientis*: microbial manipulator of arthropod reproduction. *Annu. Rev. Microbiol.* **53:** 71–102.

Takenouchi, Y. (1955). A short note on the chromosomes in three species of the Bruchidae (Coleoptera). *Jpn. J. Genet.* **30:** 7–9.

Tu, Z. and Hill, J.J. (1999). MosquI, a novel family of mosquito retrotransposons distantly related to the *Drosophila* I factors, may consist of elements of more than one origin. *Mol. Biol. Evol.* **16:** 1675–1686.

Turelli, M. (1994). Evolution of incompatibility-inducing microbes and their hosts. *Evolution* **48:** 1500–1513.

Werren, J.H. (1997). Biology of *Wolbachia. Annu. Rev. Entomol.* **42:** 587–609.

Werren, J.H. and Bartos, J.D. (2001). Recombination in *Wolbachia. Curr. Biol.* **11:** 431–435.

Werren, J.H. and Windsor, D.M. (2000). *Wolbachia* infection frequencies in insects: evidence of a global equilibrium? *Proc. R. Soc. London (B)* **267:** 1277–1285.

Werren, J.H., Windsor, D., and Guo, L.R. (1995a). Distribution of *Wolbachia* among neotropical arthropods. *Proc. R. Soc. London (B)* **262:** 197–204.

Werren, J.H., Zhang, W., and Guo, L. (1995b). Evolution and phylogeny of *Wolbachia*: reproductive parasites of arthropods. *Proc. R. Soc. London (B)* **261:** 55–71.

Wilson, E.O. (1989). *Biodiversity.* National Academy Press, Washington, D.C.

Zhou, W., Rousset, F., and O'Neill, S. (1998). Phylogeny and PCR-based classification of *Wolbachia* strains using *wsp* gene sequences. *Proc. R. Soc. London (B)* **265:** 509–515.

Index

C

D

E

F

H

I

J

K

L

M

N

Q

R

S

T

U

V

W

X

Y

Z